TRP CHANNELS

METHODS IN SIGNAL TRANSDUCTION SERIES

Joseph Eichberg, Jr., Series Editor

Published Titles

TRP CHANNELS

Edited by
Michael X. Zhu

The University of Texas Health Science Center

CRC Press
Taylor & Francis Group
Boca Raton   London   New York

CRC Press is an imprint of the
Taylor & Francis Group, an **informa** business

CRC Press
Taylor & Francis Group
6000 Broken Sound Parkway NW, Suite 300
Boca Raton, FL 33487-2742

First issued in paperback 2017

© 2011 by Taylor and Francis Group, LLC
CRC Press is an imprint of Taylor & Francis Group, an Informa business

No claim to original U.S. Government works

ISBN 13: 978-1-138-11616-0 (pbk)
ISBN 13: 978-1-4398-1860-2 (hbk)

Library of Congress Cataloging-in-Publication Data

TRP channels / editor, Michael X. Zhu.
     p. ; cm. -- (Methods in signal transduction)
"A CRC title."
Includes bibliographical references and index.
Summary: "Transient Receptor Potential (TRP) channels have gained a lot of popularity in recent years due to their involvement in a wide variety of cellular functions in many different systems. The purpose of this book is to provide detailed methods for the study of TRP channels. It includes expression and functional studies of TRP channels in heterologous systems as well as structural and functional studies of TRP channels. The author also explores various methods for detailed analysis of biophysical properties of TRP channels. In addition, the text covers high-throughput screening assays for TRP channels and offers novel and effective approaches to study endogenous TRP channel function in native systems"--Provided by publisher.
ISBN 978-1-4398-1860-2 (hardback : alk. paper)
1. TRP channels. I. Zhu, Michael X., editor. II. Title. III. Series: Methods in signal transduction.
[DNLM: 1. Transient Receptor Potential Channels. QU 55.7]

QP552.T77T76 2011
572'.696--dc22

2010051119

Visit the Taylor & Francis Web site at
http://www.taylorandfrancis.com

and the CRC Press Web site at
http://www.crcpress.com

Contents

Series Preface

The concept of signal transduction at the cellular level is now established as a cornerstone of biological sciences. Cells sense and react to environmental cues by means of a vast panoply of signaling pathways and cascades. While the steady accretion of knowledge regarding signal transduction mechanisms is continuing to add layers of complexity, this greater depth of understanding has also provided remarkable insights into how healthy cells respond to extracellular and intracellular stimuli and how these responses can malfunction in many disease states.

Central to advances in unraveling signal transduction is the development of new methods and refinement of existing ones. Progress in the field relies on an integrated approach that utilizes techniques drawn from cell and molecular biology, biochemistry, biophysics, physiology, genetics, immunology, and computational biology. The overall aim of this series is to collate and continually update the wealth of methodology now available for research into many aspects of signal transduction. Each volume is assembled by one or more editors who are leaders in their specialty. Their guiding principle is to recruit knowledgeable authors who will present procedures and protocols in a critical, yet reader-friendly, format. Our goal is to assure that each volume will be of maximum practical value to a broad audience, including students, seasoned investigators, and researchers who are new to the field.

Knowledge of the vast family of transient receptor potential (TRP) channels, which is the subject of this volume, has undergone a dramatic expansion in recent decades. A quick look at PubMed shows that over 6000 articles on this subject have appeared over the past 30 years, a figure that undoubtedly considerably underestimates the broad range of phenomena in which these channels are involved, especially since other designations were sometimes used in the earlier papers. The contents of the volume present both fundamental concepts regarding the properties of TRP channels and a variety of methodologies that can be employed to study them. Techniques that include the use of proteomics, antibodies, electrophysiology, fluorescence microscopy, and genetic manipulations are applied to cell and tissue preparations, as well as mouse models and lower eukaryotes.

In the Preface, the editor provides a brief history of the discovery of TRP channels, the recognition that they have a distinct role in regulating cellular Ca^{2+} movements, and the subsequent appreciation of their diversity. It can be anticipated that information concerning these channels will continue to steadily accumulate. It is hoped that the collection of methods presented in this book will constitute a valuable reference that will aid in the advancement of the field.

Joseph Eichberg
Series Editor

Preface

As a rapidly expanding area of research, studies on transient receptor potential (TRP) channels have undergone several interesting stages of historical development. It all started in the 1960s with the identification of a mutant fruit fly strain that lacked the sustained phase of the light response,[1] as measured using electroretinography. This technique has been used to record the electrical responses of cells in the retina, including the photoreceptors, to light illumination. These responses have been referred to as the "receptor potential." The short-lived, or transient, receptor potential to prolonged illumination in the mutant fly gave rise to the name, transient receptor potential (*trp*), which describes the mutant phenotype. It was not until 1989 when the genetic sequence, which was deleted in the *trp* mutant, was determined and shown to be a membrane protein with predicted structural similarities to voltage-gated ion channels.[2,3] Soon after, the *Drosophila* TRP protein was shown to form a light-induced nonselective cation channel, with higher Ca^{2+} permeability than Na^+, in the insect's photoreceptor cells.[4]

That TRP forms a Ca^{2+}-permeable channel downstream from the light response in insect eyes attracted a great deal of attention from investigators working in the areas of mammalian signal transduction. This is because, unlike phototransduction in vertebrates, insects utilize the G_q/phospholipase C pathway for the light response. Many hormones and neurotransmitters in the mammalian system also exert their actions through $G_{q/11}$ proteins and phospholipase C and an important signaling event of this pathway is the rise in intracellular Ca^{2+} concentrations. This occurs through both Ca^{2+} release from internal stores and Ca^{2+} influx from the extracellular space. Because of the similarity between the *Drosophila* phototransduction and the mammalian phospholipase C pathway, it was thought that a mammalian homolog(s) of the TRP channel might fulfill the rule of Ca^{2+} influx downstream from the activation of phospholipase C. Indeed, the search for the mammalian TRP homolog yielded not one, but seven, nonallelic TRP (now renamed TRP canonical or TRPC) genes in mammals.[5] Results from functional studies of heterologously expressed mammalian TRPC isoforms are consistent with a role in Ca^{2+} influx following phospholipase C activation. However, controversies exist as to whether TRPCs contribute to the formation of store-operated channels and in particular the highly Ca^{2+} selective type that conducts the so-called Ca^{2+} release-activated Ca^{2+} current. Although this is not a focus of the present volume, Chapters 2 and 3 contain considerable discussion on the potential roles of TRPCs and Orai proteins in receptor and store-operated Ca^{2+} entry. Other functions of TRPC channels, whether related or not to store-operated Ca^{2+} entry, have also been demonstrated, and some examples can be found in Chapters 4, 9, 10, and 17.

More interesting spins of the TRP field came about from several independent discoveries made by either functional cloning or genetic searches for disease-causing genes in humans during the late 1990s, right after the cloning of mammalian TRPCs, and which subsequently led to the rapid expansion of the field. Among them, the identification of TRPV1, or vanilloid receptor 1, attracted much attention because of its primary involvement in thermal sensation and nociception.[6] As such, TRPV1 has been the most studied TRP channel in recent years. TRPV1 was identified as the protein essential for the Ca^{2+} response to stimulation by capsaicin, a known potent agonist that causes pain. The derived sequence showed homology to the *Drosophila* TRP and its mammalian homologs, the TRPCs. Likewise, TRPV5, TRPV6, TRPM1, TRPP2, and TRPML1 were identified on the basis of their functions and involvement in diseases. Only after the resolution of the protein sequences was it realized that they also share homology with the *Drosophila* TRP protein. However, for a few years after the elucidation of these sequences, various names were given to different TRPs and sometimes multiple names were used for the same channel. Therefore, in 2002, a collective effort was made to unify or standardize the nomenclature of these related channel proteins.[7] The term TRP superfamily was defined to include the six distantly related subfamilies (seven in invertebrates), namely, TRPC, TRPV, TRPM, TRPA, TRPP, TRPML, and TRPN (invertebrates). To date, 28 TRP members have been described for mammalian species, making it a rather large superfamily of channel proteins with diverse functions. In the meantime, studies on invertebrate TRPs (*Drosophila* and *Caenorhabditis elegans*) continue to thrive, offering important insights on the physiological functions of these channels.

The rapid expansion of the TRP field has generated much excellent original work and many review articles. However, what is lacking is a comprehensive coverage of methodology for studying these channels. This volume is intended to fill this gap by providing broad coverage of current methods and techniques commonly used in TRP channel research. Because of the functional and sequence diversity, as well as the different physiological roles they play, techniques used for studying TRP channels are very diverse ranging from single molecular analysis to behavioral animal studies. Methods in multiple areas, such as molecular biology, fluorescence imaging, electrophysiology, cell biology, genetics, proteomics, pharmacology, system physiology, and behavioral assessment, are employed to investigate various aspects of these channels. For this reason, choosing among many possible topics in these broad areas was a daunting task. Our intent is to provide coverage of the major techniques currently used for TRP channel research in diverse areas and also to present a comprehensive viewpoint on the current standing of the field. The majority of the chapters are protocol oriented, with the goal of providing clear directions for laboratory use. Because of the breadth of the TRP field, the applications of some methods are described in multiple chapters by experts working on a variety of channel types that serve different physiological functions. This was intentional as it highlights diverse views on how the methodology can be utilized. Some chapters include discussion not only on the usefulness but also on the pitfalls associated with the use of certain techniques. This should also be of value, and together with chapters that offer comprehensive reviews on the functional regulation and diverse roles of TRP channels, students and investigators new to the field should find this book particularly informative.

Finally, we would like to thank all contributors for their enthusiastic and timely response in preparing chapters. We believe that this volume will be an excellent resource for all investigators in the TRP and related fields.

REFERENCES

1. Cosens, D. J., and A. Manning. 1969. Abnormal electroretinogram from a *Drosophila* mutant. *Nature* 224: 285–287.
2. Montell, C., and G. M. Rubin. 1989. Molecular characterization of the *Drosophila* trp locus: A putative integral membrane protein required for phototransduction. *Neuron* 2: 1313–1323.
3. Wong, F., E. L. Schaefer, B. C. Roop, J. N. LaMendola, D. Johnson-Seaton, and D. Shao. 1989. Proper function of the *Drosophila* trp gene product during pupal development is important for normal visual transduction in the adult. *Neuron* 3: 81–94.
4. Hardie, R. C., and B. Minke. 1992. The trp gene is essential for a light-activated Ca^{2+} channel in *Drosophila* photoreceptors. *Neuron* 8: 643–651.
5. Zhu, M. X., M. Jiang, M. Peyton et al. 1996. trp, a novel mammalian gene family essential for agonist-activated capacitative Ca^{2+} entry. *Cell* 85: 661–671.
6. Caterina, M. J., M. A. Schumacher, M. Tominaga, T. A. Rosen, J. D. Levine, and D. Julius. 1997. The capsaicin receptor: A heat-activated ion channel in the pain pathway. *Nature* 389: 816–824.
7. Montell, C., L. Birnbaumer, V. Flockerzi et al. 2002. A unified nomenclature for the superfamily of TRP cation channels. *Molecular Cell* 9: 229–231.

Editor

Michael X. Zhu is a professor in the Department of Integrative Biology and Pharmacology at the University of Texas Health Science Center at Houston, Texas. He received his BS from Fudan University, Shanghai, China, in 1984, and his MS and PhD from the University of Houston in 1988 and 1991, respectively. He was a postdoctoral fellow in the Department of Cell Biology at Baylor College of Medicine, Houston, Texas, from 1991 to 1994 and an assistant researcher in the Department of Anesthesiology, University of California at Los Angeles, from 1994 to 1997. In 1997, he became assistant professor in the Department of Pharmacology and Neurobiotechnology Center at The Ohio State University. In 2000, he transferred his appointment to the Department of Neuroscience and Center for Molecular Neurobiology and was promoted to associate professor in 2003 and to professor in 2010. In 2010, he moved to his current position. Dr. Zhu's research interests include several aspects of cell signaling, especially those that involve heterotrimeric G proteins and ion channels that affect Ca^{2+} signaling. Dr. Zhu's main contributions include identification and characterization of multiple TRPC channels in mammalian species and determination of the molecular identity of endolysosomal Ca^{2+} release channels activated by the Ca^{2+} mobilizing messenger, nicotinic acid adenine dinucleotide phosphate. Dr. Zhu has nearly 100 publications and is an associate editor of *The Journal of Cellular Physiology* and an editorial board member of *Pflügers Archiv European Journal of Physiology.*

Contributors

Indu S. Ambudkar
Secretory Physiology Section
Molecular Physiology and Therapeutics
    Branch
National Institute of Dental and
    Craniofacial Research
National Institutes of Health
Bethesda, Maryland

Michael Bandell
Genomics Institute of the Novartis
    Research Foundation
San Diego, California

Jonathan Berrout
Department of Integrative Biology
    and Pharmacology
University of Texas Health Science
    Center at Houston
Houston, Texas

Jun Chen
Neuroscience Research
Global Pharmaceutical Research
    and Development
Abbott Laboratories
Abbott Park, Illinois

Wanlu Du
Institute of Neuroscience
Shanghai Institutes for Biological
    Sciences
State Key Laboratory of Neuroscience
Chinese Academy of Sciences
Shanghai, China

Veit Flockerzi
Department of Experimental
    and Clinical Pharmacology
    and Toxicology
Saarland University
Homburg, Germany

Marc Freichel
Department of Experimental
    and Clinical Pharmacology
    and Toxicology
Saarland University
Homburg, Germany

Jorg Grandl
Department of Cell Biology
Scripps Research Institute
La Jolla, California

Klaus Groschner
Institute of Pharmaceutical Sciences–
    Pharmacology and Toxicology
University of Graz
Graz, Austria

Gerhard Held
José-Carreras-Center of Immuno and
    Gene Therapy
Clinic for Internal Medicine I
Saarland University Hospital
Homburg, Germany

Michael Hoschke
Department of Experimental and
    Clinical Pharmacology and
    Toxicology
Saarland University
Homburg, Germany

Hongzhen Hu
Department of Integrative Biology
　and Pharmacology
University of Texas Health Science
　Center at Houston
Houston, Texas

Junbo Huang
Institute of Neuroscience
Shanghai Institutes for Biological
　Sciences
State Key Laboratory of Neuroscience
Chinese Academy of Sciences
Shanghai, China

Min Jin
Department of Integrative Biology
　and Pharmacology
University of Texas Health Science
　Center at Houston
Houston, Texas

Shailen Joshi
Neuroscience Research, Global
　Pharmaceutical Research and
　Development
Abbott Laboratories
Abbott Park, Illinois

Martin Jung
Department of Medical Biochemistry
　and Molecular Biology
Saarland University
Homburg, Germany

Kenta Kato
Department of Synthetic Chemistry
　and Biological Chemistry
Graduate School of Engineering
Kyoto University
Kyoto, Japan

Kirill Kiselyov
Department of Biological Sciences
University of Pittsburgh
Pittsburgh, Pennsylvania

Shigeki Kiyonaka
Department of Synthetic Chemistry and
　Biological Chemistry
Graduate School of Engineering
Kyoto University
Kyoto, Japan

Elkana Kohn
Department of Medical Neurobiology
Institute of Medical Research
　Israel–Canada and the Kühne
　Minerva Center for Studies of
　Visual Transduction
The Hebrew University
Jerusalem, Israel

Jeffrey A. Kowalak
Laboratory of Neurotoxicology
National Institute of Mental Health
National Institutes of Health
Bethesda, Maryland

Ulrich Kriebs
Department of Experimental
　and Clinical Pharmacology
　and Toxicology
Saarland University
Homburg, Germany

Philip R. Kym
Neuroscience Research
Global Pharmaceutical Research
　and Development
Abbott Laboratories
Abbott Park, Illinois

Kyu Pil Lee
Molecular Physiology and Therapeutics
　Branch
National Institute of Dental
　and Craniofacial Research
National Institutes of Health
Bethesda, Maryland

V'yacheslav Lehen'kyi
Equipe labellisée par la Ligue Nationale
    contre le cancer
Université des Sciences et Technologies
    de Lille
Villeneuve d'Ascq, France

Min Li
Department of Neuroscience
High Throughput Biology Center and
    Johns Hopkins Ion Channel Center
Johns Hopkins University School of
    Medicine
Baltimore, Maryland

Sabine Link
Department of Experimental
    and Clinical Pharmacology
    and Toxicology
Saarland University
Homburg, Germany

Timothy Lockwich
Secretory Physiology Section
Molecular Physiology and Therapeutics
    Branch
National Institute of Dental and
    Craniofacial Research
National Institutes of Health
Bethesda, Maryland

Anthony Makusky
Laboratory of Neurotoxicology
National Institute of Mental Health
National Institutes of Health
Bethesda, Maryland

Stefanie Mannebach
Department of Experimental
    and Clinical Pharmacology
    and Toxicology
Saarland University
Homburg, Germany

Sanford P. Markey
Laboratory of Neurotoxicology
National Institute of Mental Health
National Institutes of Health
Bethesda, Maryland

Marcel Meissner
Department of Experimental
    and Clinical Pharmacology
    and Toxicology
Saarland University
Homburg, Germany

Melissa Miller
Department of Neuroscience
High Throughput Biology Center and
    Johns Hopkins Ion Channel Center
Johns Hopkins University School of
    Medicine
Baltimore, Maryland

Baruch Minke
Department of Medical Neurobiology
Institute of Medical Research
    Israel–Canada and the Kühne
    Minerva Center for Studies of
    Visual Transduction
The Hebrew University
Jerusalem, Israel

Yasuo Mori
Department of Synthetic Chemistry and
    Biological Chemistry
Graduate School of Engineering
Kyoto University
Kyoto, Japan

Shmuel Muallem
Molecular Physiology and Therapeutics
    Branch
National Institute of Dental and
    Craniofacial Research
National Institutes of Health
Bethesda, Maryland

Bernd Nilius
Department of Molecular Cell Biology
Laboratory Ion Channel Research
Katholieke Universiteit Leuven
Leuven, Belgium

Tomohiro Numata
Department of Synthetic Chemistry and
    Biological Chemistry
Graduate School of Engineering
Kyoto University
Kyoto, Japan

Verena C. Obmann
Department of Experimental
    and Clinical Pharmacology
    and Toxicology
Saarland University
Homburg, Germany

Roger G. O'Neil
Department of Integrative Biology
    and Pharmacology
University of Texas Health Science
    Center at Houston
Houston, Texas

Matt Petrus
Genomics Institute of the Novartis
    Research Foundation
San Diego, California

Stephan E. Philipp
Department of Experimental
    and Clinical Pharmacology
    and Toxicology
Saarland University
Homburg, Germany

Michael Poteser
Institute of Pharmaceutical Sciences–
    Pharmacology and Toxicology
University of Graz
Graz, Austria

Louis S. Premkumar
Department of Pharmacology
Southern Illinois University School of
    Medicine
Springfield, Illinois

Natalia Prevarskaya
Equipe labellisée par la Ligue Nationale
    contre le cancer
Université des Sciences et Technologies
    de Lille (USTL)
Villeneuve d'Ascq, France

Feng Qin
Department of Physiology and
    Biophysics
State University of New York at Buffalo
Buffalo, New York

Regina M. Reilly
Neuroscience Research
Global Pharmaceutical Research
    and Development
Abbott Laboratories
Abbott Park, Illinois

Mohammad Samie
Department of Molecular, Cellular, and
    Developmental Biology
University of Michigan
Ann Arbor, Michigan

Nobuaki Takahashi
Department of Synthetic Chemistry and
    Biological Chemistry
Graduate School of Engineering
Kyoto University
Kyoto, Japan

Karel Talavera
Department of Molecular Cell Biology
Laboratory Ion Channel Research
Katholieke Universiteit Leuven
Leuven, Belgium

Dominik Vogt
Department of Experimental
 and Clinical Pharmacology
 and Toxicology
Saarland University
Homburg, Germany

Yizheng Wang
Institute of Neuroscience
Shanghai Institutes for Biological
 Sciences
State Key Laboratory of Neuroscience
Chinese Academy of Sciences
Shanghai, China

David Weaver
Department of Pharmacology
Vanderbilt Program in Drug Discovery
Vanderbilt Institute of Chemical
 Biology
Vanderbilt University Medical Center
Nashville, Tennessee

Petra Weißgerber
Department of Experimental
 and Clinical Pharmacology
 and Toxicology
Saarland University
Homburg, Germany

Meng Wu
Department of Neuroscience
High Throughput Biology Center and
 Johns Hopkins Ion Channel Center
Johns Hopkins University School of
 Medicine
Baltimore, Maryland

Rui Xiao
Life Sciences Institute and Department
 of Molecular and Integrative
 Physiology
University of Michigan
Ann Arbor, Michigan

Haoxing Xu
Department of Molecular, Cellular, and
 Developmental Biology
University of Michigan
Ann Arbor, Michigan

Jia Xu
Department of Neuroscience
High Throughput Biology Center and
 Johns Hopkins Ion Channel Center
Johns Hopkins University School of
 Medicine
Baltimore, Maryland

Shang-Zhong Xu
Endocrinology, Diabetes and
 Metabolism
Hull York Medical School
University of Hull
Hull, United Kingdom

X. Z. Shawn Xu
Life Sciences Institute and Department of
 Molecular and Integrative Physiology
University of Michigan
Ann Arbor, Michigan

Fan Yang
Department of Physiology and
 Membrane Biology
University of California Davis
Davis, California

Hailan Yao
Institute of Neuroscience
Shanghai Institutes for Biological
 Sciences
State Key Laboratory of Neuroscience
Chinese Academy of Sciences
Shanghai, China

Joseph P. Yuan
Department of Physiology
University of Texas Southwestern
 Medical Center at Dallas
Dallas, Texas

Jie Zheng
Department of Physiology and
    Membrane Biology
University of California Davis
Davis, California

Alexander V. Zholos
Centre for Vision and Vascular Science
School of Medicine, Dentistry and
    Biomedical Sciences
Queen's University Belfast
Belfast, Northern Ireland, United
    Kingdom

Michael X. Zhu
Department of Integrative Biology and
    Pharmacology
University of Texas Health Science
    Center at Houston
Houston, Texas

Richard Zimmermann
Department of Medical Biochemistry
    and Molecular Biology
Saarland University
Homburg, Germany

1 High-Throughput Screening of TRPC Channel Ligands Using Cell-Based Assays

Melissa Miller, Meng Wu, Jia Xu,
David Weaver, Min Li, and Michael X. Zhu

CONTENTS

1.1   INTRODUCTION

The transient receptor potential (TRP) superfamily of cation channels has emerged as cellular sensors of various extra- and intracellular environmental changes.[1] The 28 or so TRP members in mammals and their counterparts in other animal classes are involved in diverse cellular functions, many of which are mentioned and some of which are discussed in detail in subsequent chapters of this book. However, the biggest challenge that TRP channel researchers face nowadays is the lack of pharmacological tools. With a few exceptions, for instance, TRPV1, high-affinity specific agonists and/or antagonists are lacking for the mammalian TRP channels. Despite this deficiency, active research has revealed a large number of natural and synthetic compounds having stimulatory or inhibitory effects on various TRP channels. However, even some of the more commonly used drugs, e.g., menthol, 2-aminoethoxydiphenyl borate (2-APB), allyl isothiocyanate (mustard oil), and 1-oleoyl-acetyl-sn-glycerol (OAG), may lack the specificity required for more rigorous studies, and the concentrations employed are typically in high micromolar ranges,[2-5] limiting their use in physiological experiments when the identity of the channel(s) is ambiguous. Thus there are unmet needs for more specific and potent activators and/or inhibitors for many TRP channels.

Molecular cloning and functional expression of TRP channels in heterologous systems, as well as recent advancements in high-throughput screening (HTS) technology, have made it possible to screen TRP ligands using cell-based assays. In this chapter, we will describe method development, validation, and execution of cell-based HTS assays useful for screening ligands for TRPC channels. TRPC channels are typically activated downstream from the activation of phospholipase C (PLC). This occurs through activation either of receptors that couple to $G_{q/11}$ proteins or of receptor tyrosine kinases. However, which step(s) or constituent(s) of the PLC pathway is critical for the activation of TRPC channels remains a matter of current debate, and different mechanisms exist for the different subtypes of TRPC channels.[6] Thus, nearly all steps of the PLC pathway have been shown to regulate TRPC channel activity, often in subtype-specific manners. These include the breakdown of phosphatidylinositol 4,5-bisphosphate (PIP$_2$), production of diacylglycerols and inositol 1,4,5-trisphosphate (IP$_3$), activation of IP$_3$ receptors, cytosolic Ca^{2+} increase, Ca^{2+}–calmodulin binding, internal Ca^{2+} store depletion, and protein kinase C (PKC) activation. For more detailed discussion on these topics, readers are referred to Chapters 2 and 3. In addition, physiological functions of TRPC channels have been revealed through studying TRPC knockout mice and other gene-manipulating approaches, as well as by studies on human genetic diseases (see Chapters 2 and 8). The highlights of the TRPC channel function include the involvement of TRPC6 in proteinuric kidney diseases,[7] multiple TRPC channels in cardiac hypertrophy,[8] TRPC4 and TRPC6 in neurogenic responses of intestinal smooth muscles,[9] TRPC3 in motor coordination[10] and pancreatitis,[11] and TRPC5 in neurite outgrowth[12] and fear behaviors.[13] However, these studies are time consuming, and the results have been shown to be limited because of long-term compensatory effects. For example, TRPC3 was upregulated in smooth muscle cells from TRPC6 knockout mice.[14]

All studies could benefit from having subtype-specific and high-potency drugs. Flufenamic acid has been shown to enhance the activity of TRPC6 but inhibit that of TRPC3 and C7.[15,16] However, this drug has multiple actions on several ion channels.[17] Antagonists for TRPCs are also rare and not specific. SKF96365 and 2-APB are frequently used, but they suffer from the same specificity issue as flufenamate.[18,19] In addition, the effect of 2-APB on some TRPCs (e.g., TRPC3) has been quite inconsistent.[20–22] Currently, the only "specific" TRPC drug is Pyr3, a pyrazole compound that selectively inhibits TRPC3.[23] This should encourage the search for other TRPC-specific inhibitors. Direct activators will also be beneficial because TRPC channel activation typically relies on indirect actions through receptor signaling, complicating experimental designs and data interpretation.

1.2 ESTABLISHMENT OF STABLE CELL LINES

A cell line that overexpresses the channel of interest in the stable form is a must for a successful HTS effort. This is often achieved by transfection of a plasmid vector that contains the cDNA for the channel into a host cell line and then selecting for transfected cells in the presence of antibiotics that otherwise kill all nontransfected ones. Those that survive are subjected to functional tests to confirm the expression and function of the channel proteins and then clonally purified through the process of "limiting dilution." Each of the "monoclonal" cell lines should be further tested and the stability confirmed after going through multiple passages and freeze and thaw processes. Here, we describe the expression vectors, host cell lines, transfection methods, and functional assays used for the establishment of TRPC cell lines.

1.2.1 EXPRESSION VECTORS

There are many commercially available mammalian expression vectors. We found that bicistronic vectors (e.g., pIRESneo and pIREShyg2, both from Clontech) are very convenient to use. These vectors have the coding sequence for the gene of interest and that for the selection marker (encoding a protein that can resist the antibiotics) placed in tandem and under the control of a single promoter. In between the coding sequences is the internal ribosome entry site (IRES) that allows translation initiation of the second coding sequence, which is usually the antibiotic resistance gene.[24] Stabilization usually involves some random breakage of the circular plasmid and insertion of the linear DNA fragments into chromosomes of the host cell. In order for the selection marker to be expressed and to function, the entire transcriptional unit, including the promoter, the coding sequence of the channel, the IRES site, and the coding sequence of the antibiotic-resistant gene, has to be integrated together. Therefore, when selected for antibiotic resistance, the surviving clones often expressed the gene of interest as well. As such, after selection by the antibiotics, bicistronic vectors typically give rise to >90% clones that are positive in expression of gene of interest, whereas conventional vectors, which place the gene of interest and selection marker under the control of separate promoters, usually yield approximately 30% positive clones. The success rate tends to be much lower if the expression of the gene of interest has even some minor detrimental effect on cell survival, which is often the case for TRP channels because of Ca^{2+} influx and/or membrane depolarization inherently associated with the channel function.

There are also vectors for inducible expression. These are especially useful when the expression of the gene of interest is cytotoxic. Ideally, all expression should be controlled in an inducible manner so that the uninduced cells could serve as perfect negative controls. However, to achieve inducible expression, a transcriptional activa-tor or repressor protein, e.g., tTA or rtTA of the tetracycline-inducible mammalian system, needs to be co-expressed with the gene of interest. This is technically more difficult to do since two proteins need to be stably coexpressed in the cell line. In addition, clonal selection is essential for inducible expression because leaky expres-sion, i.e., low levels of the expression of the gene of interest under uninduced condi-tions, is common. Rigorous tests of multiple monoclones will ensure that only those with no or minimal leaky expression but robust induction are selected. Excellent inducible expression vectors for mammalian cells are available commercially (e.g., Clontech and Invitrogen).

1.2.2 HOST CELLS

Human embryonic kidney (HEK) 293 cells and Chinese hamster ovary (CHO) cells are commonly used in cell-based HTS assays. These established cell lines are quite stable over long-term cultures and passages and are easily transfected with high efficiencies. However, because of this durability, the HEK293 and CHO cell lines maintained in different laboratories can vary significantly in terms of growth con-ditions, growth rate, physical appearance, and expression of endogenous proteins. Therefore, it is always desirable to use parental cells from which the stable clones were generated as negative controls when assessing the function of the expressed proteins. Ideally, stable clones transfected with the vector alone, or the same vector that expresses an unrelated protein, should be made and maintained in parallel, but this will require additional efforts and cost. The choice between HEK293 and CHO cells is mostly arbitrary, although most screen centers prefer CHO cells because of the better adherence to culture plates of these cells over HEK293 cells. On the other hand, HEK293 cells tend to give stronger fluorescence signals than CHO cells. This difference likely results from the higher anion transporter activity, which causes more dye loss from loaded cells for CHO cells as compared with those for HEK293 cells. Consequently, high concentrations of anion transport inhibitors are needed to block dye transport in CHO cells. Other host cells should also work as long as rea-sonable transfection efficiency can be achieved and established clones remain stable for a long period of time.

Host cells that stably express a transcriptional activator or repressor for inducible expression are commercially available. These are often tested and selected to minimize leaky expression. They can be directly used to express the gene of interest under the control of an inducible promoter. However, it is very important to match the vector with the specific cell line that expresses the required transcription activator or repressor.

1.2.3   TRANSFECTION AND SELECTION

Many transfection methods are available to introduce DNA into mammalian host cells. The traditional calcium phosphate method works well for HEK293 cells and CHO cells. Electroporation and its variant, nucleofection,[25] are effective over a broad range of cell types. However, this requires special equipment. Viral transduction is also very effective, but making viral vectors and producing viruses requires additional work. The more common method of choice at present is the use of cationic lipids to aid DNA transfection. These are the bases of many commercial transfection reagents. For stable TRPC-expressing cell lines, we have used Lipofectamine 2000, which allows high-efficiency transfection of HEK293 cells in wells of a 96-well plate. This has dramatically reduced the amounts of cells and DNA used for transfection and therefore is cost-effective as compared with other methods. Here, we only describe a protocol for transfecting HEK293 cells using Lipofectamine 2000. The method is applicable to CHO cells as well.

HEK293 cells are grown to 50–70% confluence on the day of transfection. The culture dish can be of any size (35- to 100-mm diameter) as long as it provides enough cells for transfection. To help cell adherence to the 96-well plate, 50 µL of polyornithine (Sigma, 20 µg/mL in sterile H_2O) is added to each well to be used. The plate is incubated at 37°C for >15 min, followed by aspiration of the solution and rinsing the wells once with ~100-µL Hank's Balanced Salt Solution (HBSS) without Mg^{2+} and Ca^{2+}. For each well, 0.2 µg of the plasmid DNA is added to 25 µL of Opti-MEM medium (Invitrogen) in a 1.5 mL Eppendorf tube; 0.4-µL Lipofectamine 2000 is diluted to 25 µL of Opti-MEM in another Eppendorf tube. Both tubes are vortexed briefly and allowed to stand at room temperature for less than 5 min before the contents are mixed together and briefly vortexed again. This procedure can be scaled up depending on the number of transfections. The DNA–lipofectamine mixture can stay at room temperature for up to 2 hours, which gives time for making multiple transfection mixtures and preparing cells. After trypsinization, the cells are counted and the desired number of cells (~150,000 cells/well × number of wells × 1.2) is collected in a suitable size tube and centrifuged at 500 × g for 5 min. After removing the supernatant, the cell pellet is resuspended in the culture medium (DMEM plus 10% heat-inactivated fetal bovine serum [ΔFBS]) without antibiotics at 1,500,000 cells/mL. Following addition of 50 µL of the DNA–lipofectamine mixture to a polyornithine-treated well, 100-µL cells are also added and the plate is incubated at 37°C, 5% CO_2 incubator for 12–16 hours. Transfection efficiency can be monitored by parallel transfection of a vector that expresses green fluorescence protein (GFP) and can vary from 20 to 70%.

To select clones, cells are trypsinized and diluted in 2 mL of the complete culture medium (DMEM, with 10% ΔFBS, 100 unit/mL penicillin, and 100 μg/mL streptomycin) supplemented with the antibiotics for the selection marker. We use 400 μg/mL G418 (Invitrogen) for channels expressed using the pIRESneo vector and 100 μg/mL hygromycin B (Roche) for the pIREShyg2 vector. Optimal antibiotic concentrations can be established by testing untransfected cells at different concentrations of the antibiotics to construct a "kill curve." The concentration chosen should be just high enough to kill all untransfected cells in approximately 2 weeks, but not too high to waste reagent or inhibit cell growth even when the selection marker is expressed.

There are several selection strategies. Because the majority of cells express the gene of interest when using a bicistronic vector, a polyclonal cell line can be established by growing the transfected cells in a 35-mm dish, or a well of a 6-well plate, for 2 to 3 weeks in the presence of the selection antibiotics, with medium change every 3–4 days. The cells can be expanded to larger and/or multiple dishes when they grow out of the capacity of the original dish/well. The dish/plate does not need polyornithine coating as long as the medium change is carried out carefully with minimal disturbance to cell attachment. In scenarios when the expression of the exogenous protein has very little detrimental effect on the growth of cells, the obtained polyclonal cell line may work just fine for all subsequent analyses including HTS. However, this is not always the case because the presence of the exogenous protein may present a growth disadvantage so that the nonexpressers grow better and faster than the expressers. Eventually, the entire population may be taken over by the nonexpressers leading to a loss of the expression. Therefore, clonal selection would be necessary. For many transfections, clonal selection is done by limiting dilution (see below) soon after the polyclonal cell line is ready. This reduces the chance of losing expression and obtaining multiple clones from the same original transfectant(s); both are consequences of long-term cultures. Alternatively, clonal selection may begin right after transfection. Thus, the day after transfection, the trypsinized cells may be diluted at a low enough density to be plated in wells of a 96-well plate. Because the ratio of successful transfection and chromosomal integration is hard to predict, it may be advisable to plate the cells out at three to four different densities. For example, the first plate receives ~2000, the second 500, and the third 125 cells/well in a volume of 150 to 200 μL. This increases the chance of obtaining clones that arise from single colonies. Again, all cells are grown under the selection pressure of the corresponding antibiotics, with medium change once every 3–4 days. The immunowasher (Nunc) is a convenient tool for removing the medium without cross contamination. Medium addition may be accomplished with the use of a multichannel pipettor or the Costar Transtar-96 Cartridges.

To identify clones that arise from a single transfectant (monoclones), each well is visually inspected using an inverted cell culture microscope with a 10× or 20× phase-contrast lens. If most cells in the well are confined to a region that appears to arise from one single cell, then it may be monoclonal. The purpose at this point is to eliminate wells that contain multiple clones and those with none. Therefore, the inspection needs to be done early (10 to 14 days after seeding), when the colonies just become visible. If it is done too late, the cell colonies will merge, making it difficult to tell whether they grow out from a single or multiple areas. This crude assessment does not guarantee that the apparent single colonies are monoclonal, but it will eliminate those that are clearly polyclonal. Another helpful hint is to see in general how many colonies are present in each well of a particular plate. If there are from zero to three colonies, then there is a good chance that some wells have signal colonies. If there are typically more than five colonies per well, then the chance of obtaining a monoclone from the plate is very slim. The identified wells should be marked and cells allowed to grow continuously until they achieve confluency.

Normally, about 12 potential monoclones should be identified and tested. If there are not enough monoclones, then wells with two or three colonies may also be selected. After reaching confluency, the selected clones are expanded to wells of 6-well plates and then further expanded to large dishes, if more cells are needed for functional analysis. For each clone, a replica should be prepared at this stage, and depending on how long it takes to confirm the functional expression of the protein/channel, cell aliquots should be frozen and stored (see Section 1.2.6). For stable TRPC-expressing cell lines, we have conducted functional tests right after the clones reached confluency at the 6-well stage.

1.2.4 FUNCTIONAL SCREENING

Functional expression is of utmost importance for a stable clone, especially for HTS. The method for functional screening should be based on the functional properties of the protein, as well as on the convenience and throughput of the assay. For TRPC channels, we have used the FLIPR membrane potential (FMP) dye (Molecular Devices, Sunnyvale, California) in combination with the FlexStation, a fluid handling an integrated fluorescence plate reader (Molecular Devices). Cells are seeded in wells of a 96-well plate pretreated with polyornithine as described above at ca. 120,000 cells/well in 150 μL. Alternatively, commercial plates coated with poly-D-lysine or other synthetic charged surfaces may be used (e.g., BD Biosciences, San Jose, California). Each clone should be seeded in multiple wells for duplicates and/or testing multiple drugs or concentrations, but enough backup cells should be kept in separate plates/dishes in order to maintain the cell line for storage and future use. Multiple clones can be seeded in the same plate. After incubation at 37°C, 5% CO_2 for 16–20 hours, cells are inspected under a microscope to ensure that the confluency is at 70–100%. If too low, cells may be grown for a longer time period.

For the FMP assay, either the diluted dye may be added directly to the cells in the culture medium or the cells can be rinsed with an assay buffer first and then loaded with the dye. To do the latter, the culture medium is dumped into a waste container by placing the plate upside down without the lid and shaking briefly. The residual liquid is absorbed by pressing the open side of the plate against a stack of paper towels. Then added to the cells is 80 μL of an assay buffer consisting of (in mM): 140 NaCl, 5.4 KCl, 1.8 $CaCl_2$, 1 $MgCl_2$, 10 glucose, 15 HEPES, pH 7.4 by NaOH. The plate is swirled to rinse off the residual medium, the buffer is decanted, and excess liquid is absorbed as before. A fresh buffer is then added. This process may be repeated once, followed by addition of a 40-μL assay buffer to each well. Then an equal volume of the FMP dye, diluted in the same assay buffer, is added.

The FMP dye relies on its partitioning to the inner side of the plasma membrane as the membrane depolarizes. A membrane impermeable quencher is included in the dye mixture to eliminate fluorescence signals from the extracellular dye; thus, only the internalized dye molecules are detected. Essentially, this dye does not need time for loading, but ~10 min is required for the dye signal to stabilize. Importantly, the loading time affects the degree of fluorescence response (M. X. Zhu, unpublished results), and because the FlexStation only reads one column (8 vertical wells of a 96-well plate) at a time, it is advisable to space the loading time between columns to that required to finish reading one column. For example, if it takes 3 min to finish reading column 1, column 2 should be loaded 3 min after column 1. In this case, it will take 36 min to finish loading the entire plate (all 12 columns), so the load time may be set to 40 min.

After loading, the plate is read in the FlexStation with excitation at 530 nm and emission at 565 nm at 0.67 Hz. From the compound plate, the integrated robotic 8-channel pipettor is programmed to deliver either 40 µL carbachol (300 µM diluted in the assay buffer) or the buffer alone (no drug control) at 30 s. This gives a final carbachol concentration of 100 µM in the stimulated wells. Carbachol is a muscarinic agonist that triggers PLC activation via M3 muscarinic receptors natively expressed in HEK293 cells that are coupled to $G_{q/11}$ proteins. TRPC-expressing cells respond to carbachol stimulation with a robust depolarization, i.e., fluorescence increase, which is absent or very small in cells that do not overexpress TRPC (Figure 1.1). Cell clones that respond to carbachol with a strong depolarization are considered positive and are subjected to further analyses.

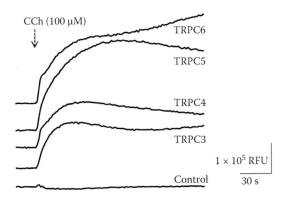

FIGURE 1.1 Detection of TRPC-mediated membrane depolarization in microtiter plates using the FLIPR membrane potential (FMP) dye. HEK293 cells stably expressing TRPC3, TRPC4, TRPC5, or TRPC6 were grown in wells of 96-well plates overnight. Cells were rinsed with an extracellular solution and loaded with the FMP dye as explained in the text. Fluorescence signals were read in the FlexStation plate reader with excitation at 530 nm and emission at 565 nm at 0.67 Hz. Carbachol (CCh, 100 µM) was added at 30 s. Note: in cells expressing TRPCs, the increase in fluorescence intensity, which indicates depolarization, was fast and strong, while the fluorescence in the control HEK293 cells decreased initially and then slowly increased. RFU: random fluorescence unit.

In the absence of a convenient high-throughput functional assay, this step can be carried out by ELISA or Western blotting to detect the expressed protein, if suitable antibodies are available. Often, a fluorescence protein tag can be added to the protein under study, if it does not significantly alter the function. In this case, screening is done at the 96-well stage when transformants are selected. In the absence of all these available tools, RT-PCR may still be used to select clones that express the exogenous transcript. The identified positive clones may then be subjected to manual patch clamping or Ca^{2+} imaging assays to confirm that the clones are functional.

1.2.5   LIMITING DILUTION

The identified positive clones are further purified by limiting dilution (normally two rounds). This is time-consuming, and therefore, only a few clones are subjected to this procedure. Briefly, cells are trypsinized and counted. About 200 cells are diluted in a 15-mL complete culture medium containing the selection antibiotics and evenly distributed to wells (150 µL/well) of a 96-well plate. Correct counting and dilution should give rises to between 0 and 4 cells/well, and this should be seen as 0 to 4 colonies when inspected 10 to 14 days after seeding. Medium change and inspection are carried as described above for transfection and selection. Those with single colonies should be marked and transferred to 6-well plates when they reach confluency, and then functionally tested. Here, the most critical factor is cell counting and dilution; too few or too many cells will all lead to failure to obtain single colonies.

1.2.6   FREEZING AND DEFROSTING

At all stages of stable cell line generation, backup cells are frozen and stored either in −80°C freezers or in liquid nitrogen tanks. There are a number of ways to do this, resulting in different efficiencies in preserving cell viability. A simple method is to grow cells in two 10-cm dishes to 50–70% confluency. This should happen within 1 or 2 days after seeding in order to ascertain the log phase growth. If it takes too long to reach the desired density, the cells should be split again and reseeded at a higher density so that they will be ready in 1 or 2 days. Then the cells from the two dishes are trypsinized and collected in a 15-mL tube, centrifuged (300 × g, 5 min), and the cell pellet resuspended in 3 mL of the freezing medium (a complete culture medium supplemented with additional 10% ΔFBS and 10% DMSO). The resuspended cells are transferred evenly (1 mL each) to 3 cryovials (use 1.5- or 2-mL capacity to ensure enough air space in the vial). It is always a good idea to label the vials clearly and to label the caps as well in order to ease future identification. The vials are placed in a −20°C freezer for 1 hour before being transferred to −80°C. Although most cell lines are stable in −80°C freezers for years, it is safer to store them in liquid nitrogen tanks, which can maintain temperature for weeks during power outage. Each cell line should be frozen down at least twice on different days in order to minimize the chance of losing the cell line.

To defrost cells, a 15-mL centrifuge tube containing 10 mL of the culture medium should be prepared. The frozen vial is warmed in a 37°C water bath with constant swirling. This ensures quick defrosting and even heat distribution within the vial. Immediately after the content is melted, the outside of the vial is wiped clean with an alcohol swab, and the entire contents are diluted into the 10-mL medium in the 15-mL tube. This process should be done quickly in order to minimize the exposure of cells to the high DMSO concentration in the freezing medium. The tube is centrifuged ($300 \times g$, 5 min), and the supernatant solution is removed. The cell pellet is resuspended in a 10-mL fresh complete culture medium containing the selection antibiotics, transferred to a 10-cm dish, and cultured overnight at $37°C$, 5% CO_2. The medium is changed the next day. Depending on the quality of freezing and defrosting, the recovery may take anywhere from 1 to 4 weeks.

1.3 ASSAY DEVELOPMENT

1.3.1 STUDYING TRPC CHANNEL FUNCTION USING FLIPR MEMBRANE POTENTIAL DYE

Methods for the functional analysis of heterologously expressed TRPC channels normally include fluorescence-based measurements of intracellular Ca^{2+} concentration ($[Ca^{2+}]_i$) changes and patch clamp recordings of transfected cells. The Ca^{2+} assay is often obstructed by the endogenous Ca^{2+} responses (both Ca^{2+} release and Ca^{2+} influx) of host cells. Thus, low concentrations of lanthanides have been used to block the endogenous store-operated Ca^{2+} entry (SOCE).[18,26] Ba^{2+} and Sr^{2+} have been used in place of Ca^{2+} in extracellular solutions to help isolate the TRPC-mediated cation influx, which is less Ca^{2+} selective than the endogenous SOCE.[18,27] However, for these to work, one has to assume that (1) TRPC channels are not inhibited by lanthanides at the concentrations used, and (2) they are permeable to Ba^{2+} and Sr^{2+}, but endogenous channels are not. These conditions are not always met. In addition, neither method blocks receptor-activated Ca^{2+} release, which masks the initial phase of Ca^{2+} entry. In most cases, divalent cation influx has to be experimentally separated from Ca^{2+} release by stimulating cells in the absence of extracellular Ca^{2+} and then reintroducing divalent cations in the bath after the release signal has subsided.[18,26] This is not convenient for HTS. Another alternative is to use Mn^{2+}, which also enters the cell through TRPC channels and quenches the Fura-2 signal. The rate of decrease in the Fura-2 signal at its isosbestic point (~360 nM) is used to quantify the relative activity of Ca^{2+} entry channels.[28] However, the relatively high and variable levels of basal Mn^{2+} leak, especially in the microplate format, makes its use a poor method for HTS. Furthermore, the use of Ba^{2+}, Sr^{2+}, and Mn^{2+} has only been validated for Fura-2, a ratiometric Ca^{2+} indicator excited by UV light, which is not always available for HTS. Direct recording of ionic currents mediated by TRPCs may be the best way to characterize these channels. However, high-throughput methods for whole-cell patch clamp recording of TRPC channels are still under development.[29]

The FMP assay kit (Molecular Devices) uses an oxonol-derived voltage-sensitive indicator mixed with fluorescent quenchers, which absorb the emitted light of the extracellular dye, eliminating the washing step after dye loading. This design is tailored specifically for the HTS of ion channels. Because the inward current of TRPCs is mainly carried by Na^+, leading to depolarization, changes in membrane potentials should provide a measure of the channel activity. We have compared the FMP dye, Fluo-4, and Fura-2 in detecting the activities of TRPC channels expressed in HEK293 cells in 96-well plates using the FlexStation. Both transient and stable expression systems may be used. The transfection method described in Section 1.2.3 is suitable for transient expression and functional testing of TRPC channels in the same 96-well plate. However, stable cell lines give more reproducible and robust signals because of the high percentage of cells expressing the channels.

We have tested HEK293 cells that stably express TRPC3, C4, C5, and C6. The cells are seeded in wells of 96-well plates at 120,000 cells/well and assayed 20 hours later. As shown in Figure 1.1, upon stimulation by carbachol (CCh), cells expressing TRPC3–6 show robust depolarization, whereas nontransfected cells or vector-transfected control cells show a slight hyperpolarization followed by a very small and slow depolarization. By contrast, TRPC-mediated increase in Ca^{2+} influx is quite subtle and difficult to distinguish from the endogenous Ca^{2+} influx using either Fluo-4 or Fura-2 in the plate reader assay (data not shown). As described above, La^{3+}, Sr^{2+}, and Ba^{2+} have been used to facilitate the detection of activity mediated by heterologously expressed TRPC channels. However, the utilities of these cations are not universal for all TRPC isoforms. Using the FlexStation, we found that 10 µM La^{3+} only facilitated the detection of TRPC5 and, to a lesser extent, TRPC3 and C6; Sr^{2+} and Ba^{2+} were only useful for TRPC3 and C6 (M. X. Zhu, unpublished results). The main problem with both Fluo-4 and Fura-2 is the prominent and somewhat variable endogenous responses. This is not a problem for the FMP assay. The small endogenous response in the FMP assay is advantageous, as it provides a method to study the TRPC function with little interference of endogenous background activity. On the other hand, caution should be taken with the FMP assay because quencher dyes present in the mixture are not necessarily pharmacologically neutral, and their potential effects on the channels should be considered.

1.3.2 Ca^{2+} Assay for TRPC4β via Activation of $G_{i/o}$-Coupled Receptors

Using the FMP assay, we found that stimulation of co-expressed µ opioid receptor (µOR), which couples to pertussis toxin (PTX)-sensitive $G_{i/o}$ proteins rather than $G_{q/11}$, activates TRPC4 and C5 but not TRPC3, C6, and C7 (data not shown). Interestingly, when examining the effect of µOR stimulation on $[Ca^{2+}]_i$ changes, we found that DAMGO, a µ agonist, was unable to elicit a Ca^{2+} signal unless TRPC4 or C5 was co-expressed. Among them, TRPC4β gives the most robust $[Ca^{2+}]_i$ increase when µOR is activated (Figure 1.2). Similar results were obtained with TRPC4 and C5 when they were co-expressed with other $G_{i/o}$-coupled receptors, e.g., 5-HT_{1A} serotonin and D_2 dopamine receptors. Therefore, in the HEK293 cells that we use, stimulation of $G_{i/o}$ pathway does not usually activate PLC and the consequent Ca^{2+} signals. On the other hand, $G_{i/o}$ protein activation is coupled to the opening of TRPC4 and C5, especially TRPC4β, channels, allowing Ca^{2+} entry. Although the detailed mechanism is unknown, this phenomenon provides an excellent assay for the HTS of TRPC4 ligands using the Ca^{2+} assay because of the very low background signal.

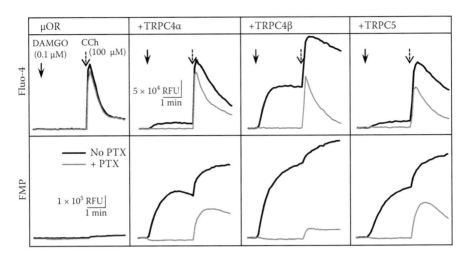

FIGURE 1.2 $G_{i/o}$-mediated activation of TRPC4 and TRPC5, with increases in intracellular Ca^{2+} concentrations and membrane depolarization. HEK293 cells stably co-expressing μ opioid receptor (μOR) and TRPC4α, TRPC4β, or TRPC5 were grown overnight in wells of 96-well plates. Cells were rinsed with an extracellular solution and loaded with Flou-4 AM followed by the FMP dye. Fluorescence signals were read in the FlexStation plate reader with excitation/emission pairs at 530/565 nm for the FMP dye and 494/525 nm for Fluo-4. The reading frequency was 6 s per data point. The μ agonist, DAMGO (1 μM) was added at 30 s. Note: in cells expressing μOR alone, DAMGO did not induce any Ca^{2+} signal (or Fluo-4 fluorescence increase) or membrane depolarization. Co-expression of TRPC4α, TRPC4β, or TRPC5 leads to increases in intracellular Ca^{2+} levels and membrane depolarization in response to DAMGO, with TRPC4β giving the most robust response. These responses were blocked by pre-incubating the cells with pertussis toxin (PTX, 200 ng/mL, 16 hours, gray lines), demonstrating the involvement of $G_{i/o}$ proteins.

1.3.3 CONFIGURATION FOR HTS

Cell-based HTS assays typically use a "mix and measure" formula, in which test compounds are added to samples without washing off the existing medium. The FMP and Ca^{2+} assays have all been formulated in such a manner. The signal-to-background ratios of both assays are quite high. Z′ factors are commonly used to judge the suitability of the assay for HTS. The Z′ factor is determined from the means (μ) and standard deviations (δ) of positive (p) and negative (n) controls from multiple wells of the same plate with the formula: $Z' = 1 - (3\delta_p + 3\delta_n)/|\mu_p - \mu_n|$.[30] Typically, a Z′ factor > 0.5 is considered a good HTS assay. The Z′ factors of FMP assays for TRPC channels typically reach 0.5 to 0.6, as determined by the responses to a saturating concentration of CCh (p) and to no ligand (n) controls. The Z′ factors of the Ca^{2+} assay for TRPC4β in response to stimulation by DAMGO via μOR also approach 0.6. Therefore, these assays are well suited for the HTS of TRPC ligands.

To this end, we have chosen TRPC4β and TRPC6 as primary targets for the HTS of TRPC ligands. The Ca^{2+} assay is used for TRPC4β utilizing the stable cell line that co-expresses mouse TRPC4β with μOR. An attempt was made to uncover activators and allosteric modulators, as well as inhibitors of TRPC4 in a single HTS, using a three-addition protocol. The test compounds are added first. Any potential activator will cause a fluorescence increase in the corresponding well. At 2 min after the addition of the test compounds, DAMGO at EC_{20} concentration is added to all wells. An allosteric modulator is expected to enhance the effect of the low concentration of DAMGO and thereby causing a stronger fluorescence increase. The third addition is DAMGO at the concentration of EC_{Max}. If a blocker is present, the response will be inhibited.

For TRPC6, we have used the FMP assay with a stable HEK293 cell line that expresses mouse TRPC6. The channel is activated through stimulation of the endogenous M3 muscarinic receptors by acetylcholine. An attempt was made to uncover both activators and inhibitors of TRPC6 in a single HTS using a two-addition protocol. Upon adding the test compounds, any potential activator will cause a fluorescence increase in the corresponding well. At 2 min after the addition of the test compounds, acetylcholine at its EC_{Max} concentration is added to all wells. If a blocker is present, the response will be inhibited.

1.4 HIGH-THROUGHPUT SCREENING

1.4.1 TRPC4

1.4.1.1 Cell Preparation and Loading

In preparation for the screen, a large back stock of cells should be grown and stored in liquid nitrogen, as discussed earlier. The development of a large cell stock ensures that all cells assayed will respond similarly in the assay, as they have undergone identical cell culture manipulation. If a cell stock is grown in multiple batches, a quality control assay should be performed to guarantee that each batch of cells responds similarly to control stimuli. Great care should be taken to assure that passage number and technique, along with freezing procedures, are held constant between batches, as deviations from standard protocols could affect cell response in screening assays.

To prepare cells for assay, frozen stocks of cells are defrosted as described above and resuspended in a culture medium supplemented with 10% ΔFBS, 50 IU/mL penicillin, and 50 µg/mL streptomycin. Note that antibiotics are not added to this medium, as their presence may interfere with a test compound effect. Cells are routinely counted using a hemocytometer, and a cell suspension for plating is prepared in a culture medium containing the supplements as described above at a concentration of 300,000 cells/mL. An aliquot of this cell suspension is then added to each well of 384-well assay plates (at 50 µL/well) using either a multichannel pipette or a multidrop dispenser. Cell plates are placed at 37°C, 5% CO_2 overnight until initiation of the screening assay.

1.4.1.2 Compound Library and Screening Protocol

In the current HTS screen, we have utilized the Molecular Libraries Small Molecule Repository (MLSMR) as our compound library, the makeup of which includes natural products, known drugs and toxins, along with compounds having specific protein targets, and structural diversity compounds. A more detailed description of the library makeup can be found on the National Institutes of Health (NIH) Web site (http://mli.nih.gov/mli/compound-repository/mlsmr-compounds/). Stock solutions of library compounds (5 μM in 100% DMSO) are stored at −80°C in sealed 384-well polypropylene microplates until one day prior to assay.

To monitor channel activity, Fluo-4, a calcium-sensitive dye, is used. Fluo-4 stock solutions are made as follows: 1 mg Fluo-4 AM, 400 μL anhydrous DMSO, and 200 μL Pluronic F-127 (20% in DMSO). This stock solution can be stored for up to one week at −20°C without detectable differences in fluorescence intensity. On the day of the assay, stock Fluo-4 AM solution is diluted to 0.25 mg/mL in a 1× assay buffer (1× HBSS, 20 mM HEPES at pH 7.4). Water-soluble probenecid (0.77mg/mL) is also added to the dye solution to inhibit organic anion transporters within the cells. A cell culture medium is manually removed from cell plates and replaced with 20 μL of Fluo-4 AM dye solution using a multichannel pipet or automated multidrop dispenser to ensure homogeneous dye transfer. Cells are incubated for 45 min at room temperature, after which time, the dye is removed and cells washed with 40 μL of the 1× assay buffer to remove any dye that is not associated with cells. This wash solution is then removed, and 20 μL of the 1× assay buffer is added to wells prior to loading on a Hamamatsu FDSS 6000 kinetic imaging plate reader. During the 45-min incubation, test compounds are diluted in the 1× assay buffer to a concentration of approximately 75 μM and DAMGO added to eight control wells (EC_{Max}) at a concentration of 1.8 μM.

After washing, cell plates are loaded into the FDSS where 4 μL/well of test compounds and controls is added to the cell plates, reaching a concentration of around 10 μM for each compound. This protocol is applicable for the detection of TRPC4 activators. For the identification of TRPC4 allosteric potentiators, cells are incubated for 110 s prior to application of 6 μL EC_{20} DAMGO (3 nM). To identify TRPC4 inhibitors, cells are incubated for another 110 s after which 6 μL of DAMGO corresponding to a concentration of EC_{Max} (1.8 μM) is added.

1.4.1.3 Results

Data for all modulator types, including activators, allosteric potentiators, and inhibitors, are processed similarly. For activators, the integrated fluorescence ratio within the window of 1 s after application of the test compound and 5 s prior to application of the EC_{20} concentration of DAMGO is used to determine the compound activity. Compounds that cause instantaneous (in <2 s) fluorescence increase are considered to be autofluorescent and are removed from further analysis. Allosteric potentiator activity is evaluated by measurement of the integrated fluorescence ratio within the window of 2 s after application of EC_{20} and 5 s prior to application of EC_{Max}. Finally, measurement of the integrated fluorescence ratio within the window of 2 s after application of maximally activating DAMGO and the assay end point is used to determine the inhibitor activity. The quality of the data obtained for all three modulator types is evaluated by calculation of the Z′ factor and the signal-to-noise ratio of each assay plate.

1.4.1.4 Hit Selection

For hit selection, integrated ratios are processed using the B score method of normalization.[31,32] In short, this method uses a two-way median polish to account for row and column effects, residuals of which are divided by the median absolute deviation to account for plate-to-plate biases. B score normalization gives a very intuitive score; whereas those compounds that increase the fluorescence signal are given a positive score, those compounds that decrease the fluorescence signal are given a negative score. Hit criteria for activators and allosteric potentiators of TRPC4 are defined as any compounds with a B score greater than that of the mean of all test compounds plus three standard deviations within their respective data set. For inhibitors, hits are defined as those compounds with B score less than the mean of all test compounds minus three standard deviations.

1.4.1.5 Initial Validation

To validate that compounds identified as hits in the primary HTS give reproducible modulation of the fluorescence signal, hit compounds are obtained from MLSMR library, and their activity is assayed using an identical approach as the primary screen. In this assay, however, test compounds are run in duplicate, and their effect on the fluorescence signal is compared to control wells. For activators, a compound must increase the fluorescence signal greater than the mean of the buffer control plus three standard deviations in both repeat plates to be validated as a TRPC4 activator. Similarly, allosteric potentiators are validated if they too are active in both repeat plates increasing the fluorescence signal greater than the mean of the EC_{20} control plus three standard deviations. For inhibitors, the compounds must result in an inhibition of fluorescence increase elicited by EC_{Max} to less than the mean of the EC_{Max} control wells minus three standard deviations in both duplicates.

1.4.2 TRPC6

1.4.2.1 Cell Preparation and Loading

To prepare cells for assay, frozen stocks of cells are defrosted as described above and resuspended in a culture medium that has been supplemented with 10% ΔFBS, 50 IU/mL penicillin, and 50 µg/mL streptomycin. As stated earlier, antibiotics are not added to this medium, as their presence may interfere with test compound effects. Cells are counted using a hemocytometer and a cell suspension for plating prepared using a medium that has been supplemented as described above. A concentration of 300,000 cells/mL is prepared, and 50 µL of this is then added to 384-well assay plates using either a multichannel pipet or a multidrop dispenser. Cell plates are cultured at 37°C, 5% CO_2 overnight.

1.4.2.2 Compound Library and Screening Protocol

TRPC6 channel activity is assayed using the FMP dye. To accomplish this, the FMP dye is diluted in a 1× assay buffer (1× HBSS, 20 mM HEPES pH 7.4). The cell culture medium is manually removed from cell plates and replaced with 20 µL diluted FMP using either a multichannel pipet or an automated multidrop dispenser to ensure the delivery of uniform volumes of dye. Cell plates are then incubated for 45 min at room temperature prior to measurement on a Hamamatsu FDSS 6000 kinetic imaging plate reader. The high-throughput assay for TRPC6 also utilizes the MLSMR compound library as described above for TRPC4. During the 45-min incubation, test compounds are diluted in the 1× assay buffer to a concentration of approximately 75 µM, and acetylcholine is added to eight control wells (EC_{Max}) at a concentration of 60 µM.

After incubation, cell plates are loaded onto the FDSS where 4 μL of test compounds and controls are added to cell plates, reaching a concentration of 10 μM for test compounds and 8 μM acetylcholine for positive controls. This protocol is applicable for the detection of TRPC6 activators. For the identification of TRPC6 inhibitors, cells are incubated with test compounds for 2 min prior to application of 6 μL acetylcholine corresponding to a concentration of EC_{Max} (roughly 60 μM with a final concentration of 10 μM).

1.4.2.3   Results

For activators, the integrated fluorescence ratio within the window of 1 s after application of the test compound and 5 s prior to application of acetylcholine is used to determine the compound activity. The quality of these data is evaluated by determination of the Z' factor and the signal-to-noise ratio of each assay plate. The inhibitor activity is evaluated by measurement of the integrated fluorescence ratio within the window of 2 s after application of maximally activating acetylcholine and the assay end point. The quality of these data is also evaluated by the Z' factor and the signal-to-noise ratio of each assay plate.

1.4.2.4   Hit Selection

As in the TRPC4 assay, the TRPC6 assay uses B score normalization of integrated ratios for hit selection. We expect that compounds that activate the TRPC6 channel will result in a depolarization of membrane potential and therefore increase the fluorescence signal of the voltage-sensitive dye. The hit criterion for activators of TRPC6 is defined as any compounds that lie outside of the mean of all test compounds plus three standard deviations. As for the TRPC4 assay, compounds that cause instantaneous fluorescence increase are not further considered. For inhibitors, we expect them to inhibit membrane depolarization caused by acetylcholine, and therefore, hits are defined as those compounds that result in diminished fluorescence increase to lower than the mean obtained for all test compounds minus three standard deviations.

1.4.2.5   Initial Validation

To validate that those compounds identified as hits in the primary HTS give reproducible modulation of the fluorescence signal, hit compounds are assayed again using an identical approach as the primary screen. In this assay, however, test compounds are run in duplicate, and their effect on the fluorescence signal is compared to control wells. For activators, a compound must increase the fluorescence signal to greater than the mean of the buffer control plus three standard deviations in both repeat plates to be validated as a TRPC6 activator. In the case of inhibitors, compounds must result in an inhibition of the fluorescence signal greater than the mean of the EC_{Max} control wells minus three standard deviations.

1.5   VALIDATION AND OPTIMIZATION

All primary hits need to be further evaluated using several secondary screen and counter screen assays to confirm the activity and the specificity on the channel. Given the different degrees of similarities both in sequences and mechanisms of regulation among TRP channels, it is likely that some compounds will act similarly on several TRP channels, while others are more specific to a particular isoform or subgroup. These drugs are all useful, as they will provide researchers with a choice of manipulating either just one particular type of channel at a time or all related ones at once.

1.5.1 Counter Screens

Cherry-picked compounds are tested for their effects on parental HEK293 cells and stable cell lines that express channels different from the target channel. For example, the potential TRPC4 ligands are tested on cells that express TRPC3, TRPC5, TRPC6, TRPV3, or TRPM8 as well as the parental HEK293 cells. These can be carried out using the FMP assay. It is important to confirm that the compound is negative in the parental cells in order to rule out any nonspecific depolarization effect by the hit compound. Those that affect all TRPCs are candidates for common TRPC activators or blockers, whereas those that only affect TRPC4 are likely to be specific to this TRPC isoform. In addition, alternative assays, such as the FMP assay, are carried out for TRPC4β; this will rule out any assay-dependent effect by the compound. Likewise, cell lines that co-express TRPC4β with other receptors, e.g., $5HT_{1A}$ serotonin, D_2 dopamine, or M_2 muscarinic receptor, are excellent reagents to confirm that the drug is actually targeting the channel, but not the μOR used in the primary screen. Furthermore, the use of the stable cell line expressing TRPC4β alone or co-expressing TRPC4β with a $G_{q/11}$-coupled receptor and stimulating the channel through the $G_{q/11}$-PLC pathway provides another alternative screen to rule out the possibility that the drugs exert their activities through modulation of $G_{i/o}$ signaling. For stimulating TRPC4β via the $G_{q/11}$-PLC pathway, the FMP assay is used.

For TRPC6, counter screens include testing, in a Ca^{2+} assay, the effect of compounds on acetylcholine-induced $[Ca^{2+}]_i$ rise in parental HEK293 cells. This will help exclude those drugs that target the muscarinic receptor. The drugs can then be tested on other TRP channels (C3, C4, C5, V3, and M8) to determine the specificity.

1.5.2 Secondary Screening

Alternative assays other than the ones used for primary screening are used to confirm the effects of the compounds. For TRPC4β, these include the FMP assay, followed by whole-cell patch clamping studies. For selected compounds, full dose response curves are generated. For TRPC6, Ca^{2+} assays can be used with the receptor ligand added in the absence of extracellular Ca^{2+} and re-addition of 1 to 2 mM Ca^{2+} in the medium 3 min later. In Fura2-loaded cells, Ba^{2+}, Sr^{2+}, and Mn^{2+} entry can also be studied.[18] Furthermore, TRPC6 is activated by addition to the medium of diacylglycerols, such as 1-oleoyl-2-acetyl-sn-glycerol (OAG), providing an alternative way to activate the channel without stimulating receptor signaling. Finally, selected drugs are tested in electrophysiological studies.

1.5.3 Compound Optimization

More often than not, the identified compound(s) needs optimization. The details of this procedure are beyond the scope of this book, and optimization should be done in collaboration with experienced medicinal chemists. Any new compounds identified or synthesized on the basis of the analysis by the chemists should be tested through the counter and secondary screening methods discussed above.

1.6 CONCLUDING REMARKS

TRP channels have emerged as important cellular sensors for internal and external changes, with broad physiological and pathophysiological implications. However, the studies of TRP channels are hampered by the poor pharmacological tools available for these channels. Molecular identification, functional expression, development of cell-based assays for HTS, and availability of large compound libraries now offer an excellent opportunity to screen for the ligands of TRP channels. In this chapter, we have described the methods used for HTS of agonists and antagonists of TRPC4 and TRPC6. Primary screening at the NIH-supported Molecular Libraries Probe Centers Network Center at Johns Hopkins University has validated the methodology and yielded a large number of potential channel activators, allosteric modulators, and inhibitors. Continued efforts into the subsequent phases of validation and compound optimization will likely generate useful drugs or probes for TRPC4, TRPC6, and related channels. Similar approaches should be applicable to other channels.

ACKNOWLEDGMENTS

This work is supported by NIH grants NS056942 (to MXZ), 5U54MH074427-03 (to DW), and GM078579 and MH084691 (to ML).

REFERENCES

1. Clapham, D. E. 2003. TRP channels as cellular sensors. *Nature* 426: 517.
2. Hofmann, T., A. G. Obukhov, M. Schaefer, C. Harteneck, T. Gudermann, and G. Schultz. 1999. Direct activation of human TRPC6 and TRPC3 channels by diacylglycerol. *Nature* 397: 259.
3. McKemy, D. D., W. M. Neuhausser, and D. Julius. 2002. Identification of a cold receptor reveals a general role for TRP channels in thermosensation. *Nature* 416: 52.
4. Hu, H. Z., Q. Gu, C. Wang et al. 2004. 2-aminoethoxydiphenyl borate is a common activator of TRPV1, TRPV2, and TRPV3. *Journal of Biological Chemistry* 279: 35,741.
5. Jordt, S. E., D. M. Bautista, H. Chuang et al. 2004. Mustard oils and cannabinoids excite sensory nerve fibres through the TRP channel ANKTM1. *Nature* 427: 260.
6. Putney, J. W. 2005. Physiological mechanisms of TRPC activation. *Pflügers Arch* 451: 29.
7. Möller, C. C., J. Flesche, and J. Reiser. 2009. Sensitizing the slit diaphragm with TRPC6 ion channels. *Journal of American Society of Nephrology* 20: 950.
8. Guinamard, R., and P. Bois. 2007. Involvement of transient receptor potential proteins in cardiac hypertrophy. *Biochimica et Biophysica Acta* 1772: 885.
9. Tsvilovskyy, V. V., A. V. Zholos, T. Aberle et al. 2009. Deletion of TRPC4 and TRPC6 in mice impairs smooth muscle contraction and intestinal motility in vivo. *Gastroenterology* 137: 1415.
10. Hartmann, J., E. Dragicevic, H. Adelsberger et al. 2008. TRPC3 channels are required for synaptic transmission and motor coordination. *Neuron* 59: 392.

11. Kim, M. S., J. H. Hong, Q. Li et al. 2009. Deletion of TRPC3 in mice reduces store-operated Ca²⁺ influx and the severity of acute pancreatitis. *Gastroenterology* 137: 1509.

12. Greka, A., B. Navarro, E. Oancea, A. Duggan, and D. E. Clapham. 2003. TRPC5 is a regulator of hippocampal neurite length and growth cone morphology. *Nature Neuroscience* 6: 837.

13. Riccio, A., Y. Li, J. Moon et al. 2009. Essential role for TRPC5 in amygdala function and fear-related behavior. *Cell* 137: 761.

14. Dietrich, A., Y. Mederos, M. Schnitzler et al. 2005. Increased vascular smooth muscle contractility in TRPC6⁻/⁻ mice. *Molecular Cell Biology* 25: 6980.

15. Inoue, R., T. Okada, H. Onoue et al. 2001. The transient receptor potential protein homologue TRP6 is the essential component of vascular α1-adrenoceptor-activated Ca²⁺-permeable cation channel. *Circulation Research* 88: 325.

16. Jung, S., R. Strotmann, G. Schultz, and T. D. Plant. 2002. TRPC6 is a candidate channel involved in receptor-stimulated cation currents in A7r5 smooth muscle cells. *American Journal of Physiology - Cell Physiology* 282: C347.

17. Schultheiss, G., M. Frings, G. Hollingshaus, and M. Diener. 2000. Multiple action sites of flufenamate on ion transport across the rat distal colon. *British Journal of Pharmacology* 130: 875.

18. Zhu, M. X., M. Jiang, and L. Birnbaumer. 1998. Receptor-activated Ca²⁺ influx via human Trp3 stably expressed in HEK293 cells: evidence for a non-capacitative Ca²⁺ entry. *Journal of Biology and Chemistry* 273: 133.

19. Xu, S. Z., F. Zeng, G. Boulay, C. Grimm, C. Harteneck, and D. J. Beech. 2005. Block of TRPC5 channels by 2-aminoethoxydiphenyl borate: a differential, extracellular and voltage-dependent effect. *British Journal of Pharmacology* 145: 405.

20. Ma, H. T., R. L. Patterson, D. B. van Rossum, L. Birnbaumer, K. Mikoshiba, and D. L. Gill. 2000. Requirement of the inositol trisphosphate receptor for activation of store-operated Ca²⁺ channels. *Science* 287: 1647.

21. Ma, H. T., K. Venkatachalam, K. E. Rys-Sikora, L. P. He, F. Zheng, and D. L. Gill. 2003. Modification of phospholipase C γ-induced Ca²⁺ signal generation by 2-aminoethoxydiphenyl borate. *Biochemical Journal* 376: 667.

22. Lievremont, J. P., G. S. Bird, and J. W. Putney, Jr. 2005. Mechanism of inhibition of TRPC cation channels by 2-aminoethoxydiphenylborane. *Molecular Pharmacology* 68: 758.

23. Kiyonaka, S., K. Kato, M. Nishida et al. 2009. Selective and direct inhibition of TRPC3 channels underlies biological activities of a pyrazole compound. *Proceedings of the National Academy Sciences of the United States of America* 106: 5400.

24. Pelletier, J., and N. Sonenberg. 1988. Internal initiation of translation of eukaryotic mRNA directed by a sequence derived from poliovirus RNA. *Nature* 334: 320.

25. Leclere, P. G., A. Panjwani, R. Docherty, M. Berry, J. Pizzey, and D. A. Tonge. 2005. Effective gene delivery to adult neurons by a modified form of electroporation. *Journal of Neuroscience Methods* 142: 137.

26. Liao, Y., C. Erxleben, J. Abramowitz et al. 2008. Functional interactions among Orai1, TRPCs, and STIM1 suggest a STIM-regulated heteromeric Orai/TRPC model for SOCE/Icrac channels. *Proceedings of the National Academy Sciences of the United States of America* 105: 2895.

27. Vazquez, G., B. J. Wedel, M. Trebak, G. St John Bird, and J. W. Putney, Jr. 2003. Expression level of the canonical transient receptor potential 3 (TRPC3) channel determines its mechanism of activation. *Journal Biological Chemistry* 278: 21649.

28. Schaefer, M., T. D. Plant, A. G. Obukhov, T. Hofmann, T. Gudermann, and G. Schultz. 2000. Receptor-mediated regulation of the nonselective cation channels TRPC4 and TRPC5. *Journal of Biological Chemistry* 275: 17517.

29. Zheng, W., R. H. Spencer, and L. Kiss. 2004. High throughput assay technologies for ion channel drug discovery. *Assay Drug Development Technology* 2: 543.

30. Zhang, J. H., T. D. Chung, and K. R. Oldenburg. 1999. A simple statistical parameter for use in evaluation and validation of high throuput screening assays. *Journal of Biomolecular Screening* 4: 67.
31. Brideau, C., B. Gunter, B. Pikounis, and A. Liaw. 2003. Improved statistical methods for hit selection in high-throughput screening. *Journal of Biomolecular Screening* 8: 634.
32. Malo, N., J. A. Hanley, S. Cerquozzi, J. Pelletier, and R. Nadon. 2006. Statistical practice in high-throughput screening data analysis. *Nature Biotechnology* 24: 167.

2 Methods to Study TRPC Channel Regulation by Interacting Proteins

Kirill Kiselyov, Kyu Pil Lee, Joseph P. Yuan, and Shmuel Muallem

CONTENTS

2.1 INTRODUCTION

TRPC (transient receptor potential, canonical) is a family of ion channels that mediate changes in plasma membrane (PM) cation permeability induced by stimulation of phospholipase C (PLC). PLC is a cellular integrator that receives input from G protein coupled receptors (GPCR) and from receptor tyrosine kinases (RTK). A central cellular function of TRPC channels is to regulate Ca^{2+} influx across the PM induced by stimulation of these receptors by hormones, neurotransmitters, and growth factors. Specific TRPC channels may also be activated by other means such as membrane stretch and changes in cell volume. Physiological, biochemical, and molecular studies revealed that TRPC channel activity is regulated by their interaction with many auxiliary proteins. Although many channels contain functional subunits, TRPC channels are unique in that their activity is modulated rather than enabled by the auxiliary proteins. Such context dependence of TRPC channel activity necessitates the development of tools that enable enumeration and exploration of the TRPC channel auxiliary proteins and their roles in the channel function. This chapter focuses on these tools and on the results and conclusions made possible because of the use of these tools.

2.2 CONTEXT-SPECIFIC REGULATION OF TRPC CHANNELS

TRPC channels are nonselective, Ca^{2+} permeable cation channels that include six closely related members in humans (seven in mice) that are also members of the large superfamily of TRP ion channels.[1-3] TRP is a strikingly versatile family of channels responsible for a wide array of sensory functions,[1-3] as well as for less understood aspects of membrane permeability. Several TRP members, including TRPC channels, have been implicated in pathological processes;[4] TRPC family members are involved in cell growth and differentiation,[5-9] kidney function,[10-12] blood pressure and vascular tone regulation,[13-17] pancreatic[18,19] and salivary gland secretion,[20] and likely other organism-level functions.

All TRPC channels are activated downstream of PLC stimulation, which, in turn, is induced by stimulation of plasma membrane receptors.[1-3] The GPCR subtypes that couple to the $G\alpha_{q/11}$ type of heterotrimeric G proteins activate PLCβ, while stimulation of RTK activates PLCγ. Both PLC isoforms cleave the minor membrane lipid phosphatidylinositol 4,5-bisphosphate (PIP_2), yielding two intracellular second messengers: diacylglycerol (DAG) and inositol 1,4,5-trisphosphate (IP_3). Both PLC isoforms are regulated by other inputs, and thus, they serve as cellular integrators that translate various extracellular stimuli into DAG and IP_3 production.[21-23]

DAG remains in the membrane and activates (or its products do) a set of signaling molecules that includes several TRPC channels (TRPC2, TRPC3, TRPC6, and TRPC7).[1-3] IP_3 diffuses into the cytoplasm and induces opening of IP_3 receptors (IP_3R), which are ligand-gated ion channels that reside in the endoplasmic reticulum (ER).[24,25] The ER is the main source of intracellular Ca^{2+} that penetrates the entire cytoplasm and allows the delivery of Ca^{2+} to various parts of the cell without compromising the cellular homeostasis. Binding of IP_3 and subsequent activation of IP_3R leads to the release of Ca^{2+} from the ER into the cytoplasm, where it regulates a vast array of cellular processes. Because the PM Ca^{2+} pumps extrude some of the ER Ca^{2+} that is released during stimulation out of the cell, prolonged cell stimulation leads to depletion of the ER Ca^{2+}.[21,26] The ER Ca^{2+} content is gauged by a recently discovered ER Ca^{2+} sensor STIM1[27,28] and is reported to the PM in the form of activation of the PM Ca^{2+} influx channels. The Ca^{2+} influx mediated by these channels makes available the Ca^{2+} that is necessary to maintain the sustained Ca^{2+} response and to restore the ER Ca^{2+} content between Ca^{2+} spikes and at the end of the stimulation period.[21,26] Two sets of channels are ascribed the role of store-operated Ca^{2+} channels (SOCs): the recently found Orai proteins[29-31] and a subset of TRPC channels.[32,33]

The structural determinants underlying the two modes of TRPC channel activation are only beginning to emerge. The DAG-sensitive TRPC subset is prominently regulated by delivery to the PM.[18,34-37] Several lines of evidence show that the presence of these channels on the PM is increased following cell stimulation.[18,34-38] It has been suggested that some TRPC channels exist in a preactivated mode in delivery vesicles underneath the PM and that the vesicles fuse with the PM upon stimulation. The resulting net increase in TRPC channels in the PM is responsible for the increase in the PM ion permeability.[18,34–37] There is little evidence that the store-sensitive TRPC channels behave in this manner. Pair-wise amino acid comparison of DAG-activated and store-operated TRPC channels, together with a large-scale computational approach, indicates an amino acid stretch in TRPC channels termed TRPC_2 as a possible structural determinant of TRPC sensitivity and their regulation by delivery to the PM.[36]

This apparent dichotomy of TRPC channel activation mechanisms is probably an oversimplification that does not accommodate complex modes of TRPC behavior in different expression systems and cellular contexts. For example, it has been shown that deletion of the DAG-sensitive TRPC3 channels leads to the loss of store-dependent Ca^{2+} influx.[9,19,39] In recombinant systems, it has been shown that the ability of TRPC3 to be activated by DAG is inversely correlated with the TRPC3 expression levels, while TRPC3 sensitivity to store depletion shows an opposite trend.[40] These and other observations gave rise to the suggestion that the TRPC channel gating modes depend on the TRPC channel expression levels and cellular context, i.e., their interaction with other proteins involved in Ca^{2+} signaling.[32,41,42]

2.3 TRPC INTERACTION WITH AUXILIARY PROTEINS

The *Drosophila* photoreceptor TRP channel, the founding member of the TRP superfamily, exists in a complex tethered by the multivalent scaffold InaD that bears multiple PDZ domains.[43,44] Interaction of TRP, PLC, and protein kinase C with selective PDZ domains in InaD brings these signaling molecules into close proximity.[43–46] The InaD mediated multivalent binding is responsible for the precise organization and localization of the light signaling proteins in the ommatidia. The functional consequence of such organization is the immensely rapid and high-fidelity signal transduction that takes place during photoreception in the *Drosophila* eye.[43–46] At present, there is fragmentary information on the role of PDZ domain scaffolds in the TRPC channel function.[47–49] Although the role of PDZ domain scaffolds deserves to be explored extensively, this chapter will focus on other TRPC interacting proteins for which better information is available.

The first indication that TRPC channels may behave differently in different cellular contexts came from the discovery that co-expression of different TRPC isoforms produces a current whose characteristics are different from any of the individual isoforms.[50–52] Targeting and insertion in the PM is also changed by co-expression of combinations of TRPC channels.[50,51,53] These observations lead to the suggestion that TRPC channels share a similar pore structure and that TRPC heteromultimer formation results in combining the channel pore sequences into shared or interdependent pores.

Calmodulin (CaM) is a versatile cytoplasmic Ca^{2+} sensor. Conformational changes induced in CaM by binding Ca^{2+} actuate various cellular processes that include gene expression and activity of various cellular switches. CaM was shown to bind several TRPC channels and regulate their activity.[54–59] Whether its interaction with the putative IP_3R binding sites in TRPC channels is the structural determinant of such regulation remains to be clarified. Nonetheless, CaM–TRPC channel interaction and the possible feedback loop provided by CaM bound to the Ca^{2+} that is brought into the cell by the activated TRPC channels create an important precedent for regulatory loops provided by the TRPC channel-interacting proteins. This is likely to be similar to the regulation of Orai1 by CaM, which appears to mediate fast Ca^{2+}-dependent inactivation of Orai1.[60]

Junctate is another protein suggested to modulate TRPC channel gating. Although it has been shown to associate with some TRPC channels and IP_3Rs,[61,62] and to modulate Ca^{2+} signaling in muscle,[63,64] little is known about its role in modulating TRPC channel activity.

Homer proteins have been cloned as a result of a search for immediate early genes whose expression changes in response to neuronal stress.[65–69] In humans, three genes code the Homer family of scaffolds.[70–73] Each gene (*Homer 1, 2,* and *3*) yields splice variants of the full-length Homer proteins bearing an EVH domain that binds the proline-rich sequences PP(X)F or the LPSSP motifs in target proteins, and a coiled-coil/leucine zipper domain responsible for multimerization.[70–75] An immediate early gene product, Homer 1a, that is upregulated during neuronal stress, lacks the coiled-coil domain and thus works as a dominant negative isoform of the full-length Homer action. In a paradigm-setting series of studies, it was established that binding of two (or more) Homers to the proline-rich sequences in IP_3R and mGluR holds the two molecules in a complex and facilitates the signal transduction between them.[67,72,76] Expression of Homer 1a or deletion of the Homer-binding sequences in mGluR retards the signal transduction between these molecules.[67,72,76] TRPC channels and IP_3Rs have proline-rich sequences capable of Homer binding, and Homer-mediated interaction of TRPC channels and IP_3Rs is responsible for correct gating of TRPCs.[74] These results constitute the first comprehensive analysis of the effect of cellular context on TRPC channel activity. Hence, it is possible that differential expression of Homer isoforms and Homer levels, perhaps as a result of cell stimulation, can modulate TRPC channel ability to respond to cellular stimuli. Indeed, Homer 1 and Homer 3 show differential membrane localization and in different poles in polarized cells,[77] which may allow differential activation of channels expressed in each membrane or cellular site.

Another recently discovered protein that regulates TRPC channel activity in response to ER Ca^{2+} store depletion is STIM1.[27,28] STIM1 was identified in siRNA screens for proteins that mediate activation of SOCs in response to depletion of ER Ca^{2+} in *Drosophila* and in human HEK293 cell line.[27,28] STIM1 is an ER Ca^{2+} sensor that bears Ca^{2+} binding EF hand in its N-terminus, which is immersed in the ER lumen.[27,28,78] As a result of ER store depletion, Ca^{2+} dissociates from STIM1 to allow formation of STIM1 clusters near the PM and STIM1 interaction with the SOCs.[27,28,32,33,79–81] It is not clear whether STIM1 commands the ER reorganization accompanying store depletion or whether it is passively transported toward the PM during the ER reorganization that is triggered by Ca^{2+} depletion. Nonetheless, a solid body of evidence shows that STIM1 is essential for activating the Ca^{2+} influx by native Ca^{2+} channels and by TRPC channels in response to ER Ca^{2+} store depletion.[82]

Another group of channels gated by STIM1 are the Orai channels.[30,31,33] Orai1 is a four-transmembrane domain protein that constitutes the channel that mediates I_{CRAC},

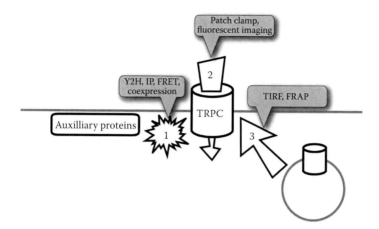

FIGURE 2.1 The key paradigms and approaches in studying TRPC regulation by auxiliary proteins. 1, TRPC channel interaction with auxiliary proteins can be probed using Y2H and verified using immunoprecipitation (IP) and FRET with the native or co-expressed proteins. 2, TRPC channel activity can be monitored by patch clamp current recording or fluorescent Ca^{2+} measurements. 3, The delivery of TRPC channels to the PM can be dynamically monitored using TIRF and FRAP.

a distinct Ca^{2+} current activated by the ER Ca^{2+} store depletion.[31,83–87] The Orai complex with STIM1 is necessary for gating of I_{CRAC} by store depletion, and formation of the complex depends on the STIM1 SOAR domain.[88] In addition, Orai interacts with TRPC channels, and this interaction is essential for the activity of Orai1 and TRPC channels in response to the ER Ca^{2+} store depletion.[89–91]

It was possible to obtain the results discussed above and to formulate the paradigm of TRPC channel regulation by auxiliary proteins owing to the use of an array of modern techniques. These techniques allowed prediction of proteins that interact with TRPC channels and mapping and testing the functional significance of such interactions. The remainder of this review will focus on the specific techniques employed in studying TRPC interacting proteins. Sample scenarios of exploration of several aspects of the regulation of TRPC channels by interacting proteins will be provided. Figure 2.1 illustrates the key paradigms of TRPC regulation by interacting proteins and the techniques that have been used to obtain the data for these paradigms.

2.4 EXPLORATION TECHNIQUES

TRPC channels appear to depend on a network of proteins that define the specific mode of their response to cell stimulation. Such context-dependent mode of activity is rare among ion channels. There are numerous implications of the context-dependent gating of TRPC channels; chief among them are expanding the repertoire of responses and fine-tuning the responses to specific stimuli. Detailed investigation of the signaling network of TRPC channels requires developing a comprehensive set of tools capable of enumerating the members of the TRPC channel protein network and delineating their function. The remainder of this chapter will focus on such techniques and examples of the experimental data obtained because of the use of these techniques.

2.4.1 PATCH CLAMP AND CA²⁺ RECORDINGS

Several cell types appear to be enriched in specific TRPC isoforms (e.g., human submandibular cell lines seem to express high levels of TRPC1[92]), while pancreatic and salivary gland cells express high levels of TRPC1, TRPC3, and TRPC6.[18] The function of native channels can also be studied by knockdown with siRNA probes, and now there are knockout mice for almost all TRPC channels.[93,94] Recombinant TRPC channels have been introduced in a vast array of cell lines using transient transfection and stable expression techniques. A combination of these conditions allows analyzing the native and recombinant TRPC channel activity. An assumption is usually made that the recombinant expression of TRPC channels leads to overwhelming presence of the TRPC construct of interest, and therefore, the resulting current represents the pure or nearly pure current mediated by the expressed TRPC isoform. This is possible with several TRPC channels that have distinctive pore properties. Nevertheless, problems inherent to the use of this assumption will be discussed at the end of this section.

Evaluation and comparison of TRPC channel activity using physiological techniques such as patch clamp and Ca^{2+} recordings using fluorescent dyes are possible because TRPC channels conduct current because of the ion flux through the open channels and because they provide an entry pathway for ions whose concentration in the cytoplasm can be easily monitored. Perhaps the best example of such ion is Ca^{2+} and its substitutes, Ba^{2+}, Sr^{2+}, and Mn^{2+}. Under resting conditions, the free cytoplasmic Ca^{2+} level is extremely low (10^{-7}–10^{-9} M), and the concentration of other divalent ions is even lower or nil, making it possible to reliably isolate minute changes in the concentration of these ions with fluorescent dyes such as Fura-2. The advantage of using Ba^{2+} and Mn^{2+} is that the cellular Ca^{2+} pumps do not transport them, which allows clear separation between Ca^{2+} influx and Ca^{2+} release and therefore isolation of the unidirectional divalent ion influx. This technique is very useful, especially when comparing between conditions and when there is a need to isolate small changes in influx activity. The technique and the protocol for using Ba^{2+} and Mn^{2+} to evaluate Ca^{2+} influx are illustrated in Figure 2.2.

A particular advantage of documenting TRPC activity using fluorescent dyes is that it provides high-throughput means of measuring channel activity before and after stimulation and between different samples. Using a plate reader can further increase the throughput. The key premise of the experiments conducted with the use of fluorescent Ca^{2+} indicators is that the TRPC channels opening results in the increased PM permeability for Ca^{2+} and other divalent cations and therefore an increase in the cellular concentration of these ions. When measuring Ca^{2+} influx, it is necessary to design the experiments in a way that allows separation of the ER Ca^{2+} release from the Ca^{2+} influx signal. The use of Ba^{2+} or Mn^{2+} as reporters of Ca^{2+} influx circumvents this problem. In this case, extracellular Ca^{2+} is removed before stimulation of cells expressing the desired TRPC channel. The stimulation triggers

Ca^{2+} release from ER stores, a process unaffected by the expression of TRPC. After the Ca^{2+} release phase is completed, Ba^{2+} or Mn^{2+} is added to the extracellular solution. The Fura-2 ratio will increase because of the Ba^{2+} influx, while the net Fura-2 fluorescence (excited at 360 nm) will decrease because of Mn^{2+} entry into the cell. Usability of this technique is somewhat limited in the cases when TRPC expression is low or if the host cell displays measurable levels of native influx activity for divalent cations. Another problem with this technique is inability to clamp the membrane potential. Therefore, changes in membrane potential during cell stimulation, whether these changes result from TRPC channel activity, introduce some uncertainty in the recordings. Thus, apparent reduction in the Ca^{2+} influx may result simply from PM depolarization because all TRPC channels also conduct Na^+ and K^+ and can thus depolarize the PM. Nevertheless, this technique has been extremely useful for delineating the key features of TRPC channel activity and regulation, such as activation by receptor stimulation or DAG and modulation by drugs or auxiliary proteins.

Whole cell current recording provides an excellent means to analyze TRPC channel activity. As channel opening results in ion current flow, the changes in TRPC activity can be easily documented as changes in current amplitude or more precisely as current density (net current amplitude per unit of cell capacitance). The latter measure is used to correct for varying cell size. The two possible modes of channel recording are the whole cell configuration to measure the current of all active channels in the cell and single channel recording to analyze the activity of single TRPC channels. The two modes with typical current–voltage relationship for the various TRPC channels and single channel properties of TRPC3 when activated by Homer 1 proteins are shown in Figure 2.3.

Each technique has advantages and limitations. The whole-cell technique allows reliable estimation of the key macro-characteristics of the TRPC channel-mediated currents that describe the key characteristics of the channel, such as selectivity and permeability. It allows cell stimulation by changing the extracellular solution. One limitation of the whole-cell configuration is that access to the cytoplasmic face of the membrane is restricted to the content of the patch pipette. Nevertheless, the whole-cell technique remains a major driving force in TRPC channel studies. Using this technique, the key characteristics of TRPC channel regulation (e.g., net current, selectivity, activation by agonists and DAG, and dependence on auxiliary proteins) have been determined along with the key features of their ion conductance characteristics (e.g., relatively low selectivity for Ca^{2+} over Na^+). These topics have been reviewed extensively.[1–3,41,42,95] As indicated before, identification of the current mediated by a particular TRPC channel is aided by their specific current–voltage relationships. These have been determined for all TRPC channels and for combination of expressed TRPC isoforms. Figure 2.3a shows the current–voltage (I/V) relationships for monovalent current most typically reported for the TRPC channels. The I/V relationships for TRPC1, TRPC3, and TRPC7 are nearly linear, while those for TRPC4, TRPC5, and TRPC6 are double rectifying. The extent of linearity and steepness of the double rectification is typical to these TRPC isoforms.

The single channel modes of the patch clamp technique, such as cell-attached and inside-out modes, allow documenting fine and selective characteristics of TRPC channel activity, and the inside-out mode allows access to the intracellular face of

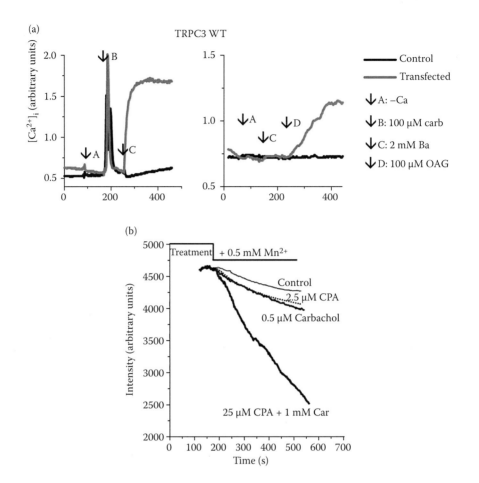

FIGURE 2.2 The use of the fluorescent dye Fura-2 to measure agonist- or OAG-induced TRPC activity: sample experimental protocols to measure the TRPC channel activity. The experimental details are described below. A complication in all these protocols is the native SOC activity. Although the native SOC activity can be inhibited by 1 μM Gd^{3+}, Gd^{3+} can also inhibit at least in part the TRPC channel activity. Therefore, all the protocols below require controls to measure the native divalent ion fluxes and subtract them from the fluxes measured in cells expressing the desired TRPC channels. (a, left) An experiment aimed to measure the TRPC channel activity induced by stimulation of GPCR or by passive Ca^{2+} store depletion using the SERCA pump inhibitor thapsigargin and Ba^{2+} as a Ca^{2+} surrogate. In the beginning of the experiment, cells are placed in a Ca^{2+} free solution in order to separate Ca^{2+} release and influx events. Next, the agonist or thapsigargin is applied to deplete ER Ca^{2+} and activate the TRPC channels. After completion of Ca^{2+} release and return of Ca^{2+} to the basal level, a Ba^{2+}-containing buffer is introduced; the amplitude of baseline deflection induced by Ba^{2+} application is a measure of the Ba^{2+} influx magnitude and, therefore, the activity of TRPC channels. Right: an experiment aimed to measure the TRPC activity stimulated by OAG using Ba^{2+} as a Ca^{2+} surrogate. In the beginning of the experiment, the cells are placed in a Ba^{2+}-containing buffer in order to measure the magnitude of the basal Ba^{2+} influx. The small upward deflection of the baseline in this panel represents the basal Ba^{2+} influx.

FIGURE 2.2 (Continued) Next, OAG is added; the amplitude of baseline deflection induced by OAG application is a measure of the Ba^{2+} influx mediated by the OAG-activated TRPC channels. (Modified from Singh, B.B. et al., *Molecular Cell* 15, 635, 2004.) (b) An experiment aimed at measuring the TRPC channel activity induced by the ER Ca^{2+} store depletion using Mn^{2+} influx. Mn^{2+}-induced quenching of Fura-2 is measured as a loss of Fura-2 fluorescence. Untreated cells were placed in a Mn^{2+}-containing buffer in order to measure the magnitude of the basal Mn^{2+} influx (thin solid line). In parallel experiments, cells were treated with carbamylcholine, cyclopiazonic acid (CPA), or a combination of the two. The rate (slope) and the magnitude of Fura-2 quench induced by the Mn^{2+} influx are the measure of the Ca^{2+} channel activity in the PM. (Modified from Yao, J. et al., *Journal of Biological Chemistry* 279, 21511, 2004.) Sample setup. *Buffers.* The standard Fura-2 loading and cell bathing solution contains (in mM) 140 NaCl, 5 KCl, 2 $CaCl_2$, 2 $MgCl_2$, 10 HEPES (pH 7.2). $CaCl_2$ is omitted in the Ca^{2+}-free buffer; the Ba^{2+} buffer contains 2 mM Ba^{2+} instead of Ca^{2+}. Buffers are applied by continuous perfusion (preferable) or by direct injection into the chamber accompanied by the immediate aspiration of the excess buffer. Each buffer application is equal to 5 to 10 chamber volumes in order to guarantee nearly complete replacement of the buffer in the chamber. *Cell loading.* The cells attached to cover slips are loaded with 1–5 µM Fura-2AM for 20–60 min at 37°C in a 5% CO_2 incubator and then incubated without Fura-2AM for 30 min to allow complete dye hydrolysis. The Fura-2 concentration and loading times differ depending on the cell type. The cells are used for experiments immediately after the loading. *Fluorescent measurements.* Experiments are performed in a dark room. The cover slip is placed in an experimental chamber and secured to the stage of an inverted microscope supplied with an objective capable of transmitting UV light, with an appropriate filter cube with a cutoff filter of <500 nm, source of excitation light, recording device, and software package. Fura-2 fluorescence is recorded at excitation wavelength of 340 and 380 nm (Ca^{2+} and Ba^{2+}) or the isosbestic point of 360 nm (Mn^{2+}). Light emitted at wavelength >500 nm is collected with a photomultiplier device or a CCD camera. Complete suitable recording systems are available from many vendors. *Data interpretation.* An increase in cytoplasmic Ca^{2+} or Ba^{2+} results in change in the Fura-2 fluorescence ratio recorded at 340/380 nm. Hence, fluorescence emitted by the sample that is being excited by a 340-nm light (excitation optima for Ca^{2+} bound Fura-2) increases, while fluorescence emitted by the sample that is being excited by a 380-nm light (excitation optima for Ca^{2+} free Fura-2) decreases. The ratio of fluorescence intensity induced by a 340-nm light over fluorescence intensity induced by a 380-nm light represents the fraction of Fura-2 bound to Ca^{2+} and, therefore, the cytoplasmic Ca^{2+} or Ba^{2+} concentrations. Consequently, plotting the ratio of fluorescence intensity against time yields the time profile of Ca^{2+} or Ba^{2+} changes in the cytoplasm. Ca^{2+} measurements obtained using this approach can be calibrated in terms of the absolute Ca^{2+} concentrations (see Chapter 15). In the experiment provided in the left panel of a, an agonist was applied while the cells were bathed in a Ca^{2+}-free buffer. A spike in the fluorescence ratios caused by agonist application represents Ca^{2+} release from the ER Ca^{2+} store, as no Ca^{2+} influx from outside the cell is possible under these conditions. The stable increase in the fluorescence ratio induced by Ba^{2+} application illustrates the Ba^{2+} influx through active channels in the PM. The signal remains stable since the SERCA and PMCA cannot pump Ba^{2+}. The magnitude of this influx can be compared between different samples either directly or normalized to the magnitude of Ca^{2+} release or Ba^{2+} influx in cells of which membranes were made permeable using the Ca^{2+} ionophore, ionomycin.

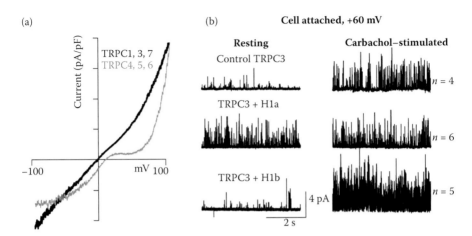

FIGURE 2.3 Recordings of TRPC-mediated whole-cell current and TRPC3 single channel activity using a cell-attached configuration. (a) Representative current–voltage characteristic obtained from cells expressing several recombinant TRPC channels activated by GPCR stimulation. (b) Single channel recording of the TRPC3 current in a cell-attached patch obtained from HEK 293 cells expressing TRPC3 and the indicated Homer 1 isoforms and stimulated with the M3 GPCR. Recording conditions are described in the following protocol. Sample setup. *Buffers.* Whole-cell mode: the standard bathing solution described above is used in the beginning of the experiment. In order to eliminate native Cl⁻ permeability, Cl⁻ in intracellular and extracellular solutions is replaced by large organic anions such as Aspartate or Methanesulfonate. The pipette solution contained (in mM): 140 Cs-aspartate (to eliminate native K$^+$ currents), 1 Mg-ATP, 0.25 BAPTA, 10 HEPES (pH 7.3); Ca^{2+} can be added and BAPTA concentration can be varied accordingly in order to set the desired cytoplasmic Ca^{2+} buffering. Pipette resistance is in the 1–4 GOhm range in order to minimize the access resistance. Membrane conductance is tested by changing the holding membrane potential using rapid alteration of membrane potential (RAMP) or step-wise protocols. Buffers are applied by continuous perfusion (preferable) or by direct injection into the chamber accompanied by the immediate aspiration of the excess buffer. Each buffer application is equal to 5 to 10 chamber volumes in order to guarantee nearly complete replacement of the buffer in the chamber. *Cell-attached mode:* In order to nullify the membrane potential and control the membrane potential on the patch, Na$^+$ in the standard bathing solution is replaced by K$^+$. The pipette solution contains (in mM) 140 Na-aspartate (or another large organic ion in order to negate Cl⁻ conductance), 1 Ca^{2+}, 10 HEPES (pH 7.6). Na$^+$ can be replaced with another ion of interest, and Ca^{2+} can be omitted from the media and the pipette solutions, as desired. Bath solution exchange can be performed as described above. Because of the low amplitudes of currents in the cell-attached mode, extra care should be taken in electrically isolating the setup from sources or electrical noise. Complete suitable recording systems for both whole-cell and cell-attached modes are available from many suppliers. *Data interpretation.* Whole-cell mode: channel activation in the whole-cell mode will result in an increase in the net current through the PM. The current amplitude can be normalized to cell capacitance in order to obtain the current density. Larger current density corresponds to increased channel activity. The current–voltage characteristic obtained using RAMP or step-wise voltage protocols in the bathing solution with different cations is used to establish the selectivity profile of the given TRPC channel to various cations. Note that TRPC channels conduct minimal Ca^{2+} current, and it is customary to measure the Na$^+$ current mediated by these channels.

the channel. The suggestions that TRPC channels may be regulated by IP_3Rs or CaM have been made using this technique.[96–99] This technique is only limited by the conductance of the channel that must be sufficiently large to reliably isolate the current. Single channel recording allows precise characterization of the channel biophysical properties and defining the channel fingerprints. The usefulness of this technique in studying channel regulation is illustrated in Figure 2.3b showing an example of single TRPC3 current, its modulation by the scaffolding protein Homer 1, and the lack of effect of Homer 1a. Co-expression of TRPC3 with Homer 1 increased the productive complexes of TRPC3–Homer 1–IP_3Rs, TRPC3 expression at the PM (not shown), and the activity and the number of active channels in the patch. On the other hand, co-expression of TRPC3 with Homer 1a resulted in no increase in the channel number, but the dissociation of the TRPC3–Homer 1–IP_3R complex by Homer 1a locked TRPC3 in a spontaneously active state.

A combination of patch-clamp and fluorescent measurements allows comprehensive description of the TRPC activity as a function of stimulation or cellular environment. They do not illuminate changes in TRPC channel localization and, when used alone, do not show to which extent the environment affects the channel activity. Therefore, the analysis of TRPC channel regulation by monitoring the channel function should be complemented by other experimental approaches that control or enumerate the channel environment.

2.4.2 YEAST TWO-HYBRID SCREEN AND BIOCHEMICAL ISOLATION OF TRPC CHANNEL COMPLEXES

As with other proteins, the quest for new proteins that interact with TRPC channels relies on inference based on the presence of sequences that are predicted to form binding motifs or ligands (as is the case with Homer) or on identifying TRPC channels interacting proteins using standard screening methods.

Yeast two-hybrid screen (Y2H) is a powerful protein interaction assay whose predictive power has been utilized in numerous biological systems. Y2H relies on the ability of yeast gene expression regulatory elements to induce the expression of reporter genes when the regulatory elements physically interact. Although Y2H screens have not been widely used to identify TRPC channel interacting proteins, their limited use identified several TRPC channel interacting proteins and mapped the interacting domains. Y2H identified the interaction of TRPC channels with RNF24,[100] Enkurin,[101] MxA,[102] and dopamine D2 receptors.[103] It helped map the sites of TRPC channel interaction with IP_3R,[97] RACK1,[104] and other TRPCs.[105] Although Y2H is prone to false positives, the use of this powerful technique will continue

FIGURE 2.3 (Continued)   Cell-attached mode: Channel activation will result in an increase in open probability, which reflects the fraction of time the channels spend in the open state. This will manifest in an increased frequency of opening events and/or increased dwell time of the opening events. Channel conductance can be estimated by recording the mean current amplitude during channel openings at different membrane potentials. Analyzing open probability at different membrane potentials provides an estimate of the channels' voltage dependence.

to enrich our understanding of TRPC channel physiology by providing novel information on the TRPC channel interaction network.

2.4.3   Co-Expression of TRPC with Auxiliary Proteins

Co-expressing TRPC channels with other proteins and analyzing the resulting change in the cellular response to stimulation at macroscopic levels have been and remain the first and primary means of studying the role of potential protein candidates in modulating the TRPC channel activity. Among other key findings on TRPC channel regulation by auxiliary proteins, heteromultimerization and regulation by Homers and STIM1 have been demonstrated with this approach.

A recent example highlighted here is the use of the co-expression approach to characterize the role of STIM1 and Orai1 in gating TRPC channels. Using a combination of co-expression, Ca^{2+} and Mn^{2+} flux measurements, and electrophysiology, it was found that co-expression of STIM1 with TRPC channels and Orai1 facilitates signal transduction from the depleted ER Ca^{2+} stores to the TRPC channels. Specifically, STIM1 was found to physically interact with TRPC1 and enhance TRPC1 response to store depletion.[33] The TRPC1–STIM1 interaction is mediated by the STIM1 ERM domain,[33] whereas gating of TRPC1 by STIM1 is mediated by the polybasic domain of STIM1.[81] The gating of TRPC1 by STIM1 was mapped down to two charged amino acids in TRPC1 and STIM1 that convey the electrostatic interaction between these molecules. Mutating these amino acids abolished the STIM1-mediated signal transduction between the ER and TRPC channels.[81] Interestingly, STIM1 seems to report the ER Ca^{2+} store depletion to the "store-insensitive" TRPC3 due to a lateral effect, whereby STIM1 interacts with TRPC1, which presents STIM1 to TRPC3 in a TRPC1–TRPC3 heteromultimer, resulting in indirect regulation of TRPC3 by STIM1.[32] Co-expression analysis also revealed that low levels of Orai1 enhance the store-dependent activity of TRPC1, TRPC3, and TRPC6,[91,106,107] which requires the STIM1-assembled Orai1/TRPC1 complex.[106,108] Surprisingly, when co-expressed at low levels in the same cell, the SOC activity of TRPC1 and Orai1 requires competent TRPC1 and Orai1 channel function.[90,106] The mutual requirement of TRPC1 and Orai1 to assemble the SOC complex can be shown biochemically. This is illustrated in Figure 2.4, which shows that when TRPC1, Orai1, and STIM1 are expressed at low levels, the interaction of STIM1 with Orai1 requires TRPC1, and conversely, interaction of STIM1 with TRPC1 requires Orai1.

The data presented above are just one example of applying the co-expression technique to studying the assembly of TRPC channel complexes and their functional consequences. In addition to the reports described above, the list of studies that successfully employed these techniques is indeed vast and includes such key findings as regulation of TRPC channel activity by interaction with PLC,[36,109] IP_3R,[97] RACK1,[104] and PM Na^+/Ca^{2+} exchanger.[110] These results illustrate the complexity of the signaling mediated by TRPC channels and delineate the molecular relationships that drive these channels' activity. They provide a critical tool for first-tier analysis of the possible candidates for direct interaction with TRPC channels. The main drawbacks of these techniques include the limited ability to match the expression levels of the potential indicators with physiologically relevant ratios and the limited control over other components of the TRPC signaling complexes.

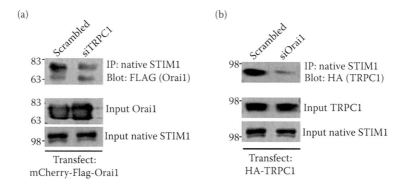

FIGURE 2.4 Mutual dependence of Orai1 and TRPC1 for interaction with STIM1. HEK cells treated with scrambled siRNA and (a) siTRPC1 or (b) siOrai1 were transfected with 100 ng (a) mCherry-Flag-Orai1 or (b) HA-TRPC1 and were used to immunoprecipitate the native STIM1 and blotted for co-IP of Orai1 (a, with anti-FFLAG) or TRPC1 (b, with anti-HA). Also shown are the inputs for Orai1, TRPC1, and the native STIM1.

2.4.4 IMMUNOPRECIPITATION AND PULL-DOWN ASSAYS

Another powerful protein–protein interaction screening technique is low- and high-volume proteomic analysis. Such analysis is a relatively high-throughput platform that allows screening candidates under different experimental conditions. In these assays, TRPC channels are immunoprecipitated from lysates prepared from native tissues or from model cells transfected with the TRPC channel of interest. Alternatively, tagged constructs representing TRPC channel domains are used to pull down proteins from lysates, and then the immunoprecipitated or pull-down proteins are analyzed by mass spectrometry. This technique has been used for immunoprecipitation (IP) of TRPC3[111] and TRPC5 and TRPC6[112] from brain extract, in which these channels are expressed at a high level. Among the TRPC5 and TRPC6 interactors identified and confirmed using IP are several cytoskeletal proteins including spectrin and neurabin, the endocytic vesicle-associated proteins dynamin and AP-2, and the PM Na[+]/K[+] ATPase.[112]

A more comprehensive analysis was performed with TRPC3. TRPC3 is reported to complex with 64 proteins, although it is not known which of the interactions are direct and which are indirect.[111] These multiple hits are usually grouped on the basis of known function and localization, as described in detail in Chapter 5.[111] TRPC3 was found to associate with the same proteins that associate with TRPC5 and TRPC6 and with additional novel proteins. Fidelity of the findings can be determined by identification of previously known interacting proteins. In the case of TRPC3, the MS/MS analysis identified Homer1, PLCβ, VAMP2, Na[+]/K[+] ATPase, SERCA2, PMCA, Gαq, and IP$_3$Rs in the TRPC3 complex.[111] All these proteins have been shown to interact with or affect the activity of TRPC3.

Novel hits are confirmed by targeted Y2H with defined domains, co-IP, and pull-down assays and are then examined for their effect on channel expression and function. So far, verification of interaction has been confirmed for few of the novel hits, and only the effect of RACK1 on the TRPC3 function has been examined.[104] This is obviously very much needed information.

Another standard use of the co-IP and pull-down assays is in determining if interactions are direct or mediated and in delineating the interacting domains. For example, STIM1 interacts with TRPC1 but not with TRPC3, but STIM1 can access TRPC3 through its interaction with TRPC1.[32] The co-IP and pull-down assays, together with molecular approaches, are indispensable in identifying the domains that mediate the protein–protein interaction. By providing domains of varying length within the original protein, it is possible to map the domains essential for TRPC channel interaction, as was done for CaM,[54–59] IP$_3$R,[54–59] Homer,[74] Junctate,[61,62] STIM1,[33] Orai1,[90,106] and other proteins.

2.4.5   FRET, TIRF, AND FRAP

Time resolution of the co-IP analysis of protein–protein interaction does not report direct interaction and the proximity between the proteins, let alone their dynamic interaction in living cells. This is accomplished by other techniques such as FRET, TIRF, and FRAP.

Fluorescent resonance energy transfer (FRET) is a quantum phenomenon that allows transfer of energy between two fluorophores in a distance of less than 20 nm. FRET requires an engineering fluorescent fusion protein pair, e.g., fusion of one protein with CFP and another one with YFP, or tagging them with fluorescent groups, e.g., by specific chemical modification or use of antibodies. Increased proximity between the proteins permits FRET, while separation of the interacting pair results in the loss of FRET. Therefore, FRET can be used to monitor the proximity between proteins and their dynamic interaction. FRET has been used successfully in determining the interaction between TRPC isoforms[35] and in studying the dynamic changes in the TRPC channel environment during cell stimulation. It was possible to demonstrate that TRPC heteromultimers assemble in the ER and to confirm the tetrameric nature of these heteromultimers.[113–115] FRET was helpful in showing that ER Ca^{2+} store depletion induces interaction of STIM1 with TRPC1 and reorganization of TRPC1 within the PM.[114,116,117] The role of ankyrin repeats in organization of TRPC heteromultimers was demonstrated using FRET as well.[113] A detailed description of FRET technology and experimental protocols is provided in Chapters 4 and 15.

The idea that the cellular localization of TRPC channels changes during cell stimulation and that the changes in TRPC channel localization are essential for their function calls for a detailed analysis of the dynamics of TRPC channel localization within cellular microdomains. Standard and confocal fluorescent microscopy does not provide the necessary spatial resolution for this task. Furthermore, these assays do not resolve the movement of individual TRPC channel-bearing compartments. This can be accomplished with TIRF and FRAP.

Total internal reflection fluorescence (TIRF) relies on creation of an evanescent (standing) wave created when light is totally reflected within the cover slip hosting the cells. Such a wave is able to excite fluorophores within a very limited (under 200 nm) space within the sample. When applied to cells attached to a cover slip, TIRF allows resolution of near-membrane fluorescence that is unobstructed by fluorescence from the rest of the cell. Therefore, TIRF is very useful for analyzing the delivery of molecules to/from the PM. TIRF microscopy showed that cell stimulation resulted in translocation of intracellular TRPC channels to the PM in a form of heteromultimers.[18,36,38,113] Furthermore, TIRF microscopy demonstrated the dynamic changes in the localization of STIM1[28,118–120] and of Homer-containing complexes of TRPC3 to the PM.[18]

Fluorescence recovery after photobleaching (FRAP) allows testing the lateral movement of molecules within the plane of the PM. It relies on bleaching of an area of the PM containing the molecule of interest attached to a fluorophore by high-intensity light. The delivery of new fluorophores into the bleached area is a function of the freedom of the lateral diffusion of the molecule of interest. A faster rate of fluorescence recovery indicates a higher mobility and/or diffusion rate. This technique has been useful in analyzing the dynamics of TRPC channel insertion into the PM.[36]

The FRET, TIRF, and FRAP platforms have no predictive power, and they rely on information inferred from other assays, such as Y2H and proteomics. Nonetheless, they have proven and will likely continue to be indispensable tools for studying the dynamic changes in TRPC channels' working environment and the molecular makeup of this environment.

2.4.6 KNOCKOUT AND KNOCKDOWN

TRPC channel knockout (KO) in model organisms (*Drosophila*, *C. elegans,* and mice) and knockdown (KD) using siRNA are very useful in analyzing the physiological role of TRPC channels in various tissues, as well as their role in cellular Ca^{2+} signaling. KO in mice showed that TRPC channels are involved in a variety of organism level functions such as pheromone recognition,[121,122] blood pressure regulation, smooth muscle motility,[13,16,123,124] and secretion.[19,20] These studies also established the SOC function of the native TRPC1, TRPC3, and TRPC4.[19,20,124] KD experiments firmly established the TRPC channels as essential components of receptor-evoked Ca^{2+} influx in all but a few tissues. Up to now, however, these approaches enjoyed only limited use in analyzing the TRPC working environment. Beyond the use of KD and KO techniques to show the essential role of STIM1 and Orai1 in TRPC channel activity, application of these techniques to other TRPC channel auxiliary proteins has been limited.

2.5 PERSPECTIVES

Integration of modern biochemical, physiological, and molecular techniques toward identifying the proteins that interact with and regulate TRPC channels has proven to be a productive approach. It led to identification of some key components of the TRPC channel interacting networks and to formulation of novel paradigms of signaling whose impact is likely to extend beyond TRPC channel physiology. It is firmly established that TRPC channels exist within a context of a signaling network that comprises a number of PM, ER, and cytoplasmic proteins. Composition of the TRPC channel signaling network defines the channel response to cell stimulation. The network can be dynamically regulated as a result of cell stimulation.

This field is likely to develop toward further specialization of our understanding of the mechanisms of the TRPC channel function as well as generalization of its key concepts. Among the likely developments are the use of the wealth of information provided by proteomic and siRNA and other genome screening techniques for a comprehensive enumeration of the components of the complexes formed by TRPC channels, and thereby solidifying the principle of "TRPCsome" as a protein network comprising the functional determinants of the TRPC channel activity. Documenting how the TRPCsome changes during the history of cell stimulation or as a function of the cell's context within the tissue will significantly advance our understanding of the integrative function of these channels and of the cell.

The current techniques are likely to remain the major tools of TRPC channel studies. However, our quest toward better understanding of these fascinating molecules is likely to be aided by new approaches, such as expanded proteomic, genomic, and computational assays.

ACKNOWLEDGMENTS

The authors would like to extend their apologies to those authors whose work has not been cited here because of manuscript length consideration. Work in the authors' labs is supported by the National Institutes of Health grants HD058577 and ES016782 to K.K. and the National Institutes of Health grants DE12309 and DK38938 and the Ruth S. Harrell Professorship in Medical Research to S.M.

REFERENCES

1. Venkatachalam, K., and C. Montell. 2007. TRP channels. *Annual Reviews of Biochemistry* 76: 387.
2. Ramsey, I. S., M. Delling, and D. E. Clapham. 2006. An introduction to TRP channels. *Annual Reviews of Physiology* 68: 619.
3. Pedersen, S. F., G. Owsianik, and B. Nilius. 2005. TRP channels: an overview. *Cell Calcium* 38: 233.
4. Nilius, B., T. Voets, and J. Peters. 2005. TRP channels in disease. *Science's STKE* 2005: re8.
5. Amaral, M. D., and L. Pozzo-Miller. 2007. TRPC3 channels are necessary for brain-derived neurotrophic factor to activate a nonselective cationic current and to induce dendritic spine formation. *Journal of Neuroscience* 27: 5179.
6. Cai, S., S. Fatherazi, R. B. Presland et al. 2006. Evidence that TRPC1 contributes to calcium-induced differentiation of human keratinocytes. *Pflugers Archiv* 452: 43.
7. Thebault, S. et al. 2005. Receptor-operated Ca^{2+} entry mediated by TRPC3/TRPC6 proteins in rat prostate smooth muscle (PS1) cell line. *Journal of Cell Physiology* 204: 320.
8. Li, Y., Y.-C. Jia, K. Cui et al. 2005. Essential role of TRPC channels in the guidance of nerve growth cones by brain-derived neurotrophic factor. *Nature* 434: 894.
9. Wu, X., T. K. Zagranichnaya, G. T. Gurda et al. 2004. A TRPC1/TRPC3-mediated increase in store-operated calcium entry is required for differentiation of H19-7 hippocampal neuronal cells. *Journal of Biological Chemistry* 279: 43392.
10. Winn, M. P., P. J. Conlon, K. L. Lynn et al. 2005. A mutation in the TRPC6 cation channel causes familial focal segmental glomerulosclerosis. *Science* 308: 1801.
11. Reiser, J., K. R. Polu, C. C. Möller et al. 2005. TRPC6 is a glomerular slit diaphragm-associated channel required for normal renal function. *Nature Genetics* 37: 739.
12. Woudenberg-Vrenken, T. E., R. J. Bindels, and J. G. Hoenderop. 2009. The role of transient receptor potential channels in kidney disease. *Nature Reviews Nephrology* 5: 441.

13. Dietrich, A. et al. 2005. Increased vascular smooth muscle contractility in TRPC6$^{-/-}$ mice. *Mol Cell Biol*, 25: 6980.

14. Dietrich, A., M. Mederos y Schnitzler, M. Gollasch et al. 2007. In vivo TRPC functions in the cardiopulmonary vasculature. *Cell Calcium* 42: 233.

15. Tiruppathi, C., G. U. Ahmmed, S. M. Vogel et al. 2006. Ca^{2+} signaling, TRP channels, and endothelial permeability. *Microcirculation* 13: 693.

16. Tiruppathi, C., M. Freichel, S. M. Vogel et al. 2002. Impairment of store-operated Ca^{2+} entry in TRPC4$^{-/-}$ mice interferes with increase in lung microvascular permeability. *Circulation Research* 91: 70.

17. Tsvilovskyy, V. V., A. V. Zholos, T. Aberle et al. 2009. Deletion of TRPC4 and TRPC6 in mice impairs smooth muscle contraction and intestinal motility in vivo. *Gastroenterology* 137: 1415.

18. Kim, J. Y., W. Zeng, J.-P. Yuan et al. 2006. Homer 1 mediates store- and inositol 1,4,5-trisphosphate receptor-dependent translocation and retrieval of TRPC3 to the plasma membrane. *Journal of Biological Chemistry* 281: 32540.

19. Kim, M. S., J. H. Hong, Q. Li et al. 2009. Deletion of TRPC3 in mice reduces store-operated Ca^{2+} influx and the severity of acute pancreatitis. *Gastroenterology* 137: 1509.

20. Liu, X., K. T. Cheng, B. C. Bandyopadhyay et al. 2007. Attenuation of store-operated Ca^{2+} current impairs salivary gland fluid secretion in TRPC1$^{-/-}$ mice. *Proceedings of the National Academy of Sciences of the United States of America* 104: 17542.

21. Parekh, A. B., and J. W. Putney, Jr. 2005. Store-operated calcium channels. *Physiological Reviews* 85: 757.

22. Harden, T. K., and J. Sondek. 2006. Regulation of phospholipase C isozymes by ras superfamily GTPases. *Annual Review of Pharmacology and Toxicology* 46: 355.

23. Balla, T. 2006. Phosphoinositide-derived messengers in endocrine signaling. *Journal of Endocrinology* 188: 135.

24. Mikoshiba, K. 2007. IP3 receptor/Ca^{2+} channel: from discovery to new signaling concepts. *Journal of Neurochemistry* 102: 1426.

25. Berridge, M. J. 2009. Inositol trisphosphate and calcium signalling mechanisms. *Biochimica et Biophysica Acta* 1793: 933.

26. Kiselyov, K., X. Wang, D. M. Shin et al. 2006. Calcium signaling complexes in microdomains of polarized secretory cells. *Cell Calcium* 40: 451.

27. Roos, J., P. J. DiGregorio, A. V. Yeromin et al. 2005. STIM1, an essential and conserved component of store-operated Ca^{2+} channel function. *Journal of Cell Biology* 169: 435.

28. Liou, J., M. L. Kim, W. D. Hoe et al. 2005. STIM is a Ca^{2+} sensor essential for Ca^{2+}-store-depletion-triggered Ca^{2+} influx. *Current Biology* 15: 1235.

29. Feske, S., Y. Gwack, M. Prakriya et al. 2006. A mutation in Orai1 causes immune deficiency by abrogating CRAC channel function. *Nature* 441: 179.

30. Vig, M., C. Peinelt, A. Beck et al. 2006. CRACM1 is a plasma membrane protein essential for store-operated Ca^{2+} entry. *Science* 312: 1220.

31. Zhang, S. L., A. V. Yeromin, X. H. Zhang et al. 2006. Genome-wide RNAi screen of Ca^{2+} influx identifies genes that regulate Ca^{2+} release-activated Ca^{2+} channel activity. *Proceedings of the National Academy of Sciences of the United States of America* 103: 9357.

32. Yuan, J. P., W. Zeng, G. N. Huang, P. F. Worley, and S. Muallem. 2007. STIM1 heteromultimerizes TRPC channels to determine their function as store-operated channels. *Nature Cell Biology* 9: 636.

33. Huang, G. N., W. Zeng, J. Y. Kim, J. P. Yan, L. Han, S. Muallem, and P. F. Worley. 2006. STIM1 carboxyl-terminus activates native SOC, I(crac) and TRPC1 channels. *Nature Cell Biology* 8: 1003.

34. Singh, B. B., T. P. Lockwich, B. C. Bandyopadhyay et al. 2004. VAMP2-dependent exocytosis regulates plasma membrane insertion of TRPC3 channels and contributes to agonist-stimulated Ca^{2+} influx. *Molecular Cell* 15: 635.

35. Hofmann, T., M. Schaefer, G. Schultz et al. 2002. Subunit composition of mammalian transient receptor potential channels in living cells. *Proceedings of the National Academy of Sciences of the United States of America* 99: 7461.

36. van Rossum, D. B., D. Oberdick, Y. Rbaibi et al. 2008. TRP_2, a lipid/trafficking domain that mediates diacylglycerol-induced vesicle fusion. *Journal of Biological Chemistry* 283: 34384.

37. Bezzerides, V. J., I. S. Ramsey, S. Kotecha, A. Greka, and D. E. Clapham. 2004. Rapid vesicular translocation and insertion of TRP channels. *Nature Cell Biology* 6: 709.

38. Smyth, J. T., L. Lemonnier, G. Vazquez, G. S. Bird, and J. W. Putney, Jr. 2006. Dissociation of regulated trafficking of TRPC3 channels to the plasma membrane from their activation by phospholipase C. *Journal of Biological Chemistry* 281: 11712.

39. Wu, X., G. Babnigg, and M. L. Villereal. 2000. Functional significance of human trp1 and trp3 in store-operated Ca^{2+} entry in HEK-293 cells. *American Journal of Physiology - Cell Physiology* 278: C526.

40. Trebak, M., G. S. Bird, R. R. McKay, and J. W. Putney, Jr. 2002. Comparison of human TRPC3 channels in receptor-activated and store-operated modes. Differential sensitivity to channel blockers suggests fundamental differences in channel composition. *Journal of Biological Chemistry* 277: 21617.

41. Kiselyov, K., and R. L. Patterson. 2009. The integrative function of TRPC channels. *Frontiers in Bioscience* 14: 45.

42. Kiselyov, K., D. M. Shin, J. Y. Kim, J. P. Yuan, and S. Muallem. 2007. TRPC channels: interacting proteins. *Handbook of Experimental Pharmacology* 2007: 559.

43. Li, H. S., and C. Montell. 2000. TRP and the PDZ protein, INAD, form the core complex required for retention of the signalplex in Drosophila photoreceptor cells. *Journal of Cell Biology* 150: 1411.

44. Chevesich, J., A. J. Kreuz, and C. Montell. 1997. Requirement for the PDZ domain protein, INAD, for localization of the TRP store-operated channel to a signaling complex. *Neuron* 18: 95.

45. Tsunoda, S., J. Sierralta, Y. Sun et al. 1997. A multivalent PDZ-domain protein assembles signalling complexes in a G-protein-coupled cascade. *Nature* 388: 243.

46. Tsunoda, S., Y. Sun, E. Suzuki, and C. Zuker. 2001. Independent anchoring and assembly mechanisms of INAD signaling complexes in Drosophila photoreceptors. *Journal of Neuroscience* 21: 150.

47. Sabourin, J., C. Lamiche, A. Vandebrouck et al. 2009. Regulation of TRPC1 and TRPC4 cation channels requires an alpha1-syntrophin-dependent complex in skeletal mouse myotubes. *Journal of Biological Chemistry* 284: 36248.

48. Obukhov, A. G., and M. C. Nowycky. 2004. TRPC5 activation kinetics are modulated by the scaffolding protein ezrin/radixin/moesin-binding phosphoprotein-50 (EBP50). *Journal of Cell Physiology* 201: 227.

49. Tang, Y., J. Tang, Z. Chen et al. 2000. Association of mammalian trp4 and phospholipase C isozymes with a PDZ domain-containing protein, NHERF. *Journal of Biological Chemistry* 275: 37559.

50. Schilling, W. P., and M. Goel. 2004. Mammalian TRPC channel subunit assembly. *Novartis Foundation Symposium* 258: 18.

51. Gudermann, T., T. Hofmann, M. Mederos y Schnitzler, and A. Dietrich. 2004. Activation, subunit composition and physiological relevance of DAG-sensitive TRPC proteins. *Novartis Foundation Symposium* 258: 103.

52. Goel, M., W. G. Sinkins, and W. P. Schilling. 2002. Selective association of TRPC channel subunits in rat brain synaptosomes. *Journal of Biological Chemistry* 277: 48303.

53. Strubing, C. et al. 2003. Formation of novel TRPC channels by complex subunit interactions in embryonic brain. *Journal of Biological Chemistry* 278: 39014.
54. Kwon, Y., T. Hofmann, and C. Montell. 2007. Integration of phosphoinositide- and calmodulin-mediated regulation of TRPC6. *Molecular Cell* 25: 491.
55. Zhu, M. X. 2005. Multiple roles of calmodulin and other Ca^{2+}-binding proteins in the functional regulation of TRP channels. *Pflügers Archiv* 451: 105.
56. Shimizu, S., T. Yoshida, M. Wakamori et al. 2005. Ca^{2+}/calmodulin dependent myosin light chain kinase is essential for activation of TRPC5 channels expressed in HEK293 cells. *Journal of Physiology* 570: 219.
57. Shi, J., E. Mori, Y. Mori, M. Mori, J. Li, Y. Ito, and R. Inoue. 2004. Multiple regulation by calcium of murine homologues of transient receptor potential proteins TRPC6 and TRPC7 expressed in HEK293 cells. *Journal of Physiology* 561: 415.
58. Wedel, B. J., G. Vazquez, R. R. McKay, G. St. J. Bird, and J. W. Putney, Jr. 2003. A calmodulin/inositol 1,4,5-trisphosphate (IP3) receptor-binding region targets TRPC3 to the plasma membrane in a calmodulin/IP3 receptor-independent process. *Journal of Biological Chemistry* 278: 25758.
59. Singh, B. B., X. Liu, J. Tang, M. X. Zhu, and I. S. Ambudkar. 2002. Calmodulin regulates Ca^{2+}-dependent feedback inhibition of store-operated Ca^{2+} influx by interaction with a site in the C terminus of TrpC1. *Molecular Cell* 9: 739.
60. Mullins, F. M., C. Y. Park, R. E. Dolmetsch, and R. S. Lewis. 2009. STIM1 and calmodulin interact with Orai1 to induce Ca^{2+}-dependent inactivation of CRAC channels. *Proceedings of the National Academy of Sciences of the United States of America* 106: 15495.
61. Stamboulian, S., M. J. Moutin, S. Treves et al. 2005. Junctate, an inositol 1,4,5-triphosphate receptor associated protein, is present in rodent sperm and binds TRPC2 and TRPC5 but not TRPC1 channels. *Developmental Biology* 286: 326.
62. Treves, S., C. Franzini-Armstrong, L. Moccagatta et al. 2004. Junctate is a key element in calcium entry induced by activation of InsP3 receptors and/or calcium store depletion. *Journal of Cell Biology* 166: 537.
63. Hong, C. S., S. J. Kwon, and M. C. Cho et al. 2008. Overexpression of junctate induces cardiac hypertrophy and arrhythmia via altered calcium handling. *Journal of Molecular and Celularl Cardiology* 44: 672.
64. Divet, A., S. Paesante, C. Grasso et al. 2007. Increased Ca^{2+} storage capacity of the skeletal muscle sarcoplasmic reticulum of transgenic mice over-expressing membrane bound calcium binding protein junctate. *Journal of Cellular Physiology* 213: 464.
65. Tadokoro, S., T. Tachibana, T. Imanaka, W. Nishida, and K. Sobue. 1999. Involvement of unique leucine-zipper motif of PSD-Zip45 (Homer 1c/vesl-1L) in group 1 metabotropic glutamate receptor clustering. *Proceedings of the National Academy of Sciences of the United States of America* 96: 13801.
66. Xiao, B., J. C. Tu, R. S. Petralia et al. 1998. Homer regulates the association of group 1 metabotropic glutamate receptors with multivalent complexes of homer-related, synaptic proteins. *Neuron* 21: 707.
67. Tu, J. C., B. Xiao, J. P. Yuan et al. 1998. Homer binds a novel proline-rich motif and links group 1 metabotropic glutamate receptors with IP3 receptors. *Neuron* 21: 717.
68. Kato, A., F. Ozawa, Y. Saitoh, Y. Fukazawa, H. Sugiyama, and K. Inokuchi. 1998. Novel members of the Vesl/Homer family of PDZ proteins that bind metabotropic glutamate receptors. *Journal of Biological Chemistry* 273: 23969.
69. Brakeman, P. R., A. A. Lanahan, R. O'Brien et al. 1997. Homer: a protein that selectively binds metabotropic glutamate receptors. *Nature* 386: 284.
70. Ehrengruber, M. U., A. Kato, K. Inokuchi, and S. Hennou. 2004. Homer/Vesl proteins and their roles in CNS neurons. *Molecular Neurobiology* 29: 213.
71. Fagni, L., P. F. Worley, and F. Ango. 2002. Homer as both a scaffold and transduction molecule. *Science's STKE* 2002: re8.

72. Xiao, B., J. C. Tu, and P. F. Worley. 2000. Homer: a link between neural activity and glutamate receptor function. *Current Opinion in Neurobiology* 10: 370.
73. Worley, P. F., W. Zeng, G. Huang et al. 2007. Homer proteins in Ca^{2+} signaling by excitable and non-excitable cells. *Cell Calcium* 42: 363.
74. Yuan, J. P., K. Kiselyov, D. M. Shin et al. 2003. Homer binds TRPC family channels and is required for gating of TRPC1 by IP3 receptors. *Cell* 114: 777.
75. Beneken, J., J. C. Tu, B. Xiao et al. 2000. Structure of the Homer EVH1 domain-peptide complex reveals a new twist in polyproline recognition. *Neuron* 26: 143.
76. Tu, J. C., B. Xiao, S. Naisbitt et al. 1999. Coupling of mGluR/Homer and PSD-95 complexes by the Shank family of postsynaptic density proteins. *Neuron* 23: 583.
77. Shin, D. M., M. Dehoff, X. Luo et al. 2003. Homer 2 tunes G protein-coupled receptors stimulus intensity by regulating RGS proteins and PLCbeta GAP activities. *Journal of Cellular Biology* 162: 293.
78. Stathopulos, P. B., L. Zheng, G. Y. Li, M. J. Plevin, and M. Ikura. 2008. Structural and mechanistic insights into STIM1-mediated initiation of store-operated calcium entry. *Cell* 135: 110.
79. Spassova, M. A., J. Soboloff, L.-P. He et al. 2006. STIM1 has a plasma membrane role in the activation of store-operated Ca^{2+} channels. *Proceedings of the National Academy of Sciences of the United States of America* 103: 4040.
80. Zhang, S. L., Y. Yu, J. Roos et al. 2005. STIM1 is a Ca^{2+} sensor that activates CRAC channels and migrates from the Ca^{2+} store to the plasma membrane. *Nature* 437: 902.
81. Zeng, W., J. P. Yuan, M. S. Kim et al. 2008. STIM1 gates TRPC channels, but not Orai1, by electrostatic interaction. *Molecular Cell* 32: 439.
82. Cahalan, M. D. 2009. STIMulating store-operated Ca^{2+} entry. *Nature Cell Biology* 11: 669.
83. McNally, B. A., M. Yamashita, A. Engh, and M. Prakriya. 2009. Structural determinants of ion permeation in CRAC channels. *Proceedings of the National Academy of Sciences of the United States of America* 106: 22516.
84. Yamashita, M., L. Navarro-Borelly, B. A. McNally, and M. Prakriya. 2007. Orai1 mutations alter ion permeation and Ca^{2+}-dependent fast inactivation of CRAC channels: evidence for coupling of permeation and gating. *Journal of General Physiology* 130: 525.
85. Yeromin, A. V., S. L. Zhang, W. Jiang, Y. Yu, O. Safrina, and M. D. Cahalan. 2006. Molecular identification of the CRAC channel by altered ion selectivity in a mutant of Orai. *Nature* 443: 226.
86. Prakriya, M., S. Feske, Y. Gwack et al. 2006. Orai1 is an essential pore subunit of the CRAC channel. *Nature* 443: 230.
87. Vig, M., A. Beck, J. M. Billingsley et al. 2006. CRACM1 multimers form the ion-selective pore of the CRAC channel. *Current Biology* 16: 2073.
88. Yuan, J. P., W. Zeng, M. R. Dorwart, Y. J. Choi, P. F. Worely, and S. Muallem. 2009. SOAR and the polybasic STIM1 domains gate and regulate Orai channels. *Nature Cell Biology* 11: 337.
89. Kim, M. S., W. Zeng, J. P. Yuan, D. M. Shin, P. F. Worley, and S. Muallem. 2009. Native store-operated Ca^{2+} influx requires the channel function of Orai1 and TRPC1. *Journal of Biological Chemistry* 284: 9733.
90. Cheng, K. T., X. Liu, H. L. Ong, and I. S. Ambudkar. 2008. Functional requirement for Orai1 in store-operated TRPC1-STIM1 channels. *Journal of Biological Chemistry* 283: 12935.
91. Liao, Y., C. Erxleben, J. Abramowitz et al. 2008. Functional interactions among Orai1, TRPCs, and STIM1 suggest a STIM-regulated heteromeric Orai/TRPC model for SOCE/Icrac channels. *Proceedings of the National Academy of Sciences of the United States of America* 105: 2895.

92. Liu, X., B. B. Singh, and I. S. Ambudkar. 2003. TRPC1 is required for functional store-operated Ca^{2+} channels. Role of acidic amino acid residues in the S5-S6 region. *Journal of Biological Chemistry* 278: 11337.

93. Birnbaumer, L. 2009. The TRPC class of ion channels: a critical review of their roles in slow, sustained increases in intracellular Ca^{2+} concentrations. *Annual Review of Pharmacology and Toxicology* 49: 395.

94. Abramowitz, J., and L. Birnbaumer. 2009. Physiology and pathophysiology of canonical transient receptor potential channels. *FASEB Journal* 23: 297.

95. Schaefer, M. 2005. Homo- and heteromeric assembly of TRP channel subunits. *Pflügers Archiv* 451: 35.

96. Kiselyov, K., X. Xu, G. Mozhayeva et al. 1998. Functional interaction between InsP3 receptors and store-operated Htrp3 channels. *Nature* 396: 478.

97. Zhang, Z., J. Tang, S. Tikunova et al. 2001. Activation of Trp3 by inositol 1,4,5-trisphosphate receptors through displacement of inhibitory calmodulin from a common binding domain. *Proceedings of the National Academy of Sciences of the United States of America* 98: 3168.

98. Tang, J., Y. Lin, Z. Zhang, S. Tikunova, L. Birnbaumer, and M. X. Zhu. 2001. Identification of common binding sites for calmodulin and inositol 1,4,5-trisphosphate receptors on the carboxyl termini of trp channels. *Journal of Biological Chemistry* 276: 21303.

99. Kiselyov, K., G. A. Mignery, M. X. Zhu, and S. Muallem. 1999. The N-terminal domain of the IP3 receptor gates store-operated hTrp3 channels. *Molecular Cell* 4: 423.

100. Lussier, M. P., P. K. Lepage, S. M. Bousquet, and G. Boulay. 2008. RNF24, a new TRPC interacting protein, causes the intracellular retention of TRPC. *Cell Calcium* 43: 432.

101. Sutton, K. A., M. K. Jungnickel, Y. Wang, K. Cullen, S. Lambert, and H. M. Florman. 2004. Enkurin is a novel calmodulin and TRPC channel binding protein in sperm. *Developmental Biology* 274: 426.

102. Lussier, M. P., S. Cayouette, P. K. Lepage et al. 2005. MxA, a member of the dynamin superfamily, interacts with the ankyrin-like repeat domain of TRPC. *Journal of Biological Chemistry* 280: 19393.

103. Hannan, M. A., N. Kabbani, C. D. Paspalas, and R. Levenson. 2008. Interaction with dopamine D2 receptor enhances expression of transient receptor potential channel 1 at the cell surface. *Biochimica et Biophysica Acta* 1778: 974.

104. Bandyopadhyay, B. C., H. L. Ong, T. P. Lockwich et al. 2008. TRPC3 controls agonist-stimulated intracellular Ca^{2+} release by mediating the interaction between inositol 1,4,5-trisphosphate receptor and RACK1. *Journal of Biological Chemistry* 283: 32821.

105. Liu, X., B. C. Bandyopadhyay, B. B. Singh, K. Groschner, and I. S. Ambudkar. 2005. Molecular analysis of a store-operated and 2-acetyl-sn-glycerol-sensitive non-selective cation channel. Heteromeric assembly of TRPC1-TRPC3. *Journal of Biological Chemistry* 280: 21600.

106. Kim, M. S., W. Zeng, J. P. Yuan et al. 2009. Native store-operated Ca^{2+} influx requires the channel function of Orai1 and TRPC1. *Journal of Biological Chemistry* 284: 9733.

107. Liao, Y., C. Erxleben, C. Yildirim, J. Abramowitz, D. L. Armstrong, and L. Birnbaumer. 2007. Orai proteins interact with TRPC channels and confer responsiveness to store depletion. *Proceedings of the National Academy of Sciences of the United States of America* 104: 4682.

108. Ong, H. L., X. Liu, K. Tsaneva-Atanasova et al. 2007. Relocalization of STIM1 for activation of store-operated Ca^{2+} entry is determined by the depletion of subplasma membrane endoplasmic reticulum Ca^{2+} store. *Journal of Biological Chemistry* 282: 12176.

109. van Rossum, D. B., R. L. Patterson, S. Sharma et al. 2005. Phospholipase Cgamma1 controls surface expression of TRPC3 through an intermolecular PH domain. *Nature* 434: 99.

110. Rosker, C., A. Graziani, M. Lukas et al. 2004. Ca^{2+} signaling by TRPC3 involves Na^+ entry and local coupling to the Na^+/Ca^{2+} exchanger. *Journal of Biological Chemistry* 279: 13696.

111. Lockwich, T., J. Pant, A. Makusky et al. 2008. Analysis of TRPC3-interacting proteins by tandem mass spectrometry. *Journal of Proteome Research* 7: 979.

112. Goel, M., W. Sinkins, A. Keightley, M. Kinter, and W. P. Schilling. 2005. Proteomic analysis of TRPC5- and TRPC6-binding partners reveals interaction with the plasmalemmal Na^+/K^+-ATPase. *Pflügers Archiv* 451: 87.

113. Schindl, R., I. Frischauf, H. Kahr et al. 2008. The first ankyrin-like repeat is the minimum indispensable key structure for functional assembly of homo- and heteromeric TRPC4/TRPC5 channels. *Cell Calcium* 43: 260.

114. Alfonso, S., O. Benito, S. Alicia et al. 2008. Regulation of the cellular localization and function of human transient receptor potential channel 1 by other members of the TRPC family. *Cell Calcium* 43: 375.

115. Amiri, H., G. Schultz, and M. Schaefer. 2003. FRET-based analysis of TRPC subunit stoichiometry. *Cell Calcium* 33: 463.

116. Pani, B., H. L. Ong, S. C. Brazer et al. 2009. Activation of TRPC1 by STIM1 in ER-PM microdomains involves release of the channel from its scaffold caveolin-1. *Proceedings of the National Academy of Sciences of the United States of America* 106: 20087.

117. Pani, B., H. L. Ong, X. Liu, K. Rauser, I. S. Ambudkar, and B. B. Singh. 2008. Lipid rafts determine clustering of STIM1 in endoplasmic reticulum-plasma membrane junctions and regulation of store-operated Ca^{2+} entry (SOCE). *Journal of Biological Chemistry* 283: 17333.

118. Tamarina, N. A., A. Kuznetsov, and L. H. Philipson. 2008. Reversible translocation of EYFP-tagged STIM1 is coupled to calcium influx in insulin secreting beta-cells. *Cell Calcium* 44: 533.

119. Wu, M. M., J. Buchanan, R. M. Luik, and R. S. Lewis. 2006. Ca^{2+} store depletion causes STIM1 to accumulate in ER regions closely associated with the plasma membrane. *Journal of Cellular Biology* 174: 803.

120. Luik, R. M., M. M. Wu, J. Buchanan, and R. S. Lewis. 2006. The elementary unit of store-operated Ca^{2+} entry: local activation of CRAC channels by STIM1 at ER-plasma membrane junctions. *Journal of Cell Biology* 174: 815.

121. Liman, E. R., D. P. Corey, and C. Dulac. 1999. TRP2: a candidate transduction channel for mammalian pheromone sensory signaling. *Proceedings of the National Academy of Sciences of the United States of America* 96: 5791.

122. Stowers, L., T. E. Holy, M. Meister et al. 2002. Loss of sex discrimination and male-male aggression in mice deficient for TRP2. *Science* 295: 1493.

123. Freichel, M., S. Philipp, A. Cavalié, and V. Flockerzi. 2004. TRPC4 and TRPC4-deficient mice. *Novartis Foundation Symposium* 258: 189.

124. Freichel, M., S. H. Suh, A. Pfeifer et al. 2001. Lack of an endothelial store-operated Ca^{2+} current impairs agonist-dependent vasorelaxation in TRP4$^{-/-}$ mice. *Nature Cell Biology* 3: 121.

125. Yao, J., Q. Li, J. Chen, and S. Muallem. 2004. Subpopulation of store-operated Ca^{2+} channels regulate Ca^{2+}-induced Ca^{2+} release in non-excitable cells. *Journal of Biological Chemistry* 279: 21511.

3 Activation of TRP Channels in Mammalian Systems

Tomohiro Numata, Shigeki Kiyonaka, Kenta Kato, Nobuaki Takahashi, and Yasuo Mori

CONTENTS

3.1   INTRODUCTION

Cation channels encoded by the gene superfamily of *transient receptor potentials* (*TRPs*) are characterized by a wide variety of activation triggers that act from outside and inside the cell.[1,2] Because of the molecular identification of *Drosophila* TRP,[3] functional characterizations of recombinant TRP homologues and gene knockout mice have indeed revealed a plethora of stimuli that activate TRP channels (Table 3.1). In addition, respective TRP homologues are susceptible to multiple activation triggers and can function as multimodal sensors. This diversity in activation properties likely contributes to widening the body's ability to react and adapt to different forms of environmental change by sensing and integrating information. In this chapter, we provide an overview of the activation triggers for mammalian TRP channels and discuss the physiological significance of activation sensitivity. Because we concentrate on "mammalian" TRPs here, readers who are interested in activation of invertebrate TRPs are referred to Chapters 20 and 21 of this book and other excellent reviews.[4–6]

3.2   PLC-COUPLED RECEPTORS AND DOWNSTREAM SIGNALING CASCADES

Stimulation of receptors at the plasma membrane (PM) activates phospholipase C (PLC) via GTP-binding (G) proteins and tyrosine kinases which generates the messengers inositol 1,4,5-trisphosphate (IP_3) and diacylglycerol (DAG) from phosphatidylinositol-4,5-bisphosphate (PIP_2). This induces an elevation of cytosolic Ca^{2+} concentration, which is controlled by two closely coupled components, IP_3-induced Ca^{2+} release via IP_3 receptors (IP_3Rs) from the intracellular Ca^{2+} store, the endoplasmic reticulum (ER), and Ca^{2+} influx across the PM. Association of TRPs with this mode of Ca^{2+} influx has been a major research area, on the basis of the knowledge that PLC-mediated light response is upstream of TRP channels in *Drosophila* photoreceptor cells.[4–6]

3.2.1   RECEPTOR PROTEINS (DIRECT COUPLING)

Interesting protein complexes have been reported for TRP channels such that TRPC3 is associated with the receptor TrkB for brain-derived nerve growth factor.[7] In protein

TABLE 3.1

Activation Triggers of TRP Channels in Mammalian Systems

	TRPC	TRPV	TRPM	TRPA	TRPP	TRPML
	PLC-Coupled Receptors and Downstream Signaling Cascades					
Receptor protein (directly)	TrkB: (C3)[7]					
G proteins	activation by G_i: (C4)[8]		Suppression by G_o: (M1)[9]			
PLC	PLCγ1, PLCγ2: (C3, C4)[10–12]		PLCs: (M7)[14]			
IP$_3$ and Ca^{2+} store depletion	Store depletion: (C1)[29–31,39,40], (C2)[34], (C3)[31,35], (C4)[32,36,38], (C5)[35], (C6)[35], (C7)[37] Interact with STIM: (TRPCs)[44–48] Interact with STIM1 and Orai: (TRPCs)[49–52] STIM1 modulation: (C1,C3)[47] Interact with IP$_3$Rs: (TRPCs)[55–58]	Passive and active store depletion: (V6)[42]	Involvements in SOC influx: (M5)[43]			

(continued)

TABLE 3.1 (Continued)
Activation Triggers of TRP Channels in Mammalian Systems

	TRPC	TRPV	TRPM	TRPA	TRPP	TRPML
Ca^{2+} and CaM	Ca^{2+} (Ba^{2+}, Sr^{2+}, La^{3+}): (C3)[67], (C5)[69–72], (C6)[73,74], (C7)[75] Ca^{2+} (inhibit): (C1, C3)[68] Ca^{2+}–CaM: (C5)[109–112], (C6)[74,113,114] Ca^{2+}–CaM (inhibit): (C1)[100,101] (C2)[102,103], (C3)[104]	Ca^{2+} (inhibit): (V4)[94], (V6)[95] Ca^{2+}–CaM (inhibit): (V1)[117–119], (V3)[121], (V4)[122], (V6)[123,124]	Ca^{2+}: (M2)[76–82], (M4)[83], (M5)[84,85] Ca^{2+} (inhibit): (M7)[96–98], (M8)[99] Ca^{2+}–CaM: (M2)[79–80], (M4)[127]	Ca^{2+}: (A1)[86–89]	Ca^{2+}: (P2)[90], (P3)[91,92]	Ca^{2+}: (ML1)[93]
PIP$_2$ and other phosphoinositides	PIP$_2$: (C5)[140], (C7)[143] PIP$_2$ (inhibit): (C4)[141], (C5)[140], (C6)[142] PIP$_3$: (C6)[115,146]	PIP$_2$: (V1)[119,130,131], (V5)[132,133], (V6)[139] PIP$_2$ (inhibit): (V1)[128,129] PIP$_3$: (V2)[145]	PIP$_2$: (M4)[134,135], (M5)[85,132], (M7)[15], (M8)[132]	PIP$_2$: (A1)[138] PIP$_2$ (inhibit): (A1)[136,137]	PIP$_2$ (inhibit): (P2)[144]	
DAG (directly)	DAG: (C2)[148], (C3,C6)[73,147,149], (C7)[75]					

Phosphorylation	PKC (inhibit): (C3,C4,C5)[157], (C5)[150], (C6)[149], (C7)[75] PKG (inhibit): (C3)[158], (C6)[159] Src: (C3)[62,63], (C4)[107], (C6)[160]	PKA: (V1)[154], (V4)[155] PKC: (V1)[150–153], (V4)[155], (V6)[123] Src: (V1)[161], (V4)[164,165]	PKC: (M4)[127,156]

Receptor-Induced Protein Translocation

Receptor-induced protein translocation	via RhoA: (C1)[173] via VAMP2: (C3)[174] EGF: (C4)[107], (C5)[167]	NGF: (V1)[130,169] IGF-I: (V2)[166] via Rab11a: (V5,V6)[171]	via PKA/PLC signaling: (A1)[172]	with PKD1: (P2)[175] via PACS: (P2)[177] via TRPC3: (P2)[178]

Redox Status Changes and Covalent Cysteine Modification

Sensitivity to redox status changes mediated by nicotinamide adenine dinucleotide (NAD+) and its metabolites	H_2O_2: (M2)[76] NAD^+: (M2)[76] ADPR, cADPR: (M2)[179–182] NAADP: (M2)[184]

(continued)

TABLE 3.1 (Continued)

Activation Triggers of TRP Channels in Mammalian Systems

	TRPC	TRPV	TRPM	TRPA	TRPP	TRPML
Cysteine oxidation, reduction, and modification	H_2O_2: (C1,C4,C5)[188] NO: (C1,C4,C5)[188], (C5)[211] tert-Butylhydroperoxide: (C3)[191] Thioredoxin, DTT: (C5)[209]	H_2O_2, NO: (V1, V3, V4)[188], (V1)[190] Diamide, chloramine-T: (V1)[189] Pungent compounds: (V1)[207] DTT: (V1)[208]		H_2O_2: (A1)[192–195] NO: (A1)[192,195,198] 15-Deoxy-$\Delta^{12,14}$-prostaglandin J₂: (A1)[193,195-197] Cinnamaldehyde, allyl isothiocyanate, allicin: (A1)[199–203] 4-Oxononenal and 4-hydroxynonenal, nitrosylated oleic acid: (A1)[204–206] H_2S: (A1)[210]		
Anoxia and hypoxia	Hypoxic condition: (C6)[214]		OGD: (M7)[212,213]			
Ultraviolet (UV)				(A1)[215]		
Other modifications	N-linked protein glycosylation: (C3, C6)[217]	N-glycosylation : (V1)[218], (V4)[219] Klotho: (V4)[221], (V5)[220]	N-glycosylation: (M8)[222]			

(continued)

Noncovalent Action of Chemicals and Signaling Mediators

Endogenous activation modulators	Arachidonic acid: (C2)[223] 5',6'-EET: (C1)[227] 20-HETE: (C6)[228,229] Sphingosine-1-phosphate: (C5)[233] Lysophospholipid: (C5)[234]	Anandamide: (V1)[225], (V4)[226] 12-HPETE, Leukotriene B4: (V1)[230] ATP: (V1)[119,237]	Arachidonic acid: (M2)[76], (M5)[224] Sphingosine: (M3)[232] Lysophosphilipid: (M8)[235] Pregnenolone sulfate: (M3)[236] AMP (inhibit): (M2)[182], (M4)[239] ADP (inhibit): (M4)[239] ATP (inhibit): (M4)[127,239], (M7)[238] Mg^{2+} (inhibit): (M3)[243], (M6)[242], (M7)[238]	Pyrophosphate, pyrotriphosphate: (A1)[240], Zn^{2+}: (A1)[241]
Exogenous activation modulators	Hyperforin: (C6)[270] GsMTx peptide (inhibit): (C6)[273] 2-APB (inhibit): (C5)[280], (C6)[277,280] Pyrazol (inhibit): (TRPCs)[282,283] Pore region antibodies: (C1)[286], (C5)[287] La^{3+}, Gd^{3+}: (C4, C5)[70,289] La^{3+}, Gd^{3+} (inhibit): (C6)[70]	Capsaicin, resiniferatoxin: (V1)[247–249] Piperine: (V1)[250] Gingerols, shogaols: (V1)[251] Vamphor: (V1)[252], (V3)[253,254] Menthol: (V3)[254,259] Eugenol: (V1)[262] Carvacrol, thymol: (V3)[263] Psychoactive terpenoid: (V2)[266] Bisandrographolide A: (V4)[267] Vanillotoxin peptides: (V1)[271]	Menthol, eucalyptol: (M8)[255–257] Icillin: (M8)[255,257] Carvacrol (inhibit): (M7)[265] 2-APB: (M6, M7)[279] 2-APB (inhibit): (M2)[281], (M3)[280], (M8)[277] BTP2: (M4)[284] Ethanol (inhibit): (M8)[285] Pore region antibody: (M3)[288]	Camphor: (A1)[252] Icilin: (A1)[259,260] Menthol: (A1)[261] Carvacrol, thymol, alkyl phenol: (A1)[263,264] Caffeine: (A1)[268] Nicotine: (A1)[269] 2-APB: (A1)[202]

TABLE 3.1 (Continued)
Activation Triggers of TRP Channels in Mammalian Systems

	TRPC	TRPV	TRPM	TRPA	TRPP	TRPML
		Acyl polyamine toxins (inhibit): (V1)[272] 4α-PDD: (V4)[274,275] 2-APB: (V1, V2 (rat and mouse), V3)[266,276–278] Ethanol: (V1)[285]				
pH						
Acidification	(C4, C5)[291,292]	(V1)[247,290]	(M6, M7)[96–98,279]	(A1)[195]	(PKD1L3 and P3)[293]	(ML1, ML2, ML3)[294]
Suppression of activation by acidification	(C5)[299], (C6)[291]	(V5)[295,296]	(M2)[298], (M5)[297]			
Alkalization			(M7)[302]	(A1)[300]		
Proton permeability			(M7)[303]		(P3)[301]	(ML1)[304]
Temperature						
Heat		>41.5°C–43°C, $Q_{10} = 14.8$–25.6: (V1)[247,290,306–309] >51.6°C–55°C, $Q_{10}{\sim}100$: (V2)[310–312]				

	TRPC	TRPV	TRPM	TRPA	TRPP
Warm		31°C–39°C, Q_{10} = 21.6: (V3)[316–318] 25°C–33.6°C, Q_{10} = 9.9–19.1: (V4)[322–325]	33.6°C–35°C, Q_{10} = 8.5: (M2)[327] 15°C–35°C, Q_{10} = 8.5–10.3: (M4, M5)[326]		(P2)[90,352–354], (P3)[301]
Cold		22°C–26°C, Q_{10} = 23–35: (M8)[255,256,260,328,329]	**Membrane Voltage** (M3)[232], (M4, M5)[83–85,326,339,340], (M8)[314,315], (M7)[303,345,346] **Mechanostimulation** (M4)[358] (M7)[345,346,357]	8°C–28°C: (A1)[260,261,335,336]	
Membrane voltage	(C5)[337]	(V1)[314,315,341], (V3)[343], (V5)[338]	(M3)[365], (M7)[345,346]	(A1)[87,203,261]	
Membrane tension	(C1)[356], (C6)[229,273]	(V2)[359]		(A1)[367]	
Hypotonicity-induced cell swelling	(C5)[366], (C6)[229,273,364]	(V2)[359], (V4)[164,165,319–321,360–363]			
Hypertonicity-induced cell shrinkage		(V1)[370,371]			
Shear stress	(C6)[229]	(V4)[361–363]	(M7)[345,357,372]		
Mechano-induced protein translocation		(V2)[374], (V4)[319]	(M7)[372]		(P1 and P2)[373]

complex signalsomes, close interactions between PM receptors and TRP channels are critical for enhancing the efficiency of their coupling, although clear evidence has not been provided to show that the interactions are indeed "direct."

3.2.2   G PROTEINS

Activation by G proteins is not widely recognized in TRP channels. Only activation of the Gi protein of TRPC4 has been reported.[8] Conversely, constitutively activated TRPM1 currents are suppressed by activated Go proteins, when cells are internally dialyzed with the Go protein activated by the nonhydrolyzable GTP analogue from the patch pipette in HEK293 cells.[9] This suppression mechanism mediates the ON pathway of retinal bipolar cells. Considering the relatively slow time course (60 s) for complete suppression by the purified Goα preparation, additional factors may be involved in TRPM1 suppression.

3.2.3   PLC

PLC proteins themselves are important regulators of TRP channel activity. PLCγ1 and PLCγ2 interact with TRPC3 and TRPC4.[10,11] In neuron-like pheochromocytoma PC12 cells, smooth muscle A7r5 cells, or DT40 B cells, knockdown of PLCγ1 and PLCγ2 abolishes Ca^{2+} entry induced by agonist stimulation. In TRPC3–PLCγ1 interactions, a partial pleckstrin homology (PH) domain—a consensus lipid- and protein-binding sequence—in PLCγ1 and a complementary PH-like domain in TRPC3 have been reported to interact to elicit lipid binding and cell surface expression of TRPC3.[12] However, structural analysis does not necessarily support this intercellular interaction but rather suggests an intracellular assembly of the partial PH domains in PLCγ1.[13] The kinase domain of TRPM7 directly associates with the C2 domain of PLCs.[14] In cardiac cells and heterologous expression systems, this interaction allows inhibition of TRPM7 activity by hydrolysis of PIP_2 through activation of PLCs upon stimulation of Gq-linked receptors and tyrosine kinase receptors (see also Section 3.2.7).[15]

3.2.4   IP$_3$ AND CA^{2+} STORE DEPLETION

Store-operated Ca^{2+} (SOC) channels are responsible for Ca^{2+} influx activated upon IP_3-induced Ca^{2+} release and the consequent depletion of ER Ca^{2+} stores. Signaling mechanisms triggered by ER store depletion that activates SOC channels at the PM have been a mystery for a long time. STIM1, the protein that was originally identified in a screen of cell adhesion molecules,[16] has recently emerged as the first of a long-sought molecules essential for the transduction of a store depletion signal.[17–19] Ca^{2+} release-activated Ca^{2+} (CRAC) channels are SOC channels that show prominent Ca^{2+}-selective permeation properties[20] and are formed by proteins termed Orai or CRAC modulators (CRACMs).[21–25] STIM proteins interact with Orai/CRACMs to activate CRAC channels.[26–28] Since molecular identification of the first mammalian TRP, TRPC1,[29,30] involvements in SOC influx have been demonstrated for TRPC channels, using recombinant systems[31–37] and gene knockout strategies.[38–41] Involvements of TRPV6[42] and TRPM5[43] in SOC channels have also been suggested. It has been demonstrated that STIM1 interacts and activates TRPCs.[44,45] STIM1 binds to TRPC1, TRPC4, and TRPC5 and determines their function as SOC channels; however, it associates with TRPC3 and TRPC6 indirectly through heteromultimerization with TRPC1 and TRPC4.[46] STIM1 gates TRPC1 and TRPC3 channels by intermolecular electrostatic interaction, which is not required for STIM1-mediated activation of CRAC channels.[47] Furthermore, lipid raft localization of TRPC1 can be regulated by STIM1, converting TRPC1 from a receptor-operated channel to an SOC channel.[48] Assembly of TRPCs with not only STIMs but also Orai1 proteins has been suggested to play an important role in the formation of SOC channels.[49–51] In rat mast cell line RBL-2H3, activation of TRPC5 channels depends on Orai1 and STIM1 upon store depletion and provides optimal Ca^{2+} or Sr^{2+} entry for degranulation.[52] Conversely, it must be noted that several lines of evidence suggest a TRPC channel activation mechanism that is independent of STIM1 and Orai1.[53,54]

Other than STIM1 and Orai proteins, a number of important associated proteins such as IP_3Rs have been reported in activating the SOC channel activity of TRPCs. Although these topics have been presented in Chapter 2, we wish to point out an important interaction here. For example, a C-terminal site of TRPCs has been identified to bind to IP_3Rs[55–57] through biochemical binding studies.[58] This site is shared by Ca^{2+}-dependent calmodulin (CaM) binding and is therefore known as the CaM- and IP_3R-binding (CIRB) domain. It will be discussed in Section 3.2.6.

3.2.5   Inositol 1,3,4,5-Tetrakisphosphate (IP_4)

Production of IP_4 by the phosphorylation of IP_3 has been shown in mammalian cells upon PLC-linked receptor stimulation.[59–61] Although several lines of evidence have suggested that IP_4 is an activator/positive regulator of Ca^{2+} entry,[62–64] the molecular entity corresponding to an IP_4-regulated Ca^{2+}-permeable channel is yet to be identified.[65]

3.2.6 Ca^{2+} and Calmodulin

First examples for activation of TRP channels independent of store depletion were TRPC6[66] and Ca^{2+}-induced activation of TRPC3.[67] Conversely and puzzlingly, opposite suppressive effects of intracellular Ca^{2+} have been reported for TRPC3 as well as TRPC1; oligomeric channels formed by TRPC1 and TRPC3 showed high sensitivity to Ca^{2+} compared to homomeric channels.[68] In the case of TRPC5, Ca^{2+} acting from both the outside and inside is important for activation.[69–72] An action site of extracellular Ca^{2+} is shared by La^{3+},[70] while inhibition by the Ca^{2+} chelator BAPTA, but not by EGTA, suggests the action site of intracellular Ca^{2+} is beneath the PM.[72] The action of extracellular Ca^{2+} is stimulatory for TRPC6, which is mimicked by Ba^{2+} and Sr^{2+},[73,74] but it is inhibitory for TRPC7.[75] Intracellular Ca^{2+} shows bell-shaped effects on TRPC6 and TRPC7: Ca^{2+} is important to support channel activity, which peaks at around 100–200 nM but turns inhibitory at higher concentrations.[74,75] Ca^{2+} applied from either the extracellular or the intracellular side has been shown to be critical for TRPM2 channel activation.[76–79] In particular, intracellular Ca^{2+} alone is very likely sufficient for TRPM2 activation without a requirement for other activation triggers (see Section 3.4).[80,81] Interestingly, intracellular Ca^{2+} induces an aberrant current inactivation in a TRPM2 variant of Guamanian amyotrophic sclerosis and Parkinsonism-dementia.[82] TRPM4[83] and TRPM5[84,85] are Ca^{2+}-activated cation channels that are responsible for cell membrane depolarization without significant Ca^{2+} permeation. The action of Ca^{2+} is likely direct, but the responsible site is unclear yet in these channel proteins. Likewise, TRPA1 is activated by the direct action of Ca^{2+} from the intracellular side through an EF hand-like motif.[86–89] TRPP2[90] and TRPP3[91] are positively regulated by Ca^{2+}, and they have EF hand-like domains, although the one in TRPP3 is unlikely to be the determinant of Ca^{2+}-induced activation.[92] An increase in single channel open probability by intracellular Ca^{2+} has been reported for the type IV mucolipidosis-associated protein TRPML1.[93] For a number of TRP channels, Ca^{2+} is a negative modulator. In TRPV4 and TRPV6, Ca^{2+} from outside suppresses and accelerates the decay of the current, and Ca^{2+} from inside suppresses the current at much lower concentrations.[94,95] Extracellular Ca^{2+} suppresses TRPM7 currents by blocking the ion-conducting pore.[96–98] On the other hand, TRPM8 is suppressed by extracellular Ca^{2+} and other divalent cations, as well as protons through surface charge screening.[99]

CaM is the most important protein that mediates the effects of Ca^{2+} in TRP channels. The CIRB domain identified on the C-termini of TRPC channels through biochemical binding studies is the site for Ca^{2+}-dependent binding of CaM.[58] CaM bound to this site is suggested to contribute to the delay between Ca^{2+} release and Ca^{2+} influx via TRPC1.[100] In TRPC2, Ca^{2+} influx strongly inhibits channel activity via the formation of Ca^{2+}-CaM to induce effective sensory adaptation of pheromone responses of vomeronasal sensory neurons.[101] The N-terminus of TRPC2 has been also shown to bind CaM in a Ca^{2+}-dependent manner.[102] In TRPC3, it has been proposed that displacement of Ca^{2+}-CaM, which sets the channel under resting conditions, by IP_3Rs at the common binding site leads to channel activation.[103] Of note, a role of the common site is also suggested in targeting TRPC3 proteins to the PM in a CaM/IP_3R-indpendent process.[104] In TRPC4 that also binds Ca^{2+}-CaM,[105] the spectrin cytoskeleton has been reported to interact with the CIRB domain to influence the epidermal growth factor (EGF) induced surface expression of the protein (see also Section 3.3).[106,107] Ca^{2+}-CaM binds to the second site on the C-terminal side of the CIRB domain to regulate Ca^{2+}-dependent facilitation in TRPC5 channels.[108] In addition, myosin light chain kinase activated by Ca^{2+}-CaM is critical for the activation of TRPC5.[109–111] Binding of Ca^{2+}-CaM plays an important role in the activation of TRPC6 as well.[112,113] The Ca^{2+}-CaM binding site in the C-terminal CIRB domain of TRPC6 has been shown to accommodate phosphoinositides, particularly phosphatidylinositol 3,4,5-trisphosphate (PIP_3) with the highest potency, and to enhance channel activity by antagonizing the suppressive effects of Ca^{2+}-CaM.[114] Comparative binding experiments have revealed that this site is present in other TRPCs and TRPV1 as well as in the voltage-dependent K^+ channels KCNQ1 and voltage-dependent Ca^{2+} channel $Ca_V 1.2$.[114] In mediating the action of extracellular Ca^{2+}, which is critical for the activation of TRPC6, activation of CaM-dependent kinase II by Ca^{2+}-CaM also plays an important role.[74] In sperm, TRPC1, TRPC2, and TRPC5, but not TRPC3, interact and colocalize with the CaM-binding protein enkurin.[115] In addition to the TRPC family, members of the TRPV family, in particular TRPV1, are susceptible to Ca^{2+}-CaM regulation. Ca^{2+} permeated through TRPV1 forms Ca^{2+}-CaM that binds to the C-terminus[116] and the ankyrin-like repeat of the TRPV1 protein[117] to induce desensitization (feedback inhibition) of its channel activity. Detailed structural analysis has shown that the Ca^{2+}-CaM binding site accommodates ATP but not PIP_2 in the N-terminal ankyrin-like repeat in TRPV1[118] while the Ca^{2+}-CaM binding site in the C-terminal CIRB domain, which accommodates PIP_3 with the highest potency, is present as well in TRPV1, as mentioned above.[114] Interestingly, Ca^{2+}-CaM formed by Ca^{2+} influx through TRPV1 causes a profound inhibition of voltage-dependent Ca^{2+} channels in sensory neurons.[119] Major roles have been suggested for the N-terminal CaM-binding site in the Ca^{2+}-dependent inhibition of TRPV3[120] and for the C-terminal CaM-binding site in Ca^{2+}-dependent potentiation of TRPV4.[121] Ca^{2+}-CaM accelerates current inactivation in human and rat TRPV6.[122,123] However, puzzlingly, Ca^{2+}-insensitive mutants of CaM suppressed mouse TRPV6 activity.[124] In TRPM2, activation induced by Ca^{2+} is mediated by Ca^{2+}-CaM,[79,80] whose binding site is located at the IQ-like motif in the N-terminus.[125] In TRPM4, although Ca^{2+}-CaM significantly facilitates activation, Ca^{2+} can still activate the channel without binding CaM, although much more weakly.[126] Protein kinase C (PKC) also represents an important pathway of Ca^{2+} signaling and is discussed in Section 3.2.9. Thus Ca^{2+} regulates TRP channels from outside and inside the cell through multiple modes of regulatory pathways.

3.2.7 PIP₂ AND OTHER PHOSPHOINOSITIDES

The effect of PIP_2 was originally reported as channel inhibition in TRPV1 channels.[127,128] However, recent understanding of the PIP_2 effect suggests that it is involved in the positive modulation of channel activation:[129] PIP_2, as well as ATP, has been shown to prevent desensitization of TRPV1 to repeated applications of capsaicin (tachyphylaxis), on the basis of structural and functional analyses.[118] This is supported by the identification of Pirt protein as a TRPV1 regulatory subunit that binds PIP_2.[130] The importance of PIP_2 has also been reported for TRPM7, in which PIP_2 hydrolysis by PLCs inhibits channel activation.[15] PIP_2 exerts a strong positive modulatory action on TRPM5,[85,131] TRPM8,[131] TRPV5,[131,132] and TRPM4,[133,134] which likely proceeds through suppression of desensitization. Like TRPV1, the inhibition of channel activity by PIP_2 was also reported for TRPA1;[135,136] however, a later report showed a promoting action of PIP_2 on TRPA1 activity.[137] Support of channel activation by PIP_2 has also been implicated as well for TRPV6[138] and TRPC5.[139] Among TRPC channels, inhibition by PIP_2 has been implicated for TRPC4,[140] TRPC6,[141] and, puzzlingly, for TRPC5,[139] while the lipid has a robust activating effect on TRPC7.[142] PIP_2-mediated inhibition has been also reported for TRPP2.[143] Because PIP_2 is degraded by PLC, the action of PIP_2 may underlie the response of some TRP channels to stimulation of PLC-coupled PM receptors, as mentioned above for TRPM7.

As mentioned in Section 3.3 below, most studies suggest that phosphatidylinositol 3-kinase (PI(3)K), which is responsible for the generation of PIP_3, is essential for eliciting channel flux activity through protein translocation of TRP channels to the PM. However, a direct action by PIP_3 also plays a role in supporting the activation of

TRPV2[144] and TRPC6.[145] In TRPC6, PIP_3 disrupts the association of CaM with the channel protein and enhances the channel activity.[114]

3.2.8 DAG (DIRECT ACTIVATION)

Activation by DAG through direct action in a membrane-delimited fashion was first described for TRPC3 and TRPC6,[146] which was followed immediately by a report for TRPC7[75] and later by that for TRPC2 in vomeronasal organs using knockout mice.[147] The action of DAG is independent of the serine/threonine kinase, PKC, because inhibitors or activators for PKCs failed to enhance activity of these TRPC channels.[73,75,146] In TRPC6, the action site of DAG has been mapped at the N-terminus.[148] Lipids, including arachidonic acid and its metabolites downstream of DAG, are important regulators of TRP channels, and are discussed in Section 3.5 below.

3.2.9 PHOSPHORYLATION

In addition to direct membrane-delimited action, "classical" PKC activation mediates regulatory actions of DAG on TRP channels. TRPV1 activation by PKC upon cellular stimulation by the pro-inflammatory peptide bradykinin, the endogenous ligand anandamide, and ATP is known to induce a pain sensation.[149,150] In this metabolic pathway-mediated mode of TRPV1 activation, TRPV1 proteins are directly phosphorylated by a PKCε subtype, which has been suggested to control pain sensation and hyperalgesia.[151,152] TRPV1 channel activity is also regulated through a reduction of desensitization by protein kinase A (PKA)-mediated protein phosphorylation.[153] Similar to TRPV1, TRPV4 channel activity is enhanced by PKA- and PKC-mediated protein phosphorylation.[154] CaM-dependent inactivation is counteracted by phosphorylation through PKC in TRPV6.[122] Ca^{2+} sensitivity of Ca^{2+}-activated TRPM4 is enhanced by PKC,[126] contributing to the control of the myogenic tone.[155] By contrast, PKC is inhibitory to TRPC channels.[75,148,149,156] TRPC3, TRPC4, and TRPC5 channels are strongly inhibited by DAG-induced PKC activation.[156] PKC exerts an inhibitory action on TRPC7 as well.[75] Phosphorylation by protein kinase G (PKG) also suppresses the activities of TRPC3[157] and TRPC6.[158] This mode of inhibition may act downstream of nitric oxide (NO) to contribute to vasorelaxation (see Section 3.4).

In addition to serine/threonine kinases, the tyrosine kinase, Src, plays important roles in TRP channel activation in response to growth factor stimulation. Src makes a significant contribution to the activation of TRPC6,[159] TRPV1,[160] TRPC3,[161,162] and TRPC4.[106] With regard to hypotonicity-induced activation of TRPV4, contradictory results have been reported for the involvement of Src.[163,164] Furthermore, phosphorylation-induced translocation by PKC or Src controls TRP channel activity, which is discussed in Section 3.3.

3.3 RECEPTOR-INDUCED PROTEIN TRANSLOCATION

Protein translocation regulated by receptor activation cannot be regarded as a conventional activation mechanism that involves channel "gating," but it plays a critical role in controlling TRP channel activities at the PM. TRPV2 was the first TRP homologue for which the involvement of this mechanism was demonstrated.[165] Insulin-like growth factor induces translocation of TRPV2, which is termed "growth factor-regulated channel" in the paper, from intracellular compartments to PM to mediate Ca^{2+} influx, when the channel is expressed recombinantly in CHO cells and endogenously in MIN6 insulinoma cells. Recombinant TRPC5 channels are susceptible to rapid vesicular translocation and insertion mediated by PI(3)K, Rho GTPase Rac1, and phosphatidylinositol 4-phosphate 5-kinase (PIP(5)Kα) upon activation of EGF receptors.[166] In neurons, translocated TRPC5 contributes to control of neurite extension and growth cone morphology through interaction with stathmin proteins.[167] Insertion of TRPV1 channels to PM is induced by nerve growth factor (NGF) in dorsal root ganglion (DRG) sensory neurons via TrkA receptors, PI(3)K, and direct tyrosine phosphorylation by Src kinase,[168] which might underlie NGF-induced sensitization or hyperalgesia.[169] Physical interaction between PI(3)K and TRPV1 may underlie this mechanism.[129] The small G protein Rab11a targets TRPV5 and TRPV6, which mediates the rate-limiting luminal influx step of transcellular Ca^{2+} transport to the PM.[170] The importance of PM translocation is also suggested in the sensitization of TRPA1-mediated nocifensive behavior through activation of PKA and PLC.[171] In addition, PM translocation is involved in Ca^{2+} entry through TRPC1 regulated via RhoA in endothelial cells,[172] as well as in Ca^{2+} entry through TRPC4 via Src family tyrosine kinases, such as Fyn, upon EGF stimulation in COS-7 cells (see also Section 3.2.6 for the involvement of spectrin).[106] VAMP-dependent exocytosis regulates PM insertion of TRPC3 and contributes to agonist-induced Ca^{2+} entry.[173] Notably, in pathogenesis of polycystic kidney disease (PKD) due to mutation of TRPP2, protein trafficking to the PM is induced through assembly of TRPP2 with polycystin-1 (PKD1), the other causative gene for PKD,[174] TRPC1,[175] or phosphofurin acidic cluster sorting proteins PACS.[176] More recently, we have raised a possibility that a pathogenic C-terminus-truncated TRPP2 is mislocalized at the apical membrane of kidney epithelial cells, where it enhances growth factor-induced Ca^{2+} influx through misassembly with TRPC3.[177] Thus, accumulating evidence strongly suggests a contribution of protein translocation to functional regulation of TRP channels, although we cannot exclude the possibility that receptor stimulation, which evokes PM translocation, may, in parallel, lead to production of messenger molecules that modify "gates" of the translocated channels.

3.4 CHANGES IN REDOX STATUS AND COVALENT CYSTEINE MODIFICATION

3.4.1 SENSITIVITY TO REDOX STATUS CHANGES MEDIATED BY NICOTINAMIDE ADENINE DINUCLEOTIDE (NAD$^+$) AND ITS METABOLITES

We have demonstrated that activation of TRPM2 is potently triggered by changes in the redox status, specifically by oxidative stress.[76] Application of H_2O_2 or stimulation with tumor necrosis factor α evokes Ca^{2+} and/or Na^+ influx via native TRPM2 channels, leading to cell death in insulinoma cells and monocytes. NAD$^+$ was proposed as a mediator of the sensitivity.[76] However, it is more likely that some metabolites of NAD$^+$, such as ADP-ribose (ADPR) and cyclic ADP-ribose (cADPR), which are

TRPM2 activators reported independently of our work,[178,179] are direct activation triggers of TRPM2.[180,181] TRPM2 activation independent of ADPR has also been proposed.[182] Nicotinic acid adenine dinucleotide phosphate (NAADP) has been reported to activate TRPM2.[183] More recently, TRPM2 has been demonstrated to release Ca^{2+} from intracellular organelles such as lysosomes.[184] In this context, it is interesting to study relationships between TRPM2 and two-pore channels (TPCs) in regulating NAADP-induced Ca^{2+} mobilization; the latter release Ca^{2+} from endo-lysosomes in response to NAADP at much lower concentrations than TRPM2.[185] TRPM2-activating NAD^+ metabolites are produced by multiple pathways.[186] For example, ADPR can be generated by the action of poly(ADPR) glycohydrolase from poly(ADPR), which is made from NAD^+ by poly(ADPR) polymerase. The ADP-ribosyl cyclase, CD38, also produces ADPR, cADPR, and NAADP. Notably, poly(ADPR) production is tightly linked to DNA damage caused by oxidants and chemicals; however, the link of oxidative stress mediated via reactive oxygen species (ROS) to CD38 and other NAD^+-hydrolyzing enzymes still remains elusive.

3.4.2 CYSTEINE OXIDATION, REDUCTION, AND MODIFICATION

Cysteine oxidation has emerged as a prominent mechanism underlying activation of various TRP channels. Direct oxidative modifications of cysteine residues by H_2O_2, nitric oxide (NO), and reactive disulfides have been demonstrated for TRPC5.[187] NO nitrosylates and reactive disulfides directly modify cysteine residues located on the N-terminal side of the pore-forming region between S5 and S6 transmembrane helices. The corresponding cysteine sites of TRPC1, TRPC4, TRPV1, TRPV3, and TRPV4 are also targets of nitrosylation that leads to channel activation. Sensitizing effects of oxidizing agents such as diamide and chloramine-T support the existence of the counterpart cysteine in TRPV1.[188] More recently, the C-terminal cytoplasmic cysteine residues that sensitize TRPV1 activation have been shown upon oxidative challenges.[189] TRPC3 shows insensitivity or, if at all, only marginal sensitivity to H_2O_2 or NO donors and reactive disulfides[187] but is sensitive to tert-butylhydroperoxide at relatively low concentrations, although the mechanism is unclear.[190] Using recombinant expression systems and isolated sensory neurons from wild-type and knockout mice, H_2O_2, NO donors, and 15-deoxy-$\Delta^{12,14}$-prostaglandin J_2 have been shown to activate TRPA1, of which a number of N-terminal cysteines are likely to mediate the effect of these activators.[191–197] Originally, TRPA1 was functionally identified as a channel activated by pungent natural compounds, including cinnamaldehyde, allyl isothiocyanate, and α,β-unsaturated aldehydes (from plants such as mustard, onion, cinnamon, and wasabi), and the garlic pungent compound, allicin. These compounds are potentially susceptible to the nucleophilic attack of the sulfhydryl group of cysteine residues, as well as by cold temperature, receptor stimulation, and cannabinoids.[198–200] Later examinations of various noxious compounds finally led to a conclusion that electrophilic compounds that covalently modify cysteine residues through mechanisms such as Michael addition are indeed potent activators of TRPA1 channels in recombinant and native systems.[201,202] Other α,β-unsaturated aldehyde derivatives (endogenous autacoids, 4-oxononenal, and 4-hydroxynonenal), as well as nitrosylated oleic acid, also show activating effects on TRPA1 channels.[203–205]

Interestingly, TRPV1 also shows sensitivity to pungent compounds from onion and garlic, such as allicin, through covalent modification of a single cysteine residue located in the N-terminal region.[206]

Chemical agents that reverse or reduce oxidatively modified cysteine residues can induce activation of several TRP channels. Dithiothreitol (DTT), an agent that maintains the sulfhydryl group of cysteines in a reduced state, was first reported to facilitate heat- or capsaicin-activated currents in cultured DRG neurons.[207] Activation of TRPC5 by extracellular thioredoxin and DTT,[208] which act on the same cysteine residues as oxidative modification and nitrosylation,[187] and activation of TRPA1 by the hydrogen sulfide (H_2S) donor, sodium hydrogen sulfide (NaHS), have been reported.[209]

Except for sensing functions in sensory neurons, the physiological significance of activation via cysteine modification still remains obscure in TRP channels. In the case of TRPC5, activation induced by NO via nitrosylation should enhance the Ca^{2+} influx responsible for augmentation of NO production by endothelial type NO synthase, forming a positive feedback loop of NO production upon vasodilator stimulation.[186,210] By contrast, in smooth muscle cells, phosphorylation by PKG, which is activated downstream of NO, suppresses activities of TRPC3[157] and TRPC6[158] (see also Section 3.2). This mode of channel inhibition in smooth muscle cells might contribute, together with NO-activated TRPC5 channels and enhanced NO production in endothelial cells, to full vasorelaxation (see Section 3.11).

3.4.3 ANOXIA AND HYPOXIA

Anoxia induced by oxygen-glucose deprivation (OGD) activates cationic currents that correspond to TRPM7 currents in cultured central neurons.[211] This OGD-activated TRPM7 current is responsible for anoxic neuronal death in vitro and in vivo.[212] ROS, or reactive nitrogen species, are involved in this TRPM7 activation. In smooth muscle cells from precapillary pulmonary arteries, hypoxia (1% O_2) induces cation influx, and consequently hypoxic pulmonary vasoconstriction, both of which were shown to be mediated by TRPC6.[213] Accumulation of DAG is likely to be responsible for the TRPC6 activation.

3.4.4 ULTRAVIOLET (UV) LIGHT

UV light has been shown to activate TRPA1 channels presumably through generation of ROS.[214] Therefore, TRPA1, as well as other ROS-activated TRP channels, are likely candidates for Ca^{2+}-permeable cation currents observed in different mammalian cell types.[215]

3.4.5 OTHER MODIFICATIONS

Glycosylation plays important roles in TRP activation. In TRPC3 and TRPC6, N-linked protein glycosylation has a major dominant effect on basal channel activity.[216] Mutation of the second extracellular glycosylation, which is found in TRPC6 but not in TRPC3, converted tightly receptor-regulated TRPC6 into constitutively active channels like TRPC3, while introduction of the second site markedly reduced the constitutive activity of TRPC3. N-glycosylation is also known to enhance ligand-binding and gating properties of TRPV1,[217] while disruption of N-glycosylation enhanced membrane trafficking and hypotonic stress-induced activity in TRPV4 (see also Section 3.9 for osmoregulation).[218] The β-glucuronidase, Klotho, regulates accumulation of TRPV5 and stimulates the activity of TRPV5 through extracellular hydrolysis of TRPV5-associated N-linked oligosaccharides.[219] Taking into consideration that Klotho and TRPV5 colocalize in the distant part of a nephron, it must be noted that TRPV5 regulates secretion of Klotho.[220] In TRPM8, N-glycosylation localizes the protein in lipid rafts, which suppresses menthol- and cold-mediated responses.[221]

3.5   NONCOVALENT ACTION OF CHEMICALS AND SIGNALING MEDIATORS

3.5.1   ENDOGENOUS ACTIVATION MODULATORS

Because this topic is discussed partially in Section 3.2 for signaling molecules regulated downstream of PLC-coupled receptors and in Section 3.4 for redox-related molecules, we discuss here mainly the remaining important signaling molecules that regulate activation of TRP channels.

Lipid metabolites are potent endogenous activators of TRP channels. Arachidonic acid is freed from a phospholipid molecule by the enzyme phospholipase A2 (PLA_2), which cleaves off the fatty acid, but can also be generated from DAG by DAG lipase. Activation by arachidonic acid was first shown for TRPM2 as an enhancement of its NAD^+ activation sensitivity.[76] In the vomeronasal organ, TRPC2 can be involved in arachidonic acid-activated currents.[222] TRPM5 expressed abundantly in taste receptor cells is potently activated by arachidonic acid.[223] The endocannabinoid anandamide activates TRPV1.[224] Anandamide and its metabolite, arachidonic acid, activate TRPV4 via cytochrome P450 epoxygenase-dependent formation of 5′, 6′-epoxyeicosatrienoic acid (5′, 6′-EET) in endothelial cells.[225] It is also suggested that 5′, 6′-EET plays roles in store depletion-induced activation of TRPC1.[226] Another arachidonic acid metabolite, 20-hydroxyeicosatetraenoic acid (20-HETE), which is produced via cytochrome P450, is a potent activator of TRPC6,[227] suggesting that some of the reported effects of arachidonic acid are mediated by 20-HETE. The 20-HETE sensitivity of TRPC6 underlies its mechanosensitivity under receptor stimulation.[228] In addition to 20-HETE, lipoxygenase metabolites of arachidonic acid, such as 12-(S)-hydroxyperoxyeicosatetraenoic acid (12-HPETE) and leukotriene B_4, are potent activators of TRPV1,[229] which may be responsible for inflammatory hyperalgesia.[230]

Among sphingolipids, which are known to inhibit many ion channels, sphingosine but not sphingosine-1-phosphate has been, for the first time, demonstrated to activate TRPM3 channels.[231] Sphingosine-1-phosphate, by contrast, is an effective activator of TRPC5 channels.[232] Common endogenous lysophospholipids, including lysophosphatidylcholine (LPC) produced by PLA_2, also activate TRPC5[233] and TRPM8.[234] On the other hand, the neurosteroid, pregnenolone sulfate (PS), is capable of directly activating TRPM3.[235] It is interesting to note that localization to lipid rafts is important for TRPC1 and TRPM8 activity.[48,221]

Other than NAD[+] and its metabolites, nucleotide phosphates show interesting modulatory action on TRP channel activation. ATP enhances the activity of TRPV1[236] via binding to the ankyrin-like repeats.[118] By contrast, TRPM7 is inhibited by ATP and GTP complexed with Mg^{2+}.[237] Paradoxically, TRPM4 is inhibited and facilitated by ATP[126,238] and is inhibited by AMP and ADP.[238] TRPM2 is inhibited by AMP.[181]

Organic phosphates, such as pyrophosphate and pyrotriphosphate, support activation of TRPA1 by pungent chemicals.[239] IP_3 and IP_4 also show partial effects.

Other than Ca^{2+}, several endogenous inorganic ions are important regulators of TRP channels. Zn^{2+} activates TRPA1 channels, which underlies the regulatory role of Zn^{2+} in sensory transmission.[240] The "so-called" Mg^{2+}-absorption channels, TRPM7,[237] TRPM6,[241] and TRPM3,[242] are inhibited and tightly regulated by intracellular Mg^{2+}. In case of TRPM7, Mg^{2+} cooperates physiologically with ATP and GTP to inhibit channel activity.[237] Clear lines of evidence have been obtained for these actions using X-ray structure analysis and biochemical analyses.[243,244]

3.5.2　EXOGENOUS ACTIVATION MODULATORS

Various natural and artificial exogenous compounds are activators of TRP channels. For those readers particularly interested in herbal compounds and toxins, it is highly recommended to read the excellent review by Vriens et al.[245] A milestone of this research area is the identification of the capsaicin receptor (or vanilloid receptor), which was later designated as TRPV1.[246] The regions responsible for binding exogenous vanilloid compounds such as capsaicin and resiniferatoxin, an extremely potent capsaicin analogue from Europhobia plants, have been identified on the cytoplasmic side of the membrane[247,248] and are shared at least in part by endovanilloids, such as anandamide.[247] Other pungent compounds, such as piperine from black pepper[249] and both pungent and nonpungent compounds, gingerols, and shogaols from ginger, are also potent activators of TRPV1.[250]

Monoterpenes, such as camphor or menthol, are well-known plant-derived natural products, which, like capsaicin, show sensory effects. Camphor activates TRPV1[251] and TRPV3[252,253] but blocks TRPA1.[251] The "cold" receptor TRPM8 channel is activated strongly by natural cooling compounds, such as monoterpenes menthol and eucalyptol, as well as by the synthetic super-cooling agent icilin.[254–256] A high-throughput random mutagenesis screen has revealed that the second transmembrane region S2 is the determinant for menthol sensitivity of TRPM8.[257] Menthol also activates TRPV3.[253,258] It has been reported that TRPA1, another cold receptor channel, is activated by icilin but is insensitive to (or inhibited by) menthol.[258,259] However, a recent dose response study has revealed a bimodal action of menthol on TRPA1: submicromolar to low-micromolar concentrations cause channel activation, whereas higher concentrations lead to a reversal block.[260] Eugenol, the major component of essential oil of cloves, activates TRPV1.[261] This compound and others from oregano, clove, and thyme such as carvacrol and thymol, which are known as skin sensitizers and allergens and induce a warm sensation on the tongue, strongly activate and sensitize TRPV3.[262] Interestingly, carvacrol, thymol, and other alkyl phenols activate

TRPA1,[262,263] while carvacrol desensitizes TRPA1[262] and inhibits TRPM7.[264] The main psychoactive terpenoid from the cannabis plant is a potent activator for TRPV2,[265] while bisandrographolide A isolated from a Chinese herb activates TRPV4.[266] TRPA1 shows responses to caffeine[267] and nicotine.[268] The former sensitivity underlies the sensation of bitter taste, while the latter mediates irritation of mucosa and skin. A bicyclic polyprenylated acylphloroglucinol derivative, hyperforin, extracted from St. John's wort, activates TRPC6,[269] which underlies its antidepressant profile.

Several spider toxins have been identified that affect TRP channels. Vanillotoxin peptides isolated from the venom of a tarantula activate TRPV1,[270] whereas acyl polyamine toxins from the venom of a funnel web spider block TRPV1.[271] GsMTx peptide from the tarantula venom inhibits pressure-induced activation of TRPC6.[272]

The synthetic phorbol ester, 4α-phorbol 12,13-didecanoate (4α-PDD), has been identified as a TRPV4-selective activator.[273] The determinant of 4α-PDD sensitivity is located in the S3 and S4 transmembrane domains of TRPV4.[274] By contrast, the effects of another synthetic compound, 2-aminoethoxyphenyl borate (2-APB), are variable among TRP channels. 2-APB activates thermosensor TRP channels, TRPV1, TRPV2 (rat and mouse but not human orthologs), TRPV3,[265,275–277] TRPA1,[201] TRPM6, and TRPM7,[278] whereas it inhibits TRPC6, TRPM8,[276] TRPC5, TRPM3,[279] and TRPM2.[280] Although pyrazole compounds are now widely accepted as selective inhibitors of TRPCs,[281,282] it has been reported that BTP2, a synthetic derivative of pyrazole compounds, shows activating effects on TRPM4 to inhibit Ca^{2+} entry through membrane depolarization.[283] Ethanol inhibits TRPM8 but activates TRPV1 by weakening the PIP_2 interaction.[284] Antibodies that recognize the N-terminal side of the pore region show selective inhibitory effects on TRPC5, TRPC1, and TRPM3[285–287] (see Chapters 6 and 7 for more detailed discussion). Thus, the development of non-natural agents will provide promising tools to selectively modify respective TRP channel subtypes.

Polyvalent inorganic cations, La^{3+} and Gd^{3+}, potentiate TRPC4 and TRPC5 currents but suppress TRPC6 currents.[70,288] The action site located at the extracellular side of the pore mouth can be shared by Ca^{2+}, which also potentiates TRPC5.

3.6 pH

3.6.1 Acidification

Among TRP channels, activation by acidification was first shown for TRPV1.[246,289] The concentration–response curve of TRPV1 for protons shows a half-maximal response at pH 5.4. Acidification also potentiates capsaicin- and heat-activated TRPV1 currents. Interestingly, acidification-sensitive activation is a characteristic of TRPV1 and is only seen in the TRPV subfamily.

TRPM6 and TRPM7 respond to acidification-induced current potentiation.[96–98,278] The 50% potentiation pH is 4.5 for TRPM7. Glu1047 in the pore-forming region is critical for the pH sensitivity of mouse TRPM7, while two residues Glu1052 and Asp1054 are critical for that of human TRPM7. In potentiating TRPM7 activity, it is likely that extracellular acidification markedly reduces the blocking effects of Mg^{2+} and Ca^{2+} on the Cs^+ currents by decreasing their binding affinities; IC_{50} values of Mg^{2+} and Ca^{2+} are increased by 510- and 447-fold, respectively, at pH 4.0. The 50% potentiation pH is 4.3 for TRPM6. Pore glutamates, Glu1024 and Glu1029, are key determinants of pH sensitivity in the mouse TRPM6 channel. The magnitude of increase by acidification of TRPM6 currents is smaller than for TRPM7 currents (9.9- and 3.8-fold for TRPM7 and TRPM6, respectively, at pH 4.0). Interestingly, heteromeric TRPM6/TRPM7 channels show enhancement of proton sensitivity by approximately one pH unit.[278]

Channel activities of TRPC4 and TRPC5 are potentiated by decreases in extracellular pH.[290,291] Lowering the pH increased both G protein-activated and spontaneous TRPC5 currents. Both effects were already observed upon small reductions in pH (from 7.4 to 7.0). The pH dependency is bell-shaped with the peak around pH 6–6.5; at higher concentrations, such as at pH 5.5, the activating effects of protons become inhibitory. The potentiating effect of protons is dependent on extracellular Na^+. Mutation of Glu543 and Glu595 responsible for lanthanoid sensitivity of mouse TRPC5 modified the potentiation by protons (see also La^{3+} sensitivity in Section 3.5 and Ca^{2+} sensitivity in Section 3.2).

With regard to other subfamilies, acid-induced responses are observed only after the removal of an acid solution of less than pH 3.0 for the channel complex formed by PKD1L3 and TRPP3.[292] The activity of TRPML1, as well as that of its homologues, TRPML2 and TRPML3, is enhanced at low pH (e.g., pH 4.6).[293]

External acidification also induces human TRPA1 activation, which is critically impaired by mutation of Cys421 in the N-terminal region involved in activation by other triggers.[194]

3.6.2 SUPPRESSION OF ACTIVATION BY ACIDIFICATION

The activity of some TRP channels is suppressed by acidification. External protons block TRPV5 with an IC_{50} of pH 6.55, while internal protons inhibit TRPV5 with an IC_{50} of pH 7.31. In the case of extracellular proton-induced inhibition, a decrease in single-channel conductance was observed. Amino acid residues at the outer mouth of the channel pore are responsible for external pH sensitivity, and Lys607 at the intracellular proximal C-terminus is responsible for intracellular pH sensitivity in rabbit TRPV5.[294,295] For mouse TRPM5 channels, external acidification has two effects: a fast reversible block of the current (IC_{50} of pH 6.2) and a slower irreversible enhancement of current inactivation.[296] Glu830 located at the linker between S3 and S4 and His934 located at the linker between S5 and S6 each plays a role in the acid block, while His896 located at the linker between S5 and S6 has a role in recovery from acid-enhanced inactivation. External protons block human TRPM2 with an IC_{50} of pH 5.3 by decreasing single-channel conductance, whereas internal protons inhibit TRPM2 with an IC_{50} of pH 6.7.[297] Three titratable residues, His958, Asp964, and Glu994, at the outer vestibule of the channel pore are responsible for extracellular pH sensitivity, whereas Asp933 located at the C-terminus of the S4–S5 linker is responsible for intracellular pH sensitivity. TRPC6 is inhibited by external protons with an IC_{50} of pH 5.7.[290] As mentioned above, TRPC5 shows a bell-shaped pH dependency with the peak of enhancing effects around pH 6–6.5 in the presence of extracellular Na^+; at high proton concentrations, such as pH 5.5, the effect of protons is inhibitory.[298] In the presence of extracellular Cs^+ instead of Na^+, external acidification inhibits TRPC5.

3.6.3 ALKALIZATION

TRPA1 is activated by external NH_4^+-induced intracellular alkalization with an EC_{50} around pH 8.[299] TRPP3 activity is increased by extracellular alkalization (pH 9) and is decreased by acidification (pH 2.7).[300] This is in contrast to the report by Inada et al. (see Section 3.6.1).[292] TRPM7 currents are activated by cytosolic alkalization and inhibited by acidification; they can be reactivated by PIP_2 following rundown in inside-out patches.[301] TRPM7 channels can be regulated by a charge screening mechanism and might function as sensors of intracellular pH.

3.6.4 PROTON PERMEABILITY

TRPM7 shows proton-dependent enhancement of channel activity with an EC_{50} value of pH 3.9.[302] It is conceivable that the proton dependency of TRPM7 currents is attained through stabilization of channel opening by protons and permeation of protons. Asp1054, located at the selectivity filter of the pore, is an essential determinant of the proton conductivity of human TRPM7. There might be coupled movement of the regulatory ions and permeant ions through the pore domain of TRPM7 channels. Proton permeation has been shown for TRPML1, which also shows acidification-sensitive channel opening.[303]

3.7 TEMPERATURE

The ability to detect temperature is important for survival, both for maintenance of homeostasis and for avoidance of noxious temperatures. Many enzyme reactions have Q_{10} values of ~3, as does the gating of many ion channels.[304] The Q_{10} values of nine mammalian temperature-sensitive TRP channels (TRPV1, TRPV2, TRPV3, TRPV4, TRPM2, TRPM4, TRPM5, TRPM8, and TRPA1) far surpass the temperature dependence of the gating processes characterized by other ion channels with regular Q_{10} values. The temperature-sensitive TRP channels can be categorized into three groups with distinct thermal activation thresholds and Q_{10} values: heat-, warm-, and cold-activated channels.

3.7.1 HEAT

A milestone of this specific research area was the identification of TRPV1,[246] as remarked for TRP channels activated by exogenous modulators in Section 3.5. TRPV1 is activated by heat as well as by vanilloid compounds and protons.[246,289,305] The temperature range of the activation threshold of rat TRPV1 is 41.5°C–43°C, and Q_{10} is 20.6–25.6.[289,305–307] Human TRPV1 responds to a temperature greater than 42°C.[308] TRPV2 has a threshold of 51.6°C–55°C[309–311] and a Q_{10} value of ~100.[311] For TRPV1, the region responsible for temperature sensitivity is suggested to lie in the C-terminal tail. A TRPV1 mutant with a deletion of 72 residues from the C-terminus end showed a lowered activation thermal threshold (from 41.5° to 28.6°C) and slowing of the activation rate of heat-evoked membrane currents (Q_{10} from 25.6 to 4.7).[307] Moreover, replacing the C-terminus of TRPV1 with that of cold-sensitive channel TRPM8 (see Section 3.7.3) resulted in cold temperature sensitivity, making the channel cold-sensitive instead of heat-sensitive. Vice versa, TRPM8 could be turned into a heat-sensitive channel by replacing its C-terminus with that of TRPV1.[312] On the basis of a biophysical analysis of the whole-cell gating kinetics of the two prototype thermoTRPs, the cold receptor TRPM8 and the heat receptor TRPV1, a general principle for temperature-sensitive channel gating has been proposed,[313,314] which is discussed in Section 3.8.

3.7.2 WARM

TRPV3 and TRPV4, which are expressed in multiple tissues including DRG neurons and skin keratinocytes, are activated at warm temperatures.[315–317] The reported activation threshold of TRPV3 is variable ranging from 31° to 39°C,[315–317] and TRPV3 has a Q_{10} of 21.6.[317] TRPV4, originally identified as an osmosensor ion channel,[318–320] is activated at warm temperatures ranging from 25° to 33.6°C with a Q_{10} of 9.9–19.1.[321–324]

Ca^{2+}-activated cation channels TRPM4 and TRPM5 are activated by heating in the temperature range of 15°C–35°C: Q_{10} values are 8.5 for TRPM4 at –25 mV and 10.3 for TRPM5 at –75 mV. The activation of these channels is underlain by a temperature-dependent shift of the voltage-dependent activation curve,[325] which is discussed in detail in Section 3.8.

TRPM2 is known to be activated by ADPR or H_2O_2.[76,178,179] Electrophysiological data show that TRPM2 can be robustly potentiated by exposure to warm temperatures apparently via direct heat-evoked channel gating with a threshold of 33.6°C–35°C and a Q_{10} of 32.3–44.4 in the presence of an agonist (ADPR, cADPR, or NAD$^+$), and activated with a threshold of 33.9°C and a Q_{10} of 15.6 in the absence of an agonist at –60 mV.[326] This implies synergistic effects of temperature and agonists on TRPM2 activation.

3.7.3 COLD

In contrast to the above seven thermosensitive TRP channels, two other TRP channels, TRPM8 and TRPA1, are activated by cold stimuli.[252,254,256] Cold-activated TRPM8 currents have a thermal threshold of 22°–26°C[254,255,259] and a Q_{10} of 35 (Ref. 327) or 23 (Ref. 328). The threshold temperature for activation is elevated by menthol[254,255,313,327] and by higher pH.[327] TRPM8 is expressed in multiple tissues, including sensory nerves.[253,255,259] Responses to cold are also observed in DRG[254,259,329,330] and trigeminal neurons.[331,332] In TRPM8-deficient mice, thermotaxis data suggest that innocuous cold thermosensation is severely disrupted.[330–332] These reports clearly suggest that TRPM8 plays an important role in cold nociception.

TRPA1 is activated by cold with a lower threshold temperature than TRPM8 but with a broad range (8°C–28°C), which gives rise to an average threshold of 17°C.[259,333] In inside-out patches, when the temperature is decreased to below 18°C, unitary conductance of the channel is decreased, but the open probability is increased. This increase in the open probability of TRPA1 may underlie the increase in whole-cell currents induced by deep cooling.[333] TRPA1 is expressed in DRG neurons and in trigeminal ganglion neurons.[259,333,334] Ca^{2+}-imaging experiments and patch-clamp recordings in DRG neurons and trigeminal ganglion neurons show cold sensitivity,[259,333,334] leading to the suggestion that TRPA1 is involved in cold nociception. In fact, cold plate and tail-flick experiments in TRPA1-deficient mice have revealed a TRPA1-dependent, cold-induced nociceptive behavior in mice. Thus, TRPA1 likely acts as a major sensor for noxious cold.[334]

3.8   MEMBRANE VOLTAGE

Voltage-dependent regulation of the channel activity has been implicated in a number of TRP channels at relatively early stages of their functional characterization. For example, a voltage-dependent component of activation has been reported for TRPC5[335] and TRPV5.[336] TRPM4 and TRPM5, categorized as Ca^{2+}-activated cation channels,[83–85,337] show clear voltage dependence with slow activation kinetics,[84,338] while thermosensor channel TRPV1 also exhibits time- and voltage-dependent activation/deactivation properties.[339] Nilius et al.[313] and coincidently Latorre and his colleagues[314] have established a general principle in attempts to unify multimodal gating sensitivities to temperature, membrane voltage, and chemical ligands: temperature sensing is tightly linked to voltage-dependent gating in the cold-sensitive channel TRPM8 and the heat-sensitive channel TRPV1. Both channels are activated upon depolarization, and changes in temperature result in graded shifts of their voltage-dependent activation curves, while agonist substances menthol and capsaicin shift the activation curves of TRPM8 and TRPV1, respectively, toward physiological membrane potentials. Functional analyses of mutants identified charged residues in the transmembrane region S4 and the adjacent linker region S4–S5 as a voltage sensor of TRPM8.[340] The above concept has been applied to TRPM4 and TRPM5, in which heat-induced activation is thought to be due to a shift of the voltage activation curve to hyperpolarizing potentials.[325] Physiologically, the temperature sensitivity of TRPM5 underlies the temperature effects on perceived taste in humans, including enhanced sweet perception. Voltage-dependent gating and its involvement in agonist and temperature sensitivities are widely recognized phenomena among TRP channels, including TRPV3,[341] TRPM3,[231] and TRPA1.[87,202,260] Conversely, Matta and Ahern have conducted experiments at room temperature using activators at high concentrations and raised a possibility that a voltage-sensitive mechanism is unlikely to represent a final, gating mechanism for TRPV1 and TRPM8.[342] In fact, divalent cations as well as voltage-dependent gating might be a causal factor for outward rectification of whole-cell currents of TRPM7.[302,343,344] In TRPV6, strong inward rectification[345] rather derives from voltage-dependent binding and unbinding of Mg^{2+}.[95,346,347]

Structurally, among TRP subfamilies, TRPPs are the only subfamily that carries explicit amino acid sequences reminiscent of voltage sensors in the transmembrane S4 of voltage-dependent cation channels:[348] charged residues at every third position are separated mostly by hydrophobic residues in the "canonical" voltage sensor S4.[349]

Mainly through functional analyses using reconstitution in a lipid bilayer, TRPP2 has been, in fact, demonstrated to be a voltage-gated cation channel,[90,350–352] while others have reported contradictory results demonstrating TRPP2 currents with, if any, only slight voltage dependence.[143,174,353] TRPP3, the closest relative of TRPP2, also showed regulation of activation by voltage, as well as by changes in pH and cell volume.[300]

3.9 MECHANOSTIMULATION

There is substantial evidence that TRP channels are mechanically activated or modulated. Currently, available data suggest that at least 11 mammalian TRPs are mechanosensitive. Activation of respective TRP channels directly or indirectly responding to mechanical stress has been suggested. However, this area of TRP research is as difficult to approach as the research of other mechanosensitive channels because mechanostimulations and their immediate cellular impacts are not described by one or a few molecular events.

3.9.1 MEMBRANE TENSION

A number of endogenously expressed TRP channels show activation by membrane tension. In a strategy that is based on detergent solubilization of frog oocyte membrane proteins, followed by liposome reconstitution and evaluation by patch-clamp, the TRPC1 channel activation was achieved by applying pressure (suction) pulses.[354] Single-channel recordings in cell-attached and inside-out patches have revealed that native TRPM7 channels in human epithelial HeLa cells are exquisitely sensitive to membrane stretch.[344] TRPM7 was also reported as the stretch-activated cation channel responsible for transducing mechanical signals into Ca^{2+} flickers in a human embryonic WI-38 fibroblast.[355] A patch-clamp study has suggested that negative pressure (≥ 20 mmHg, cell-attached mode) activates single TRPM4-like channels in isolated rat cerebral artery myocytes.[356] The TRPC6 channel becomes sensitive to mechanical stimuli after receptor activation in rat embryonic aortic smooth muscle cells.[228] TRPV2 in mouse vascular smooth muscle cells can be activated by membrane stretch by cell swelling through hypotonic stimulation.[357] Mechanosensitivity of these TRP channels has been also evaluated in heterologous expression systems. Transfection of human TRPC1 into CHO-K1 cells also significantly increased mechanosensitive cation channel activity in response to increasing steps of pressure (≤ 100 mmHg).[354] Heterologously expressed TRPM7 in HEK293T cells is activated by membrane stretch. In cell-free patches, membrane stretch directly augmented the single-channel activity of TRPM7 by increasing the opening probability at potentials <20 mV.[343] It was reported that in CHO-K1 cells transfected with mouse TRPV2 cDNA (TRPV2-CHO), application of membrane stretch through the recording pipette activated single non-selective cation channels.[357] Moreover, the stretch of TRPV2-expressing CHO cells cultured on an elastic silicon membrane showed significant elevation of intracellular Ca^{2+} concentration. TRPC6 was also suggested to function as a direct sensor of mechanically and osmotically induced membrane stretch when negative pressure is applied to the back of the pipette.[272]

3.9.2 HYPOTONICITY-INDUCED CELL SWELLING

The observed time courses of hypotonicity-induced mechanical responses are usually much slower than those of responses to increases in membrane tension. This implies the involvement of sequential processes converting mechanical forces into biochemical signals such as activation of enzymes, production of second messengers, and altered membrane lipid metabolisms. In fact, hypotonicity-activated cation channel TRPV4 is activated through mobilization of cellular signals.[318–320,358–360] Hypotonic stress results in Src family tyrosine kinase-dependent tyrosine phosphorylation of TRPV4 at the residue essential for the channel function.[163] Another N-terminal tyrosine residue plays a crucial role in the activation of TRPV4 by hypotonic cell swelling.[361] Cell swelling activates TRPV4 also via the PLA_2-dependent formation of arachidonic acid, which is then metabolized further by cytochrome P450 epoxygenase to form the TRPV4-activating messenger 5′, 6′-EET.[164] Activation of TRPV4 is potentiated by warm temperatures, e.g., while at room temperature (22°C–24°C), exposure of TRPV4-transfected cells to hypotonic medium (225 mOsm/L) induces only modest channel activation, at 37°C, the same medium causes more robust TRPV4 activation (also see temperature).[359]

Patch-clamp studies have suggested that whole-cell TRPC6 currents are enhanced by exposure to hypoosmotic bath solutions (250 mOsm),[228,272,362] suggesting that the mechanical sensing and chemical lipid sensing share a common molecular mechanism that may involve lateral-lipid tension[272] and PLC/DAG and PLA_2/ω-Hydroxylase/20-HETE pathways.[228] Hypotonic solution activates TRPM3, TRPM7, TRPC5, TRPV2, and TRPA1, although it is unclear whether biochemical signals are directly involved in these hypotonicity responses.[343,344,357,363–365] Swelling-activated cation channels are involved in physiological functions, such as cell volume regulation by TRPM7 in human epithelial cells,[344] by TRPC1 in mouse liver cells,[366] and by TRPV4 in human tracheal epithelial cells,[367] as well as stretch sensing by TRPV2 in rat vascular smooth muscle cells,[357] and by TRPC6 in rat artery[362] and embryonic smooth muscle cells.[228]

3.9.3 HYPERTONICITY-INDUCED CELL SHRINKAGE

Body fluid homeostasis requires the release of arginine-vasopressin (AVP) from the neurohypophysis. This release is controlled by specific and highly sensitive "osmoreceptors" in the hypothalamus. Indeed, AVP-releasing neurons in the supraoptic nucleus (SON) are directly osmosensitive, and this osmosensitivity is mediated by stretch-inhibited cation channels. An N-terminal splice variant of TRPV1, which is activated by hyperosmolality and inhibited by hypo-osmolality, is reported as the molecular identity of the osmosensory transduction channel in AVP neurons[368] and organum vasculosum lamina terminalis neurons.[369]

3.9.4 SHEAR STRESS

TRPC6, TRPV4, TRPM7, and TRPP2 reported as hypotonically activated channels are also activated in response to shear stress.[228,343,355,359–361,370,371] Patch-clamp studies show that TRPM7 is activated by shear stress.[343,370] To observe the effect of shear stress, bath solution in the chamber (135-μL volume) was perfused at a flow rate of 15 or 35 μL/s. The whole-cell TRPM7 current was reversibly increased in a manner dependent on the perfusion speed.[343] Ca^{2+} flicker activity, which steers cell migration, is coupled to membrane tension by application of shear stress by means of TRPM7 through a stretch-induced activation.[355] In TRPP1 and TRPP2, fluid shear stress (0.75 dyne/cm²) promoted cilium bending and associated Ca^{2+} signaling in mouse embryonic kidney cells. On fluid stimulation, an immediate rise in intracellular Ca^{2+} concentration, peaking roughly 10–20 s after stimulation, was detected throughout the cell population.[371] In contrast to these shear stress-activated TRP channels that are considered as direct mechanosensors, TRPC6 and TRPV4 rather integrate and synergize receptor and mechanical signals. In TRPC6, shear force (10 dyne/cm²), which was ineffective in the absence of receptor stimulation, markedly enhanced the magnitude of the whole-cell TRPC6 current activated by receptor stimulation.[228] A synergistic effect has been documented also for the epithelial TRPV4 channel.[359,360] At room temperature in TRPV4-transfected HEK293 cells, increasing the fluid shear stress from 0 to 10 dyne/cm² had little or no effect on the intracellular Ca^{2+} concentration, while at 37°C, the same fluid shear stress resulted in a marked increase in the intracellular Ca^{2+} concentration. At 37°C, the channel response was tightly regulated by the magnitude of the shear stress applied. Step elevations in the shear stress from 0 to 20 dyne/cm² elicited proportional increases in the intracellular Ca^{2+} concentrations. The shear stress-induced activation is affected not only by warm temperature but also by phosphorylation of an N-terminal tyrosine residue.[361]

3.9.5 MECHANO-INDUCED PROTEIN TRANSLOCATION

Facilitated translocation of TRPV2 has been found in myotubes prepared from sarcoglycan-deficient BIO14.6 hamsters, which is activated in response to myocyte stretch and is responsible for enhanced Ca^{2+} influx leading to cell damage as measured by creatine phosphokinase efflux. Cell stretch increases TRPV2 translocation to the sarcolemma, a process that requires entry of external Ca^{2+}.[372] The relatively normal osmosensitivity of the TRPV4 mutant lacking the ankyrin-like repeat implies that an ankyrin-mediated connection to the cytoskeleton is not essential for TRPV4 gating by osmotic stress.[318] The connection of TRPV4 to cytoskeleton via the ankyrin-like repeat may be required for rapid responsiveness. TRPM7 was found in tubulovesicular structures that were translocated to the region of the PM upon induction of shear stress.[370]

3.10 PROTEIN–PROTEIN INTERACTION

This important subject is discussed in detail in Chapters 2, 4, and 5. Some protein–protein interactions are also dealt with in other parts of this chapter depending on the relation to the subject discussed.

3.11 CONCLUSION

Many TRP channels can respond to multiple activation triggers and therefore serve as polymodal signal detectors. For example, the sensitivity of TRPV1 to noxious heat can be greatly enhanced by inflammatory agents that activate PLC signaling pathways.[150] Such integration allows TRPV1 to detect subthreshold stimuli and provides a mechanism through which tissue injury produces thermal hypersensitivity.[373] Moreover, in TRPM8, significant and maximal activation by icilin requires intracellular Ca^{2+}.[256] The codependency of activation, wherein channel gating requires the simultaneous presence of two agonists, suggests that some TRP channels serve as coincidence detectors. Another important aspect in the multimodality of the activation sensitivity of TRP channels is its role in signal integration and amplification (Figure 3.1). When a TRP channel that is part of a specific signaling cascade is activated by downstream or upstream constituents (small molecule messengers/enzymes/scaffolding proteins) of the cascade, in addition to the immediately upstream activation trigger, the cascade is equipped with positive feedback or feed-forward loops. This mechanism, which is capable of ensuring fidelity of cellular responses and minimizing the variation of their magnitude, may contribute to the synchronization of responses in neighboring cells that comprise functional domains in tissues. For example, in vascular endothelial cells upon vasodilator receptor stimulation, Ca^{2+} influx via NO-activated TRPC5 channels can amplify production of NO by endothelial type NO synthase,[187] resulting in the enhancement of NO production in

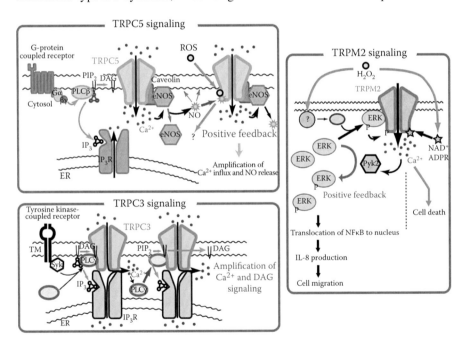

FIGURE 3.1 **(See color insert.)** Integration and amplification of calcium signals by TRP channelsome. TRPC5 signaling: TRPC5 mediates a feedback cycle of G-protein-coupled receptor-activated Ca^{2+} and NO signaling. TRPC3 signaling: TRPC3 amplifies tyrosin kinase-coupled receptor-activated Ca^{2+} and DAG signaling. TRPM2 signaling: Ca^{2+}-influx via H_2O_2-activated TRPM2 amplifies Erk activation through Pyk2-Ras phosphorylation cascade. ADPR and protein modification process may mediate TRPM2 activation and Ca^{2+} entry upon H_2O_2.

nearby endothelial cells and NO-dependent relaxation of smooth muscle cells. The intercellular amplification of NO production should eventually lead to vasodilation synchronized as a whole at vascular tissue levels. Our understanding of the activation of TRP channels is now stepping forward from functional identification of single molecules to analysis and integration of molecular systems.

REFERENCES

1. Clapham, D. E. 2003. TRP channels as cellular sensors. *Nature* 426: 517–524.
2. Voets, T., and B. Nilius. 2003. TRPs make sense. *Journal of Membrane Biology* 192: 1–8.
3. Montell, C., and G. M. Rubin. 1989. Molecular characterization of the Drosophila trp locus: a putative integral membrane protein required for phototransduction. *Neuron* 2: 1313–1323.
4. Minke, B., and B. Cook. 2002. TRP channel proteins and signal transduction. *Physiology Review* 82: 429–472.
5. Montell, C. 2005. Drosophila TRP channels. *Pflügers Archiv* 451: 19–28.
6. Minke, B., and M. Parnas. 2006. Insights on TRP channels from in vivo studies in Drosophila. *Annu. Rev. Physiol.* 68: 649–684.
7. Li, H. S., X. Z. Xu, and C. Montell. 1999. Activation of a TRPC3-dependent cation current through the neurotrophin BDNF. *Neuron* 24: 261–273.
8. Jeon, J.-P., K. P. Lee., E. J. Park et al. 2008. The specific activation of TRPC4 by Gi protein subtype. *Biochemical and Biophysical Research Commununication* 377: 538–543.
9. Koike, C., T. Obara, Y. Uriu et al. 2010. TRPM1 is a component of the retinal ON bipolar cell transduction channel in the mGluR6 cascade. *Proceedings of the National Academy of Sciences of the United States of America* 107: 332–337.
10. Patterson, R. L., D. B. van Rossum, D. L. Ford et al. 2002. Phospholipase C-γ is required for agonist-induced Ca^{2+} entry. *Cell* 111: 529–541.
11. Nishida, M., K. Sugimoto, Y. Hara et al. 2003 Amplification of receptor signalling by Ca^{2+} entry-mediated translocation and activation of PLCγ2 in B lymphocytes. *EMBO Journal* 22: 4677–4688.
12. van Rossum, D. B., R. L. Patterson, S. Sharma et al. 2005. Phospholipase Cγ1 controls surface expression of TRPC3 through an intermolecular PH domain. *Nature* 434: 99–104.
13. Wen, W., J. Yan, and M. Zhang. 2006. Structural characterization of the split pleckstrin homology domain in phospholipase C-gamma1 and its interaction with TRPC3. *Journal of Biological Chemistry* 281: 12060–12068.
14. Runnels, L. W., L. Yue, and D. E. Clapham. 2001. TRP-PLIK, a bifunctional protein with kinase and ion channel activities. *Science* 291: 1043–1047.
15. Runnels, L. W., L. Yue, and D. E. Clapham. 2002. The TRPM7 channel is inactivated by PIP_2 hydrolysis. *Nature Cell Biology* 4: 329–336.
16. Oritani, K., and P. Kincade. 1996. Identification of stromal cell products that interact with pre-B cells. *Journal of Cell Biology* 134: 771–782.
17. Roos, J., P. J. DiGregorio, A. V. Yeromin et al. 2005. STIM1, an essential and conserved component of store-operated Ca^{2+} channel function. *Journal of Cell Biology* 169: 435–445.
18. Liou, J., M. L. Kim, W. D. Heo et al. 2005. STIM is a Ca^{2+} sensor essential for Ca^{2+}-store-depletion-triggered Ca^{2+} influx. *Current Biology* 15: 1235–1241.
19. Zhang, S. L., Y. Yu, J. Roos et al. 2005. STIM1 is a Ca^{2+} sensor that activates CRAC channels and migrates from the Ca^{2+} store to the plasma membrane. *Nature* 437: 902–905.

20. Hoth, M., and R. Penner. 1992. Depletion of intracellular calcium stores activates a calcium current in mast cells, *Nature* 355: 353–356.
21. Feske, S., Y. Gwack, M. Prakriya et al. 2006. A mutation in Orai1 causes immune deficiency by abrogating CRAC channel function. *Nature* 441: 179–185.
22. Vig, M., C. Peinelt, A. Beck et al. 2006. CRACM1 is a plasma membrane protein essential for store-operated Ca^{2+} entry. *Science* 312: 1220–1223.
23. Zhang, S. L., A. V. Yeromin, X. H.-F. Zhang et al. 2006. Genome-wide RNAi screen of Ca^{2+} influx identifies genes that regulate Ca^{2+} release-activated Ca^{2+} channel activity. *Proceedings of the National Academy of Sciences of the United States of America* 103: 9357–9362.
24. Prakriya, M., S. Feske, Y. Gwack et al. 2006. Orai1 is an essential pore subunit of the CRAC channel. *Nature* 443: 230–233.
25. Yeromin, A. V., S. L. Zhang, W. Jiang et al. 2006. Molecular identification of the CRAC channel by altered ion selectivity in a mutant of Orai. *Nature* 443: 226–229.
26. Peinelt, C., M. Vig, D. L. Koomoa et al. 2006. Amplification of CRAC current by STIM1 and CRACM1 (Orai1). *Nature Cell Biology* 8: 77–73.
27. Soboloff, J., M. A. Spassova, X. D. Tang et al. 2006. Orai1 and STIM reconstitute store-operated calcium channel function. *Journal of Biological Chemistry* 281: 20661–20665.
28. Mercer, J. C., W. I. Dehaven, J. T. Smyth et al. 2006. Large store-operated calcium selective currents due to co-expression of Orai1 or Orai2 with the intracellular calcium sensor, Stim1. *Journal of Biological Chemistry* 281: 24979–24990.
29. Zhu, X., P. B. Chu, M. Peyton et al. 1995. Molecular cloning of a widely expressed human homologue for the Drosophila trp gene. *FEBS Letters* 373: 193–198.
30. Wes, P. D., J. Chevesich, A. Jeromin et al. 1995. TRPC1, a human homolog of a Drosophila store-operated channel. *Proceedings of the National Academy of Sciences of the United States of America* 92: 9652–9696.
31. Zhu, X., M. Jiang, M. Peyton et al. 1996. trp, a novel mammalian gene family essential for agonist-activated capacitative Ca^{2+} entry. *Cell* 85: 661–671.
32. Philipp, S., A. Cavalié, M. Freichel et al. 1996. A mammalian capacitative calcium entry channel homologous to Drosophila TRP and TRPL. *EMBO Journal* 15: 6166–6171.
33. Philipp, S., J. Hambrecht, L. Braslavski et al. 1998. A novel capacitative calcium entry channel expressed in excitable cells. *EMBO Journal* 17: 4274–4282.
34. Vannier, B., M. Peyton, G. Boulay et al. 1999. Mouse trp2, the homologue of the human trpc2 pseudogene, encodes mTrp2, a store depletion-activated capacitative Ca^{2+} entry channel. *Proceedings of the National Academy of Sciences of the United States of America* 96: 2060–2064.
35. Mizuno, N., S. Kitayama, Y. Saishin et al. 1999. Molecular cloning and characterization of rat trp homologues from brain. *Brain Research. Molecular Brain Research* 64: 41–51.
36. Kinoshita, M., A. Akaike, M. Satoh et al. 2000. Positive regulation of capacitative Ca^{2+} entry by intracellular Ca^{2+} in Xenopus oocytes expressing rat TRP4. *Cell Calcium* 28: 151–159.
37. Riccio, A., C. Mattei, R. E. Kelsell et al. 2002. Cloning and functional expression of human short TRP7, a candidate protein for store-operated Ca^{2+} influx. *Journal of Biological Chemistry* 277: 12302–12309.
38. Freichel, M., S. H. Suh, A. Pfeifer et al. 2001. Lack of an endothelial store-operated Ca^{2+} current impairs agonist-dependent vasorelaxation in TRP4$^{-/-}$ mice. *Nature Cell Biology* 3: 121–127.
39. Mori, Y., M. Wakamori, T. Miyakawa et al. 2002. Transient receptor potential 1 regulates capacitative Ca^{2+} entry and Ca^{2+} release from endoplasmic reticulum in B lymphocytes. *Journal of Experimental Medicine* 195: 673–681.

40. Liu, X., K. T. Cheng, B. C. Bandyopadhyay et al. 2007. Attenuation of store-operated Ca²⁺ current impairs salivary gland fluid secretion in TRPC1⁻ᐟ⁻ mice. *Proceedings of the National Academy of Sciences of the United States of America* 104: 17542–17547.
41. Kim, M. S., J. H. Hong, Q. Li et al. 2009. Deletion of TRPC3 in mice reduces store-operated Ca²⁺ influx and the severity of acute pancreatitis. *Gastroenterology* 137: 1509–1517.
42. Yue, L., J.-B. Peng, M. A. Hediger et al. 2001. CaT1 manifests the pore properties of the calcium-release-activated calcium channel. *Nature* 410: 705–709.
43. Pérez, C. A., L. Huang, M. Rong et al. 2002. A transient receptor potential channel expressed in taste receptor cells. *Nature Neuroscience* 5: 1169–1176.
44. Huang, G. N., W. Zeng, J. Y. Kim et al. 2006. STIM1 carboxyl-terminus activates native SOC, Icrac and TRPC1 channels. *Nature Cell Biology* 8: 1003–1010.
45. López, J. J., G. M. Salido, J. A. Pariente et al. 2006. Interaction of STIM1 with endogenously expressed human canonical TRP1 upon depletion of intracellular Ca²⁺ stores. *Journal of Biological Chemistry* 281: 28254–28264.
46. Yuan, J. P., W. Zeng, G. N. Huang et al. 2007. STIM1 heteromultimerizes TRPC channels to determine their function as store-operated channels. *Nature Cell Biology* 9: 636–645.
47. Zeng, W., J. P. Yuan, M. S. Kim et al. 2008. STIM1 gates TRPC channels, but not Orai1, by electrostatic interaction. *Molecular Cell* 32: 439–448.
48. Alicia, S., Z. Angélica, S. Carlos et al. 2008. STIM1 converts TRPC1 from a receptor-operated to a store-operated channel: Moving TRPC1 in and out of lipid rafts. *Cell Calcium* 44: 479–491.
49. Liao, Y., C. Erxleben, E. Yildirim et al. 2007. Orai proteins interact with TRPC channels and confer responsiveness to store depletion. *Proceedings of the National Academy of Sciences of the United States of America* 104: 4682–4687.
50. Liao, Y., C. Erxleben, J. Abramowitz et al. 2008. Functional interactions among Orai1, TRPCs, and STIM1 suggest a STIM-regulated heteromeric Orai/TRPC model for SOCE/Icrac channels. *Proceedings of the National Academy of Sciences of the United States of America* 105: 2895–2900.
51. Liao, Y., N. W. Plummer, M. D. George et al. 2009. A role for Orai in TRPC-mediated Ca²⁺ entry suggests that a TRPC:Orai complex may mediate store and receptor operated Ca²⁺ entry. *Proceedings of the National Academy of Sciences of the United States of America* 106: 3202–3206.
52. Ma, H.-T., Z. Peng, T. Hiragun et al. 2008. Canonical transient receptor potential 5 channel in conjunction with Orai1 and STIM1 allows Sr²⁺ entry, optimal influx of Ca²⁺, and degranulation in a rat mast cell line. *Journal of Immunology* 180: 2233–2239.
53. Li, J., P. Sukumar, C. J. Milligan et al. 2008. Interactions, functions, and independence of plasma membrane STIM1 and TRPC1 in vascular smooth muscle cells. *Circulation Research* 103: e97–e104.
54. DeHaven, W. I., B. F. Jones, J. G. Petranka et al. 2009. TRPC channels function independently of STIM1 and Orai1. *Journal of Physiology* 587: 2275–2298.
55. Kiselyov, K., X. Xu, G. Mozhayeva et al. 1998. Functional interaction between InsP₃ receptors and store-operated Htrp3 channels. *Nature* 396: 478–482.
56. Boulay, G., D. M. Brown, N. Qin et al. 1999. Modulation of Ca²⁺ entry by polypeptides of the inositol 1,4,5-trisphosphate receptor (IP₃R) that bind transient receptor potential (TRP): Evidence for roles of TRP and IP3R in store depletion-activated Ca²⁺ entry. *Proceedings of the National Academy of Sciences of the United States of America* 96: 14955–14960.
57. Rosado, J. A., and S. O. Sage. 2000. Coupling between inositol 1,4,5-trisphosphate receptors and human transient receptor potential channel 1 when intracellular Ca²⁺ stores are depleted. *Biochemistry Journal* 350: 631–635.

58. Tang, J., Y. Lin, Z. Zhang et al. 2001. Identification of common binding sites for calmodulin and inositol 1,4,5-trisphosphate receptors on the carboxyl termini of trp channels. *Journal of Biological Chemistry* 276: 21303–21310.
59. Batty, I. R., S. R. Nahorski, and R. F. Irvine. 1985. Rapid formation of inositol 1,3,4,5-tetrakisphosphate following muscarinic receptor stimulation of rat cerebral cortical slices. *Biochemistry Journal* 232: 211–215.
60. Irvine, R. F., A. J. Letcher, J. P. Heslop et al. 1986. The inositol tris/tetrakisphosphate pathway–demonstration of Ins(1,4,5)P_3 3-kinase activity in animal tissues. *Nature* 320: 631–634.
61. Dillon, S. B., J. J. Murray, M. W. Verghese et al. 1987. Regulation of inositol phosphate metabolism in chemoattractant-stimulated human polymorphonuclear leukocytes. Definition of distinct dephosphorylation pathways for IP_3 isomers. *Journal of Biological Chemistry* 262: 11546–11552.
62. Irvine, R. F. 1990. 'Quantal' Ca^{2+} release and the control of Ca^{2+} entry by inositol phosphates—a possible mechanism. *FEBS Letters* 263: 5–9.
63. Lückhoff, A., and D. E. Clapham. 1992. Inositol 1,3,4,5-tetrakisphosphate activates an endothelial Ca^{2+}-permeable channel. *Nature* 355: 356–358.
64. Hermosura, M. C., H. Takeuchi, A. Fleig et al. 2000. $InsP_4$ facilitates store-operated calcium influx by inhibition of $InsP_3$ 5-phosphatase. *Nature* 408: 735–740.
65. Dong, Y., D. L. Kunze, L. Vaca et al. 1995. Ins(1,4,5)P3 activates Drosophila cation channel Trpl in recombinant baculovirus-infected Sf9 insect cells. *American Journal of Physiology* 269: C1332–C1339.
66. Boulay, G., X. Zhu, M. Peyton et al. 1997. Cloning and expression of a novel mammalian homolog of Drosophila transient receptor potential (Trp) involved in calcium entry secondary to activation of receptors coupled by the Gq class of G protein. *Journal of Biological Chemistry* 272: 29672–29680.
67. Zitt, C., A G. Obukhov, C. Strübing et al. 1997. Expression of TRPC3 in Chinese hamster ovary cells results in calcium-activated cation currents not related to store depletion. *Journal of Cell Biology* 138: 1333–1341.
68. Lintschinger, B., M. Balzer-Geldsetzer, T. Baskaran et al. 2000. Coassembly of Trp1 and Trp3 proteins generates diacylglycerol- and Ca^{2+}-sensitive cation channels. *Journal of Biological Chemistry* 275: 27799–27805.
69. Okada, T., S. Shimizu, M. Wakamori et al. 1998. Molecular cloning and functional characterization of a novel receptor-activated TRP Ca^{2+} channel from mouse brain. *Journal of Biological Chemistry* 273: 10279–10287.
70. Jung, S., A. Mühle, M. Schaefer et al. 2003. Lanthanides potentiate TRPC5 currents by an action at extracellular sites close to the pore mouth. *Journal of Biological Chemistry* 278: 3562–3571.
71. Blair, N. T., J. S. Kaczmarek, and D. E. Clapham. 2009. Intracellular calcium strongly potentiates agonist-activated TRPC5 channels. *Journal of General Physiology* 133: 525–546.
72. Gross, S. A., G. A. Guzmán, U. Wissenbach et al. 2009. TRPC5 is a Ca^{2+}-activated channel functionally coupled to Ca^{2+}-selective ion channels. *Journal of Biological Chemistry* 284: 34423–34432.
73. Inoue, R., T. Okada, H. Onoue et al. 2001. The transient receptor potential protein homologue TRP6 is the essential component of vascular alpha(1)-adrenoceptor-activated Ca^{2+}-permeable cation channel. *Circulation Research* 88: 325–332.
74. Shi, J., E. Mori, Y. Mori et al. 2004. Multiple regulation by calcium of murine homologues of transient receptor potential proteins TRPC6 and TRPC7 expressed in HEK293 cells. *Journal of Physiology* 561: 415–432.
75. Okada, T., R. Inoue, K. Yamazaki et al. 1999. Molecular and functional characterization of a novel mouse transient receptor potential protein homologue TRP7. Ca^{2+}-permeable cation channel that is constitutively activated and enhanced by stimulation of G protein-coupled receptor. *Journal of Biological Chemistry* 274: 27359–27370.

76. Hara, Y., M. Wakamori, M. Ishii et al. 2002. LTRPC2 Ca^{2+}-permeable channel activated by changes in redox status confers susceptibility to cell death. *Molecular Cell* 9: 163–173.
77. McHugh, D., R. Flemming, S.-Z. Xu et al. 2003. Critical intracellular Ca^{2+} dependence of transient receptor potential melastatin 2 (TRPM2) cation channel activation. *Journal of Biological Chemistry* 278: 11002–11006.
78. Hill, K., N. J. Tigue, R. E. Kelsell et al. 2006. Characterisation of recombinant rat TRPM2 and a TRPM2-like conductance in cultured rat striatal neurones. *Neuropharmacology* 50: 89–97.
79. Starkus, J., A. Beck, A. Fleig et al. 2007. Regulation of TRPM2 by extra- and intracellular calcium. *Journal of General Physiology* 130: 427–440.
80. Du, J., J. Xie, and L. Yue. 2009. Intracellular calcium activates TRPM2 and its alternative spliced isoforms. *Proceedings of the National Academy of Sciences of the United States of America* 106: 7239–7244.
81. Olah, M. E., M. F. Jackson, H. Li et al. 2009. Ca^{2+}-dependent induction of TRPM2 currents in hippocampal neurons. *Journal of Physiology* 587: 965–979.
82. Hermosura, M. C., A. M. Cui, R. C. V. Go et al. 2008. Altered functional properties of a TRPM2 variant in Guamanian ALS and PD. *Proceedings of the National Academy of Sciences of the United States of America* 105: 18029–18034.
83. Launay, P., A. Fleig, A.-L. Perraud et al. 2002. TRPM4 is a Ca^{2+}-activated nonselective cation channel mediating cell membrane depolarization. *Cell* 109: 397–407.
84. Hofmann, T., V. Chubanov, T. Gudermann et al. 2003. TRPM5 is a voltage-modulated and Ca^{2+}-activated monovalent selective cation channel. *Current Biology* 13: 1153–1158.
85. Liu, D., and E. R. Liman. 2003. Intracellular Ca^{2+} and the phospholipid PIP2 regulate the taste transduction ion channel TRPM5. *Proceedings of the National Academy of Sciences of the United States of America* 100: 15160–15165.
86. Nagata, K., A. Duggan, G. Kumar et al. 2005. Nociceptor and hair cell transducer properties of TRPA1, a channel for pain and hearing. *Journal of Neuroscience* 25: 4052–4061.
87. Zurborg, S., B. Yurgionas, J. A. Jira et al. 2007. Direct activation of the ion channel TRPA1 by Ca^{2+}. *Nature Neuroscience* 10: 277–279.
88. Doerner, J. F., G. Gisselmann, H. Hatt et al. 2007. Transient receptor potential channel A1 is directly gated by calcium ions. *Journal of Biological Chemistry* 282: 13180–13189.
89. Wang, Y. Y., R. B. Chang, H. N. Waters et al. 2008. The nociceptor ion channel TRPA1 is potentiated and inactivated by permeating calcium ions. *Journal of Biological Chemistry* 283: 32691–32703.
90. Koulen, P., Y. Cai, L. Geng et al. 2002. Polycystin-2 is an intracellular calcium release channel. *Nature Cell Biology* 4: 191–197.
91. Chen, X.-Z., P. M. Vassilev, N. Basora et al. 1999. Polycystin-L is a calcium-regulated cation channel permeable to calcium ions. *Nature* 401: 383–386.
92. Li, Q., Y. Liu, W. Zhao et al. 2002. The calcium-binding EF-hand in polycystin-L is not a domain for channel activation and ensuing inactivation. *FEBS Letters* 516: 270–278.
93. LaPlante, J. M., J. Falardeau, M. Sun et al. 2002. Identification and characterization of the single channel function of human mucolipin-1 implicated in mucolipidosis type IV, a disorder affecting the lysosomal pathway. *FEBS Letters* 532: 183–187.
94. Watanabe, H., J. Vriens, A. Janssens et al. 2003. Modulation of TRPV4 gating by intra- and extracellular Ca^{2+}. *Cell Calcium* 33: 489–495.
95. Bödding, M. 2005. Voltage-dependent changes of TRPV6-mediated Ca^{2+} currents. *Journal of Biological Chemistry* 280: 7022–7029.
96. Jiang, J., M. Li, and L. Yue. 2005. Potentiation of TRPM7 inward currents by protons. *Journal of General Physiology* 126: 137–150.
97. Li, M., J. Du, J. Jiang, W. Ratzan et al. 2007. Molecular determinants of Mg^{2+} and Ca^{2+} permeability and pH sensitivity in TRPM6 and TRPM7. *Journal of Biological Chemistry* 282: 25817–25830.

98. Numata, T., and Y. Okada. 2008. Molecular determinants of sensitivity and conductivity of human TRPM7 to Mg^{2+} and Ca^{2+}. *Channels (Austin)* 2: 283–286.
99. Mahieu, F., A. Janssens, M. Gees et al. 2010. Modulation of the cold-activated cation channel TRPM8 by surface charge screening. *Journal of Physiology* 588: 315–324.
100. Vaca, L., and A. Sampieri. 2002. Calmodulin modulates the delay period between release of calcium from internal stores and activation of calcium influx via endogenous TRP1 channels. *Journal of Biological Chemistry* 277: 42178–42187.
101. Spehr, J., S. Hagendorf, J. Weiss et al. 2009. Ca^{2+}-calmodulin feedback mediates sensory adaptation and inhibits pheromone-sensitive ion channels in the vomeronasal organ. *Journal of Neuroscience* 29: 2125–2135.
102. Yildirim, E., A. Dietrich, and L. Birnbaumer. 2003. The mouse C-type transient receptor potential 2 (TRPC2) channel: Alternative splicing and calmodulin binding to its N terminus. *Proceedings of the National Academy of Sciences of the United States of America* 100: 2220–2225.
103. Zhang, Z., J. Tang, S. Tikunova, J. D. Johnson et al. 2001. Activation of Trp3 by inositol 1,4,5-trisphosphate receptors through displacement of inhibitory calmodulin from a common binding domain. *Proceedings of the National Academy of Sciences of the United States of America* 98: 3168–3173.
104. Wedel, B. J., G. Vazquez, R. R. McKay et al. 2003. A calmodulin/inositol 1,4,5-trisphosphate (IP3) receptor-binding region targets TRPC3 to the plasma membrane in a calmodulin/IP3 receptor-independent process. *Journal of Biological Chemistry* 278: 25758–25765.
105. Trost, C., C. Bergs, N. Himmerkus et al. 2001. The transient receptor potential, TRP4, cation channel is a novel member of the family of calmodulin binding proteins. *Biochemistry Journal* 355: 663–670.
106. Odell, A. F., J. L. Scott, and D. F. V. Helden. 2005. Epidermal growth factor induces tyrosine phosphorylation, membrane insertion, and activation of transient receptor potential channel 4. *Journal of Biological Chemistry* 280: 37974–37987.
107. Odell, A. F., D. F. V. Helden, and J. L. Scott. 2008. The spectrin cytoskeleton influences the surface expression and activation of human transient receptor potential channel 4 channels. *Journal of Biological Chemistry* 283: 4395–4407.
108. Ordaz, B., J. Tang, R. Xiao et al. 2005. Calmodulin and calcium interplay in the modulation of TRPC5 channel activity. Identification of a novel C-terminal domain for calcium/calmodulin-mediated facilitation. *Journal of Biological Chemistry* 280: 30788–30796.
109. Kim, M. T., B. J. Kim, J. H. Lee et al. 2006. Involvement of calmodulin and myosin light chain kinase in activation of mTRPC5 expressed in HEK cells. *American Journal of Physiology Cell Physiology* 290: C1031–C1040.
110. Kim, B. J., J.-H. Jeon, S. J. Kim et al. 2007. Role of calmodulin and myosin light chain kinase in the activation of carbachol-activated cationic current in murine ileal myocytes. *Canadian Journal of Physiology and Pharmacology* 85: 1254–1262.
111. Shimizu, S., T. Yoshida, M. Wakamori et al. 2006. Ca^{2+}-calmodulin-dependent myosin light chain kinase is essential for activation of TRPC5 channels expressed in HEK293 cells. *Journal of Physiology* 570: 219–235.
112. Boulay, G. 2002. Ca^{2+}-calmodulin regulates receptor-operated Ca^{2+} entry activity of TRPC6 in HEK-293 cells. *Cell Calcium* 32: 201–207.
113. Kim, J. Y., and D. Saffen. 2005. Activation of M1 muscarinic acetylcholine receptors stimulates the formation of a multiprotein complex centered on TRPC6 channels. *Journal of Biological Chemistry* 280: 32035–32047.
114. Kwon, Y., T. Hofmann, and C. Montell. 2007. Integration of phosphoinositide- and calmodulin-mediated regulation of TRPC6. *Molecular Cell* 25: 491–503.
115. Sutton, K. A., M. K. Jungnickel, Y. Wang et al. 2004. Enkurin is a novel calmodulin and TRPC channel binding protein in sperm. *Developmental Biology* 274: 426–435.

116. Numazaki, M., T. Tominaga, K. Takeuchi et al. 2003. Structural determinant of TRPV1 desensitization interacts with calmodulin. *Proceedings of the National Academy of Sciences of the United States of America* 100: 8002–8006.
117. Rosenbaum, T., A. Gordon-Shaag, M. Munari et al. 2004. Ca^{2+}/calmodulin modulates TRPV1 activation by capsaicin. *Journal of General Physiology* 123: 53–62.
118. Lishko, P. V., E. Procko, X. Jin et al. 2007. The ankyrin repeats of TRPV1 bind multiple ligands and modulate channel sensitivity. *Neuron* 54: 905–918.
119. Wu, Z.-Z., S.-R. Chen, and H.-L. Pan. 2006. Signaling mechanisms of down-regulation of voltage-activated Ca^{2+} channels by transient receptor potential vanilloid type 1 stimulation with olvanil in primary sensory neurons. *Neuroscience* 141: 407–419.
120. Xiao, R., J. Tang, C. Wang et al. 2008. Calcium plays a central role in the sensitization of TRPV3 channel to repetitive stimulations. *Journal of Biological Chemistry* 283: 6162–6174.
121. Strotmann, R., G. Schultz, and T. D. Plant. 2003. Ca^{2+}-dependent potentiation of the nonselective cation channel TRPV4 is mediated by a C-terminal calmodulin binding site. *Journal of Biological Chemistry* 278: 26541–26549.
122. Niemeyer, B. A., C. Bergs, U. Wissenbach et al. 2001. Competitive regulation of CaT-like-mediated Ca^{2+} entry by protein kinase C and calmodulin. *Proceedings of the National Academy of Sciences of the United States of America* 98: 3600–3605.
123. Derler, I., M. Hofbauer, H. Kahr et al. 2006. Dynamic but not constitutive association of calmodulin with rat TRPV6 channels enables fine tuning of Ca^{2+}-dependent inactivation. *Journal of Physiology* 577: 31–44.
124. Lambers, T. T., A. F. Weidema, B. Nilius et al. 2004. Regulation of the mouse epithelial Ca^{2+} channel TRPV6 by the Ca^{2+}-sensor calmodulin. *Journal of Biological Chemistry* 279: 28855–28861.
125. Tong, Q., W. Zhang, K. Conrad et al. 2006. Regulation of the transient receptor potential channel TRPM2 by the Ca^{2+} sensor calmodulin. *Journal of Biological Chemistry* 281: 9076–9085.
126. Nilius, B., J. Prenen, J. Tang et al. 2005. Regulation of the Ca^{2+} sensitivity of the nonselective cation channel TRPM4. *Journal of Biological Chemistry* 280: 6423–6433.
127. Chuang, H. H., E. D. Prescott, H. Kong et al. 2001. Bradykinin and nerve growth factor release the capsaicin receptor from PtdIns(4,5)P2-mediated inhibition. *Nature* 411: 957–962.
128. Prescott, E. D., and D. Julius. 2003. A modular PIP2 binding site as a determinant of capsaicin receptor sensitivity. *Science* 300: 1284–1288.
129. Stein, A. T., C. A. Ufret-Vincenty, L. Hua et al. 2006. Phosphoinositide 3-kinase binds to TRPV1 and mediates NGF-stimulated TRPV1 trafficking to the plasma membrane. *Journal of General Physiology* 128: 509–522.
130. Kim, A. Y., Z. Tang, Q. Liu et al. 2008. Pirt, a phosphoinositide-binding protein, functions as a regulatory subunit of TRPV1. *Cell* 133: 475–485.
131. Rohács, T., C. M. B. Lopes, I. Michailidis et al. 2005. PI(4,5)P2 regulates the activation and desensitization of TRPM8 channels through the TRP domain. *Nature Neuroscience* 8: 626–634.
132. Lee, J., S. K. Cha, T. J. Sun et al. 2005. PIP2 activates TRPV5 and releases its inhibition by intracellular Mg^{2+}. *Journal of General Physiology* 126: 439–451.
133. Zhang, Z., H. Okawa, Y. Wang et al. 2005. Phosphatidylinositol 4,5-bisphosphate rescues TRPM4 channels from desensitization. *Journal of Biological Chemistry* 280: 39185–39192.
134. Nilius, B., F. Mahieu, J. Prenen et al. 2006. The Ca^{2+}-activated cation channel TRPM4 is regulated by phosphatidylinositol 4,5-biphosphate. *EMBO Journal* 25: 467–478.
135. Dai, Y., S. Wang, M. Tominaga et al. 2007. Sensitization of TRPA1 by PAR2 contributes to the sensation of inflammatory pain. *Journal of Clinical Investigations* 117: 1979–1987.

136. Kim, D., E. J. Cavanaugh, and D. Simkin. 2008. Inhibition of transient receptor potential A1 channel by phosphatidylinositol-4,5-bisphosphate. *American Journal of Physiology Cell Physiology* 295: C92–C99.

137. Karashima, Y., J. Prenen, V. Meseguer et al. 2008. Modulation of the transient receptor potential channel TRPA1 by phosphatidylinositol 4,5-biphosphate manipulators. *Pflügers Archiv* 457: 77–89.

138. Thyagarajan, B., V. Lukacs, and T. Rohacs. 2008. Hydrolysis of phosphatidylinositol 4,5-bisphosphate mediates calcium-induced inactivation of TRPV6 channels. *Journal of Biological Chemistry* 283: 14980–14987.

139. Trebak, M., L. Lemonnier, W. I. DeHaven et al. 2009. Complex functions of phosphatidylinositol 4,5-bisphosphate in regulation of TRPC5 cation channels. *Pflügers Archiv* 457: 757–769.

140. Otsuguro, K., J. Tang, Y. Tang et al. 2008. Isoform-specific inhibition of TRPC4 channel by phosphatidylinositol 4,5-bisphosphate. *Journal of Biological Chemistry* 283: 10026–10036.

141. Albert, A. P., S. N. Saleh, and W. A. Large. 2008. Inhibition of native TRPC6 channel activity by phosphatidylinositol 4,5-bisphosphate in mesenteric artery myocytes. *Journal of Physiology* 586: 3087–3095.

142. Lemonnier, L., M. Trebak, and J. W. Putney. 2008. Complex regulation of the TRPC3, 6 and 7 channel subfamily by diacylglycerol and phosphatidylinositol-4,5-bisphosphate. *Cell Calcium* 43: 506–514.

143. Ma, R., W. P. Li, D. Rundle, J. Kong et al. 2005. PKD2 functions as an epidermal growth factor-activated plasma membrane channel. *Molecular and Cellular Biology* 25: 8285–8298.

144. Penna, A., V. Juvin, J. Chemin et al. 2006. PI3-kinase promotes TRPV2 activity independently of channel translocation to the plasma membrane. *Cell Calcium* 39: 495–507.

145. Tseng, P. H., H. P. Lin, H. Hu et al. 2004. The canonical transient receptor potential 6 channel as a putative phosphatidylinositol 3,4,5-trisphosphate-sensitive calcium entry system. *Biochemistry* 43: 11701–11708.

146. Hofmann, T., A. G. Obukhov, M. Schaefer et al. 1999. Direct activation of human TRPC6 and TRPC3 channels by diacylglycerol. *Nature* 397: 259–263.

147. Lucas, P., K. Ukhanov, T. Leinders-Zufall et al. 2003. A diacylglycerol-gated cation channel in vomeronasal neuron dendrites is impaired in TRPC2 mutant mice: mechanism of pheromone transduction. *Neuron* 40: 551–561.

148. Zhang, L., and D. Saffen. 2001. Muscarinic acetylcholine receptor regulation of TRP6 Ca²⁺ channel isoforms. Molecular structures and functional characterization. *Journal of Biological Chemistry* 276: 13331–13339.

149. Premkumar, L. S., and G. P. Ahern. 2000. Induction of vanilloid receptor channel activity by protein kinase C. *Nature* 408: 985–990.

150. Tominaga, M., M. Wada, and M. Masu. 2001. Potentiation of capsaicin receptor activity by metabotropic ATP receptors as a possible mechanism for ATP-evoked pain and hyperalgesia. *Proceedings of the National Academy of Sciences of the United States of America* 98: 6951–6956.

151. Numazaki, M., T. Tominaga, H. Toyooka et al. 2002. Direct phosphorylation of capsaicin receptor VR1 by protein kinase Cε and identification of two target serine residues. *Journal of Biological Chemistry* 277: 13375–13378.

152. Bhave, G., H.-J. Hu, K. S. Glauner et al. 2003. Protein kinase C phosphorylation sensitizes but does not activate the capsaicin receptor transient receptor potential vanilloid 1 (TRPV1). *Proceedings of the National Academy of Sciences of the United States of America* 100: 12480–12485.

153. Bhave, G., W. Zhu, H. Wang et al. 2002. cAMP-dependent protein kinase regulates desensitization of the capsaicin receptor (VR1) by direct phosphorylation. *Neuron* 35: 721–731.
154. Fan, H.-C., X. Zhang, and P. A. McNaughton. 2009. Activation of the TRPV4 ion channel is enhanced by phosphorylation. *Journal of Biological Chemistry* 284: 27884–27891.
155. Earley, S., S. V. Straub, and J. E. Brayden. 2007. Protein kinase C regulates vascular myogenic tone through activation of TRPM4. *American Journal of Physiology Heart and Circulatory Physiology* 292: H2613–H2622.
156. Venkatachalam, K., F. Zheng, and D. L. Gill. 2003. Regulation of canonical transient receptor potential (TRPC) channel function by diacylglycerol and protein kinase C. *Journal of Biological Chemistry* 278: 29031–29040.
157. Kwan, H. Y., Y. Huang, and X. Yao. 2004. Regulation of canonical transient receptor potential isoform 3 (TRPC3) channel by protein kinase G. *Proceedings of the National Academy of Sciences of the United States of America* 101: 2625–2630.
158. Takahashi, S., H. Lin, N. Geshi et al. 2008. Nitric oxide-cGMP-protein kinase G pathway negatively regulates vascular transient receptor potential channel TRPC6. *Journal of Physiology* 586: 4209–4223.
159. Hisatsune, C., Y. Kuroda, K. Nakamura et al. 2004. Regulation of TRPC6 channel activity by tyrosine phosphorylation. *Journal of Biological Chemistry* 279: 18887–18894.
160. Jin, X., N. Morsy, J. Winston et al. 2004. Modulation of TRPV1 by nonreceptor tyrosine kinase, c-Src kinase. *American Journal of Physiology Cell Physiology* 287: C558–C563.
161. Vazquez, G., B. J. Wedel, B. T. Kawasaki et al. 2004. Obligatory role of Src kinase in the signaling mechanism for TRPC3 cation channels. *Journal of Biological Chemistry* 279: 40521–40528.
162. Kawasaki, B. T., Y. Liao, and L. Birnbaumer. 2006. Role of Src in C3 transient receptor potential channel function and evidence for a heterogeneous makeup of receptor- and store-operated Ca^{2+} entry channels. *Proceedings of the National Academy of Sciences of the United States of America* 103: 335–340.
163. Xu, H., H. Zhao, W. Tian et al. 2003. Regulation of a transient receptor potential (TRP) channel by tyrosine phosphorylation. SRC family kinase-dependent tyrosine phosphorylation of TRPV4 on TYR-253 mediates its response to hypotonic stress. *Journal of Biological Chemistry* 278: 11520–11527.
164. Vriens, J., H. Watanabe, A. Janssens et al. 2004. Cell swelling, heat, and chemical agonists use distinct pathways for the activation of the cation channel TRPV4. *Proceedings of the National Academy of Sciences of the United States of America* 101: 396–401.
165. Kanzaki, M., Y.-Q. Zhang, H. Mashima et al. 1999. Translocation of a calcium-permeable cation channel induced by insulin-like growth factor-I. *Nature Cell Biology* 1: 165–170.
166. Bezzerides, V. J., I. S. Ramsey, S. Kotecha et al. 2004. Rapid vesicular translocation and insertion of TRP channels. *Nature Cell Biology* 6: 709–720.
167. Greka, A., B. Navarro, E. Oancea et al. 2003. TRPC5 is a regulator of hippocampal neurite length and growth cone morphology. *Nature Neuroscience* 6: 837–845.
168. Zhang, X., J. Huang, and P. A. McNaughton. 2005. NGF rapidly increases membrane expression of TRPV1 heat-gated ion channels. *EMBO Journal* 24: 4211–4223.
169. Bonnington, J. K., and P. A. McNaughton. 2003. Signalling pathways involved in the sensitisation of mouse nociceptive neurones by nerve growth factor. *Journal of Physiology* 551: 433–446.
170. van de Graaf, S. F. J., Q. Chang, A. R. Mensenkamp et al. 2006. Direct interaction with Rab11a targets the epithelial Ca^{2+} channels TRPV5 and TRPV6 to the plasma membrane. *Molecular and Cellular Biology* 26: 303–312.

171. Schmidt, M., A. E. Dubin, M. J. Petrus et al. 2009. Nociceptive signals induce trafficking of TRPA1 to the plasma membrane. *Neuron* 64: 498–509.

172. Mehta, D., G. U. Ahmmed, B. C. Paria et al. 2003. RhoA interaction with inositol 1,4,5-trisphosphate receptor and transient receptor potential channel-1 regulates Ca^{2+} entry. Role in signaling increased endothelial permeability. *Journal of Biological Chemistry* 278: 33492–33500.

173. Singh, B. B., T. P. Lockwich, B. C. Bandyopadhyay et al. 2004. VAMP2-dependent exocytosis regulates plasma membrane insertion of TRPC3 channels and contributes to agonist-stimulated Ca^{2+} influx. *Molecular Cell* 15: 635–646.

174. Hanaoka, K., F. Qian, A. Boletta et al. 2000. Co-assembly of polycystin-1 and -2 produces unique cation-permeable currents. *Nature* 408: 990–994.

175. Tsiokas, L., T. Arnould, C. Zhu et al. 1999. Specific association of the gene product of PKD2 with the TRPC1 channel. *Proceedings of the National Academy of Sciences of the United States of America* 96: 3934–3939.

176. Köttgen, M., T. Benzing, T. Simmen et al. 2005. Trafficking of TRPP2 by PACS proteins represents a novel mechanism of ion channel regulation. *EMBO Journal* 24: 705–716.

177. Miyagi, K., S. Kiyonaka, K. Yamada et al. 2009. A pathogenic C terminus-truncated polycystin-2 mutant enhances receptor-activated Ca^{2+} entry via association with TRPC3 and TRPC7. *Journal of Biological Chemistry* 284: 34400–34412.

178. Perraud, A. L., A. Fleig, C. A. Dunn et al. 2001. ADP-ribose gating of the calcium-permeable LTRPC2 channel revealed by Nudix motif homology. *Nature* 411: 595–599.

179. Sano, Y., K. Inamura, A. Miyake et al. 2001. Immunocyte Ca^{2+} influx system mediated by LTRPC2. *Science* 293: 1327–1330.

180. Perraud, A. L., C. L. Takanishi, B. Shen et al. 2005. Accumulation of free ADP-ribose from mitochondria mediates oxidative stress-induced gating of TRPM2 cation channels. *Journal of Biological Chemistry* 280: 6138–6148.

181. Kolisek, M., A. Beck, A. Fleig et al. 2005. Cyclic ADP-ribose and hydrogen peroxide synergize with ADP-ribose in the activation of TRPM2 channels. *Molecular Cell* 18: 61–69.

182. Wehage, E., J. Eisfeld, I. Heiner et al. 2002. Activation of the cation channel long transient receptor potential channel 2 (LTRPC2) by hydrogen peroxide. A splice variant reveals a mode of activation independent of ADP-ribose. *Journal of Biological Chemistry* 277: 23150–23156.

183. Beck, A., M. Kolisek, L. A. Bagley et al. 2006. Nicotinic acid adenine dinucleotide phosphate and cyclic ADP-ribose regulate TRPM2 channels in T lymphocytes. *FASEB Journal* 20: 962–964.

184. Lange, I., S. Yamamoto, S. Partida-Sanchez et al. 2009. TRPM2 functions as a lysosomal Ca^{2+}-release channel in beta cells. *Science Signaling* 2: ra23.

185. Calcraft, P. J., M. Ruas, Z. Pan et al. 2009. NAADP mobilizes calcium from acidic organelles through two-pore channels. *Nature* 459: 596–600.

186. Massullo, P., A. Sumoza-Toledo, H. Bhagat et al. 2006. TRPM channels, calcium and redox sensors during innate immune responses. *Seminars in Cell and Developmental Biology* 17: 654–666.

187. Yoshida, T., R. Inoue, T. Morii et al. 2006. Nitric oxide activates TRP channels by cysteine S-nitrosylation. *Nature Chemical Biology* 2: 596–607.

188. Susankova, K., K. Tousova, L. Vyklicky et al. 2006. Reducing and oxidizing agents sensitize heat-activated vanilloid receptor (TRPV1) current. *Molecular Pharmacology* 70: 383–394.

189. Chuang, H. H., and S. Lin. 2009. Oxidative challenges sensitize the capsaicin receptor by covalent cysteine modification. *Proceedings of the National Academy of Sciences of the United States of America* 106: 20097–20102.

190. Balzer, M., B. Lintschinger, and K. Groschner. 1999. Evidence for a role of Trp proteins in the oxidative stress-induced membrane conductances of porcine aortic endothelial cells. *Cardiovascular Research* 42: 543–549.

191. Sawada, Y., H. Hosokawa, K. Matsumura et al. 2008. Activation of transient receptor potential ankyrin 1 by hydrogen peroxide. *European Journal of Neuroscience* 27: 1131–1142.

192. Andersson, D. A., C. Gentry, S. Moss et al. 2008. Transient receptor potential A1 is a sensory receptor for multiple products of oxidative stress. *Journal of Neuroscience* 28: 2485–2494.

193. Bessac, B. F., M. Sivula, C. A. von Hehn et al. 2008. TRPA1 is a major oxidant sensor in murine airway sensory neurons. *Journal of Clinical Investigations* 118: 1899–1810.

194. Takahashi, N., Y. Mizuno, D. Kozai et al. 2008. Molecular characterization of TRPA1 channel activation by cysteine-reactive inflammatory mediators. *Channels (Austin)* 2: 287–298.

195. Maher, M., H. Ao, T. Banke et al. 2008. Activation of TRPA1 by farnesyl thiosalicylic acid. *Molecular Pharmacology* 73: 1225–1234.

196. Taylor-Clark, T. E., B. J. Undem, D. W. Macglashan et al. 2008. Prostaglandin-induced activation of nociceptive neurons via direct interaction with transient receptor potential A1 (TRPA1). *Molecular Pharmacology* 73: 274–281.

197. Miyamoto, T., A. E. Dubin, M. J. Petrus et al. 2009. TRPV1 and TRPA1 mediate peripheral nitric oxide-induced nociception in mice. *PLoS ONE* 4: e7596.

198. Jordt, S. E., D. M. Bautista, H.-H. Chuang et al. 2004. Mustard oils and cannabinoids excite sensory nerve fibres through the TRP channel ANKTM1. *Nature* 427: 260–265.

199. Bandell, M., G. M. Story, S. W. Hwang et al. 2004. Noxious cold ion channel TRPA1 is activated by pungent compounds and bradykinin. *Neuron* 41: 849–857.

200. Bautista, D. M., S. E. Jordt, T. Nikai et al. 2006. TRPA1 mediates the inflammatory actions of environmental irritants and proalgesic agents. *Cell* 124: 1269–1282.

201. Hinman, A., H.-H. Chuang, D. M. Bautista et al. 2006. TRP channel activation by reversible covalent modification. *Proceedings of the National Academy of Sciences of the United States of America* 103: 19564–19568.

202. Macpherson, L. J., A. E. Dubin, M. J. Evans et al. 2007. Noxious compounds activate TRPA1 ion channels through covalent modification of cysteines. *Nature* 445: 541–545.

203. Trevisani, M., J. Siemens, S. Materazzi et al. 2007. 4-Hydroxynonenal, an endogenous aldehyde, causes pain and neurogenic inflammation through activation of the irritant receptor TRPA1. *Proceedings of the National Academy of Sciences of the United States of America* 104: 13519–13524.

204. Taylor-Clark, T. E., M. A. McAlexander, C. Nassenstein et al. 2008. Relative contributions of TRPA1 and TRPV1 channels in the activation of vagal bronchopulmonary C-fibres by the endogenous autacoid 4-oxononenal. *Journal of Physiology* 586: 3447–3459.

205. Taylor-Clark, T. E., S. Ghatta, W. Bettner et al. 2009. Nitrooleic acid, an endogenous product of nitrative stress, activates nociceptive sensory nerves via the direct activation of TRPA1. *Molecular Pharmacology* 75: 820–829.

206. Salazar, H., I. Llorente, A. Jara-Oseguera et al. 2008. A single N-terminal cysteine in TRPV1 determines activation by pungent compounds from onion and garlic. *Nature Neuroscience* 11: 255–261.

207. Vyklický, L., A. Lyfenko, K. Susánková et al. 2002. Reducing agent dithiothreitol facilitates activity of the capsaicin receptor VR-1. *Neuroscience* 111: 435–441.

208. Xu, S. Z., P. Sukumar, F. Zeng et al. 2008. TRPC channel activation by extracellular thioredoxin. *Nature* 451: 69–72.

209. Streng, T., H. E. Axelsson, P. Hedlund et al. 2008. Distribution and function of the hydrogen sulfide-sensitive TRPA1 ion channel in rat urinary bladder. *Eur. Urol.* 53: 391–399.

210. Foster, M. W., D. T. Hess, and J. S. Stamler. 2006. S-nitrosylation TRiPs a calcium switch. *Nature Chemical Biology* 2: 570–571.

211. Aarts, M., K. Iihara, W. L. Wei et al. 2003. A key role for TRPM7 channels in anoxic neuronal death. *Cell* 115: 863–877.

212. Sun, H. S., M. F. Jackson, L. J. Martin et al. 2009. Suppression of hippocampal TRPM7 protein prevents delayed neuronal death in brain ischemia. *Nature Neuroscience* 12: 1300–1307.

213. Weissmann, N., A. Dietrich, B. Fuchs et al. 2006. Classical transient receptor potential channel 6 (TRPC6) is essential for hypoxic pulmonary vasoconstriction and alveolar gas exchange. *Proceedings of the National Academy of Sciences of the United States of America* 103: 19093–19098.

214. Hill, K., and M. Schaefer. 2009. Ultraviolet light and photosensitising agents activate TRPA1 via generation of oxidative stress. *Cell Calcium* 45: 155–164.

215. Mendez, F., and R. Penner. 1998. Near-visible ultraviolet light induces a novel ubiquitous calcium-permeable cation current in mammalian cell lines. *Journal of Physiology* 507: 365–377.

216. Dietrich, A., M. M. y. Schnitzler, J. Emmel et al. 2003. N-linked protein glycosylation is a major determinant for basal TRPC3 and TRPC6 channel activity. *Journal of Biological Chemistry* 278: 47842–47852.

217. Wirkner, K., H. Hognestad, R. Jahnel et al. 2005. Characterization of rat transient receptor potential vanilloid 1 receptors lacking the N-glycosylation site N604. *Neuroreport* 16: 997–1001.

218. Xu, H., Y. Fu, W. Tian et al. 2006. Glycosylation of the osmoresponsive transient receptor potential channel TRPV4 on Asn-651 influences membrane trafficking. *American Journal of Physiology Renal Physiology* 290: F1103–F1109.

219. Chang, Q., S. Hoefs, A. W. van der Kemp et al. 2005. The beta-glucuronidase klotho hydrolyzes and activates the TRPV5 channel. *Science* 310: 490–493.

220. Imura, A., Y. Tsuji, M. Murata et al. 2007. alpha-Klotho as a regulator of calcium homeostasis. *Science* 316: 1615–1618.

221. Morenilla-Palao, C., M. Pertusa, V. Meseguer et al. 2009. Lipid raft segregation modulates TRPM8 channel activity. *Journal of Biological Chemistry* 284: 9215–9224.

222. Spehr, M., H. Hatt, and C. H. Wetzel. 2002. Arachidonic acid plays a role in rat vomeronasal signal transduction. *Journal of Neuroscience* 22: 8429–8437.

223. Oike, H., M. Wakamori, Y. Mori et al. 2006. Arachidonic acid can function as a signaling modulator by activating the TRPM5 cation channel in taste receptor cells. *Biochimica Biophysica Acta* 1761: 1078–1084.

224. Zygmunt, P. M., J. Petersson, D. A. Andersson et al. 1999. Vanilloid receptors on sensory nerves mediate the vasodilator action of anandamide. *Nature* 400: 452–457.

225. Watanabe, H., J. Vriens, J. Prenen et al. 2003. Anandamide and arachidonic acid use epoxyeicosatrienoic acids to activate TRPV4 channels. *Nature* 424: 434–438.

226. Ben-Amor, N., P. C. Redondo, A. Bartegi et al. 2006. A role for 5,6-epoxyeicosatrienoic acid in calcium entry by de novo conformational coupling in human platelets. *Journal of Physiology* 570: 309–323.

227. Basora, N., G. Boulay, L. Bilodeau et al. 2003. 20-Hydroxyeicosatetraenoic acid (20-HETE) activates mouse TRPC6 channels expressed in HEK293 cells. *Journal of Biological Chemistry* 278: 31709–31716.

228. Inoue, R., L. J. Jensen, Z. Jian et al. 2009. Synergistic activation of vascular TRPC6 channel by receptor and mechanical stimulation via phospholipase C/diacylglycerol

and phospholipase A_2/ω-hydroxylase/20-HETE pathways. *Circulation Research* 104: 1399–1409.

229. Hwang, S. W., H. Cho, J. Kwak et al. 2000. Direct activation of capsaicin receptors by products of lipoxygenases: Endogenous capsaicin-like substances. *Proceedings of the National Academy of Sciences of the United States of America* 97: 6155–6160.

230. Shin, J., H. Cho, S. W. Hwang et al. 2002. Bradykinin-12-lipoxygenase-VR1 signaling pathway for inflammatory hyperalgesia. *Proceedings of the National Academy of Sciences of the United States of America* 99: 10150–10155.

231. Grimm, C., R. Kraft, G. Schultz et al. 2005. Activation of the melastatin-related cation channel TRPM3 by D-erythro-sphingosine. *Molecular Pharmacology* 67: 798–805.

232. Xu, S.-Z., K. Muraki, F. Zeng et al. 2006. A sphingosine-1-phosphate-activated calcium channel controlling vascular smooth muscle cell motility. *Circulation Research* 98: 1381–1389.

233. Flemming, P. K., A. M. Dedman, S.-Z. Xu et al. 2006. Sensing of lysophospholipids by TRPC5 calcium channel. *Journal of Biological Chemistry* 281: 4977–4982.

234. Andersson, D. A., M. Nash, and S. Bevan. 2007. Modulation of the cold-activated channel TRPM8 by lysophospholipids and polyunsaturated fatty acids. *Journal of Neuroscience* 27: 3347–3355.

235. Wagner, T. F. J., S. Loch, S. Lambert et al. 2008. Transient receptor potential M3 channels are ionotropic steroid receptors in pancreatic β cells. *Nature Cell Biology* 10: 1421–1430.

236. Kwak, J., M. H. Wang, S. W. Hwang et al. 2000. Intracellular ATP increases capsaicin-activated channel activity by interacting with nucleotide-binding domains. *Journal of Neuroscience* 20: 8298–8304.

237. Nadler, M. J. S., M. C. Hermosura, K. Inabe et al. 2001. LTRPC7 is a Mg·ATP-regulated divalent cation channel required for cell viability. *Nature* 411: 590–595.

238. Nilius, B., J. Prenen, T. Voets et al. 2004. Intracellular nucleotides and polyamines inhibit the Ca^{2+}-activated cation channel TRPM4b. *Pflügers Archiv* 448: 70–75.

239. Kim, D., and E. J. Cavanaugh. 2007. Requirement of a soluble intracellular factor for activation of transient receptor potential A1 by pungent chemicals: role of inorganic polyphosphates. *Journal of Neuroscience* 27: 6500–6509.

240. Hu, H., M. Bandell, M. J. Petrus et al. 2009. Zinc activates damage-sensing TRPA1 ion channels. *Nature Chemical Biology* 5: 183–190.

241. Voets, T., B. Nilius, S. Hoefs et al. 2004. TRPM6 forms the Mg^{2+} influx channel involved in intestinal and renal Mg^{2+} absorption. *Journal of Biological Chemistry* 279: 19–25.

242. Oberwinkler, J., A. Lis, K. M. Giehl et al. 2005. Alternative splicing switches the divalent cation selectivity of TRPM3 channels. *Journal of Biological Chemistry* 280: 22540–22548.

243. Yamaguchi, H., M. Matsushita, A. C. Nairn et al. 2001. Crystal structure of the atypical protein kinase domain of a TRP channel with phosphotransferase activity. *Molecular Cell* 7: 1047–1057.

244. Ryazanova, L. V., M. V. Dorovkov, A. Ansari et al. 2004. Characterization of the protein kinase activity of TRPM7/ChaK1, a protein kinase fused to the transient receptor potential ion channel. *Journal of Biological Chemistry* 279: 3708–3716.

245. Vriens, J., B. Nilius, and R. Vennekens. 2008. Herbal compounds and toxins modulating TRP channels. *Current Neuropharmacology* 6: 79–96.

246. Caterina, M. J., M. A. Schumacher, M. Tominaga et al. 1997. The capsaicin receptor: a heat-activated ion channel in the pain pathway. *Nature* 389: 816–824.

247. Jordt, S.-E., and D. Julius. 2002. Molecular basis for species-specific sensitivity to hot chili peppers. *Cell* 108: 421–430.

248. Jung, J., S.-Y. Lee, S. W. Hwang et al. 2002. Agonist recognition sites in the cytosolic tails of vanilloid receptor 1. *Journal of Biological Chemistry* 277: 44448–44454.

249. McNamura, F. N., A. Randall, and M. J. Gunthorpe. 2005. Effects of piperine, the pungent component of black pepper, at the human vanilloid receptor (TRPV1). *British Journal of Pharmacology* 144: 781–790.

250. Iwasaki, Y., A. Morita, T. Iwasawa et al. 2006. A nonpungent component of steamed ginger-[10]-shogaol-increases adrenaline secretion via the activation of TRPV1. *Nutritional Neuroscience* 9: 169–178.

251. Xu, H., N. T. Blair, and D. E. Clapham. 2005. Camphor activates and strongly desensitizes the transient receptor potential vanilloid subtype 1 channel in a vanilloid-independent mechanism. *Journal of Neuroscience* 25: 8924–8937.

252. Moqrich, A., S. W. Hwang, T. J. Earley et al. 2005. Impaired thermosensation in mice lacking TRPV3, a heat and camphor sensor in the skin. *Science* 307: 1468–1472.

253. Vogt-Eisele, A. K., K. Weber, M. A. Sherkheli et al. 2007. Monoterpenoid agonists of TRPV3. *British Journal of Pharmacology* 151: 530–540.

254. Mckemy, D. D., W. M. Neuhausser, and D. Julius. 2002. Identification of a cold receptor reveals a general role for TRP channels in thermosensation. *Nature* 416: 52–58.

255. Peier, A. M., A. Moqrich, A. C. Hergarden et al. 2002. A TRP channel that senses cold stimuli and menthol. *Cell* 108: 705–715.

256. Chuang, H., W. M. Neuhausser, and D. Julius. 2004. The super-cooling agent icilin reveals a mechanism of coincidence detection by a temperature-sensitive TRP channel. *Neuron* 43: 859–869.

257. Bandell, M., A. E. Dubin, M. J. Petrus et al. 2006. High-throughput random mutagenesis screen reveals TRPM8 residues specifically required for activation by menthol. *Nature Neuroscience* 9: 493–500.

258. Macpherson, L. J., S. W. Hwang, T. Miyamoto et al. 2006. More than cool: Promiscuous relationships of menthol and other sensory compounds. *Molecular and Cellular Neuroscience* 32: 335–343.

259. Story, G. M., A. M. Peier, A. J. Reeve et al. 2003. ANKTM1, a TRP-like channel expressed in nociceptive neurons, is activated by cold temperatures. *Cell* 112: 819–829.

260. Karashima, Y., N. Damann, J. Prenen et al. 2007. Bimodal action of menthol on the transient receptor potential channel TRPA1. *Journal of Neuroscience* 27: 9874–9884.

261. Yang, B.-H., Z. G. Piao, Y.-B. Kim et al. 2003. Activation of vanilloid receptor 1 (VR1) by eugenol. *Journal of Dental Research* 82: 781–785.

262. Xu, H., M. Delling, J. C. Jun et al. 2006. Oregano, thyme and clove-derived flavors and skin sensitizers activate specific TRP channels. *Nature Neuroscience* 9: 628–635.

263. Lee, S. P., M. T. Buber, Q. Yang et al. 2008. Thymol and related alkyl phenols activate the hTRPA1 channel. *British Journal of Pharmacology* 153: 1739–1749.

264. Parnas, M., M. Peters, D. Dadon et al. 2009. Carvacrol is a novel inhibitor of Drosophila TRPL and mammalian TRPM7 channels. *Cell Calcium* 45: 300–309.

265. Neeper, M. P., Y. Liu, T. L. Hutchinson et al. 2007. Activation properties of heterologously expressed mammalian TRPV2. *Journal of Biological Chemistry* 282: 15894–15902.

266. Smith, P. L., K. N. Maloney, R. G. Pothen et al. 2006. Bisandrographolide from Andrographis paniculata activates TRPV4 channels. *Journal of Biological Chemistry* 281: 29897–29904.

267. Nagatomo, K., and Y. Kubo. 2008. Caffeine activates mouse TRPA1 channels but suppresses human TRPA1 channels. *Proceedings of the National Academy of Sciences of the United States of America* 105: 17373–17378.

268. Talavera, K., M. Gees, Y. Karashima et al. 2009. Nicotine activates the chemosensory cation channel TRPA1. *Nature Neuroscience* 12: 1293–1299.

269. Leuner, K., V. Kazanski, M. Müller et al. 2007. Hyperforin—a key constituent of St. John's wort specifically activates TRPC6 channels. *FASEB Journal* 21: 4101–4111.

270. Siemens, J., S. Zhou, R. Piskorowski et al. 2006. Spider toxins activate the capsaicin receptor to produce inflammatory pain. *Nature* 444: 208–212.
271. Kitaguchi, T., and K. J. Swartz. 2005. An inhibitor of TRPV1 channels isolated from funnel web spider venom. *Biochemistry* 44: 15544–15549.
272. Spassova, M. A., T. Hewavitharana, W. Xu et al. 2006. A common mechanism underlies stretch activation and receptor activation of TRPC6 channels. *Proceedings of the National Academy of Sciences of the United States of America* 103: 16586–16591.
273. Watanabe, H., J. B. Davis, D. Smart et al. 2002. Activation of TRPV4 channels (hVRL-2/mTRP12) by phorbol derivatives. *Journal of Biological Chemistry* 277: 13569–13577.
274. Vriens, J., G. Owsianik, A. Janssens et al. 2007. Determinants of 4α-phorbol sensitivity in transmembrane domains 3 and 4 of the cation channel TRPV4. *Journal of Biological Chemistry* 282: 12796–12803.
275. Chung, M.-K., H. Lee, A. Mizuno et al. 2004. 2-aminoethoxydiphenyl borate activates and sensitizes the heat-gated ion channel TRPV3. *Journal of Neuroscience* 24: 5177–5182.
276. Hu, H.-Z., Q. Gu, C. Wang et al. 2004. 2-aminoethoxydipenyl borate is a common activator of TRPV1, TRPV2, and TRPV3. *Journal of Biological Chemistry* 279: 35741–35748.
277. Hu, H., J. Grandl, M. Bandell et al. 2009. Two amino acid residues determine 2-APB sensitivity of the ion channels TRPV3 and TRPC4. *Proceedings of the National Academy of Sciences of the United States of America* 106: 1626–1631.
278. Li, M., J. Jiang, and L. Yue. 2006. Functional characterization of homo- and heteromeric channel kinase TRPM6 and TRPM7. *Journal of General Physiology* 127: 525–537.
279. Xu, S.-Z., F. Zeng, G. Boulay et al. 2005. Block of TRPC5 channels by 2-aminoethoxydiphenyl borate: a differential, extracellular and voltage-dependent effect. *British Journal of Pharmacology* 145: 405–414.
280. Togashi, K., H. Inada, and M. Tominaga. 2008. Inhibition of the transient receptor potential cation channel TRPM2 by 2-aminoethoxydiphenyl borate (2-APB). *British Journal of Pharmacology* 153: 1324–1330.
281. He, L.-P., T. Hewavitharana, J. Soboloff et al. 2005 A functional link between store-operated and TRPC channels revealed by the 3,5-bis(trifluoromethyl)pyrazole derivative, BTP2. *Journal of Biological Chemistry* 280: 10997–11006.
282. Kiyonaka S., K. Kato, M. Nishida et al. 2009. Selective and direct inhibition of TRPC3 channels underlies biological activities of a pyrazole compound. *Proceedings of the National Academy of Sciences of the United States of America* 106: 5400–5405.
283. Takezawa, R., H. Cheng, A. Beck et al. 2006. A Pyrazole derivative potently inhibits lymphocyte Ca^{2+} influx and cytokine production by facilitating transient receptor potential melastatin 4 channel activity. *Molecular Pharmacology* 69: 1413–1420.
284. Benedikt, J., J. Teisinger, L. Vyklicky et al. 2007. Ethanol inhibits cold-menthol receptor TRPM8 by modulating its interaction with membrane phosphatidylinositol 4,5-bisphosphate. *Journal of Neurochemistry* 100: 211–224.
285. Xu, S.-Z., and D. J. Beech. 2000. TrpC1 is a membrane-spanning subunit of store-operated Ca^{2+} channels in native vascular smooth muscle cells. *Circulation Research* 88: 84–87.
286. Xu, S.-Z., F. Zeng, M. Lei et al. 2005. Generation of functional ion-channel tools by E3 targeting. *Nature Biotechnology* 23: 1289–1293.
287. Naylor, J., C. J. Milligan, F. Zeng et al. 2008. Production of a specific extracellular inhibitor of TRPM3 channels. *British Journal of Pharmacology* 155: 567–573.
288. Schaefer, M., T. D. Plant, A. G. Obukhov et al. 2000. Receptor-mediated regulation of the nonselective cation channels TRPC4 and TRPC5. *Journal of Biological Chemistry* 275: 17517–17526.
289. Tominaga, M., M. J. Caterina, A. B. Malmberg et al. 1998. The cloned capsaicin receptor integrates multiple pain-producing stimuli. *Neuron* 21: 531–543.

290. Semtner, M., M. Schaefer, O. Pinkenburg et al. 2007. Potentiation of TRPC5 by protons. *Journal of Biological Chemistry* 282: 33868–33878.

291. Kim, M. J., J.-P. Jeon., H. J. Kim et al. 2008. Molecular determinant of sensing extracellular pH in classical transient receptor potential channel 5. *Biochemical and Biophysical Research Communications* 365: 239–245.

292. Inada, H., F. Kawabata, Y. Ishimaru et al. 2008. Off-response property of an acid-activated cation channel complex PKD1L3-PKD2L1. *EMBO Report* 9: 690–697.

293. Dong, X.-P., X. Cheng, E. Mills et al. 2008. The type IV mucolipidosis-associated protein TRPML1 is an endolysosomal iron release channel. *Nature* 455: 992–996.

294. Yeh, B.-I., T.-J. Sun., J. Z. Lee et al. 2003. Mechanism and molecular determinant for regulation of rabbit transient receptor potential type 5 (TRPV5) channel by extracellular pH. *Journal of Biological Chemistry* 278: 51044–51052.

295. Yeh, B.-I., Y. K. Kim, W. Jabbar et al. 2005. Conformational changes of pore helix coupled to gating of TRPV5 by protons. *EMBO Journal* 24: 3224–3234.

296. Liu, D., Z. Zhang, and E. R. Liman. 2005. Extracellular acid block and acid-enhanced inactivation of the Ca^{2+}-activated cation channel TRPM5 involve residues in the S3-S4 and S5-S6 extracellular domains. *Journal of Biological Chemistry* 280: 20691–20699.

297. Du, J., J. Xie, and L. Yue. 2009. Modulation of TRPM2 by acidic pH and the underlying mechanisms for pH sensitivity. *Journal of General Physiology* 134: 471–488.

298. Kim, M. J, J.-P. Jeon, H. J. Kim et al. 2008. Molecular determinant of sensing extracellular pH in classical transient receptor potential channel 5. *Biochemical and Biophysical Research Communications* 365: 239–245.

299. Fujita, F., K. Uchida, T. Moriyama et al. 2008. Intracellular alkalization causes pain sensation through activation of TRPA1 in mice. *Journal of Clinical Investigations* 118: 4049–4057.

300. Shimizu, T., A. Janssens, T. Voets et al. 2009. Regulation of the murine TRPP3 channel by voltage, pH, and changes in cell volume. *Pflügers Archiv* 457: 795–807.

301. Kozak, J. A., M. Matsushita, A. C. Nairn et al. 2005. Charge screening by internal pH and polyvalent cations as a mechanism for activation, inhibition, and rundown of TRPM7/MIC channels. *Journal of General Physiology* 126: 499–514.

302. Numata, T., and Y. Okada. 2008. Proton conductivity through the human TRPM7 channel and its molecular determinants. *Journal of Biological Chemistry* 283: 15097–15103.

303. Soyombo, A. A., S. Tjon-Kon-Sang, Y. Rbaibi et al. 2006. TRP-ML1 regulates lysosomal pH and acidic lysosomal lipid hydrolytic activity. *Journal of Biological Chemistry* 281: 7294–7301.

304. Hille, B. 2001. *Ion Channels of Excitable Membrane*, 3rd ed. Sunderland, MA: Sinauer Associates.

305. Caterina, M. J., and D. Julius. 2001. The vanilloid receptor: a molecular gateway to the pain pathway. *Annual Review of Neuroscience* 24: 487–517.

306. Welch, J. M., S. A. Simon, and P. H. Reinhart. 2000. The activation mechanism of rat vanilloid receptor 1 by capsaicin involves the pore domain and differs from the activation by either acid or heat. *Proceedings of the National Academy of Sciences of the United States of America* 97: 13889–13894.

307. Vlachová, V., J. Teisinger, K. Susánková et al. 2003. Functional role of C-terminal cytoplasmic tail of rat vanilloid receptor 1. *Journal of Neuroscience* 23: 1340–1350.

308. Hayes, P., H. J. Meadows, M. J. Gunthorpe et al. 2000. Cloning and functional expression of a human orthologue of rat vanilloid receptor-1. *Pain* 88: 205–215.

309. Caterina, M. J., T. A. Rosen, M. Tominaga et al. 1999. A capsaicin-receptor homologue with a high threshold for noxious heat. *Nature* 398: 436–441.

310. Ahluwalia, J., H. Rang, and I Nagy. 2002. The putative role of vanilloid receptor-like protein-1 in mediating high threshold noxious heat-sensitivity in rat cultured primary sensory neurons. *European Journal of Neuroscience* 16: 1483–1489.

311. Leffler, A., R. Madalina, L. C. Nau et al. 2007. A high-threshold heat-activated channel in cultured rat dorsal root ganglion neurons resembles TRPV2 and is blocked by gadolinium. *European Journal of Neuroscience* 26: 12–22.

312. Brauchi, S., G. Orta, M. Salazar et al. 2006. A hot-sensing cold receptor: C-terminal domain determines thermosensation in transient receptor potential channels. *Journal of Neuroscience* 26: 4835–4840.

313. Voets, T., G. Droogmans, U. Wissenbach et al. 2004. The principle of temperature-dependent gating in cold- and heat-sensitive TRP channels. *Nature* 430: 748–754.

314. Brauchi, S., P. Orio, and R. Latorre. 2004. Clues to understanding cold sensation: Thermodynamics and electrophysiological analysis of the cold receptor TRPM8. *Proceedings of the National Academy of Sciences of the United States of America* 101: 15494–15499.

315. Peier, A. M., A. J. Reeve, D. A. Andersson et al. 2002. A heat-sensitive TRP channel expressed in keratinocytes. *Science* 296: 2046–2049.

316. Smith, G. D., M. J. Gunthorpe, R. E. Kelsell et al. 2002. TRPV3 is a temperature-sensitive vanilloid receptor-like protein. *Nature* 418: 186–190.

317. Xu, H., I. S. Ramsey, S. A. Kotecha et al. 2002. TRPV3 is a calcium-permeable temperature-sensitive cation channel. *Nature* 418: 181–186.

318. Liedtke, W., Y. Choe, M. A. Martí-Renom et al. 2000. Vanilloid receptor-related osmotically activated channel (VR-OAC), a candidate vertebrate osmoreceptor. *Cell* 103: 525–535.

319. Strotmann, R., C. Harteneck, K. Nunnenmacher et al. 2000. OTRPC4, a nonselective cation channel that confers sensitivity to extracellular osmolarity. *Nature Cell Biology* 2: 695–702.

320. Wissenbach, U., M. Bödding, M. Freichel et al. 2000. Trp12, a novel Trp related protein from kidney. *FEBS Letters* 485: 127–134.

321. Güler, A. D., H. Lee, T. Iida et al. 2002. Heat-evoked activation of the ion channel, TRPV4. *Journal of Neuroscience* 22: 6408–6414.

322. Watanabe, H., J. Vriens, S. H. Suh et al. 2002. Heat-evoked activation of TRPV4 channels in a HEK293 cell expression system and in native mouse aorta endothelial cells. *Journal of Biological Chemistry* 277: 47044–47051.

323. Chung, M.-K., H. Lee, A. Mizuno et al. 2004. TRPV3 and TRPV4 mediate warmth-evoked currents in primary mouse keratinocytes. *Journal of Biological Chemistry* 279: 21569–21575.

324. Chung, M.-K., H. Lee, and M. J. Caterina. 2003. Warm temperatures activate TRPV4 in mouse 308 keratinocytes. *Journal of Biological Chemistry* 278: 32037–32046.

325. Talavera, K., K. Yasumatsu, T. Voets et al. 2005. Heat activation of TRPM5 underlies thermal sensitivity of sweet taste. *Nature* 438: 1022–1025.

326. Togashi, K., Y. Hara, T. Tominaga et al. 2006. TRPM2 activation by cyclic ADP-ribose at body temperature is involved in insulin secretion. *EMBO Journal* 25: 1804–1815.

327. Andersson, D. A., H. W. N. Chase et al. 2004. TRPM8 activation by menthol, icilin, and cold is differentially modulated by intracellular pH. *Journal of Neuroscience* 24: 5364–5369.

328. Brauchi, S., G. Orta, C. Mascayano et al. 2007. Dissection of the components for PIP2 activation and thermosensation in TRP channels. *Proceedings of the National Academy of Sciences of the United States of America* 104: 10246–10251.

329. Reid, G., A. Babes, and F. Pluteanu. 2002. A cold- and menthol-activated current in rat dorsal root ganglion neurones: properties and role in cold transduction. *Journal of Physiology* 545: 595–614.

330. Dhaka, A., A. N. Murray, J. Mathur et al. 2007 TRPM8 is required for cold sensation in mice. *Neuron* 54: 371–378.

331. Bautista, D. M., J. Siemens, J. M. Glazer et al. 2007. The menthol receptor TRPM8 is the principal detector of environmental cold. *Nature* 448: 204–208.

332. Colburn, R. W., M. Lou, L. J. Stone et al. 2007. Attenuated cold sensitivity in TRPM8 null mice. *Neuron* 54: 379–386.

333. Sawada, Y., H. Hosokawa, A. Hori et al. 2007. Cold sensitivity of recombinant TRPA1 channels. *Brain Research* 1160: 39–46.

334. Karashima, Y., K. Talavera, W. Everaerts et al. 2009. TRPA1 acts as a cold sensor in vitro and in vivo. *Proceedings of the National Academy of Sciences of the United States of America* 106: 1273–1278.

335. Yamada, H., M. Wakamori, Y. Hara et al. 2000. Spontaneous single-channel activity of neuronal TRP5 channel recombinantly expressed in HEK293 cells. *Neuroscience Letters* 285: 111–114.

336. Nilius, B., R. Vennekens, J. Prenen et al. 2000. Whole-cell and single channel monovalent cation currents through the novel rabbit epithelial Ca^{2+} channel ECaC. *Journal of Physiology* 527: 239–248.

337. Prawitt, D., M. K. Monteilh-Zoller, L. Brixel et al. 2003. TRPM5 is a transient Ca^{2+}-activated cation channel responding to rapid changes in $[Ca^{2+}]_i$. *Proceedings of the National Academy of Sciences of the United States of America* 100: 15166–15171.

338. Nilius, B., J. Prenen, G. Droogmans et al. 2003. Voltage dependence of the Ca^{2+}-activated cation channel TRPM4. *Journal of Biological Chemistry* 278: 30813–30820.

339. Ahern, G. P., and L. S. Premkumar. 2002. Voltage-dependent priming of rat vanilloid receptor: effects of agonist and protein kinase C activation. *Journal of Physiology* 545: 441–451.

340. Voets, T., G. Owsianik, A. Janssens et al. 2007. TRPM8 voltage sensor mutants reveal a mechanism for integrating thermal and chemical stimuli. *Nature Chemical Biology* 3: 174–182.

341. Chung, M.-K., A. D. Güler, and M. J. Caterina. 2005. Biphasic currents evoked by chemical or thermal activation of the heat-gated ion channel, TRPV3. *Journal of Biological Chemistry* 280: 15928–15941.

342. Matta, J. A., and G. P. Ahern. 2007. Voltage is a partial activator of rat thermosensitive TRP channels. *Journal of Physiology* 585: 469–482.

343. Numata, T., T. Shimizu, and Y. Okada. 2007. Direct mechano-stress sensitivity of TRPM7 channel. *Cellular and Physiological Biochemistry* 19: 1–8.

344. Numata, T., T. Shimizu, and Y. Okada. 2007. TRPM7 is a stretch- and swelling-activated cation channel involved in volume regulation in human epithelial cells. *American Journal of Physiology Cell Physiology* 292: C460–C467.

345. Peng, J. B., X. Z. Chen, U. V. Berger et al. 1999. Molecular cloning and characterization of a channel-like transporter mediating intestinal calcium absorption. *Journal of Biological Chemistry* 274: 22739–22746.

346. Voets, T., J. Prenen, A. Fleig, R. Vennekens et al. 2001. CaT1 and the calcium release-activated calcium channel manifest distinct pore properties. *Journal of Biological Chemistry* 276: 47767–47770.

347. Voets, T., A. Janssens, J. Prenen et al. 2003. Mg^{2+}-dependent gating and strong inward rectification of the cation channel TRPV6. *Journal of General Physiology* 121: 245–260.

348. Mochizuki, T., G. Wu, T. Hayashi et al. 1996. PKD2, a gene for polycystic kidney disease that encodes an integral membrane protein. *Science* 272: 1339–1342.

349. Numa, S. 1987. A molecular view of neurotransmitter receptors and ionic channels. *Harvey Lectures* 83: 121–165.

350. Gonzalez-Perrett, S., M. Batelli, K. Kim et al. 2002. Voltage dependence and pH regulation of human polycystin-2-mediated cation channel activity. *Journal of Biological Chemistry* 277: 24959–24966.

351. Luo, Y., P. M. Vassilev, X. Li, Y. Kawanabe et al. 2003. Native polycystin 2 functions as a plasma membrane Ca^{2+}-permeable cation channel in renal epithelia. *Molecular and Cellular Biology* 23: 2600–2607.

352. Pelucchi, B., G. Aguiari, A. Pignatelli et al. 2006. Nonspecific cation current associated with native polycystin-2 in HEK-293 cells. *Journal of American Society of Nephrology* 17: 388–397.

353. Delmas, P., S. M. Nauli, X. Li et al. 2004. Gating of the polycystin ion channel signaling complex in neurons and kidney cells. *FASEB Journal* 18: 740–742.

354. Maroto, R., A. Raso, T. G. Wood et al. 2005. TRPC1 forms the stretch-activated cation channel in vertebrate cells. *Nature Cell Biology* 7: 179–185.

355. Wei, C., X. Wang, M. Chen et al. 2009. Calcium flickers steer cell migration. *Nature* 457: 901–905.

356. Morita, H., A. Honda, R. Inoue et al. 2007. Membrane stretch-induced activation of a TRPM4-like nonselective cation channel in cerebral artery myocytes. *Journal of Pharmacology Science* 103: 417–426.

357. Muraki, K., Y. Iwata, Y. Katanosaka et al. 2003. TRPV2 is a component of osmotically sensitive cation channels in murine aortic myocytes. *Circulation Research* 93: 829–838.

358. Nilius, B., J. Prenen, U. Wissenbach et al. 2001. Differential activation of the volume-sensitive cation channel TRP12 (OTRPC4) and volume-regulated anion currents in HEK-293 cells. *Pflügers Archiv* 443: 227–233.

359. Gao, X., L. Wu, and R. G. O'Neil. 2003. Temperature-modulated diversity of TRPV4 channel gating: activation by physical stresses and phorbol ester derivatives through protein kinase C-dependent and -independent pathways. *Journal of Biological Chemistry* 278: 27129–27137.

360. Wu, L., X. Gao, R. C. Brown et al. 2007. Dual role of the TRPV4 channel as a sensor of flow and osmolality in renal epithelial cells. *American Journal of Physiology Renal Physiology* 293: F1699–F1713.

361. Wegierski, T., U. Lewandrowski, B. Müller, A. Sickmann et al. 2009. Tyrosine phosphorylation modulates the activity of TRPV4 in response to defined stimuli. *Journal of Biological Chemistry* 284: 2923–2933.

362. Welsh, D. G., A. D. Morielli, M. T. Nelson et al. 2002. Transient receptor potential channels regulate myogenic tone of resistance arteries. *Circulation Research* 90: 248–250.

363. Grimm, C., R. Kraft, S. Sauerbruch et al. 2003. Molecular and functional characterization of the melastatin-related cation channel TRPM3. *Journal of Biological Chemistry* 278: 21493–21501.

364. Gomis, A., S. Soriano, C. Belmonte et al. 2008. Hypoosmotic- and pressure-induced membrane stretch activate TRPC5 channels. *Journal of Physiology* 586: 5633–5649.

365. Zhang, X.-F., J. Chen, C. R. Faltynek et al. 2008. Transient receptor potential A1 mediates an osmotically activated ion channel. *European Journal of Neuroscience* 27: 605–611.

366. Chen, J., and G. J. Barritt. 2003. Evidence that TRPC1 (transient receptor potential canonical 1) forms a Ca^{2+}-permeable channel linked to the regulation of cell volume in liver cells obtained using small interfering RNA targeted against TRPC1. *Biochemistry Journal* 373: 327–336.

367. Arniges, M., E. Vázquez, J. M. Fernández-Fernández et al. 2004. Swelling-activated Ca^{2+} entry via TRPV4 channel is defective in cystic fibrosis airway epithelia. *Journal of Biological Chemistry* 279: 54062–54068.

368. Sharif Naeini, R., M. F. Witty, P. Séguéla et al. 2006. An N-terminal variant of Trpv1 channel is required for osmosensory transduction. *Nature Neuroscience* 9: 93–98.

369. Ciura, S., and C. W. Bourque. 2006. Transient receptor potential vanilloid 1 is required for intrinsic osmoreception in organum vasculosum lamina terminalis neurons and for normal thirst responses to systemic hyperosmolality. *Journal of Neuroscience* 26: 9069–9075.

370. Oancea, E., T. W. Joshua, and D. E. Clapham. 2006. Functional TRPM7 channels accumulate at the plasma membrane in response to fluid flow. *Circulation Research* 98: 245–253.
371. Nauli, S. M., F. J. Alenghat, Y. Luo et al. 2003. Polycystins 1 and 2 mediate mechanosensation in the primary cilium of kidney cells. *Nature Genetics* 33: 129–137.
372. Iwata, Y., Y. Katanosaka, Y. Arai et al. 2003. A novel mechanism of myocyte degeneration involving the Ca^{2+}-permeable growth factor-regulated channel. *Journal of Cell Biology* 161: 957–967.
373. Julius, D., and A. I. Basbaum. 2001. Molecular mechanisms of nociception. *Nature* 413: 203–210.

4 Studying Subunit Interaction and Complex Assembly of TRP Channels

Michael Poteser and Klaus Groschner

CONTENTS

4.1 INTRODUCTION

4.1.1 Functional Significance of Multimerization and Complex Assembly

Transient receptor potential (TRP) proteins serve as subunits in diverse cation channel complexes. These signaling units are typically composed of scaffolds, adaptors, regulatory subunits, and enzymes (for review, see Refs. 1 and 2). The functional core unit within these complexes is a homo- or hetero-oligomeric TRP pore assembly that is likely to contain the essential structures for channel gating and regulation. The obvious structural similarities between TRP- and voltage-gated K^+ channel proteins have promoted the view of a tetrameric TRP pore complex,[3,4] and this concept has been confirmed experimentally.[5,6] Tight protein–protein interactions between the pore-forming, transmembrane subunits determine the biophysical and regulatory properties of the TRP cation conductances and are thus essential determinants of the (patho)physiological role of individual signal complexes.[4] Communication and cross-talk of the oligomeric TRP pore assemblies with auxiliary proteins such as adaptors, scaffolds, and regulators appear to involve a highly diverse array of molecular interactions, ranging from tight physical association to rather loose and dynamic interactions.[7,8] Both qualitative composition and stoichiometry of cation channel complexes containing a certain TRP protein are most likely dependent on the cellular host environment, as well as the cell phenotype.[9] Our current incomplete understanding of the (patho)physiological function of many TRP proteins is partly due to the lack of information on multimeric assembly and complex composition in native cellular environments, and the difficulties to assign a certain signaling process in native cells to a defined TRP complex arrangement. Therefore, in-depth analysis of the composition of native TRP complexes and identification of the molecular structures and mechanisms of protein–protein interaction within functional TRP signalosomes is prerequisite for understanding the (patho)physiological impact of these structures[10] and for uncovering their potential as therapeutic targets. Because of the remarkable plasticity of functional TRP complexes based on dynamic cellular trafficking and complex rearrangement,[7,8] this task is exceptionally challenging. Successful strategies will require refined classical biochemical, biophysical, and functional measurements complemented by cutting-edge proteome analysis. In the following, we will address the potentials as well as pitfalls and limitations of individual approaches with respect to certain aspects of the TRP complex analysis. Focus will be placed on the advantages and pitfalls of established methods complemented by a brief outline of alternative and novel techniques. Technical requirements and problems will be discussed for analysis of pore complex interactions, as well as regulatory and/or targeting interactions including standard step-by-step working procedures for key methods.

4.1.2 ANALYSIS OF PORE ASSEMBLIES, STRUCTURAL MODELS, AND LESSONS FROM K+ CHANNEL PORES

The structural similarity between TRP and K_v proteins has promoted the view of a similar oligomeric assembly of pore structures.[4] For both TRP and K_v channels, tetramerization of channel subunits is essential to generate a functional permeation pathway. Assembly of pore proteins is likely mediated by multiple interaction domains within both the cytosolic N- and C-termini,[11–13] which host important structures for tetramerization in both ion channel families. Homo- and heteromeric pore assemblies are possible, and certain limitations for heteromeric assembly of TRP proteins have been indicated in expression studies.[14] Pore assemblies are, in general, considered as fairly stable in a given cell type, and the potential components of TRP pores are limited, as mixing between distantly related members of the superfamily is, with a few exceptions, a rare phenomenon. Nonetheless, composition and stoichiometry of native TRP pores may exhibit a certain degree of plasticity in terms of adaptation to the functional needs of native cells.

Both classical biochemical methods such as co-immunoprecipitation from native cells as well as in vitro pull-down assays and heterologous expression approaches for complex reconstitution, combined with fluorescence microscopy (fluorescent resonance energy transfer [FRET] approaches), appear, in principle, well suited and have proven valuable for the analysis of TRP pore complexes. As a general limitation of standard heterologous expression systems, the potential impact of cell type-specific targeting and/or regulatory interactions, which are lacking in common expression systems, needs consideration. Nonetheless, for pore complex analysis, heterologous expression strategies bear the advantage of both qualitative and quantitative evaluation of assemblies by functional measurements as a highly valuable complementation of biochemical strategies.

4.1.3 ANALYSIS OF INTERACTION WITH REGULATORS AND SCAFFOLDS AND DYNAMIC ASPECTS OF COMPLEX ASSEMBLY

Linkage of TRP pore complexes with regulatory components and scaffolds within cellular signalplexes involves highly dynamic molecular interactions and has been proposed to include protein–protein interactions that transiently bridge different cellular membrane systems. Such regulatory interactions are typically dependent on the cellular activity and phenotype and correspond to distinct functional states of the pore complex. Moreover, rapid translocation of TRP pore complexes along with regulation of cellular activity[9,15–17] has repeatedly been reported. Experimental analysis of the dynamic features of regulatory protein–protein interactions requires specific approaches and refined experimental strategies as compared to investigations on the assembly of pore tetramers. Time and activity dependence of protein–protein interactions generates a number of constraints to standard techniques that need to be considered for data interpretation. We will address this issue specifically for the methods and strategies outlined below.

4.2 CLASSICAL METHODS, THEIR ADVANTAGES AND LIMITATIONS

The selection of an appropriate strategy to analyze protein–protein interactions within ion channel complexes is typically guided by the level of scientific hypothesis development. While a number of methods are preferentially suited for screening within a large pool of proteins, other methods are designed to test and characterize interaction between already suggested interacting partners. Both approaches are limited at certain aspects and may be used in combination to generate conclusive results.

4.2.1 YEAST TWO-HYBRID APPROACH

Yeast two-hybrid systems represent a classical approach for the analysis of specific protein–protein interactions, which is well suited for hypothesis-generating screens using protein or drug libraries.

Taking advantage of the ordered domain structure of eukaryotic transcription factors, this technique detects protein–protein interaction by bringing together a specific DNA-binding sequence and a transcription activation domain (Gal-4, LexA) that are fused to the proteins of interest and transferred using specially designed vectors. Following the classical approach, the physical interaction results in the activation of reporter constructs like lacZ, HIS3, or LEU2.

In case the method is used for fishing/screening experiments, the DNA-binding sequence usually encodes for the protein of interest (bait), while the cDNA library comprises fusion constructs of activation domains (prey).

4.2.1.1 Advantages

The yeast two-hybrid system allows for semi-quantitative analysis of even weak protein–protein interactions because of the amplifying effect of the reporter gene activation. If the general equipment and expertise in handling yeast and yeast genetics is already available, the main advantage may be given by the speed and ease of the experiments that, in addition, only require the generation of cDNA.

4.2.1.2 Requirements, Limitations, and Problems

Yeast strains have to meet the growth selection needs of the plasmids used and display deficiency of the wild-type (e.g., GAL-4) transcription factor and its binding protein (e.g., GAL80). Use of yeast two-hybrid systems requires some technical experience with yeast cell culture and genetics.

The phylogenetic distance between yeast and mammalian cells results in clear limitations of this method regarding mammalian cell-specific folding and posttranslational protein modifications like disulfide-bridge formation, glycosylation, phosphorylation, and folding, a fact that led to the development of mammalian cell-based approaches.[18] The presence of a native and functional yeast TRP channel homolog

(TRPY1) with about 20% amino acid identity and 40% overall homology indicates that at least the folding of the general TRP channel structure is most likely not compromised in yeast. A possible toxic effect of nuclear targeted TRP-channel fragments has not been reported so far but may be caused by other proteins leading to counter selection during cell growth.

Because nuclear expression of proteins, as required in common yeast two-hybrid systems, may cause toxicity and misfolding of nonnuclear proteins, cytoplasmic protein recruitment systems (Sos recruitment system, SRS, Ras recruitment system, RRS) have been developed to overcome these limitations.[19–21] In this approach, the membrane recruitment of the guanyl nucleotide exchange factor (hSos), via protein–protein interaction, has been shown sufficient to activate the Ras signaling pathway and to complement a cdc25 temperature-sensitive mutation of the yeast strain. The hybrid bait is therefore designed as C-terminally truncated hSos (SRS) or CAAX box lacking Ras (RRS) and the prey fused to a membrane localization signal, such as v-Src myristoylation. Thus, a nonnuclear alternative to classical yeast two-hybrid method is available.

Two-hybrid approaches are not applicable if the reporter protein transcription is already activated by expression of one of the proteins of interest alone, a fact that is hard to predict and has to be tested using a truncated/modified construct. Less likely, but still possible, is the activation of the reporter gene by a third unknown protein.

The yeast two-hybrid system has so far mainly been used to identify interactions between cytosolic portions of TRPC channels like N- and C-termini and loops and other intracellular proteins[22–27] or to terminal structures involved in the heteromeric channel assembly.[28–30] It has been shown that isolated subdomains may interact better than full-length clones, possibly because of altered folding properties, and one way to avoid possible "false positives" is to use full-length cDNA or confirmation of results by non yeast two-hybrid methods like co-immunoprecipitation. Park et al.[31] used TRPM4b as a bait in a protein library approach and identified TRPC3 as an interacting protein within a human fetal brain matchmaker cDNA library.

4.2.2 CO-IMMUNOPRECIPITATION

Protein-complex co-immunoprecipitation (Co-IP) using specific antibodies has become one of the most common techniques to analyze protein–protein interactions and has been already used in the wake of intensive TRP channel science.[11,32] Several modified protocols have been developed and optimized for particular tasks, all of which share the need for antibodies coupled to a solid substrate at some point of the procedure. Alternatively, the antibody directed against a protein is added directly to the mixture or lysate and protein A or G coated beads are added subsequently. Such protocols may be of advantage for the detection of target proteins at low concentration or antibody affinity. For the more common, direct approach, the antibodies are already immobilized on microbeads and are then added to the lysate. After the Protein A (or G)-antibody, or antibody–target interactions are completed, the unbound proteins are removed by washing the solid phase, leaving the purified antibody–antigen complexes bound to the matrix. The sample can then be separated by SDS-PAGE for Western blot analysis. Unequivocally, the critical step in Co-IP is specific molecular recognition of proteins by antibodies. The use of commercially available, barely suitable TRP antibodies is therefore a common source for pitfalls and problems (see below and also Chapter 6).

4.2.2.1 Advantages

This technique allows for the co-precipitation of proteins in their native conformation and posttranslational state and can be used for screening type experiments as well as for the detection of defined protein interactions. Besides the limitations given by antibody quality, immunoprecipitation is a rather selective method. In addition, Co-IP does not require heterologous (over)expression of a protein and thus avoids artifacts produced by unbalanced relative concentrations of protein and antigen.

4.2.2.2 Requirements, Limitations, and Problems

Precipitation steps are typically performed in several rounds: Epitope specificity of antibodies is limited, and confirmation by a second antibody, preferably detecting a different site of the protein, is recommended, in particular, when the quality and suitability of an antibody are unclear (see below). The same strategy may be helpful in circumventing the problem of precipitating protein–protein complexes that hide their epitopes by structural interactions. In screening approaches, a larger number of proteins will be pulled down in the first round of precipitation, and immuno-precipitation may then be repeated using newly identified proteins as a target of precipitation. However, results should be double-checked by precipitating both potential interaction partners, along with the use of different antibodies for control purposes.

In general, if available, a knockdown system (tissue from knockout animal models or genetically modified cell lines) should be used for control experiment to demonstrate the suitability of the employed antibodies.

Agarose and sepharose beads are the most common materials for the solid-phase support. These beads resemble highly porous structures with large surface to volume ratios and a very high potential binding capacity. One possible caveat arising from this point is the requirement to saturate the surface of these sponge-like beads with antibody because incomplete coverage increases the possibility of unspecific protein–bead interactions. This problem can be addressed by careful calculation of antibody and bead concentrations, by using nonporous beads, or by preclearing. In a preclearing step, beads without antibody coating are added to the protein mixture, in an effort to capture and remove all nonspecifically binding proteins from the lysate. Nonporous, spherical beads, also referred to as monodisperse beads, are utilized in magnetic and nonmagnetic forms and are mainly used for investigating proteins of larger size because there is no size limit, as is the case with agarose/sepharose pores. Magnetic beads add the possibility to retain the protein–bead complex at the tube wall during washing steps, a feature that allows for a fast and precise removal of supernatants. However, the advantages of monodisperse beads come with the price of a reduced binding capacity, as compared to agarose or sepharose.

A common problem is the coelution of antibody heavy and light chains with the antigen, which may interfere with downstream analysis. This problem can be addressed by introducing a cross-linking step, where the beads and antibodies are irreversibly linked by formation of stable amide bonds.

A general disadvantage of Co-IP is given by the fact that this approach does not allow for rapid, conclusive identification of direct molecular interactions between proteins or to exclude involvement of a third protein. Nonetheless, pull down of larger protein complexes is a suitable first step for further analysis of complex composition by identifying the components by mass spectrometry or Western blotting.

4.2.2.3   Step-by-Step Showcase Protocol

Co-IP strategy as used for identification of TRPC4 interaction partners in endothelial cells.[9]

Prepare Cell Lysates

1. Wash the cells twice with ice-cold phosphate-buffered saline (PBS, see composition below). Scrape the cells ($3–5 \times 10^6$) off in 5 mL PBS and centrifuge them with 900 rpm for 4 min at 4°C.
2. Discard the supernatant and resuspend the cell pellet in a 500-μL *mammalian cell lysis buffer* (QUIAGEN), containing 5-μL Protease Inhibitor solution. *Note:* For optimized lysis conditions (stringency, salt concentration, divalent cation concentration, and pH) and to obtain a high protein yield, it is recommended to use a commercially available, ready-to-use cell lysis buffer.
3. Shake lysate overhead for 10 min at 4°C and subsequently centrifuge at $20,000 \times g$ for 5 min at 4°C. *Tip:* If the lysate does not become clear, sonicate it for 10 s.
4. Transfer the supernatant (cleared lysate) to a new 1.5-mL vial and determine the protein concentration by any standard method.

Prepare Protein A or G Beads

5. Wash the beads with ice-cold PBS, centrifuge them with a maximum of $500 \times g$ for 1 min at room temperature, and repeat this once. *Note:* The choice of either protein A or protein G beads is based on the affinity of the antibodies for protein A or protein G. *Tips:* When pipetting the beads, it is advisable to cut the end of the pipette tips off to prevent damage to the beads. To avoid beads sticking on the lid of the vial, it is advisable not to vortex.

Preclear Samples

6. Divide the cleared lysate in aliquots corresponding to 500 μg proteins and add 500 μL PBS to each sample.
7. Add 50 μL of protein A or protein G beads, and rotate for 45 min at 4°C.
8. Centrifuge gently (maximum at $500 \times g$) for 1 min at 4°C.
9. Transfer the supernatant (precleared lysate) to a new 1.5-mL vial. *Note:* The preclearing step reduces the background that is due to undesirable adhesion of sample components to the beads. Samples can be incubated either with protein A or protein G beads or with an irrelevant antibody of the same species of origin and same Ig subclass.

Incubate to Form Antibody–Antigen Complexes

10. Add 1–3 µg of antibody to each vial, except the one for control, and gently rotate the tubes overnight at 4°C or for 2h at room temperature.
11. Add 60 µl of washed beads and incubate them with rotation for 2h at room temperature.

Purify Protein Complexes

12. After incubation, centrifuge to remove the supernatant and wash the beads with a fresh buffer (PBS containing 1% Triton-X-100) for 4 times, 5 min each at 4°C with rotation. *Tip:* Centrifuge gently (500 × g, 2 min) each time to prevent any damage to the beads.
13. Finally, add 25–50 µL of 2× SDS loading buffer to the beads.
14. Boil at 95°C for 5 min to denature the proteins and separate them from the beads, and then centrifuge at a maximum speed for 2 min.
15. The supernatant, now containing the protein of interest, can be stored at –20°C, but it is preferable to run the sample immediately on an SDS-PAGE gel.

Detect Proteins by Western Blot

16. Transfer of the separated proteins to a nitrocellulose membrane may be performed in a standard blotting chamber or with the Dry-Blot-System (Invitrogen, USA).
17. After transfer, block the membrane overnight at 4°C with 5% nonfat dry milk in PBST (0.1% Tween-20 in PBS) and TBST (0.1% Tween-20 in Tris-buffered saline, TBS) under gentle agitation. *Note*: The choice of washing buffer (PBST or TBST) depends mainly on the antibody being used. Blocking of the membrane and dilution of the antibodies have to be done with the same buffer.
18. On the next day, incubate the membrane with the primary antibody for 1 hour at room temperature under gentle agitation.
19. Wash the membrane 4 times with buffer (PBST or TBST) for 10 min with agitation.
20. Add the secondary antibody and incubate for 1 hour at room temperature with agitation.
21. Repeat the washing procedure (step 19).
22. Incubate the membrane with ChemiGlow West Chemiluminescence Substrate Sample Kit (Alpha Innotech, USA) for 5 min.
23. Protein bands can be detected by densitometric analysis.

Reagents/buffers needed
   10× TBS stock solution: pH 7.6

NaCl	80.0 g/L
Tris Base	24.4 g/L
in distilled water	

10× PBS stock solution: pH 6.9

NaCl	80.0 g/L
KCl	24.0 g/L
NaHPO$_4$	144.0 g/L
KH$_2$PO$_4$	2.0 g/L

in distilled water

4.2.3   Fluorescence Resonance Energy Transfer (FRET) Microscopy

FRET is a microscopy-based method for the detection of molecular interactions and utilizes a photophysical phenomenon analogous to "near-field communication." A donor fluorophore in its excited state may be able to transfer some of its energy to an acceptor fluorophore in close proximity (typically less than 10 nm), resulting in enhanced fluorescence of the acceptor at its specific emission wavelength. For monitoring interactions by FRET, each protein needs to be tagged with a fluorophore that fulfills certain spectral properties (FRET pair). The most common labels are the GFP variants CFP (cyan fluorescent protein) and YFP (yellow fluorescent protein), which require the generation of fusion constructs for each protein of interest. The technically least pretentious but also slightly more error-prone method is analysis of sensitized emission by taking images of cells expressing exclusively donor and acceptor, and a combination of donor and acceptor to calculate the energy transfer using an adequate image analysis software. With these procedures, the FRET images are corrected for background fluorescence and spectral bleed-through using an algorithm and values from control experiments. More robust are approaches that take advantage of changes in the fluorescence of donor or acceptor during or after controlled photobleaching[33,34] or fluorescence lifetime measurements (FLIM; fluorescence lifetime imaging microscopy).[35–37]

FRET has become one of the most common techniques to study TRP-channel protein interactions and has been used to analyze TRPC subunit multimerization[12] and stoichiometry[38] as well as association of TRP channels with other proteins.[39–41]

4.2.3.1   Advantages

Perhaps the most striking advantage of FRET is the possibility for real-time observation of the dynamics of protein–protein interaction in an intact cellular system. Once appropriate fusion proteins are generated, FRET results can be obtained relatively fast by employing standard expression systems. FRET experiments may be combined with advanced, high-resolution techniques of fluorescence microscopy, like confocal laser scanning, multiphoton microscopy (two-photon microscopy), or total internal reflection fluorescence (TIRF) to combine the strengths of each of these methods.

Specifically, TIRF-FRET strategies potentially allow for the detection of rapid changes in protein–protein interactions within channel complexes residing in the plasma membrane.[42] This advantage is, however, limited for TRP complexes by the rapid dynamics of channel recruitment, retrieval, as well as lateral trafficking. Therefore, TIRF-FRET analysis of the TRP complex assembly may be performed only on a sufficiently "stable" set of TRP complexes (Figure 4.1). A sample protocol for TIRF-FRET measurement of TRPC3 homomeric interactions in the membrane of HEK293 cells is outlined below.

FRET analysis is very sensitive, is well scalable, and allows for quantification from the subcellular, organelle level, and single cells up to large (mL) volume photometry with cell populations. In addition, FRET-based approaches are relatively cost-effective when compared to other techniques used for detecting protein–protein interactions. FRET may be used not only for studies on interactions of distinct proteins but also for monitoring conformational changes of single proteins (intramolecular FRET[43]).

4.2.3.2 Requirements, Limitations, and Problems

One limitation of FRET approaches is the fact that false-negative results may be produced simply by the lack of suitability of the molecular (dipole) orientation of the chromophore labels, a molecular property that is generally hard to predict.[44] Thus, under common experimental circumstances, only a positive FRET signal represents potentially conclusive information.

Spectral bleed-through or cross-talk is a major problem in FRET microscopy and often unavoidable because of a given spectral overlap of the fluorophores used. Spectral bleed-through-associated problems may be minimized by optimizing the optical system of the microscope, employing bleaching-based FRET techniques, or by careful calculation and subtraction of the contaminating component of fluorescence. The principle of computed spectral unmixing[45,46] requires the analysis of bleed-through of each of the fluorophores in each channel when expressed alone (spectral profile). On the basis of this profile, the image that results from the overlapping spectra of the fluorophores can be corrected. This method may be used for

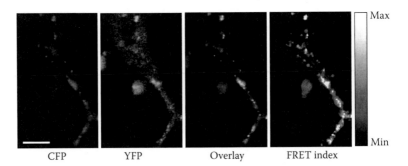

CFP   YFP   Overlay   FRET index

FIGURE 4.1 **(See color insert.)** TIRF-FRET demonstration of TRPC3–TRPC3 homomerization in HEK293 cells. Typical sensitized-emission FRET obtained by TRPC3 homomerization in HEK293 cells transfected to express about equal levels of CFP- and YFP-tagged TRPC3 (N-terminal fusions). The fusion proteins colocalize in distinct clusters within the evanescent field, i.e., in or close to the plasma membrane. FRET generated in homomultimers that are localized within fairly stable clusters in a region of the basolateral membrane is clearly detectable. Shown are images of CFP and YFP fluorescence as well as of an overlay and the calculated FRET image using color-coding of nFRET values (index). Note that FRET in channels that are localized in highly mobile vesicles will be missed. Scale bar represents 5 μm.

simple subtractive correction but can also be utilized in combination with advanced techniques like multiphoton microscopy and Fourier spectroscopy.[47]

Although false-positive FRET results are unlikely when data analysis is carefully performed, control experiments in which interaction between the fluorophores is prevented by excessive expression of an untagged protein or interacting fragment[48] are strongly recommended.

As FRET analysis typically requires heterologous expression of tagged proteins, unbalanced expression levels of the fluorophores and membrane crowding may occur and impair analysis. As a matter of fact, expression levels of equally transfected cells are often found to vary and statistical analysis of a sufficient number of cells is necessary, although, in principle, FRET does not require absolutely balanced amounts of fluorescent protein pairs. Moreover, high expression levels of fluorophores (GFP variants) may interfere with physiological pathways, a phenomenon that needs consideration.[49]

Similar to approaches utilizing fluorescence labeling of proteins, expression of the FRET chromophores may induce phototoxicity, a still not well-understood phenomenon that leads to selective death of fluorophore-expressing cells upon illumination. The severity of this problem depends on the experimental system (e.g., fluorophore, cell type, illumination) and tends to be correlated with light intensity and exposure times.

In most types of FRET experiments, the general, technical goal will be to achieve maximum emission fluorescence using minimal excitation intensities and exposure times. This will not only reduce phototoxicity but also minimize unwanted bleaching of the fluorophore. The bioluminescence resonance energy transfer (BRET) technique has been developed to avoid problems induced by external illumination by using bioluminescent luciferase as a replacement for CFP to enable photon emission by a donor without excitation light. BRET has also been considered for screening approaches.[50]

Another problem of fluorescence microscopy that may also affect FRET experiments arises from cellular autofluorescence. Autofluorescence is often produced by mitochondrial or lysosomal proteins[51] and may become a severe obstacle when examining labeled specimen of dim fluorescence. One possibility to cope with this problem is to switch to a less affected excitation wavelength and/or fluorophore.

4.2.3.3   Step-by-Step Showcase Protocol

FRET strategy as used for demonstration of TRPC multimerization in the plasma membrane of HEK293 cells.[48]

Prepare TRP-expressing HEK293 Cells:

1. Seed cells (HEK293) on cover slips at low density to allow growth to approximately 70–80% confluency after overnight culture.
2. Transfect cells using a standard transfection reagent and transfection protocol (48) with 2–3 µg cDNA/10^6 cells. The cDNA amounts should be adjusted to obtain fairly equal expression levels of proteins and to avoid membrane overcrowding.

3. Prepare single transfections of CFP- and YFP-fusion proteins as well as cotransfections to express both fusion proteins and, if applicable, controls (i.e., triple transfection including an unlabelled competing species).

4. At 12 hours after transfection, start to test for appropriate expression by fluorescence microscopy. *Note:* Depending on the cell type and culture conditions, cells may be used for 2–3 days after transfection. Coating cover slips with poly-lysine or similar substances may improve cell adhesion and morphological integrity of attached cells but cannot be employed for TIRF experiments.

Record FRET (3-cube, Sensitized Emission Measurements) in TRP-expressing HEK293 Cells:

5. Place cells on the stage of a microscope equipped with appropriate optics for fluorescence detection (e.g., Axiovert 200M, Zeiss, Jena, Germany), digital imaging (e.g., Photometrics CoolSNAP fx-HQ monochrome camera, Roper Scientific, Tucson, AZ, USA), and a beam-splitting device (e.g., MultiSpec Imager, Visitron Systems, Gmbh., Puchheim, Germany). Alternatively, a dual-camera system may be used. Filter sets suitable for CFP and YFP excitation and emission as well as dichroic mirrors (485 nm, 535 nm, Chroma Technology Corp., Bellows Falls, VT, USA) and a light source able to supply the required excitation wavelengths and intensities (445 nm, 488 nm, ModuLaser, Centerville, UT, USA) are required. *Notes:* (1) While excitation of fluorophores using laser light provides sufficient intensities for FRET, it also may facilitate the generation of artifacts owing to bleaching, autofluorescence, and overexposure. (2) For evaluation, alignment of images with pixel-level accuracy is required. Reference grid-based prerecording alignment of beam-splitter channels facilitates later manipulations. Whenever technically possible, images should be recorded into prealigned stack images because offline, manual alignment of images may be challenging.

6. Take images from single fluorophore-protein transfected cells to estimate the spectral bleed-through. To analyze the contribution of each fluorophore emission to the FRET channel image in the absence of FRET, pairs of donor (CFP) channel and FRET channel images, as well as acceptor (YFP) channel and FRET channel images, have to be recorded. Use donor/acceptor double-transfected cells for generating FRET images. Three images are required for each FRET analysis: donor image (CFP channel), acceptor image (YFP channel), and FRET (CFP excitation/YFP emission) channel image. *Note:* Carefully select exposure time parameters to avoid overexposed/saturated images. If a beam splitter is used, the two channels of the image may differ significantly in intensity, and selected exposure time has to be optimized for sufficient excitation of the dim fluorophore while avoiding overexposure of the brighter one. However, the exposure time and other parameters that influence image intensity have to be constant for all images used in analysis.

Record FRET (3-cube, Sensitized Emission Measurements) in TIRF Mode:

7. TIRF requires special objectives (Zeiss 100×/1.45 α-PLAN FLUO), laser excitation (Modu-Laser, 445 nm, 488 nm), and mechanical light beam controller/condenser (VisiTirf Dual-Port Kondensor VS T1, Visitron Systems). After selecting a well-attached cell in epifluorescence, carefully adjust the focal plane and the laser beam angle until only basal structural elements of the cellular membrane are visible (e.g., vesicles, microdomains, pseudopodia). *Note:* TIRF recordings are highly sensitive to bleaching, while typical TIRF objectives require high excitation intensity. Thus, combination of TIRF-FRET requires high sensitivity of the detection system and efficient optical lenses and filters. Typically, fluorescence microscopes for TIRF measurements are controlled by specialized software to acquire images (e.g., MetaVue, Universal Imaging Corp., Downingtown, PA, USA).

8. Take images for sensitized emission FRET analysis as outlined for epifluorescence.[6]

9. Analyze images using suitable software (MetaMorph, Molecular Devices, Sunnyvale, CA). Most software products include special tools for manual alignment of images. However, as image alignment quality is critical for FRET analysis, this step has to be performed with maximum care. *Note:* The technical feature of the microscope determines the minimum time required for acquisition of images for a single set of interaction analysis and therefore the time resolution. Be aware that if proteins are targeted to relatively small clusters (microdomains) and display a certain degree of lateral mobility, correct alignment may be impossible owing to insufficient time resolution (e.g., result of TRPC3/TRPC3 TIRF-FRET below).

10. Correct FRET images for cross-talk between optical channels. The corrected FRET image (nFRET) has to be calculated, after background subtraction, using the sensitized emission correction method according to nFRET = rawFRET − corrD*I_{donor} − corrA*$I_{acceptor}$. The calibration/correction factors (corrD, corrA) represent the relative cross-talk contribution of the donor and acceptor channel to FRET, as obtained from experiments where only CFP or YFP was expressed (corrD = 0.4, corrA = 0.1) and multiplied by the actual donor and acceptor channel intensity (I_{donor}, $I_{acceptor}$).

11. For statistical analysis, the mean cellular nFRET intensity of a sufficient number of cells has to be determined.

4.2.4 FUNCTIONAL APPROACHES TO ANALYZE PORE COMPLEX ASSEMBLY

The assembly of TRP proteins into multimers and the existence of heteromeric TRP pores of defined subunit composition were recognized early on in TRP research. This concept was suggested by dominant negative effects of nonfunctional TRP fragments or loss of function mutations,[11,52,53] as well as by the generation of unique properties by co-expression of different subunits.[48,54] Thus, electrophysiological experiments using defined expression of pore proteins provide important information on the interactions and stoichiometry within ion channel complexes. Experimental approaches to test for a specific functional assembly of pore subunits and to analyze subunit stoichiometry include both the independent co-expression of potential subunits and the expression or co-expression of concatenated subunits (concatamers consisting of a defined combination of subunits).[55-57] In both cases, solid knowledge on the functional properties of the individual subunits is a prerequisite for such strategies. Typically, species that exhibit easily detectable functional features (e.g., mutants with altered permeation of gating) serve as reporters for the inclusion of the proteins into a channel complex.

One classical strategy to confirm interactions between TRP channel proteins is to test whether loss-of-function mutations of a particular species, preferentially proteins that lack a functional pore structure, are able to prevent currents through the potential heteromerization partner.[58,59]

Disruption of the pore region by either point mutations or deletions within the SS5 region generates nonfunctional proteins that, upon integration into homo- or heteromeric pore assemblies, eliminate conductivity of the heteromeric complex.[59,60] So far, a gradual modification of channel properties, depending on the subunit stoichiometry and composition of hybrid TRP channels, has not been reported. The effectiveness of pore mutants to inhibit TRP conductances indicates that inclusion of a single mutant molecule may be sufficient to eliminate channel function.

4.2.4.1 Advantages

Both concepts, i.e., dominant negative suppression of channel function as well as transfer of mutant properties (e.g., modified blocker sensitivity/affinity) to a heteromeric channel complex, will, in principle, allow not only to test for an interaction at a qualitative level but also to determine subunit stoichiometry and to test certain concepts relating pore properties and stoichiometry. Formation of hybrid channels is expected to result in distinct, composition (stoichiometry)-dependent channel properties. Assuming that (1) wild-type and mutant channels express equally in the respective cell system and (2) both proteins are equally well integrated into channel complexes with random aggregate formation, the fraction of functionally modified channels and thus the remaining cellular current or signaling response for co-expression of a wild-type TRP protein and a mutant can be predicted by assuming a binomial distribution of channel complexes.

$$F_i = \binom{n}{i} f_{mut}^i * f_{wt}^{n-i}$$

with F_i as the channels that are of i-type, f_{mut} and f_{wt} the fractions of mutant and wild-type subunits, respectively, and n as the subunit stoichiometry.

A formalism described by MacKinnon[61] can be used to predict the subunit stoichiometry of hybrid channels composed of wild type and a blocker insensitive mutation, with an approximation of n given by

$$\frac{1}{\ln(f_{mut})} * \ln\left(\frac{U_{mix}}{U_{mut}}\right) \to n$$

using an approach in which the ratio of unblocked current response hybrid channels (U_{mix}) and blocker insensitive mutant monomers (U_{mut}) is determined at high blocker concentrations.[61,62] Thus, mutants with eliminated sensitivity to blocking agents are expected to be particularly useful tools to investigate assembly of pore complexes.

Alternatively, mutations that generate a gain in blocker sensitivities may be employed. We have recently demonstrated that TRPC mutants, which contain an artificial antibody sensitivity due to insertion of an extracellularly located haemagglutinin (HA) epitope, are able to confer this sensitivity to hybrid channels (see the showcase protocol below). Measurements of the anti-HA sensitivity of channels generated by co-expression of wild-type and mutant proteins at different expression ratios yield results consistent with the current concept of tetrameric TRPC pore complexes.

4.2.4.2 Requirements, Limitations, and Problems

As interpretation of data depends critically on the suitability of the mutant protein, the construct used as an analysis tool needs to be characterized thoroughly in initial experiments. Pitfalls may arise from dominant negative effects of the mutant proteins that are independent of integration into a pore complex such as competition of the nonfunctional mutant and the wild-type proteins for an essential regulatory component. The use of fluorescent fusions of the mutants is helpful to test and confirm proper expression and targeting of the proteins. This will also allow for concomitant evaluation of pore complex interactions and subunit stoichiometry by FRET approaches as outlined above.

For strategies employing concatamers of multiple pore-forming subunits, assembly of the proteins into higher aggregates needs to be considered as a typical source of inconsistent results and misinterpretations. This issue may be addressed by blue native PAGE analysis of protein aggregation.[55]

In general, functional measurements using dominant negative TRP mutants are difficult to interpret in terms of pore assembly and bear a high risk of misinterpretation unless properties that are measured are tightly linked to the pore architecture. A classical example is the sensitivity to blocking agents as a typical property of the pore complex itself.

4.2.4.3 Step-by-Step Showcase Protocol

Conditional dominant negative strategy used for identification of TRPC3/TRPC4 interactions.[9,48]

Prepare TRP-expressing HEK293 Cells:

1. Seed HEK293 cells on glass cover slips suitable for electrophysiological measurements to obtain a density of approximately 40–50% within 24 hours of culture.
2. After 24 hours, transfect cells with equal amounts of cDNA (approximately 2–3 μg cDNA/10^6 cells) encoding a YFP- and a CFP-tagged TRPC species, one of which needs to display unique antibody or blocker sensitivity (e.g., mutants that carry an extracellular epitope, e.g., YFP-exoHA-TRPC4).[48] Prepare cells that are transfected with both constructs as well as cells that are, in a similar manner, transfected to express either one of the constructs.

3. Check expression of constructs 48 hours after transfection by fluorescence microscopy and perform electrophysiological measurements if expression is sufficient and fairly equal for both constructs according to fluorescence intensities.

Measure TRP Currents in HEK293 Cells in the Absence and Presence of a Blocking Agent or Antibody:

4. Transfer cells to a standard patch-clamp setup to measure currents through the generated TRPC channels (see composition of solutions below). Currents can be activated with common agonists of Gq-coupled receptors expressed in the HEK293 cell line, e.g., muscarinic receptors using carbachol (100 μM). *Note:* Currents are typically transient, and the peak current densities can be used for quantification and analysis. This requires a careful determination of cell capacitance.

5. Perform experiments to obtain an appropriate measure of mean currents in agonist-stimulated cells in the absence and presence of an appropriate blocking agent (e.g., anti-HA; 1:200) for channels generated at approximately equal expression of both species as well as by each of the species alone. *Note:* Effective block may require sufficient time for equilibration and thus pre-incubation of the antibody or blocker with the cells.

6. Evaluate the functional interaction of the two species based on the expected blocker sensitivities (e.g., anti-HA sensitivities). Channels generated by HA-TRPC4 are fully sensitive to anti-HA. TRPC3 itself is fairly insensitive to anti-HA. Mixed expression of both channel types without heteromerization is expected to generate a conductance corresponding to a combination of both individual conductances and in a HA-insensitive component that is indistinguishable from that obtained by TRPC3 only. For heteromerization and dominant transfer of HA sensitivity, almost complete block of the mixed conductance by anti-HA is predicted from the concept outlined above. The insensitive component calculated for tetrameric channels is <10% (6.25%). *Note:* The existence of a HA-insensitive, agonist-induced background conductance needs to be considered, and therefore, additional controls to characterize the employed HEK293 cell line in the absence of heterologous TRPC expression, ideally expressing a GFP variant only (vector control), are required. Moreover, unspecific blocking effects of the anti-HA product, which will be evident in the experiments with the epitope-deficient (wild-type) species, need consideration. Importantly, one should use only antibody preparations that are devoid of any potentially channel blocking components, such as stabilizers.

Reagents/buffers needed for electrophysiology
  Bath solution pH 7.4:

NaCl	140 mM
KCl	5.4 mM

HEPES   10 mM
CaCl$_2$ 2 mM
MgCl$_2$ 1 mM
in distilled water

Pipette solution pH 7.2:

Cs methansulfonate   110 mM
CsCl$_2$ 25 mM
HEPES   30 mM
MgCl$_2$ 5 mM
EGTA   3 mM
in distilled water

4.3   COMPLEMENTARY AND ALTERNATIVE APPROACHES

4.3.1   CHEMICAL CROSS-LINKING AND LABEL TRANSFER

Cross-linkers are reagents with at least two reactive groups, permitting the establishment of inter- as well as intramolecular cross linkages. Interacting proteins may be cross-linked and then subjected to SDS-PAGE, in-gel digestion, and liquid chromatography or mass spectrometry (MS) for identification of the interacting partners.

In vivo cross-linking is possible using either formaldehyde to cross-link proteins to labeled constructs (Strep-Protein Interaction Experiment, SPINE) or photo reactive amino acid analogs (diazirine) that are incorporated to cross-link protein domains that interact and/or colocalize within some angstroms.

For label transfer, one of the interacting proteins is labeled with a cross-linking agent. After an interaction that induces the agent to bind also to the second protein, the original link between the primary protein and the label is cleaved, resulting in the transfer of the label to the second protein. This method often utilizes sulfhydryl-directed biotin labels to cysteine residues of a protein in the primary step.[63]

4.3.2   TANDEM AFFINITY PURIFICATION (TAP)

A fusion protein carrying a TAP tag, a label that incorporates calmodulin binding protein and Protein A, is introduced to the target system. In the first affinity purification step, the protein and its possible interaction partners are bound to IgG-coated beads. In the second step, the tag is broken enzymatically, and a certain part of the tag is used for detection of the complex in a subsequent affinity purification step. After washing steps, the proteins of interest are analyzed using MS or other techniques to identify the interactions.

4.3.3   SURFACE PLASMON RESONANCE, DUAL-POLARIZATION
INTERFEROMETRY, AND STATIC LIGHT SCATTERING

Surface plasmon resonance (SPR) based approaches utilize optical properties of transparent optical media interfaces: light coming from the media with the higher refractive index is partly refracted and reflected at the interface. Similar to TIRF microscopy, light is introduced to the specimen at certain angles, resulting in total internal reflection and the propagation of an evanescent wave into the medium of the lower refractive index. For SPR, the media interface is coated with a layer of gold, causing monochromatic polarized light to be reduced in intensity because of resonance energy transfer between the evanescent wave and surface plasmons. As a linear relationship between protein or other biomolecule mass and resonance energy has been found, association and dissociation can be analyzed.[64]

Dual-polarization interferometry (DPI) is a technique where molecular scale layers that are adsorbed to the surface of a waveguide are probed by an evanescent wave. For DPI, laser light is split into two waveguides, one with the exposed surface and another one for reference. Polarization-dependent interference reveals information about surface chemistry and molecular interactions.[65]

Static light scattering (SLS) is another biophysical method that uses changes in Rayleigh scattering (small particle induced scattering of light as commonly observed in gases) induced by protein complexes and allows for detecting interactions without modification of the protein.[66]

4.4 PERSPECTIVES

Extensive combination of complementary strategies, including techniques that allow for analysis of protein–protein interactions at a high time and spatial resolution, will help to gain further insight into the composition of TRP signalplexes, as well as into the dynamics of protein–protein interactions within these complexes and their significance for cellular control of TRP-mediated ion transport. This will provide a clue to understanding the (patho)physiological functions of TRP channels.

ACKNOWLEDGMENT

The authors thank Dr. C. Romanin for helpful comments and discussions.

REFERENCES

1. Venkatachalam, K., and C. Montell. 2007. TRP channels. *Annual Review of Biochemistry* 76: 387–417.
2. Owsianik, G., D. D'Hoedt, T. Voets, and B. Nilius. 2006. Structure-function relationship of the TRP channel superfamily. *Reviews of Physiology, Biochemistry, and Pharmacology* 156: 61–90.
3. Schindl, R., and C. Romanin. 2007. Assembly domains in TRP channels. *Biochemical Society Transactions* 35: 84–85.
4. Lepage, P. K., and G. Boulay. 2007. Molecular determinants of TRP channel assembly. *Biochemistry Society Transactions* 35: 81–83.
5. Zhang, P., Y. Luo, B. Chasan et al. 2009. The multimeric structure of polycystin-2 (TRPP2): structural-functional correlates of homo- and hetero-multimers with TRPC1. *Human Molecular Genetics* 18: 1238–1251.
6. Fujiwara, Y., and D. L. Minor, Jr. 2008. X-ray crystal structure of a TRPM assembly domain reveals an antiparallel four-stranded coiled-coil. *Journal of Molecular Biology* 383: 854–870.

7. Eder, P., and K. Groschner. 2008. TRPC3/6/7: topical aspects of biophysics and pathophysiology. *Channels (Austin)* 2.
8. Eder, P., M. Poteser, and K. Groschner. 2007. TRPC3: a multifunctional, pore-forming signalling molecule. *Handbook of Experimental Pharmacology* 2007: 77–92.
9. Graziani, A., M. Poteser, W. M. Heupel et al. 2009. Cell-cell contact formation governs Ca^{2+} signaling by TRPC4 in the vascular endothelium: evidence for a regulatory TRPC4β-catenin interaction. *Journal of Biological Chemistry* 285: 4213–4223.
10. Nilius, B. 2007. TRP channels in disease. *Biochimica et Biophysica Acta* 1772: 805–812.
11. Xu, X. Z., H. S. Li, W. B. Guggino, and C. Montell. 1997. Coassembly of TRP and TRPL produces a distinct store-operated conductance. *Cell* 89: 1155–1164.
12. Schindl, R., I. Frischauf, H. Kahr et al. 2008. The first ankyrin-like repeat is the minimum indispensable key structure for functional assembly of homo- and heteromeric TRPC4/TRPC5 channels. *Cell Calcium* 43: 260–269.
13. Erler, I., D. M. Al-Ansary, U. Wissenbach, T. F. Wagner, V. Flockerzi, and B. A. Niemeyer. 2006. Trafficking and assembly of the cold-sensitive TRPM8 channel. *Journal of Biological Chemistry* 281: 38396–38404.
14. Schaefer, M. 2005. Homo- and heteromeric assembly of TRP channel subunits. *Pflügers Archiv* 451: 35–42.
15. Smyth, J. T., L. Lemonnier, G. Vazquez, G. S. Bird, and J. W. Putney, Jr. 2006. Dissociation of regulated trafficking of TRPC3 channels to the plasma membrane from their activation by phospholipase C. *Journal of Biological Chemistry* 281: 11712–11720.
16. Graziani, A., C. Rosker, S. D. Kohlwein et al. 2006. Cellular cholesterol controls TRPC3 function: evidence from a novel dominant-negative knockdown strategy. *Biochemistry Journal* 396: 147–155.
17. Bezzerides, V. J., I. S. Ramsey, S. Kotecha, A. Greka, and D. E. Clapham. 2004. Rapid vesicular translocation and insertion of TRP channels. *Nature Cell Biology* 6: 709–720.
18. Fearon, E. R., T. Finkel, M. L. Gillison et al. 1992. Karyoplasmic interaction selection strategy: a general strategy to detect protein–protein interactions in mammalian cells. *Proceedings of the National Academy of Sciences of the United States of America* 89: 7958–7962.
19. Aronheim, A. 2001. Membrane recruitment systems for analysis of protein–protein interactions. *Methods in Molecular Biology* 177: 319–328.
20. Aronheim, A. 2001. Ras signaling pathway for analysis of protein–protein interactions. *Methods in Enzymology* 332: 260–270.
21. Hubsman, M., G. Yudkovsky, and A. Aronheim. 2001. A novel approach for the identification of protein–protein interaction with integral membrane proteins. *Nucleic Acids Research* 29: E18.
22. Mery, L., F. Magnino, K. Schmidt, K. H. Krause, and J. F. Dufour. 2001. Alternative splice variants of hTrp4 differentially interact with the C-terminal portion of the inositol 1,4,5-trisphosphate receptors. *FEBS Letters* 487: 377–383.
23. Sutton, K. A., M. K. Jungnickel, Y. Wang, K. Cullen, S. Lambert, and H. M. Florman. 2004. Enkurin is a novel calmodulin and TRPC channel binding protein in sperm. *Developmental Biology* 274: 426–435.
24. Lussier, M. P., S. Cayouette, P. K. Lepage et al. 2005. MxA, a member of the dynamin superfamily, interacts with the ankyrin-like repeat domain of TRPC. *Journal of Biological Chemistry* 280: 19393–19400.
25. Lussier, M. P., P. K. Lepage, S. M. Bousquet, and G. Boulay. 2008. RNF24, a new TRPC interacting protein, causes the intracellular retention of TRPC. *Cell Calcium* 43: 432–443.
26. Odell, A. F., D. F. Van Helden, and J. L. Scott. 2008. The spectrin cytoskeleton influences the surface expression and activation of human transient receptor potential channel 4 channels. *Journal of Biological Chemistry* 283: 4395–4407.

27. Hannan, M. A., N. Kabbani, C. D. Paspalas, and R. Levenson. 2008. Interaction with dopamine D2 receptor enhances expression of transient receptor potential channel 1 at the cell surface. *Biochimica et Biophysica Acta* 1778: 974–982.

28. Engelke, M., O. Friedrich, P. Budde et al. 2002. Structural domains required for channel function of the mouse transient receptor potential protein homologue TRP1beta. *FEBS Letters* 523: 193–199.

29. Liu, X., B. C. Bandyopadhyay, B. B. Singh, K. Groschner, and I. S. Ambudkar. 2005. Molecular analysis of a store-operated and 2-acetyl-sn-glycerol-sensitive non-selective cation channel. Heteromeric assembly of TRPC1-TRPC3. *Journal of Biological Chemistry* 280: 21600–21606.

30. Lepage, P. K., M. P. Lussier, F. O. McDuff, P. Lavigne, and G. Boulay. 2009. The self-association of two N-terminal interaction domains plays an important role in the tetramerization of TRPC4. *Cell Calcium* 45: 251–259.

31. Park, J. Y., E. M. Hwang, O. Yarishkin et al. 2008. TRPM4b channel suppresses store-operated Ca^{2+} entry by a novel protein–protein interaction with the TRPC3 channel. *Biochemical and Biophysical Research Communications* 368: 677–683.

32. Huber, A., P. Sander, A. Gobert, M. Bahner, R. Hermann, and R. Paulsen. 1996. The transient receptor potential protein (Trp), a putative store-operated Ca^{2+} channel essential for phosphoinositide-mediated photoreception, forms a signaling complex with NorpA, InaC and InaD. *EMBO Journal* 15: 7036–7045.

33. Kenworthy, A. K. 2001. Imaging protein–protein interactions using fluorescence resonance energy transfer microscopy. *Methods* 24: 289–296.

34. Karpova, T., and J. G. McNally. 2006. Detecting protein–protein interactions with CFP-YFP FRET by acceptor photobleaching. *Current Protocols in Cytometry* Chapter 12:Unit 12.7.

35. Pepperkok, R., A. Squire, S. Geley, and P. I. Bastiaens. 1999. Simultaneous detection of multiple green fluorescent proteins in live cells by fluorescence lifetime imaging microscopy. *Current Biology* 9: 269–272.

36. Levitt, J. A., D. R. Matthews, S. M. Ameer-Beg, and K. Suhling. 2009. Fluorescence lifetime and polarization-resolved imaging in cell biology. *Current Opinion in Biotechnology* 20: 28–36.

37. Lleres, D., S. Swift, and A. I. Lamond. 2007. Detecting protein–protein interactions in vivo with FRET using multiphoton fluorescence lifetime imaging microscopy (FLIM). *Current Protocols in Cytometry* Chapter 12: Unit 12.10.

38. Amiri, H., G. Schultz, and M. Schaefer. 2003. FRET-based analysis of TRPC subunit stoichiometry. *Cell Calcium* 33: 463–470.

39. Gervasio, O. L., N. P. Whitehead, E. W. Yeung, W. D. Phillips, and D. G. Allen. 2008. TRPC1 binds to caveolin-3 and is regulated by Src kinase – role in Duchenne muscular dystrophy. *Journal of Cellular Science* 121: 2246–2255.

40. Alicia, S., Z. Angelica, S. Carlos, S. Alfonso, and L. Vaca. 2008. STIM1 converts TRPC1 from a receptor-operated to a store-operated channel: moving TRPC1 in and out of lipid rafts. *Cell Calcium* 44: 479–491.

41. Zhang, F., S. Jin, F. Yi, and P. L. Li. 2008. TRP-ML1 Functions as a lysosomal NAADP-sensitive Ca^{2+} release channel in coronary arterial myocytes. *Journal of Cellular and Molecular Medicine* 13: 3174–3185.

42. Riven, I., E. Kalmanzon, L. Segev, and E. Reuveny. 2003. Conformational rearrangements associated with the gating of the G protein-coupled potassium channel revealed by FRET microscopy. *Neuron* 38: 225–235.

43. Berger, W., H. Prinz, J. Striessnig, H. C. Kang, R. Haugland, and H. Glossmann. 1994. Complex molecular mechanism for dihydropyridine binding to L-type Ca^{2+}-channels as revealed by fluorescence resonance energy transfer. *Biochemistry* 33: 11875–11883.

44. Ivanov, V., M. Li, and K. Mizuuchi. 2009. Impact of emission anisotropy on fluorescence spectroscopy and FRET distance measurements. *Biophysical Journal* 97: 922–929.

45. Zimmermann, T. 2005. Spectral imaging and linear unmixing in light microscopy. *Advances in Biochemical Engineering/Biotechnology* 95: 245–265.
46. Zimmermann, T., J. Rietdorf, and R. Pepperkok. 2003. Spectral imaging and its applications in live cell microscopy. *FEBS Letters* 546: 87–92.
47. Tsurui, H., H. Nishimura, S. Hattori, S. Hirose, K. Okumura, and T. Shirai. 2000. Seven-color fluorescence imaging of tissue samples based on Fourier spectroscopy and singular value decomposition. *Journal of Histochemistry and Cytochemistry* 48: 653–662.
48. Poteser, M., A. Graziani, C. Rosker et al. 2006. TRPC3 and TRPC4 associate to form a redox-sensitive cation channel. Evidence for expression of native TRPC3-TRPC4 heteromeric channels in endothelial cells. *Journal of Biological Chemistry* 281: 13588–13595.
49. Sekar, R. B., E. Kizana, R. R. Smith et al. 2007. Lentiviral vector-mediated expression of GFP or Kir2.1 alters the electrophysiology of neonatal rat ventricular myocytes without inducing cytotoxicity. *American Journal of Physiology. Heart and Circulatory Physiology* 293: H2757–H2770.
50. Bacart, J., C. Corbel, R. Jockers, S. Bach, and C. Couturier. 2008. The BRET technology and its application to screening assays. *Biotechnology Journal* 3: 311–324.
51. Monici, M. 2005. Cell and tissue autofluorescence research and diagnostic applications. *Biotechnology Annual Review* 11: 227–256.
52. Groschner, K., S. Hingel, B. Lintschinger et al. 1998. Trp proteins form store-operated cation channels in human vascular endothelial cells. *FEBS Letters* 437: 101–106.
53. Tsuruda, P. R., D. Julius, and D. L. Minor, Jr. 2006. Coiled coils direct assembly of a cold-activated TRP channel. *Neuron* 51: 201–212.
54. Lintschinger, B., M. Balzer-Geldsetzer, T. Baskaran et al. 2000. Coassembly of Trp1 and Trp3 proteins generates diacylglycerol- and Ca^{2+}-sensitive cation channels. *Journal of Biological Chemistry* 275: 27799–27805.
55. Nicke, A., J. Rettinger, and G. Schmalzing. 2003. Monomeric and dimeric byproducts are the principal functional elements of higher order P2X1 concatamers. *Molecular Pharmacology* 63: 243–252.
56. Isacoff, E. Y., Y. N. Jan, and L. Y. Jan. 1990. Evidence for the formation of heteromultimeric potassium channels in Xenopus oocytes. *Nature* 345: 530–534.
57. Shapiro, M. S., and W. N. Zagotta. 1998. Stoichiometry and arrangement of heteromeric olfactory cyclic nucleotide-gated ion channels. *Proceedings of the National Academy of Sciences of the United States of America* 95: 14546–14551.
58. Greka, A., B. Navarro, E. Oancea, A. Duggan, and D. E. Clapham. 2003. TRPC5 is a regulator of hippocampal neurite length and growth cone morphology. *Nature Neuroscience* 6: 837–845.
59. Strubing, C., G. Krapivinsky, L. Krapivinsky, and D. E. Clapham. 2003. Formation of novel TRPC channels by complex subunit interactions in embryonic brain. *Journal of Biological Chemistry* 278: 39014–39019.
60. Liu, X., B. B. Singh, and I. S. Ambudkar. 2003. TRPC1 is required for functional store-operated Ca^{2+} channels. Role of acidic amino acid residues in the S5-S6 region. *Journal of Biological Chemistry* 278: 11337–11343.
61. MacKinnon, R. 1991. Determination of the subunit stoichiometry of a voltage-activated potassium channel. *Nature* 350: 232–235.
62. Cooper, E., S. Couturier, and M. Ballivet. 1991. Pentameric structure and subunit stoichiometry of a neuronal nicotinic acetylcholine receptor. *Nature* 350: 235–238.
63. Fancy, D. A. 2000. Elucidation of protein–protein interactions using chemical cross-linking or label transfer techniques. *Current Opinion in Chemical Biology* 4: 28–33.
64. van der Merwe, P. A., and A. N. Barclay. 1996. Analysis of cell-adhesion molecule interactions using surface plasmon resonance. *Current Opinion in Immunology* 8: 257–261.

65. Swann, M. J., L. L. Peel, S. Carrington, and N. J. Freeman. 2004. Dual-polarization interferometry: an analytical technique to measure changes in protein structure in real time, to determine the stoichiometry of binding events, and to differentiate between specific and nonspecific interactions. *Analytical Biochemistry* 329: 190–198.
66. Senisterra, G. A., and P. J. Finerty, Jr. 2009. High throughput methods of assessing protein stability and aggregation. *Molecular Biosystems* 5: 217–223.

5 Proteomic Analysis of TRPC Channels

Timothy Lockwich, Anthony Makusky,
Jeffrey A. Kowalak, Sanford P. Markey,
and Indu S. Ambudkar

CONTENTS

5.1 INTRODUCTION

Ca^{2+} is a ubiquitous and fundamental signaling component that is utilized by cells to regulate a diverse range of critical cellular functions. Typically, cells respond to a Ca^{2+} signal that is generated inside the cell in response to activation of a wide variety of cell surface receptors, including those involved in neurotransmitter, hormonal, and sensory signaling. In most cases, the initial Ca^{2+} signal generated in the cell is a specific increase in cytoplasmic $[Ca^{2+}]$ ($[Ca^{2+}]_i$) resulting from release of Ca^{2+} from internal Ca^{2+} stores (mainly the endoplasmic reticulum [ER]) or entry of Ca^{2+} from the external medium across the plasma membrane. Both routes involve movement of Ca^{2+} through Ca^{2+} channels that are localized within these cellular membranes. While intracellular Ca^{2+} release from ER occurs via channels activated by inositol 1,4,5-trisphosphate (IP_3), cyclic ADP-ribose, or Ca^{2+} itself, Ca^{2+} influx across the plasma membrane is achieved via numerous types of Ca^{2+} channels, including voltage-gated Ca^{2+} channels and store-operated Ca^{2+} channels, as well as a variety of ligand-gated cation channels, although the type of channels can vary depending on the cell type.[1–3] Among these, the transient receptor potential (TRP) superfamily of ion channels have been described to be involved in a diverse array of signaling mechanisms that regulate critical sensory functions in cells as well as other processes such as secretion, proliferation, neuronal guidance, cell death, and development.

This article will focus on the TRPC subfamily of Ca^{2+}-permeable channels that are activated in response to stimulation of G-protein-coupled receptors linked to phosphatidylinositol 4,5-bisphosphate (PIP_2) hydrolysis.[4,5] Ca^{2+} entry via TRPC channels regulates key cellular functions in a variety of cell types, including those from skeletal and cardiac muscle, exocrine, endocrine, endothelial, neuronal, epithelial, and smooth muscle cells.[6] Given the importance of these channels in cell functions, it is critical to resolve the mechanism(s) involved in the regulation of the TRPC channel function. TRPC channels can be activated by several different mechanisms following neurotransmitter stimulation of cells. Diacylglycerol generated from PIP_2 hydrolysis serves as a messenger to activate TRPC3, TRPC6, and TRPC7. Ca^{2+} itself has been shown to directly activate TRPC4 and TRPC5. Furthermore, TRPC3, TRPC5, and TRPC6 are dynamically recruited to the plasma membrane following stimulation.[4] TRPC1 and TRPC6 have also been reported to be activated by cell stretching.[5] Another mechanism proposed for TRPC1, TRPC3, and TRPC4 is by internal Ca^{2+} store depletion. In this mechanism, release of Ca^{2+} from intracellular Ca^{2+} stores is sensed by the ER-Ca^{2+} sensor protein STIM1, which translocates to peripheral regions of the cells where it interacts with and activates specific channels. Examples of these channels include CRAC channels (a major component of which is Orai1) and TRPC channels (e.g., TRPC1, TRPC4, and TRPC3). Indeed, STIM1–TRPC1 complex is formed when cells are stimulated.[6] Orai1 has also been detected in this complex.[6] Thus, the assembly of store-operated channels is a highly complicated, spatially/temporally coordinated process that involves several different functionally distinct proteins. It is highly likely that there are other as yet unidentified components of this complex that are critical for the activation and regulation of store-operated Ca^{2+} entry (SOCE).

Much of the initial insights into TRPC protein interactions have been derived from studies with the *Drosophila* TRP channel, which belongs to the TRPC subfamily and was the first TRP channel to be identified.[3] The channel is localized in the *Drosophila* eye where it has a critical role in phototransduction. It was shown that this TRP channel resides in a multiprotein signalplex. Both TRP–TRP interactions and TRP interactions with other proteins in the signaling complex are important for proper channel activity and regulation of phototransduction.[3] The *Drosophila* TRP forms a dynamic complex with scaffolding (e.g., INAD) and signaling proteins (e.g., phospholipase and calmodulin). INAD forms the core of this complex because it has the ability, via multiple PDZ domains, to bind to numerous signaling proteins and serve as a platform for their interaction with TRP and regulation of the channel function. Critical protein sequences, conserved in TRP channel families, appear to be involved in these various specific protein–protein interactions. These include the coiled–coiled domain (TRP–TRP interaction) as well as the ankyrin-like repeat region (TRP-signaling protein interactions), calmodulin- and lipid-binding domains, as well as other less well-characterized protein sequences. Because mammalian TRPC proteins share many of the same structural signatures with the *Drosophila* TRP channel, it was hypothesized that these proteins also share the property of homomeric or heteromeric interactions with other TRPC channels and signaling proteins. It is now well established that a number of proteins interacting with mammalian TRPC channels are crucial for their function and regulation.[7] Qualitative and quantitative differences in the protein components associated with different TRPC channels have not yet been clearly described.

Initial studies for assessing protein–protein interactions involving TRPC channels mainly utilized yeast two-hybrid analysis, GST-fusion protein interactions, co-immunoprecipitations (co-IP), and immunolocalizations. Although the search for a PDZ-domain containing scaffolding protein did not lead to identification of an INAD-like protein, several other scaffolding proteins have been identified that interact with specific TRPC members. These include NHERF (TRPC4), Homer (TRPC1, TRPC3), junctaphilin (TRPC3), and RACK1 (TRPC3).[4,8] Other proteins that have been noted to be associated with a number of TRPC channels include key Ca^{2+} signaling proteins such as plasma membrane Ca^{2+}-ATPases (PMCA), G proteins, phospholipase C (PLC), sarco/endoplasmic reticulum Ca^{2+} ATPases (SERCA), and IP_3 receptors (IP_3Rs). Complicating our understanding of these protein–protein interactions is the detection of these interacting proteins residing in other organelles and structures within the cell, e.g., ER,[9] cytoskeletal scaffolding structures,[3] and subpopulations of mitochondria (peripheral) located near the plasma membrane.[10] Thus, defining a physiological relevance for these interactions has been a challenge. Furthermore, it is also clear that the subcellular environment of the TRP proteins is dynamic. For example, there is an increase in the interaction of TRPC3 with IP_3Rs and G proteins following stimulation of cells. This interaction appears to be required for plasma membrane localization of TRPC3 and its function. The complex also modulates IP_3R function.[11] Homer interactions with TRPC3, TRPC1, as well as IP_3Rs are dynamically regulated by cell stimulation, which impacts on the channel function.[7] Association of TRPC5 with the exocyst complex is involved in its trafficking.[12] Partitioning of TRPC1 into lipid raft domains increases upon store depletion, where it colocalizes and associates with STIM1 following store depletion.[13,14] Further, interaction of TRPC1 with caveolin-1 (Cav1) is required for plasma membrane localization of the channel. However, in response to stimulation, the channel dissociates from Cav1 and binds to STIM1.[15] Clearly, in addition to the basic "resting" complex of proteins associated with TRPC channels, the changes triggered by activation also need to be elucidated.

Attempts to delineate the protein–protein interactions in TRPC channel complexes using such techniques as co-IP and immunolocalization have revealed an extensive list of proteins associated with TRPC channels.[7] However, there are inherent limitations in these methods because these are primarily based on *hypothesis*-driven approaches. Hypothesis-driven approaches require prior insight to identify candidate proteins that are then screened for interactions with the target (TRPC) protein based on other observations, e.g., functional data. Although this might be a useful approach for confirming "proposed" interacting proteins, a *"discovery*-based" method can be expected to yield a much more comprehensive and unbiased survey of interacting proteins. New applications of existing technologies (mass spectrometry and separation sciences combined with high-throughput data processing) have allowed the possibility of discovery-driven approaches designed to provide an unbiased and complete spectrum of binding partners without the need for prior preliminary data. Discovery-driven approaches used before, e.g., yeast two-hybrid and GST-fusion protein interaction analysis, have their own critical drawbacks, i.e., only binary interactions are detected, there is a high false-positive rate, and the high-throughput data management methods, including genome and bioinformatics data, are not readily available.

In this article, we describe how a discovery-based proteomic approach can be employed to identify novel protein–protein interactions associated with TRPC channels. We will also discuss attempts made in our laboratory in this direction including the pitfalls that we encountered. Most of the previously reported studies with TRPC channels report partial or selective analysis of proteins. In particular, many studies have been done with heterologous expression systems that have their own major drawbacks. We will discuss these and also compare them with studies involving endogenous proteins from tissues. The studies we have carried out represent the only high-throughput proteomic analyses of endogenous TRPC family members. Our results confirmed many previously proposed protein interactions with TRPC1 and TRPC3, as well as identified new, previously unreported interactions.[16] Because physiological responses involving these channels require changes in the molecular components associated with them, the method used for proteomic analysis should allow quantitative assessment of the relative levels of proteins in these complexes. In this report, we give an overview of new methodologies that are being applied to quantify the proteins in the signalplex in the resting state and also applications to quantitatively assess changes in the proteome associated with channel activation.

5.2 PROTEOMIC ANALYSIS OF TRPC-ASSOCIATED PROTEIN COMPLEX

5.2.1 BIOCHEMICAL CONSIDERATIONS

5.2.1.1 Proteomic Analysis of Membrane Proteins

Unlike soluble proteins, special biochemical considerations are required for proteomic analysis of integral membrane proteins, owing to the fact that these proteins are hydrophobic and associated with a specific lipid environment. These consider-ations also apply to analysis of ion channel proteins such as TRPC channels, which are primarily associated with plasma membrane Ca^{2+} signaling mechanisms. The main points that need to be taken into account in the experimental strategy used for proteomic analysis of membrane proteins are as follows:

1. Membrane proteins need to be solubilized using detergents in order to release them from the lipid environment. It is imperative that the condi-tions used preserve the functional integrity of the protein. For example, pro-tein interactions, such as those that occur via hydrophobic domains, can be altered by the type of a detergent. Of course, care should be taken to avoid denaturation or aggregation of the protein because this will also adversely affect the interacting proteins. Typically, nonionic detergents such as octyl-glucoside, which has a high critical micelle concentration, are suitable for this procedure. Inclusion of lipids (similar to the endogenous milieu) as well as glycerol (dehydrating reagent) helps to maintain the integrity of protein complexes.

2. The availability of trypsin cleavage sites is usually low owing to the hydro-phobic nature of integral membrane proteins. This is a hindrance for pro-teomic analyses because of the limitations of MS/MS analysis, i.e., large peptides are difficult to fragment using conventional collision-induced dis-sociation (CID). Peptides with an upper limit of 25–30 amino acid residues in length typically yield high-quality MS/MS spectra, although exceptions to this length guideline do indeed exist. Additionally, the ability to extract large peptide fragments from the polyacylamide gel (in-gel digests) is lim-ited.[17] Together, this results in a low yield of peptides that has a signifi-cant impact on the detection of proteins present in low abundance, such as plasma membrane ion channels.

3. Ideally, an in-solution digest of proteins or protein complexes would yield the most thorough results in a shotgun proteomic analysis, as losses asso-ciated with sample handling and poor recovery from the gel matrix are minimized. However, many detergents, while efficient in solubilizing the target protein, are unsuitable for in-solution digest owing to incompatibility issues in the downstream HPLC and electrospray ionization-based mass spectroscopy.

4. Integral membrane proteins of the plasma membrane are usually present in low abundance. This drawback requires alternate enrichment strategies[18] to achieve a sufficient material to analyze. A consequence of this scaling up, unfortunately, can also result in a proportional increase in nonspe-cifically bound proteins. An alternative approach would be to carry out an initial enrichment step prior to purification, e.g., one can start with a purified plasma membrane preparation rather than whole tissue or crude membrane fractions. However, availability of tissue could be a serious limi-tation to this. Despite such preliminary enrichment steps, control experi-ments are required to sort out putative specific and nonspecific interactions with the target protein. Examples of appropriate control experiments for

immunoaffinity-based isolations would be preclearing the lysate using buffer conditioned solid support (e.g., Protein A derivatized magnetic beads) and/or preclearing using an immobilized nonspecific immunoglobulin to minimize the detection of nonspecific interactions. Proteins identified as potential interactors in discovery-based assays still require further assessment using orthogonal biochemical and functional assays (see below).

The following sections will discuss methodologies that have been applied to successfully perform shotgun proteomic analysis on integral membrane proteins and that have been used for TRPC channels. Additionally, new procedures that can enhance proteomic discoveries will be introduced (e.g., new gel-free platforms using organic acid-cleavable detergents that have been developed to solubilize membranes[18]). These detergents perform their traditional role up to the HPLC step but can be removed by acid-induced cleavage to allow in-solution digest and analysis of the extracted target protein.

5.2.1.2   Use of Immunoprecipitation: Limitations and Advantages

Once solubilized from its membrane environment, the target protein needs to be isolated in sufficient quantities prior to proteomic analysis. One approach used for this is affinity purification. Although it is well established that with proper optimization detergents do not interfere with antibody-antigen interactions,[19] the affinity and specificity that a particular antibody has toward its antigen vary considerably. Another major drawback is the commercial availability of reliable antibodies. Most commercially available antibodies are developed for use as diagnostic reagents, e.g., for Western blotting or immunohistochemical staining. These antibodies must be evaluated for their properties as preparative affinity reagents. Another critical concern is that the immunoprecipitation (IP) reagents (e.g., beads and the Fc portion of the antibody) provide numerous sites for nonspecific binding to occur. Thus, it is important to design the IP protocol to minimize these problems, as well as to control and account for them. Further, because the concentration of the antibody is relatively high compared to the protein sample to be tested, one can predict a high level of contamination from IgG peptides. One modification that has been successfully applied is covalent conjugation of the antibody to the matrix. However, one must demonstrate that the antibody–matrix cross-linking has not impaired the ability of the antibody to capture and release the appropriate bait protein. Thus, it is possible to selectively resolve biologically relevant protein–protein interactions from experimentally introduced contaminants.[19]

If an ideal antibody targeted toward the native protein is not available, a heterologously expressed epitope-tagged protein can be used. In this case, the antibody directed against the tag is used for affinity purification of the protein complex. However, if the tagged bait protein expression level is too high, the assembly of the protein complex may not occur in a physiological/stoichiometric manner.[20] Overexpression systems also lead to "backup" of membrane trafficking pathways, resulting in a large amount of protein in other intracellular compartments, including ER and Golgi, or in extreme cases trapped in inclusion bodies. In the latter case, the protein is often resistant to solubilization. Isolation of proteins by IP would include proteins associated with the target localized in ER and Golgi, and these are generally chaperones, heat shock, and other trafficking-related proteins that might not be relevant to the mature functional protein that ultimately assembles in the plasma membrane.

Finally, another important consideration while implementing IP for isolation of a membrane protein complex is the stringency of the wash. This wash is performed after immobilization of the protein on the beads, prior to elution. The purpose of this procedure is to wash off no-specifically/loosely attached proteins so that a reproducible core group of proteins that are tightly associated with the target protein can be retained. Specificity of this protein complex can be determined by comparing samples obtained using control IgG with those obtained using antibody targeted to the target protein. Cells in which the target protein has been knocked down are also a very useful control, although this would largely be determined by the efficiency of the knockdown. In all these cases, the wash conditions critically determine the outcome. If the wash is too mild (i.e., lacking in ionic strength or detergent), too many nonspecific proteins will remain associated with the target protein. Conversely, if the wash buffer is too stringent (e.g., a RIPA buffer with a high SDS concentration), it can result in disruption of transient and/or low affinity (but nonetheless important) interactions between the target protein and putative interacting proteins.[19] Thus, a considerable effort is required to ascertain that the IP and wash conditions are yielding a consistent and reproducible array of proteins. Polyacrylamide gel electrophoresis (PAGE) and silver staining can be used to examine the panel of proteins in the sample at every step.

5.2.1.3 Biochemical Confirmation of Proteomic Data

The experimental approaches discussed above are aimed toward controlling critical variables that can impact immunopurification of membrane proteins for proteomic analysis. Despite these efforts, it is important to remember that the proteins detected in the final sample after proteomic analysis can only be defined as "putative interacting" proteins until they are confirmed using biochemical, molecular, or functional techniques. This confirmation is especially important when identification of a protein is based on a single peptide. Although, in principle, a single unique peptide is sufficient to identify a protein,[21] the reliability of the identification increases exponentially with the number of peptides used to identify the protein. Some of the more commonly used techniques to confirm the identifications detected by MS/MS include Western blotting, direct yeast two-hybrid complement and interactions, immunolocalization/colocalization, cross-linking, and multiepitope-ligand cartography (for a review, see Ref. 19). The putative interacting protein can also be modulated in cells (overexpression or knockdown), and the effect on the target protein function can be assessed. Together, these methods provide a comprehensive assessment of the proteins that associate with the target protein.

5.3 OVERVIEW OF TRPC PROTEOMIC ANALYSIS

5.3.1 Previous Proteomic Studies of TRPC Proteins

Despite the extensive amount of work on deducing the protein–protein interactions involving TRPC proteins using traditional methods such as yeast two-hybrid analysis and co-IP, there is very little published work regarding proteomic analysis of TRPC channels. In one of the few publications, TRPC5 and TRPC6 immunocomplexes were analyzed by spot-picking after SDS-PAGE.[22] After sequencing using MS/MS and subsequent identification, a partial list of interacting proteins was obtained from this approach, including several cytoskeletal proteins such as spectrin, myosin, actin, drebrin, tubulin, and neurabin. Additionally, endocytic vesicle-associated proteins such as clathrin, adapter-related protein complex AP-2, and a dynamin-1-like protein were identified, as well as the plasmalemmal Na^+/K^+-ATPase. Although several important observations were achieved in this study, two important shortcomings were evident. (1) The analysis was not based on a comprehensive approach since only select bands were analyzed. (2) More importantly, no control experiments were done to eliminate nonspecific binding. In many cases, quantitative comparisons between a common control and TRPC-IP peptide are needed to distinguish between specific and nonspecific interactions.

Another study used proteomic techniques to map autophosphorylation sites on TRPM6 and TRPM7.[23] This paper, while somewhat unrelated to the determination of protein–protein interactions in TRPC channels, nevertheless shows how proteomic techniques can be applied to better understand protein modifications (e.g., posttranslation modifications) that might affect function.

5.3.2   CHOICE OF PROTEOMIC TECHNIQUES IN THE ANALYSIS OF TRPC PROTEINS

We will discuss two principal methodologies that can be utilized for initial discovery-based proteomic analysis. The first, geLC–MS/MS, utilizes one-dimensional (1D) SDS-PAGE to separate the protein components of the TRPC channel complex isolated by affinity purification (Figure 5.1a). This 1D PAGE is followed by sectioning (~1 mm each, see Figure 5.1b) the entire length of each gel lane as opposed to selected bands. Because the whole lane is sampled, visualization by protein staining is optional, although routinely employed as documentation. Individual gel bands are subjected to in-gel enzymatic proteolysis and subsequent extraction of the peptide hydrolysate, followed by tandem mass spectral analysis of the peptides and submission of the data to a database search engine in order to determine the identity of the peptides and infer the identity of their precursor proteins. Some of the benefits of the geLC–MS/MS approach are the facile ability of working with detergent-containing samples and the significant resolving power (theoretical plates of separation) of SDS-PAGE. Importantly, all peptides produced from enzymatic digestion of proteins in a given gel slice remain in a single fraction, unless the underlying precursor protein spans a slice junction. This fact substantially simplifies inference of the correct precursor protein during subsequent bioinformatic analysis of the data because the approximate molecular weight of the precursor can be determined from molecular weight standards. Limitations of the geLC–MS/MS approach are principally related to the partial and variable recovery of peptides in the extraction step.

The second approach, termed two-dimensional liquid chromatography–mass spectrometry (LC/LC–MS/MS), begins with solution phase digestion of the affinity purified mixture of proteins in aggregate, followed by peptide fractionation using, for

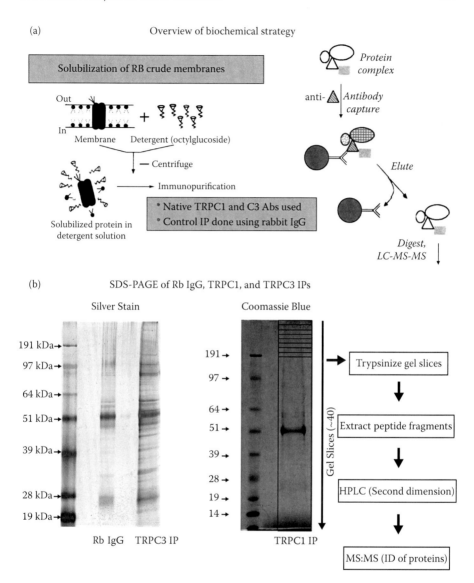

FIGURE 5.1 Overview of the biochemical strategy of isolating "specific TRPC1 or TRPC3 binding partners" from the rat brain. (a) After isolating rat brain crude membranes and solubilizing the membranes with octylglucoside, immunoprecipitates were obtained using native TRPC1 or TRPC3 antibodies. (b) The concentrated immunopurified fractions (with appropriate control samples) were then separated using SDS-PAGE. Subsequent gel slices (~40 slices) were treated with trypsin, and the resultant peptide fragments were extracted from the gel, separated using HPLC, and sequenced using MS/MS.

example, strong cation exchange (SCX) and reversed-phase (RP) liquid chromatographies coupled with tandem mass spectral analysis. Alternate chemistries can and have been employed as variations on the multidimensional LC theme. Two-dimensional LC can be performed while directly coupled to the mass spectrometer,[24,25] although there are advantages to decoupling the chromatographic steps. Performing offline SCX allows the practitioner to have the ability to increase sample loading capacity, employ organic modifiers in the SCX solvents, and take advantage of increased chromatographic resolution by employing a linear elution gradient as compared to step elution mandated by online 2D LC–MS/MS.

5.3.3 INSTRUMENTATION AND DETAILS OF MS/MS ANALYSIS

Electrospray ionization (ESI) is an ideal interface to couple LC systems directly online to modern tandem mass spectrometers, e.g., ion traps, ion trap-Orbitrap, and triple quadrupoles (QqQ). As the name implies, nanoflow LC utilizes submicroliter-per-minute flow rates; 200–400 nL min^{-1} is typical. At the time the TRPC experiments were performed, our laboratory utilized an automated 1D HPLC system that was constructed from Shimadzu LC-VP Series components (Kyoto, Japan), which allowed the use of both an autosampler and an HPLC at submicroliter-per-minute flow rates. For the uninitiated, it is difficult to accommodate the system volume of an autosampler with nanoflow rates used with 300-μm ID HPLC columns. The optimal flow rate for autosampler use greatly exceeds that required for microbore HPLC. Conversely, at flow rates optimal for microbore HPLC, autosampler loading times would be prohibitively long. Our system utilized three LC-10ADVP pumps with microflow control kits, an SIL-10ADVP automatic injector, an SCL-10AVP Controller, and a Cheminert CN2 nanovolume switching valve (Valco Instruments Company Inc., Houston, TX, USA).

One of the pumps (Pump C) delivered Buffer A at 40 μL min^{-1} to the autosampler. Samples were selected and loaded onto a peptide CapTrap (0.5 mm × 2 mm, PLRP-S 5μ 100Å; Michrom BioResource Inc., Auburn, CA, USA), which was connected to the nanovolume switching valve. This valve was used as a solvent selector for trapping, desalting, and loading samples onto the reverse-phase column. The other two LC-10ADVP pumps (A and B) were operated in conventional mode developing a linear gradient (10% B [3 min], a linear gradient of 10–60% B [40 min], 60–80% B [10 min], 80% B [2 min]) at 10 μL min^{-1}. The LC eluant was connected to a micro splitter valve (Upchurch Scientific Inc., Oak Harbor, WA, USA) to deliver an operating flow rate of 400 nL min^{-1} to the switching valve. After extensive washing with Buffer A, the LC gradient was brought in-line, and the peptide samples were eluted sequentially from the Peptide CapTrap onto a PicoTip fused silica HPLC column (BetaBasic C18, 0.075 × 100 mm, 360-μm OD 15-μm spray tip) and into a ThermoFinnigan LCQ Classic ESI-ion trap mass spectrometer (San Jose, CA, USA).

Mobile phase buffers were RP-A, water/acetonitrile/formic acid = 95.0/5/0.1 (v/v/v); RP-B, water/acetonitrile/formic acid = 20/80/0.1 (v/v/v); and RP-C, water/formic acid = 100/0.1 (v/v). The LCQ was operated in positive ion mode with a dynamic exclusion set to repeat count = 1, exclusion duration = 0.5 min, exclusion mass width = 3 amu. Spectra were acquired in a data-dependent manner with the top five most intense ions in the precursor scan selected for CID. Normalized collision energy was 35.0, and the minimum precursor ion intensity required was 2× baseline at equilibrium.

For quantitative measurements, our laboratory utilizes a true, splitless, nanoflow HPLC system (Eksigent Technologies, Inc., Dublin, CA, USA). The HPLC columns, buffers, and gradient are the same as described above. In order to insure ion beam stability under all LC conditions, we use an Advion Nanomate (Ithaca, NY, USA) ESI ion source that is mounted on an LTQ-Orbitrap mass spectrometer (ThermoFisher Scientific, San Jose, CA, USA). This instrument is operated in data-dependent mode. Survey MS scans are acquired in the Orbitrap in profile mode with the resolution set to a value of 60,000. Up to five of the most intense ions per scan are fragmented and analyzed in the linear ion trap.

5.3.4 SOFTWARE REQUIREMENTS FOR MS/MS PROTEOMIC ANALYSIS

The Center for Information Technology (CIT) at the National Institutes of Health maintains a computer cluster hosting the Mascot Search Engine (Matrix Science, Inc., Boston, MA, USA). The NIH Mascot cluster consists of four nodes: one head node and three computational nodes. Each node is a dual-core dual-socket (four cores total) 2.6-GHz Opteron with 8 GB of RAM, running 64-bit Centos 5.4. The nodes are connected by 1-Gb/s ethernet to each other and to the fileserver on which the data reside. File storage is on a Netapp FAS 960 file server. CIT performs regularly scheduled updates of the genomic sequence libraries used in MS-based protein identification. Individual users remotely connect to the cluster using the Mascot Daemon. In addition to defining the configuration of search parameters, the Daemon automates conversion of the raw MS/MS data into a concatenated list of mass versus intensity pairs for each data file (.mgf), transmission of the .mgf files to the search engine, and management of the search result files.

The following parameters were used to search the TRPC data described herein: database = SwissProt (derived from UniProt); taxonomy = All; enzyme = trypsin; maximum number of missed cleavages = 2; peptide charge states = +1 - +4; mass measurement = monoisotopic; low resolution ms: peptide mass tolerance = 1.2 Da; fragment ion mass tolerance = 0.6 Da; high resolution ms: peptide mass tolerance = 50 ppm, fragment ion mass tolerance = 0.6 Da; fixed modification = carbamidomethyl-Cysteine; variable modification = methionine oxidation (Met). Subsequent searches were performed to detect common posttranslational modifications (e.g., protein N-acetylation) and chemical artifacts (e.g., pyro-Glutamic acid formation and pyro-carbamidomethyl-Cysteine from N-terminal Gln and camCys, respectively). In these searches, taxonomy was restricted to *Rattus*. While a detailed discussion of the scoring algorithm used by the Mascot search engine is beyond the scope of this paper, it is important to point out that because the number of rat proteins is very small compared to the total number of protein sequences in the SwissProt database, the rate of false-positive identifications will be greatly increased. Practitioners are well advised to check the more general mammalian taxonomy search results to compare with *Rattus* taxonomy and to manually inspect new identifications made as a result of subsequent searches on a subset of total database records.

5.4   VALIDATION OF PROTEOMIC DATA

5.4.1   Experimental Design to Compensate for Nonspecific Interactions

To illustrate didactically how various experimental parameters affect a membrane protein proteomic study, a brief outline of our initial work on the analyses of TRPC1 and TRPC3 proteomes follows. In our initial attempts to increase the copy number of TRPCs in our system and to simplify the extraction step, we transfected HEK293 cells with FLAG-tagged TRPC1 or TRPC3 and used untransfected HEK293 cells as a control. After obtaining a crude membrane fraction, octylglucoside solubilizates were immunoprecipitated with anti-FLAG-linked agarose beads and the IP fraction (of FLAG-TRPC3 eluted with free FLAG peptide) tested for Ca^{2+} permeability in proteoliposomes. After confirming an active FLAG-TRPC3 complex,[16] we proceeded to scale up the IP, and to maximize the chances of retaining as many proteins as possible, we used a low stringency IP wash buffer (TBS which consisted of 50 mM Tris HCl, 150 mM NaCl, pH 7.4) before eluting the IP fraction. Subsequent MS/MS analysis of the IP fractions of both FLAG-TRPC1 and FLAG-TRPC3 revealed an extremely large set of proteins for both (>5000). To obtain a more manageable set of proteins, we repeated the experiment, but instead of washing with mild TBS, we washed the IP fraction with a more stringent IP wash buffer that contained (among other things) 0.5 M NaCl and 0.5% NP-40 (for a complete formulation, see Ref. 16). MS/MS of these IP fractions resulted in a more manageable number of proteins (944 for TRPC1 and 256 for TRPC3). However, it became evident that a sizeable number of the identified proteins were a result of overexpression (Table 5.1). A relatively high percentage of proteins associated with protein synthesis and trafficking, including those in the ER and Golgi, were detected. As discussed above, this is a common problem with overexpression systems because the immunopurification pulls down all proteins, including those in intracellular compartments.[20] Fully functional TRPC

TABLE 5.1
Proteomic Results Using Overexpression of FLAG-Tagged TRPCs

	FLAG-TRPC1	FLAG-TRPC3
Unfiltered	944	256
Functional breakdown (filtered)	460	102
Signaling/metabolic	185	56
Mitochondrial	75	12
Structural	31	6
Overexpression	61	10
DNA/RNA functions	75	12
Unknown	33	6
Total	460	102
% Overexpression (FLAG)	13.3%	9.8%
% Overexpression (native)	5.3%	2.6%

channels are normally present in a complex where each component is assembled in a specific stoichiometric ratio with available natural binding partners. Overabundance of newly synthesized or partially assembled TRPC proteins will result in exposure to chaperone proteins (for eventual destruction) or their congregation in sorting vesicles waiting for natural binding partners to become available. In either case, these other protein interactions will be higher relative to the interactions of the mature protein within its functional milieu, resulting in difficulty in detecting the latter interactions. On the basis of these observations and considerations, we next pursued a more realistic proteomic assessment using native TRPC channels obtained from natural tissue sources. In native systems, TRPC channels have a low turnover. Thus, we can expect most of the protein to be in its mature state in the plasma membrane.

We used rat brain for our initial studies with native TRPC channels because this is a well-characterized tissue source that has been previously shown to substantially express TRPC1 and TRPC3.[26] The procedure we utilized was to homogenize quick-frozen rat brains followed by differential centrifugation to obtain a crude membrane fraction. The membrane fraction was solubilized using a buffer containing octyl-glucoside/lipid/glycerol mixture, and the solubilized fraction was used to immuno-precipitate either endogenous TRPC1 or TRPC3 by their respective antibodies. To account for nonspecific interactions, we precleared the solubilizates with Protein A Sepharose CL-4B beads and also performed a control IP using rabbit IgG (because the native anti-TRPC antibodies are derived from rabbit). The flowchart of our final experimental design is shown in Figures 5.1a and 5.1b; this figure depicts a multistep approach that included IP, SDS-PAGE followed by in-gel digest, extraction, separation, and analysis by MS/MS. This approach required considerable scale-up of the protein preparation because of the lower amount of TRPC proteins in a native system. We determined that a 10-fold higher amount of crude membrane protein from brain (as compared to membranes from cells expressing FLAG-tagged protein) was needed to obtain an IP sample that was concentrated enough for detection of proteins after in-gel digest and extraction. Despite these additional efforts that were required, the approach was successful, and significantly more functionally relevant proteins were identified in the TRPC sample with a lower percentage of irrelevant interactions (see Table 5.2 and compare with Table 5.1).

TABLE 5.2
Functional Overlap of the TRPC1 and TRPC3 Proteomes

Functional Overlap	TRPC1	TRPC3	Both	Both (%)
Chaperone	4	2	0	0
Endoctosis	1	1	5	71
Metabolic	17	8	9	26
Neural growth	3	5	2	20
Signaling	3	13	4	20
Structural	4	2	5	45
Transport	3	3	8	57
Vesicle fusion	4	6	3	23

5.4.2 Criteria for the Determination of Valid Identification

5.4.2.1 Importance of Multipeptide Identification

Our evaluation strategy to assign a candidate protein as a specific TPPC-interacting protein is based on matching sequenced peptide fragments from the MS/MS analysis to known proteins via database searching. This process, termed protein inference, has been reviewed extensively, and a generalized list of guidelines to help eliminate false positives has evolved (for a review, see Refs. 27 and 28). In our analysis, several layers of filtering mechanisms were used. For example, known contaminating proteins (e.g., IgG's and keratins) were initially filtered out because these proteins are obvious artifacts. After initial filtering of known contaminants, candidate proteins then had to meet the following criteria to be putatively identified as a relevant protein:

1. The candidate protein has two or more sibling peptides (multihit), and it is present only in the TRPC-IPs.
2. Multihit proteins appearing in both the TRPC-IPs and control IgG-IP (which also had one or more shared peptides that could be used for quantitative assessment) must demonstrate at least a twofold quantitative increase between TRPC-IP and control-IP.
3. Proteins identified by a single peptide could be assigned as a TRPC-interacting protein only if additional verification (e.g., Western blotting or previous functional studies) could confirm the relevance or presence of the protein in TRPC function or regulation.

In all cases, the existence of at least one unique (distinct) peptide (that could only be present in the candidate's protein sequence) was required. To analyze a family of proteins, cluster analysis of shared peptides was performed using MassSieve[29] in order to determine if unique sequences were present in individual family members. Only family members with at least one unique sequence were included into any of the "specific TRPC interacting protein" category (the selection strategy is detailed in Figure 5.2).

5.4.2.2 Quantitative Analysis of Shared Peptides

If a protein was identified in both the control and TRPC runs and a common peptide sequence was used to identify both, quantitative data could be used to differentiate between specific and nonspecific binding. The stability of the LC–MS/MS ion current makes it useful to retrieve label-free quantitative information such as retention time, peak intensity, and integrated area from the MS1 data, which affords the ability to compare the same putative precursor protein between experiments. For quantification, there are multiple software tools with the capability to integrate MS1 data corresponding to identified peptides from MS2 scans. We used an in-house software tool, DBParser, to extract retention time and peak intensity from MS1 raw data based on the precursor mass identified by Mascot. The peak area and the number of scans were calculated from the selected ion chromatogram. Mass tolerance for extraction of the ion current is user-selected and instrument-dependent. In the case of the LCQ classic employed in these studies, a mass tolerance of ±0.6 Da was used. In

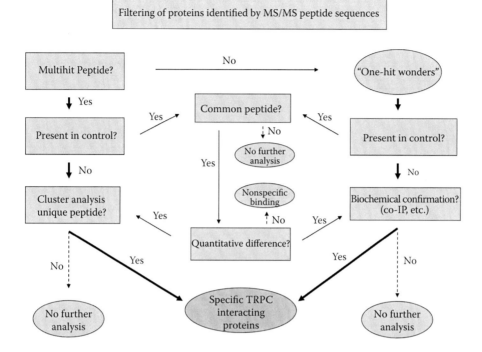

FIGURE 5.2 Flowchart of considerations used to assign a candidate protein a "specific TRPC1 or TRPC3 binding partner." After filtering of obvious artifacts such as keratins and immune proteins, "Multiple Peptide Hits" proteins were distinguished from "One Peptide Identification" proteins by DBParser reports, which assigned peptides to individual proteins. Common peptides (needed for quantitative analysis) were also derived from the DBParser reports. To determine the uniqueness of the peptides assigned to the identified proteins, cluster analysis using Mass Sieve was performed. Only proteins with unique peptides were considered for designation as a "specific TRPC1 or TRPC3 binding partner."

addition to peptide sequence, m/z, Ions Score, Homology Score, and Identity Score from Mascot search, quantification reports include retention time, peak intensity, area, and number of scans. Quantitative differences of the same protein identified in both the control and TRPC samples were analyzed by comparing the sum intensity ratio (TRPC/control) of a unique peptide present in both runs (when possible). If a quantitative ratio of more than twofold resulted, we interpreted this as a quantitative difference and not resulting from a mere nonspecific binding event. When possible, Western blot analysis of equal amounts of control-IP and TRPC3-IP fractions was analyzed to confirm quantitative differences.

5.4.3 TRPC3 PROTEOME

In our initial attempt to elucidate the functional components and regulation of TRPC-associated Ca^{2+} influx, we utilized a solubilization–reconstitution approach that we had previously used to assess Ca^{2+} permeability pathways (as gauged by $^{45}Ca^{2+}$ uptake assays) in plasma membranes from salivary glands.[30] FLAG-TRPC3 immunocomplexes were released using a free FLAG peptide and subsequently reconstituted into proteoliposomes and function gauged by $^{45}Ca^{2+}$ uptake assays. Immunopurified FLAG-TRPC3 displayed 1-oleoyl-2-acetyl-sn-glycerol (OAG) -stimulated Ca^{2+} influx following reconstitution in a proteoliposomal system. Western blots verified the presence of FLAG-TRPC3 in the reconstituted vesicles. Proteoliposomes containing FLAG-TRPC3 displayed an approximately twofold higher $^{45}Ca^{2+}$ uptake compared to proteoliposomes prepared from immunoprecipitates obtained from control nontransfected HEK293 cells.[16] Further, OAG, but not the inactive analog 1,3-dioctanoyl-sn-glycerol, increased $^{45}Ca^{2+}$ uptake in the FLAG-TRPC3 proteoliposomes. Interestingly, a TRPC3 mutant with mutation in its pore region displayed reduced $^{45}Ca^{2+}$ uptake into proteoliposomes. These data demonstrate that our solubilization and IP conditions yield a functional TRPC3 protein. Despite the preservation of function, proteomic studies of FLAG-TRPC3 revealed the presence of a considerable number of artificially induced interactions in the immuno-purified FLAG-TRPC3 complex (see table above). Thus, a native anti-TRPC3 antibody was used to immunoprecipitate TRPC3 from solubilized rat brain crude membranes using the same conditions we have previously used that allowed retention of function. Proteins in the TRPC3 and control (using rabbit IgG) immunoprecipitates were analyzed, and analysis of the sequence data revealed the presence of 76 specific TRPC3-associated proteins (as determined by the selection criteria described above). Figure 5.3 shows the functional distribution of the proteins contained in the TRPC3 proteome.

5.4.4   TRPC1 PROTEOME

As described above, we have also used a native anti-TRPC1 antibody (from rabbit) to immunoprecipitate TRPC1 from solubilized rat brain crude membranes using the same experimental approach that preserved the TRPC3 function. Proteins in the TRPC1 and control (using rabbit IgG) immunoprecipitates were separated by SDS-PAGE and subsequent gel slices trypsinized as outlined above. After the extraction of the peptides, the peptide fragments were separated and sequenced using HPLC and MS/MS techniques, respectively. Analysis of the sequence data revealed the presence of 59 specific TRPC1-associated proteins that can be initially included into the TRPC1 proteome, based on multiple peptide identification (and not present in the control). Additionally, eight proteins present in both the TRPC1 and control immunoprecipitates were determined to have quantitative differences between the TRPC1-IP fraction and the control (IgG) nonspecific binding fraction. After biological confirmation, eight additional proteins initially identified by one peptide in the TRPC1-IP (but not in the control IP) are also included in the TRPC1 proteome (making a total of 75 validated proteins associated with TRPC1). Figure 5.3 shows the functional distribution of the proteins contained in the TRPC1 proteome.

5.4.5   FUNCTIONAL OVERLAP OF THE TRPC1 AND TRPC3 PROTEOMES

Interestingly, we observed many of the same proteins in both the TRPC1 and TRPC3 proteomes (approximately 50%). Moreover, when we analyzed the functional

(a) TRPC1 functional pie chart

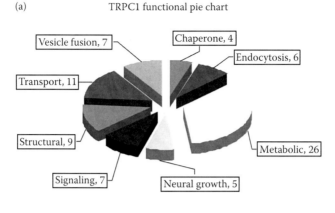

(b) TRPC3 functional pie chart

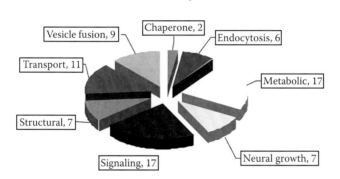

FIGURE 5.3 Functional classifications of the (a) TRPC1 and (b) TRPC3 proteomes constructed from proteins deemed to be specific TRPC1 or TRPC3 binding partners as described in Figure 5.2. (From Lockwich, T., J. et al., *Journal of Proteome Research* 7(3), 979–989, 2008 Mar, Epub 2008 Jan 19. PMID: 18205297 [PubMed–indexed for MEDLINE]. ACS. With permission.)

similarities among common proteins, a very asymmetrical distribution was evident (Table 5.2). The functional comparison between the two reveals a greater overlap in certain functional groups, e.g., protein involved in endocytosis. Hybrid TRPC1/TRPC3 channels have been proposed to exist,[8] and, consistent with this observation, TRPC3 co-immunoprecipitates with TRPC1-IP from rat brain extracts (unpublished observations). It is tempting therefore to speculate that, in certain functional areas, TRPC1 and TRPC3 coexist as heteromultimers and thus share the same protein partners. In contrast, the functional areas where little overlap in binding partners is observed could indicate independent roles for TRPC1 and TRPC3. Further studies will be needed to elucidate the complexes where TRPC1 and TRPC3 interact independently as opposed to complexes where both are present. To guide us in this attempt, one approach is to examine the distribution of the channels in thin sections of the rat brain. Mapping these regions together with localization of specific proteins might provide clues as to the relevance of the findings obtained from TRPC1 and TRPC3 proteomic analysis.

5.5 QUANTITATIVE PROTEOMIC TECHNIQUES

5.5.1 SILAC (Stable Isotope Labeling by Amino Acids in Cell Culture)

The plasma membrane is a dynamic system, and the composition of membrane proteins can dramatically change after contact with other cells, signal molecules, or simply the maturation of a cell. Some proteins are only temporarily exposed to the cell surface, but, nevertheless, these transient changes are important to understanding how signals are relayed to the interior of the cell to elicit responses. SILAC is a technique that exploits the fact that mammalian cells cannot synthesize a number of amino acids.[31] Isotopically labeled analogs of these amino acids can be synthesized and are commercially available. Either naturally occurring or "heavy" amino acids (added to amino acid deficient cell culture media) can then be incorporated into all proteins as they are synthesized. Because there is no chemical difference between the light and heavy amino acids, two groups of cells can be grown that should behave exactly alike. To ensure complete incorporation of the heavy amino acid into the experimental cell culture group, at least six passages of cell culture are allowed to proceed before testing for complete incorporation. The experimental cell population can then be treated in a specific way, such as activation by thapsigargin or specific agonists. Equal amounts of protein from both sets of cell culture conditions can then be mixed and fractionated, based on binding to a target protein (in this case, IP). Changes in the levels of binding partners (between the control and treated cell cultures) can be quantitatively analyzed at the level of the peptide mass spectrum or peptide fragment mass spectrum. This type of analysis can determine if the amount of a binding partner decreases, remains the same, or increases as a result of the specific treatment.[32] Such an approach can be used for detecting dynamic changes in the TRPC complex. For example, control cells can be labeled with naturally occurring lysine and arginine, while stimulated (or otherwise treated) cells can be labeled with heavy amino acids $^{13}C_6$-lysine and $^{13}C_6,^{15}N_4$-arginine. Solubilization and immunopurification of the protein of interest followed by MS analysis should reveal important quantitative changes on the interacting partners. Such studies have not yet been carried out with TRPC channels and should provide invaluable data as to the regulatory components of these channels.

5.5.2 Quick LC–MS (Quantitative IP Combined with Knockout)

One of the most difficult aspects of proteomic analysis when methods such as IP are used is the presence of nonspecifically bound proteins. Despite experimental designs to minimize these interactions (e.g., preclearing of the extract and use of control antibodies), interference invariably results from proteins that nonspecifically bind to any of the components used to isolate the target protein. The prevailing method to eliminate nonspecific interactions is to compare proteins identified in a specific extraction with proteins extracted from a lysate that does not contain the target protein. Clearly, this is not possible in cases that target ubiquitously expressed proteins. Therefore, our previous methods to correct for nonspecific binding can only be partially effective. Recently, a new strategy has been developed to better discriminate specifically and nonspecifically bound members of an extracted protein complex. This new strategy, named QUICK,[33] combines SILAC with RNA interference (RNAi) to knock down target protein expression in one of the cell cultures. After IP of equal protein samples and subsequent analysis by LC–MS/MS, doublets should be observed, which appear either in equal amounts (nonspecific) or in much greater amount in the non-RNAi treated culture (indicative of a specific interaction). One obvious drawback with QUICK is that this technique cannot be used with native tissue because it requires cell culture. These and other limitations due to the nonspecific interactions need to be resolved in future studies.

5.6 CONCLUSION

The recent application of high-throughput informational processing combined with more established techniques such as HPLC and mass spectrometry has allowed the emergence of proteomics, a powerful new tool in determining protein–protein interactions. Previously undreamed of capabilities that allow the complete identification of a complex mixture of proteins are now possible. With this new capability come important limitations, however. The constant reassessment of data to confirm validity and specificity is a necessary caveat that must be respected to allow the field of proteomics to evolve into a mainstream biochemical technique. Through our experiences analyzing the proteomes of several components of the Ca^{2+} entry pathway, we have identified several important areas that have drastic effects on the quality of the results. The first critical area is experimental design. Although the use of epitope-tagged proteins has revolutionized the ease of amplifying and extracting a target protein, it does promote interactions that are not physiologically relevant. In our estimation, it is advisable to scale up a natural system rather than to attempt a proteomic analysis on an overexpressed epitope-tagged system. Another important consideration is prefractionation. As discussed earlier, membrane proteins are usually present in low copy numbers, and therefore, every attempt should be made to concentrate the target protein before the actual proteomic analysis. For example, in our studies on the TRPC1 and TRPC3 proteomes, a crude membrane fraction from the rat brain homogenate was secured before proceeding to solubilization. Other variations of this fractionation strategy (i.e., isolating lipid raft domains before analysis) have allowed others a suitable starting point for proteomic analyses of membrane proteins.[34]

For shotgun proteomic analyses involving immunoprecipitation to isolate the target protein and associated proteins, important consideration should be given to the stringency of the washing buffer used prior to elution. As reported earlier, a wash buffer too low in ionic strength and detergent concentration results in an overly complex immunocomplex, whereas an overly harsh wash buffer can strip away important interactions. The selection of an appropriate wash buffer must be determined experimentally. In our opinion, a good starting point would be a *functionally active* complex that includes the minimal number of components. The ability to control antibody leaching into the system should also be an important consideration. Incorporation of covalently conjugated antibody–matrix systems in our proteomic projects has resulted in improved detection of natural protein interactions with the target protein. Other new and developing methods, as well as improved bioinformatics tools, should provide greater ease and specificity for carrying out successful proteomic methodology.

We hope that this review of our proteomic analysis of components of the Ca^{2+} entry pathway can help others to contemplate a proteomic approach to other membrane proteins. Although a proteomic analysis consists of many steps (complete with their inherent pitfalls), we feel that the discovery-based results obtained from such an approach are more than an ample reward for the risk and effort needed to obtain them.

REFERENCES

1. Birnbaumer, L. 2009. The TRPC class of ion channels: a critical review of their roles in slow, sustained increases in intracellular Ca^{2+} concentrations. *Annual Review of Pharmacology and Toxicology* 49: 395.
2. Ramsey, I. S., M. Delling, and D. E. Clapham. 2006. An introduction to TRP channels. *Annual Review of Physiology* 68: 619.
3. Montell, C. 2005. TRP channels in *Drosophila* photoreceptor cells. *Journal of Physiology* 567: 45.
4. Ambudkar, I. S. 2007. Trafficking of TRP channels: determinants of channel function. *Handbook of Experimental Pharmacology* 179: 541.
5. Spassova, M. A., T. Hewavitharana, W. Xu, J. Soboloff, and D. L. Gill. 2006. A common mechanism underlies stretch activation and receptor activation of TRPC6 channels. *Proceedings of the National Academy of Science of the United States of America* 103: 16,586.
6. Ambudkar, I. S., H. L. Ong, X. Liu, B. C. Bandyopadhyay, and K. T. Cheng. 2007. TRPC1: the link between functionally distinct store-operated calcium channels. *Cell Calcium* 42: 213.
7. Kiselyov, K., J. Y. Kim, W. Zeng, and S. Muallem. 2005. Protein-protein interaction and function TRPC channels. *Pflügers Archiv* 451: 116.
8. Ambudkar, I. S., and H. L. Ong. 2007. Organization and function of TRPC channelosomes. *Pflügers Archiv* 455: 187.
9. Delmas, P., N. Wanaverbecq, F. C. Abogadie, M. Mistry, and D. A. Brown. 2002. Signaling microdomains define the specificity of receptor-mediated InsP₃ pathways in neurons. *Neuron* 34: 209.
10. Park, M. K., M. C. Ashby, G. Erdemli, O. H. Peterson, and A. V. Tepikin. 2001. Perinuclear, perigranular and sub-plasmalemmal mitochondria have distinct functions in the regulation of cellular calcium transport. *EMBO Journal* 20: 1863.
11. Singh, B. B., T. P. Lockwich, B. C. Bandyopadhyay et al. 2004. VAMP2-dependent exocytosis regulates plasma membrane insertion of TRPC3 channels and contributes to agonist-stimulated Ca^{2+} influx. *Molecular Cell* 15(4): 635.
12. Bezzerides, V. J., I. S. Ramsey, S. Kotecha, A. Greka, and D. E. Clapham. 2004. Rapid vesicular translocation and insertion of TRP channels. *Nature Cell Biology* 6: 709.
13. Lockwich, T. P., X. Lui, B. B. Singh, J. Jadlowiec, S. Weiland, and I. S. Ambudkar. 2000. Assembly of Trp1 in a signaling complex associated with caveolin-scaffolding lipid raft domains. *Journal of Biological Chemistry* 275: 11,934.
14. Lockwich, T., B. B. Singh, X. Liu, and I. S. Ambudkar. 2001. Stabilization of cortical actin induces internalization of transient receptor potential 3 (Trp3)-associated caveolar

Ca^{2+} signaling complex and loss of Ca^{2+} influx without disruption of Trp3-inositol triphosphate receptor association. *Journal of Biological Chemistry* 276: 42,401.

15. Pani, B., H. L. Ong, S. C. Brazer, X. Liu, K. Rauser, B. B. Singh, and I. S. Ambudkar. 2009. Activation of TRPC1 by STIM1 in ER-PM microdomains involves release of the channel from its scaffold caveolin-1. *Proceedings of the National Academy of Science of the United States of America* 106: 20,087.

16. Lockwich, T., J. Pant, A. Makusky et al. 2007. Analysis of TRPC3-interacting proteins by Tandem mass spectrometry. *Journal of Proteome Research* 7: 979.

17. Tan, S., H. T. Tan, and C. M. Chung. 2008. Membrane proteins and membrane proteomics. *Proteomics* 8: 3924.

18. Lu, B., D. B. McClatchy, J. Y. Kim, and J. R. Yates III. 2008. Strategies for shotgun identification of integral membrane proteins by tandem mass spectrometry. *Proteomics* 8: 3947.

19. Zhou, M., and T. D. Veenstra. 2007. Proteomic analysis of protein complexes. *Proteomics* 7: 2688.

20. Kabbani, N. 2008. Proteomics of membrane receptors and signaling. *Proteomics* 8: 4146.

21. Rohrbough, J. G., L. Breci, N. Merchant, S. Miller, and P. A. Haynes. 2006. Verification of single-peptide protein identifications by the application of complementary database search algorithms. *Journal of Biomolecular Techniques* 17(5): 327.

22. Goel, M., W. Sinkins, A. Keightley, M. Kinter, and W. P. Schilling. 2005. Proteomic analysis of TRPC5- and TRPC6-binding partners reveals interaction with the plasmalemmal Na^{+}/K^{+}-ATPase. *Pflügers Archiv* 451(1): 87.

23. Clark, K., J. Middlebeek, N. A. Morrice, C. G. Figdor, E. Lasonder, and F. N. van Leewen. 2008. Massive autophosphorylation of the Ser/Thr-rich domain controls protein kinase activity of TRPM6 and TRPM7. *PLoS ONE* 3(3): e1876, doi:10.1371/journal.pone.0001876.

24. Masuda, J., D. M. Maynard, M. Nishimura, T. Ueda, J. A. Kowalak, and S. P. Markey. 2005. Fully automated micro- and nanoscale one- or two-dimensional high-performance liquid chromatography system for liquid chromatography–mass spectrometry compatible with non-volatile salts for ion exchange chromatography. *Journal of Chromatography A* 1063: 57.

25. Maynard, D. M., J. Masuda, X. Y. Yang, J. A. Kowalek, and S. P. Markey. 2004. Characterizing complex peptide mixtures using a multi-dimensional liquid chromatography–mass spectrometry system: *Saccharomyces cerevisiae* as a model system. *Journal of Chromatography B* 810: 69.

26. Eder, P., M. Poteser, and K. Groschner. 2007. TRPC3: a multifunctional, pore-forming signalling molecule. In *Transient Receptor Potential (TRP) Channels*, vol. 179, eds. V. Flockerzi and B. Nilius, 77. Berlin Heildelberg: Springer.

27. Nesvizhskii, A. I. 2006. Protein identification by tandem mass spectrometry and sequence database searching. In *Methods in Molecular Biology*, vol. 367, ed. R. Matthiesen, 87. Totowa, NJ: Humana Press.

28. Lu, B., T. Xu, S. K. Park, and J. R. Yates III. 2009. Shotgun protein identification and quantification by mass spectrometry. In *Proteomics, Methods in Molecular Biology*, vol. 564, eds. J. Reinders and A. Sickman, 261. Totowa, NJ: Humana Press.

29. Slotta, D. J., M. A. McFarland, A. J. Makusky, and S. P. Markey. 2007. MassSieve: A new visualization tool for mass spectrometry-based proteomics. Paper presented at *55th Annual Conference on Mass Spectrometry and Allied Topics*, Indianapolis, IN, 2007, June 4–7 (American Society for Mass Spectrometry).

30. Lockwich, T., J. Chauthaiwale, S. V. Ambudkar, and I. S. Ambudkar. 1995. Reconstitution of a passive Ca^{2+}-transport pathway from the basolateral plasma membrane of rat parotid gland acinar cells. *Journal of Membrane Biology* 148: 277.

31. Ong, S., B. Blagoev, I. Kratchmarova et al. 2002. Stable isotope labeling by amino acids in cell culture, SILAC, as a simple and accurate approach to expression proteomics. *Molecular and Cellular Proteomics* 1(5): 376.

32. Mann, M. 2006. Functional and quantitative proteomics using SILAC. *Nature Reviews. Molecular Cell Biology* Dec. (7): 952.

33. Selbach, M., and M. Mann. 2006. Protein interaction screening by quantitative immuno-precipitation combined with knockdown (QUICK). *Nature Methods* 3: 981.

34. Josic, D., and J. G. Clifton. 2007. Mammalian plasma membrane proteomics. *Proteomics* 7: 3010.

6 Lessons of Studying TRP Channels with Antibodies

Marcel Meissner, Verena C. Obmann,
Michael Hoschke, Sabine Link, Martin Jung,
Gerhard Held, Stephan E. Philipp,
Richard Zimmermann, and Veit Flockerzi

CONTENTS

6.1 INTRODUCTION

Members of the transient receptor potential (TRP) family are membrane proteins; they constitute cation channels that are involved in a vast variety of physiological processes in mammals, flies, and worms. Rather than going into specific functions of single TRP proteins, this contribution describes different procedures used to generate antibodies specific for TRP proteins, the characterization of these antibodies by Western blotting and immunoprecipitation, and common pitfalls that have to be considered when pursuing these applications. The antibodies for TRPC3, TRPC4, TRPM3, and TRPM4, which are described, have been generated in the authors' laboratory.

Generating a research-grade antibody for a membrane protein is a painstaking effort that occupies months of laborious bench work and needs money for

135

synthesizing, expressing, and purifying the antigen, for hosting the suitable animals to be used for immunization, for generating and selecting hybridoma cells, and for purifying and characterizing the final product. Especially, the validation process, including analysis of the requisite efficacy and/or specificity of the antibody, is crucial and a major ordeal that in a few cases is accomplished by many commercial suppliers that have proliferated and offer for sale antibodies directed at a wide range of proteins. Unfortunately, those commercial antibodies fail in many cases even the most basic tests of activity and/or specificity (e.g., selectively interacting with the target protein in protein lysates prepared from cells overexpressing the target protein versus lysates from cells that do not express the target protein at all), and it appears to be the rule rather than the exception that vendors pass the burden of antibody validation/quality control to the end user. The end user has to decide either to simply use such an antibody as it is for her/his intended research goals, with the assumption that the vendor has performed adequate quality control to demonstrate activity and specificity, or, instead, to spend considerable funds and time-consuming experiments to critically evaluate the antibody before use and eventual publication of results.

In some fields of TRP channel research, the poor quality of commercially available antibody reagents has caused considerable frustration among investigators and has led to publication and perpetuation of erroneous research results (compare[1]). A recent issue of *Naunyn-Schmiedeberg's Archives of Pharmacology* (volume 379, pages 385–434) deals with the poor quality of commercial antibodies for G-protein coupled receptors; several other important articles and editorials also discuss the use of antibodies and their crucial validation in general.[2–8]

In this chapter, we discuss the in-house generation of different antibodies, including polyclonal, monoclonal, and Fabs (Fragments antigen binding) generated against peptides and recombinant fragments derived from TRPC3, TRPC4, TRPM3, and TRPM4. Because excellent recipes for the generation of antibodies do exist (e.g., Refs. 9 and 10), we do not include here step-by-step protocols but rather point to the problems in obtaining specific antibodies especially for TRP proteins.

6.2   GENERATION OF POLYCLONAL ANTISERA FOR TRPM4 IN RABBITS USING SYNTHETIC PEPTIDES AS ANTIGENS

Antibodies that recognize intact proteins can be produced through the use of short (~10–25 amino acid residues) synthetic peptides derived from the primary structure, without first having to isolate the protein. An antibody produced in response to a simple linear peptide will most likely recognize a linear epitope in a protein. Furthermore, that epitope must be solvent-exposed to be accessible to the antibody. The general features of a protein that correspond to these criteria are turns or loop structures that are generally found on the protein surface connecting other elements of secondary structure and areas of high hydrophilicity, especially those containing charged residues. Accordingly, computer algorithms that predict protein hydrophilicity and tendency to form turns are useful (e.g., http://www.expasy.org/tools/protscale.html). In general, which prediction method to use is not crucial because there tends to be a high level of agreement among them. In addition, hydrophilic protein stretches obeying the above rules are far from abundant in TRP proteins, which, as integral membrane proteins, are rather hydrophobic (Figures 6.1a and 6.2a). None of the prediction methods will identify the one single sequence guaranteed to produce an effective antibody against any given protein. Rather, several sequences will be identified that have a higher than average probability of being an effective antigen.

Figure 6.1a shows a Kyte and Doolittle plot[11] for the mouse TRPM4 protein. The hydrophobic stretches (positive values) predicted to represent transmembrane domains are indicated, as well as the positions of a protein fragment used for generating monoclonal antibodies and the peptides 578, 732, 733, 734, 735, and 679 likely to be antigenic according to the criteria listed above. After designing these six peptides, they were synthesized, coupled to a carrier protein, and used for immunizing six rabbits.

Rabbits are the usual animal of choice because they are genetically divergent from the mouse (and human) sources of the proteins studied; they provide as much as 25 mL of serum from each bleed without significant harmful effects. One has to consider that even in genetically identical animals, a single preparation of antigen will elicit different antibodies. Because most laboratory rabbits are outbred, these differences are more pronounced.[9] The synthesized peptides, the haptens, have to be coupled to a carrier protein, a relatively large molecule capable of stimulating an immune response independently. Haptens themselves are too small and cannot elicit antibody responses on their own because they cannot cross link B-cell receptors and they cannot recruit T-cell help. When coupled to a carrier protein, however, they become immunogenic because the protein will carry multiple hapten groups that can now cross link B-cell receptors. In addition, T-cell-dependent responses are possible because T cells can be primed to respond to peptides derived from the protein. The most commonly used carrier protein is keyhole limpet hemocyanin, which is usually preferred over bovine or rabbit serum albumin because it tends to elicit a stronger immune response and is evolutionarily more remote from mammalian proteins.

The speed of developing a specific antibody depends on priming and boosting immunizations, but the actual amounts of specific antibody produced will vary considerably, depending on the immunogenicity of the antigen. Booster immunizations are started 4–8 weeks after the priming immunization and continued at 2- to 3-week intervals. Prior to the priming immunization and following the primary and each booster immunization, blood is taken and serum prepared. The pre-immune serum—from blood taken prior to immunization—is a critical control to ensure that the antibody activity detected in later bleeds is due to the immunization.

The presence of specific antibodies will then be determined using an appropriate technique such as Western blot of protein samples containing the target protein. In the case of TRPM4, we used protein lysates from COS cells that do not express TRPM4 ("C" in Figure 6.1b, c, d, f) and COS cells that are transfected with the mTRPM4 cDNA ("C-M4" in Figure 6.1b, c, d, f). Figure 6.1b and c show the corresponding Western blots using the sera obtained from bleeds 1, 2, and 3 after immunizations with peptides 578 and 679. A number of protein bands are recognized by these sera in lysates from both control COS cells (C) and TRPM4-expressing COS cells (C-M4), indicating that they are not related to TRPM4. However, the ~135-kDa TRPM4 protein (arrowhead in Figure 6.1b and c) is clearly recognized by sera from the second and

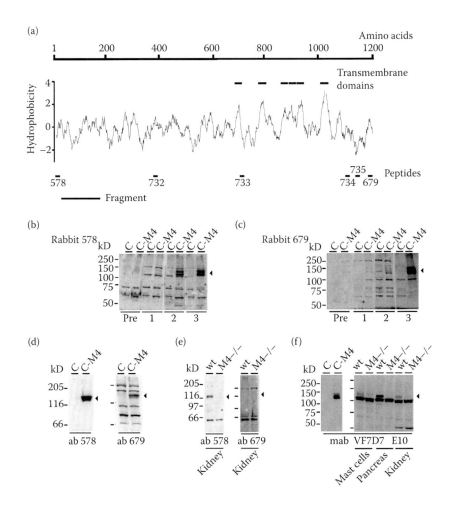

FIGURE 6.1 Characterization of TRPM4 antibodies. (a) Hydropathy blot of the TRPM4 protein sequence according to Kyte and Doolittle.[11] Predicted transmembrane domains and the positions of peptides (peptides 578, 732, 733, 734, 735, 679) and a protein frag- ment, which were used as antigens, are indicated. (b, c) Immunoblots using pre-immune sera (pre), sera from the first (1), second (2), and third (3) bleeding after immunization with peptide (b) 578 and (C) 679, and protein lysates from COS cells (C) and COS cells expressing the TRPM4 cDNA (C-M4) which were separated on 7% SDS polyacrylamide gels and transferred onto nitrocellulose membranes. The arrowhead indicates the expected size of the TRPM4 protein (~135 kDa). (d) Immunoblots using the affinity-purified ab 578 and ab 679 and protein lysates from COS cells (C) and COS cells expressing the TRPM4 cDNA (C-M4). Both abs recognize the ~135-kDa TRPM4 protein, but ab 679 also detects several other proteins that are not related to TRPM4 because they are also present in the nontransfected COS cells (C). (e) Immunoblots of microsomal membrane proteins (75 μg per lane) from mouse kidney prepared from wild-type mice (wt) and TRPM4- deficient mice[12] (M4−/−) and run on 7% SDS gels were probed with ab 578 and ab 679.

TABLE 6.1

Peptides Derived from the Mouse TRPM4 Sequence Used as Antigens to Immunize Rabbits

Peptide	Amino acid Residues	Number of Identical Amino Acids with the Sequence of		Specific/Sensitive in Western Blot for TRPM4 in	
		mTRPM4	mTRPM5	COS Cells	Primary Tissue
578	19	19	—	Yes	Yes
679	17	17	—	(Yes)	No
732	18	18	10	No	No
733	9	9	1	No	No
734	18	18	7	No	No
735	14	14	6	No	No

Note: For the relative position of the sequences within the TRPM4 primary sequence, see Figure 6.1. The TRPM4 and TRPM5 amino acid sequences are closely related, but TRPM5 was not detectable by any of the sera.

third bleeds of rabbit 578 and serum from the third bleed of rabbit 679. After affinity purification, the purified ab 578 nicely recognizes only TRPM4, whereas the ab 679 recognizes TRPM4 and at least five additional proteins both in lysates from COS controls (C), as well as TRPM4-expressing COS cells (C-M4) (Figure 6.1d). Although the affinity purification was repeated, the specificity of ab 679 was not improved.

Immunization of rabbits with the remaining four TRPM4-derived peptides did not yield any antibodies at all (Table 6.1), showing that only two out of the six immunizations yielded antibodies capable to decorate the TRPM4 protein overexpressed in COS cells. In another series of Western blots (Figure 6.1e) using microsomal protein fractions from mouse kidney, only ab 578 recognized the ~135-kDa TRPM4 protein (Figure 6.1e, wt, wild type), whereas ab 679 did not. As a control, the protein fractions from the same type of tissues from TRPM4-deficient mice[12] were used (Figure 6.1e, M4$^{-/-}$).

A monoclonal antibody (mab VF7D7E10), prepared in parallel using the indicated N-terminal TRPM4 fragment for immunization (Figure 6.1a), also nicely recognizes the TRPM4 protein overexpressed in COS cells (Figure 6.1f, left panel, C-M4 versus C), as well as the endogenous protein present in protein fractions from wild-type mice (Figure 6.1f, right panel). In contrast to ab 578 (Figure 6.1e), it also

FIGURE 6.1 (Continued) (f) Immunoblots using the monoclonal antibody (mab) VF7D7 E10 and (left panel) protein lysates from COS cells (C) and COS cells expressing the TRPM4 cDNA (C-M4) and membrane proteins (150 μg per lane) from mast cells, pancreas, and kidney prepared from wild-type mice (wt) and TRPM4-deficient mice (M4−/−) (right panel). Note that in addition to the ~135-kDa TRPM4 protein, the mab VF7D7 E10 also recognizes a protein of ~120 kDa, which is expressed at similar levels in the three cells/tissues; its identity is not known, but it is not related to TRPM4 because it is also present in protein fractions from TRPM4-deficient mice.

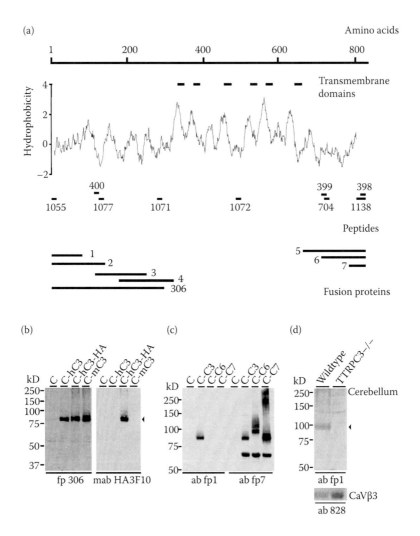

FIGURE 6.2 Characterization of TRPC3 antibodies. (a) Hydropathy blot of the TRPC3 protein sequence according to Kyte and Doolittle.[11] Predicted transmembrane domains and the positions of peptides (peptides 1055, 400, 1070, 1071, 1072, 399, 704, 1138, 398) and the protein fragments 1, 2, 3, 4, 5, 6, 7, and 306, which were used as antigens, are indicated. (b) Immunoblots using protein lysates of COS cells (C) and COS cells expressing the cDNA of the human (h), mouse (m), or HA-tagged human TRPC3 proteins separated on a 10% SDS gel, blotted on nitrocellulose membranes, and incubated with ab fp 306 (left) or the anti-HA mab (right). The human and mouse proteins run very similarly and are detectable by ab fp 306, whereas the anti-HA mab only recognizes the HA-tagged TRPC3 protein. (c) Immunoblots of protein lysates from COS cells, which were transfected with the cDNA of mTRPC3 (C-C3), mTRPC6 (C-C6), and mTRPC7 (C-C7), separated on SDS polyacrylamide gels, blotted, and incubated with polyclonal ab fp1 (left) or ab fp7 (right). The ab fp1 only recognizes TRPC3 (left), whereas ab fp7 recognizes TRPC3, TRPC6, and TRPC7.

recognizes additional proteins of ~120 kDa, which are also present in the fractions obtained from TRPM4-deficient mice[12] (Figure 6.1f, M4[−/−]) and therefore are not related to TRPM4. This example demonstrates the importance of testing the antibody on native tissues and using samples from the gene knockout mice, as without the TRPM4 knockout mice being available, it would have been difficult to decide which of the proteins recognized by this mab in protein fractions from wild-type animals is, in fact, TRPM4.

In summary, our efforts on producing TRPM4 antibodies have yielded one sensitive and specific polyclonal anti-TRPM4 antibody out of six immunizations using different potential antigenic peptides (ab 578, Table 6.1) and, in addition, one sensitive but less specific monoclonal antibody (mab VF7D7E10) from a purified N-terminal fragment.

6.3   GENERATION OF POLYCLONAL ANTISERA FOR TRPC3 IN RABBITS USING RECOMBINANT PROTEIN FRAGMENTS AS ANTIGENS

TRPC3, together with TRPC6 and TRPC7, constitutes a structurally related subgroup within the TRPC subfamily of proteins.[13] Overall, amino acid sequence identity among the three proteins is 69.4% with TRPC3 being more closely related to TRPC7 (81.0% identity) than to TRPC6 (71.2% identity). The sequence similarities among the three proteins add further problems to the generation of specific antibodies as shown below. In order to make TRPC3-specific antibodies, we first immunized rabbits with synthetic peptides derived from the TRPC3 primary structure (Figure 6.2a, peptides 398, 399, 400, 704, 1055, 1070, 1071, 1072, and 1138) as described above. However, only two of the nine immunizations using nine rabbits yielded antibodies (ab 1070 and 1071) that specifically recognized the TRPC3 protein overexpressed in COS or human embryonic kidney (HEK) 293 cells. However, neither of them recognized the TRPC3 protein in Western blots using protein fractions from mice.

We therefore prepared His-tagged TRPC3 fusion proteins (fp) 1 to 7 and a TRPC3-maltose binding protein (MBP) fusion protein (fp306)[14] (Figure 6.2a) using fragments derived from the N- and C-terminal sequences of the TRPC3 protein. The fusion constructs were expressed as recombinant proteins in *Escherichia coli*, affinity-purified, and used for immunization. Then the sera obtained from consecutive bleedings were tested for the presence of antibodies against TRPC3. Only three out of eight immunized rabbits produced antibodies that recognize TRPC3 expressed in COS cells: fp306 (Figure 6.2b, left panel) and fp1 (Figure 6.2c, left panel), both encompassing parts of the TRPC3 N-terminus, and fp7 (Figure 6.2c, right panel), covering the C-terminal end. The antibodies were affinity-purified

FIGURE 6.2 (Continued)   (d) Immunoblots using the ab fp1 and membrane proteins (150 μg per lane) from cerebella prepared from wild-type and TRPC3-deficient mice (TRPC3−/−).[15] The TRPC3 protein (arrowhead) is only recognized in the protein fraction from wild-type but not TRPC3-deficient mice. To estimate protein loading per lane of the gel, the blot was stripped thereafter and incubated in the presence of the antibody for CaVβ3.[24]

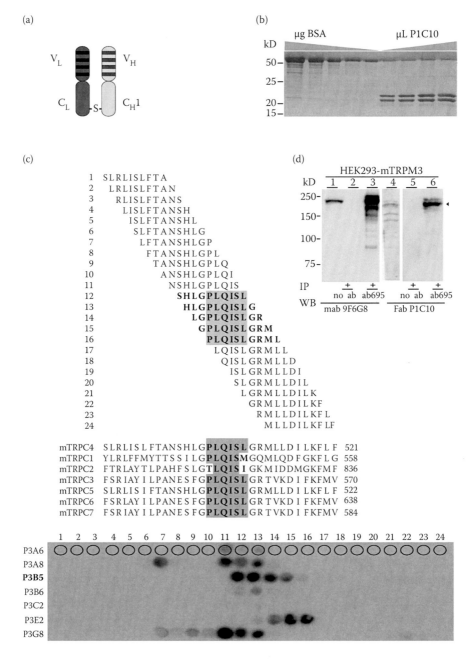

FIGURE 6.3 Characterization of TRPC4 and TRPM3 Fab fragments. (a) The Fab fragment consists of the variable (V) and constant (C) domains of the immunoglobulin light (L) and heavy (H) chains linked by a disulfide bridge. (b) Coomassie-stained 6.5%/15% SDS polyacrylamide gel run under reducing conditions to quantify the amount of recombinant Fab P1C10 (10, 15, 20, and 25 µL) using defined amounts of bovine serum albumin (10, 8, 4, 2, and 1 µg) as standards.

using the respective TRPC3-protein fragments fused to GST (fp1, fp7) or, in the case of ab fp306, by chromatography on MBP-sepharose followed by a second purification step on TRPC3-MBP-sepharose to remove the anti-His-/anti-MBP-antibodies.[14] The antibody fp306 nicely recognized both the mouse (Figure 6.2B left panel, mC3) and the human TRPC3 (Figure 6.2b, left panel, hC3) expressed in COS cells, but not in COS cells that had not been transfected with the respective cDNA (Figure 6.2b, "C"). As an additional control, an HA-tagged hTRPC3 cDNA was expressed (Figure 6.2b, C-hC3-HA) to demonstrate that the antibodies for the HA-tag (mab HA3F10, Figure 6.2b, right panel) and for TRPC3 (fp 306, Figure 6.2b, left panel) decorate proteins of the same molecular weight.

As shown in Figure 6.3c, the affinity-purified ab fp1 only recognizes TRPC3 (calculated Mr 95,672), whereas ab fp7 recognizes TRPC3 and, in addition, the slightly larger TRPC6 (calculated Mr 106,733) and TRPC7 (calculated Mr 99,475), indicating that the epitope recognized by ab fp7 is common to all three proteins. Accordingly, we used ab fp1 to identify the TRPC3 protein in microsomal membrane protein fractions prepared from various mouse tissues. It appeared that TRPC3 was detected in protein fractions prepared from cerebellum (Figure 6.2d, wild type), corresponding well to our previous findings that TRPC3 is needed for mGluR-dependent signaling in mouse cerebellar Purkinje cells.[15] As a control, we used protein fractions of the same tissue from TRPC3-deficient mice[15] (Figure 6.2d, TRPC3$^{-/-}$).

6.4 GENERATION OF RECOMBINANT FRAGMENT ANTIGEN BINDING (FAB FRAGMENTS) FOR TRPC4 AND TRPM3 USING SYNTHETIC PEPTIDES AS ANTIGENS

The large size of standard immunoglobulins comprising two heavy protein chains and two light chains that are intricately folded may impose practical limitations,

---

FIGURE 6.3 (Continued) The recombinant Fab purified by Talon Beads from bacterial lysates consists of a heavy and a light chain, which, depending on amino acid sequence composition, run at different sizes in the electrophoresis. (c) Epitope mapping of recombinant Fabs for TRPC4 using dot blots. The 24 peptides, each comprising 10 amino acid residues and covering aa 489 to 521 of the TRPC4 sequence (c, top panels), were spotted by a robot to nitrocellulose and incubated with the recombinant Fabs P3A6, P3A8, P3B5, P3B6, P3C2, P3E2, and P3G8 (c, bottom panel). After washing, specifically bound Fabs were detected by the peroxidase coupled secondary anti human Fab at a dilution of 1:100,000. The peptide spots recognized by Fab P3B5 are highlighted as is the minimal epitope PLQISL recognized by this Fab; this epitope is present in all TRPCs except TRPC1 and TRPC2. (d) Immunoblot of TRPM3 expressed in HEK293 cells detected by Fab P1C10 and mab 9F6G8,[25] which was used as a control. Cell lysates (lane 1, 15 µL and lane 4, 30 µL) and proteins retained after immunoprecipitation using the TRPM3 ab 695[25] (lane 3, 10 µL and lane 6, 25 µL) were run on a 4%/7% SDS polyacrylamide gel and incubated in the presence of Fab P1C10 (at 100 µg/mL) and mab 9F6G8 (at 5 µg/mL) as indicated. As a control, the precipitation procedure was performed in the absence of the primary ab 695 (no ab, lane 2, 10 µL and, lane 5, 25 µL). The specifically bound mab 9F6-G8 and Fab P1C10 were detected using the secondary anti-rat (1:80.000) and anti-human-Fab antibody (1:100.000), respectively.

especially when the antibody should gain access to hard-to-reach regions of the target protein, for example, to the pore region of ion channels. For such a condition, simpler and smaller proteins might perform better than full-size immunoglobulins. Especially for the cryptic epitopes of peptides displayed in an MHC class I context, antibody fragments were created by chopping off the stem of the Y-shaped immunoglobulin leaving just one "hand" to perform the chemical duty of the antibody:[16] These antibody fragments, so called Fragments antigen binding or Fabs (~50 kDa), comprise a complete light chain paired with the V_H and C_H1 domains of a heavy chain (Figure 6.3a). These fragments cannot recruit other effector molecules and cells in the same way as the full-size antibodies (~150 kDa) do because they lack the protein stem that performs such task, but they might be able to sneak into domains of an ion channel protein, which line the ion-conducting pore or which are critical for channel gating and thereby might interfere with the channel function.

For isolation and selection of Fabs specific for mouse TRPC4 and mouse TRPM3, a phagemid library expressing a nonimmune, semisynthetic human Fab repertoire of 3.7×10^{10} independent fragments[17] was screened with peptides (13–57 amino acid residues in length) derived from mouse TRPC4 and mouse TRPM3. The peptides were immobilized and incubated with the phages. Specifically bound phages were eluted after several washing steps and used to infect *E. coli* to obtain recombinant Fabs. These recombinant Fabs were purified from bacteria, and their authenticity was confirmed by Coomassie Blue-staining, which yielded one band of 50 kDa under nonreducing conditions and two bands of ~23 kDa under reducing conditions (Figure 6.3b). The specificity of the recombinant Fabs was determined by ELISA using the respective peptides/protein fragments as antigens. Sixty-three clones out of 96 were specific for TRPC4 and 50 out of 96 for TRPM3. DNA sequencing of those positive clones yielded 9 and 12 independent Fabs for TRPC4 and TRPM3, respectively.

Further characterization of the recombinant antibodies included immunoblot analyses (1) of peptides (10- to 19mers) representing the antigenic TRP protein sequences and that were spotted on nitrocellulose membranes (dot blots, Figure 6.3c) and (2) of lysates of cells expressing TRPM3 and TRPC4, respectively (Western blots, Figure 6.3d). The epitope mapping by the dot blot of seven independent Fabs for TRPC4 is shown at the bottom panel of Figure 6.3c. Each spot, 1 to 24, represents a peptide decamer immobilized on the nitrocellulose membrane. The 24 peptides cover amino acid residues (aa) 489 to 521 of the TRPC4 primary structure, with sequences shifted by one amino acid residue per spot (Figure 6.3c, top panel). The Fab P3B5 (Figure 6.3c, bottom, bold) decorates spots 12 to 16 (Figure 6.3c, bottom panel), representing aa 500 to 513 (Figure 6.3c, top panel, bold). The smallest epitope present in all five spots is highlighted in gray and most probably represents the minimal epitope recognized by this Fab. Similar approaches revealed that the shortest and longest sequences recognized by the recombinant Fabs cover 4 and 10 aa, respectively. Interestingly, the 6-aa epitope recognized by Fab P3B5 is present not only within the 33-aa TRPC4 protein fragment used to screen the phagemid library but also in all other TRPCs except TRPC1 and TRPC2 (Figure 6.3c).

For detecting the full-length proteins in Western blots by the Fabs, TRPM3 and TRPC4 were initially enriched by immunoprecipitation, and then Fabs were used at concentrations up to 100 μg/mL (which corresponds to ~2 μM) to detect the TRP proteins in cell lysates. The Western blots in Figure 6.3D summarize the results obtained with the TRPM3 Fab P1C10. It recognizes the TRPM3 protein in cell lysate (Figure 6.3d, lane 4), as well as among the proteins retained after immunoprecipitation with ab 695, an anti-peptide antibody generated for TRPM3 (Figure 6.3d, lane 6, arrowhead). As a control, we used the monoclonal antibody 9F6G8 (Figure 6.3d, lanes 1 to 3) generated against a TRPM3 fragment.

Recombinant Fabs are efficiently synthesized in bacteria, for example, P1C10 can be easily tailored to produce a concentration of at least 1.2 mg/L—one might rig them to tow fluorescent proteins—and because of their smaller size compared to full-length immunoglobulins, they might be suitable tools for immunocytochemistry and for interfering with the TRP protein function.

6.5 GENERAL COMMENTS ON THE VALIDATION OF ANTIBODIES

As shown above, the specificity controls are crucial before antibodies should be used. From our experience, the following approaches appear to be essential to evaluate the quality of any antibody:

1. Western blot: Selective decoration of the target protein in protein lysates prepared from cells overexpressing the target protein versus lysates from cells that do not express the target protein at all (compare Figures 6.1 and 6.2). Ideally, if different antibodies raised against different epitopes of the same target protein are available, they should label the same protein, as in the example shown in Figure 6.2b for ab fp306 and the anti-HA-antibody.
2. Immuncytochemistry: Selective staining of cells overexpressing the target protein versus those that do not express the target protein at all.

One should consider that HEK293 cells, like COS cells and Chinese hamster ovary (CHO) cells, have originally been generated and selected by many investigators because they very efficiently overexpress foreign proteins.[18] Antibodies recognizing proteins overexpressed in these cells do not necessarily have the sensitivity to recognize the same proteins expressed under their endogenous promotors in primary cells and tissues. Therefore, the most rigorous approaches include comparing results obtained using the equivalent cells/tissue sections or protein fractions from wild-type animals and knock out animals in which the target antigen has been genetically deleted. This deletion should be demonstrated by independent techniques such as Southern blots and PCR.

Reduction of antibody staining intensity upon knock-down approaches such as siRNA might be useful as long as independent controls—i.e., controls that do not use the antibody to be validated—have demonstrated that the expression of the protein or mRNA is really knocked down.

Finally, one should be aware that the widely used "absorption control," the competition of antibodies generated against synthetic peptides with excess peptide, determines only the specificity of the antibody for the incubating peptide but does not prove the specificity of the antibody for the target protein in the tissue.

6.6 COMPARING PROTEIN EXPRESSION DATA
WITH RNA EXPRESSION DATA

To study whether a given gene is expressed in specific cells or tissues, mRNA/transcript expression can be analyzed by Northern blots and RT-PCR. By pursuing these approaches, one only measures the expression levels of the mRNA, and the results do not substitute for protein expression data. Although extensive work on RT-PCR techniques is going on, little attention appears to have been paid to the relation between the mRNA expression level and corresponding protein abundance in eukaryotes. The few studies that addressed this question yield amazing results:[19–22] In order to obtain an estimate of the overall relationship between mRNA and protein abundances in human liver, Anderson and Seilhamer[20] found a correlation coefficient of 0.48 between them. These results, which are halfway between a perfect correlation and no correlation at all, were confirmed by studies comparing protein and mRNA abundances for one gene product across 60 human cell lines[19] or for more than 150 gene products in the yeast *Saccharomyces cerevisiae.*[23] Figure 6.4 shows extracted data from the latter study,[23] which concluded that the correlation between mRNA and protein levels is insufficient to predict protein expression levels from quantitative mRNA data. It was shown for some genes that while the mRNA levels were almost the same, the protein levels varied by more than 20-fold (Figure 6.4, data taken from table 1 of Ref. 23).

Surely, antibodies are not always available, nor are knock out mice or knock-down approaches that can be readily applied to evaluate an antibody. The results summarized above reveal that, for analysis of gene expression, simple deduction from mRNA transcript analysis is insufficient and that approaches using antibodies or antibody-based enrichment of the proteins of interest, combined with state-of-the-art mass spectrometry-based proteomics using stable isotope labeling, comprise the currently available approaches for quantitative analysis of proteins present in a given tissue or cell.

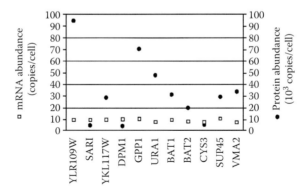

FIGURE 6.4   Correlation between protein expression and mRNA abundance. Data of the 11 yeast proteins are taken from table 1 from Ref. 23. The mRNA abundance (copies per cell, open squares) ranges from 8.9 copies (URA1, BAT2, CYS3, VMA2) to 11.9 (SSUP45), and the protein abundance (10^3 copies per cell, solid circles) varies from 5.4×10^3 (SARI) to 94.4×10^3 (YLR109W).

6.7 CONCLUSION

Antibody-based strategies for characterization, localization, and isolation of target proteins are among the most crucial and widely used techniques in molecular and cellular science. Although generation of antibodies is assumed by many merely to represent a routine procedure, evaluation and appropriate use of antibodies remain a sophisticated and challenging ordeal. This applies especially for low abundance membrane proteins like the TRPs.

ACKNOWLEDGMENTS

The authors thank Michael X. Zhu for the cDNA of the HA-tagged human TRPC3; Christine Wesely, Ute Soltek, Christin Matka, and Kerstin Fischer for their valuable assistance; and Sabine Pelvay, Ramona Gölzer, and Elisabeth Ludes for synthesizing peptides, immunizing, and bleeding rabbits. This work was supported by Homburger Forschungsförderungsprogramm (HOMFOR) (M.M., S.L., M.J., S.E.P., G.H., R.Z., and V.F.), by Forschungskommission der Universität des Saarlandes (S.E.P., V.F.), by Deutsche Forschungsgemeinschaft (M.J., S.E.P., R.Z., and V.F.), and by Fonds der Chemischen Industrie (V.F.).

REFERENCES

1. Flockerzi, V., C. Jung, T. Aberle et al. 2005. Specific detection and semi-quantitative analysis of TRPC4 protein expression by antibodies. *Pflügers Archiv* 451: 81–86.
2. Swaab, D. F., C. W. Pool, and F. W. Van Leeuwen. 1977. Can specificity ever be proved in immunocytochemical staining. *Journal of Histochemistry and Cytochemistry* 25: 388–391.
3. Willingham, M. C. 1999. Conditional epitopes. Is your antibody always specific? *Journal of Histochemistry and Cytochemistry* 47: 1233–1236.
4. Burry, R. W. 2000. Specificity controls for immunocytochemical methods. *Journal of Histochemistry and Cytochemistry* 48: 163–166.
5. Saper, C. B., and P. E. Sawchenko. 2003. Magic peptides, magic antibodies: guidelines for appropriate controls for immunohistochemistry. *Journal of Comparative Neurology* 465: 161–163.
6. Gown, A. M. 2004. Unmasking the mysteries of antigen or epitope retrieval and formalin fixation. *American Journal of Clinical Pathology* 121: 172–174.
7. Lipman, N. S., L. R. Jackson, L. J. Trudel, and F. Weis-Garcia. 2005. Monoclonal versus polyclonal antibodies: distinguishing characteristics, applications, and information resources. *ILAR Journal* 46: 258–268.
8. Rhodes, K. J., and J. S. Trimmer. 2008. Antibody-based validation of CNS ion channel drug targets. *Journal of General Physiology* 131: 407–413.
9. Harlow, E., and D. Lane. 1988. *Antibodies a Laboratory Manual*. Cold Spring Harbor, NY: Cold Spring Harbor Laboratory Press.
10. Harlow, E., and D. Lane. 1999. *Using Antibodies*. Cold Spring Harbor, NY: Cold Spring Harbor Laboratory Press.
11. Kyte, J., and R. F. Doolittle. 1982. A simple method for displaying the hydropathic character of a protein. *Journal of Molecular Biology* 157: 105–132.
12. Vennekens, R., J. Olausson, M. Meissner et al. 2007. Increased IgE-dependent mast cell activation and anaphylactic responses in mice lacking the calcium-activated nonselective cation channel TRPM4. *Nature Immunology* 8: 312–320.

13. Flockerzi, V. 2007. An introduction on TRP channels. *Handbook of Experimental Pharmacology* 2007: 1–19.
14. Philipp, S., B. Strauss, D. Hirnet et al. 2003. TRPC3 mediates T-cell receptor-dependent calcium entry in human T-lymphocytes. *Journal of Biological Chemistry* 278: 26,629–26,638.
15. Hartmann, J., E. Dragicevic, H. Adelsberger et al. 2008. TRPC3 channels are required for synaptic transmission and motor coordination. *Neuron* 59: 392–398.
16. Held, G., M. Matsuo, M. Epel et al. 2004. Dissecting cytotoxic T cell responses towards the NY-ESO-1 protein by peptide/MHC-specific antibody fragments. *European Journal of Immunology* 34: 2919–2929.
17. de Haard, H. J., N. van Neer, A. Reurs et al. 1999. A large non-immunized human Fab fragment phage library that permits rapid isolation and kinetic analysis of high affinity antibodies. *Journal of Biological Chemistry* 274: 18,218–18,230.
18. Eaton, D. L., W. I. Wood, D. Eaton et al. 1986. Construction and characterization of an active factor VIII variant lacking the central one-third of the molecule. *Biochemistry* 25: 8343–8347.
19. Tew, K. D., A. Monks, L. Barone et al. 1996. Glutathione-associated enzymes in the human cell lines of the National Cancer Institute Drug Screening Program. *Molecular Pharmacology* 50: 149–159.
20. Anderson, L., and J. Seilhamer. 1997. A comparison of selected mRNA and protein abundances in human liver. *Electrophoresis* 18: 533–537.
21. Anderson, N. L., and N. G. Anderson. 1998. Proteome and proteomics: new technologies, new concepts, and new words. *Electrophoresis* 19: 1853–1861.
22. Pradet-Balade, B., F. Boulme, H. Beug, E. W. Mullner, and J. A. Garcia-Sanz. 2001. Translation control: bridging the gap between genomics and proteomics? *Trends in Biochemical Science* 26: 225–229.
23. Gygi, S. P., Y. Rochon, B. R. Franza, and R. Aebersold. 1999. Correlation between protein and mRNA abundance in yeast. *Molecular Cell Biology* 19: 1720–1730.
24. Link, S., M. Meissner, B. Held et al. 2009. Diversity and developmental expression of L-type calcium channel β2 proteins and their influence on calcium current in murine heart. *Journal of Biological Chemistry* 284: 30129–30137.
25. Wagner, T. F., S. Loch, S. Lambert et al. 2008. Transient receptor potential M3 channels are ionotropic steroid receptors in pancreatic beta cells. *Nature Cell Biology* 10: 1421–1430.

7 Assessing TRPC Channel Function Using Pore-Blocking Antibodies

Shang-Zhong Xu

CONTENTS

7.1   INTRODUCTION

There are ~23,000 human protein-coding genes. One third of them encode membrane proteins including ion channels, receptors, transporters, and exchangers.[1,2] To understand the physiological function of individual genes, the development of specific tools targeting the gene of interest is essential. The technology by gene modification, such as antisense oligonucleotides, small interfering RNA (siRNA), gene knockout, or transgenic animals, has provided useful approaches to reveal individual gene function; however, some limitations of these techniques are inevitable. Among them, the knockdown of protein expression takes time (at least days) and will likely cause compensatory changes that obscure the interpretation of the research findings. The study of cell surface receptors and ion channels has benefited greatly from the use of small molecule blockers and activators, which cause acute modulation of the protein function, allowing direct comparison of the sample before and after the drug treatment. This often provides a clear answer as to whether the target protein is involved in the physiological process being studied without a concern of the compensatory effect.

The traditional approach for identifying a specific tool for ion channel study is mainly based on screening synthetic chemicals or natural compounds, which, during the past few decades, has provided some reliable research tools and effective therapeutic drugs, such as tetrodotoxin that blocks voltage-gated Na^+ channels, dihydropyridines and cone snail toxins for voltage-gated Ca^{2+} channels, and many receptor agonists or antagonists. Screening chemical compounds is undoubtedly useful; however, the process is laborious and time-consuming, and there is no guarantee that a specific and potent drug will be found. Therefore, more targeted approaches would appear to be necessary in order to match the fast pace of new ion channel discovery.

Antibodies are renowned for their exquisite specificity and unlimited diversity.[3] Therefore, we have tried to develop a new class of antibodies targeting ion channel pore regions, for example, the third extracellular loop of TRPC channels, called E3-targeting antibodies, in order to functionally interfere with channel properties.[4,5] Not only can these E3-targeting antibodies be used as ordinary antibodies for protein labeling, Western blotting, immunostaining, and immunoprecipitation, but they are also useful as specific pharmacological tools for in vitro or even in vivo functional studies. Since our reports on the methodology and applications,[4–8] several groups have tried the method, and some pore-blocking antibodies have been successfully developed or reproduced, such as the pore-blocking antibodies for TRPC channels,[9,10] Eag1 potassium channels,[11] TRPV1 channels,[12] TRPM3 channels,[13] and CaV1.2 channels.[14]

Given the huge potential use of pore-blocking antibodies in ion channel research, this chapter describes the generation of functional antibodies and their applications including how to design a functional antibody, how to screen the antibodies, how the antibody can be used as a pharmacological tool (especially for assessing the TRPC function), and what the advantages and limitations are.

7.2 PORE-BLOCKING ANTIBODY AND GENERATION

Antibodies that can alter the ion channel function via direct binding or interaction with the ion channel protein are functional antibodies. Besides the pore-blocking antibodies we reported,[4] functional modulation by antibodies targeting the cytoplasmic protein regions has also been described, such as the C-terminal antibodies for a potassium channel ($K_V 1.2$)[15] and the inositol 1,4,5-trisphosphate receptor (IP_3R),[16] as well as the N-terminal antibody for the stromal interaction molecule 1 (STIM1).[17] Antibodies targeting the alpha subunit of G-proteins[18] or T-tube membrane[19] also indirectly change the ion channel function.

The development of pore-blocking antibodies could be a simple and straightforward approach for stopping ion flow via a specific ion channel and useful for studying individual channel function in native cells, especially for these ion channels lacking specific blockers.[20] The E3-targeting methodology is for ion channels that contain six transmembrane segments and three extracellular loops for each subunit, such as TRPCs and shaker potassium channels.[4] The strategy can be easily adopted for other ion channel families with three extracellular loops, such as TRPVs, TRPMs, TRPA1, and cyclic nucleotide-gated (CNG) ion channels,[20] or even for other ion channels with just one or two extracellular loops, such as inwardly rectifying potassium channels and ORAI channels.[17,21] In order to help researchers achieve a high success rate for pore-blocking antibody generation, some principles are given as follows.

7.2.1 ION CHANNEL TOPOLOGY ANALYSIS

The structure of TRPCs is similar to the well-described Shaker potassium channels including six-membrane spanning segments (S1–S6), a putative channel pore region located between S5 and S6, and the intracellularly located N- and C-termini. The hydropathy plot and the structure prediction software, e.g., the Expert Protein Analysis System (ExPASy) on the server of the Swiss Institute of Bioinformatics, are helpful for understanding a new ion channel topology.

7.2.2 SELECTION OF TARGET EPITOPE

The selection of antigenic peptide is a critical step for successful generation of a pore-blocking antibody. Sequence alignment with other related isoforms is essential for the selection of isoform-specific epitopes, such as those used for TRPC1[5] and TRPC5.[4] These E3 epitopes are very close to or even partially overlap with the putative ion selectivity filter (Figure 7.1). There is software, such as Lasergene, to help predict peptide antigenicity. The following principles should be considered.

7.2.2.1 Antibody Accessibility

The selected target region for antibody generation should be accessible by IgG and is potentially important for the ion channel function. For example, the third extracellular

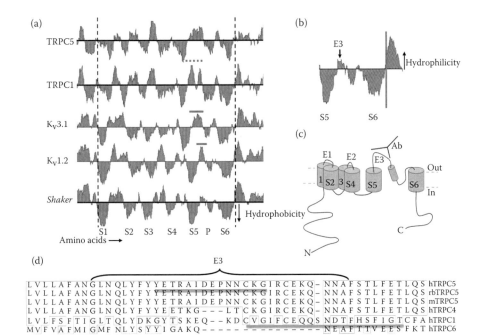

FIGURE 7.1 E3-targeting methodology. (a) Hydropathy plots of ion channel subunits with six transmembrane segments (S1–S6). Shaker is a *Drosophila* K+ channel, and $K_V1.2$ and $K_V3.1$ are mammalian homologs. TRPC1 and TRPC5 show a similarity structure to the K+ channels. The E3-targeting regions are indicated by a solid line or dashed line. (b) Kyte–Doolittle pattern for the S5–S6 region of Shaker. (c) Predicted membrane topology for TRPC channels and the hydrophilic region of the third extracellular loop (E3) as the selected target for pore-blocking antibody. (d) Alignment of TRPC channel S5–S6 regions and the epitopes (underlined) for pore-blocking antibody generation. (Adapted from Xu, S.Z. et al. *Nature Biotechnology*, 23, 1289, 2005. With permission.)

loop near the channel pore region is important for the TRPC channel function. The cysteine[8] and negatively charged glutamate (Glu)[22] residues are functionally important for TRPC5 channel activity. Some ion channels may be glycosylated, and therefore, the potential glycosylated site should be avoided if possible because antibodies targeted to the peptide sequences may not recognize the modified native protein.

7.2.2.2 Hydrophilicity and Flexibility

The hydrophilic regions tend to reside on the surface of membrane proteins, while the hydrophobic regions tend to be found hidden in the interior of the protein structure or in the membrane. Therefore, the hydropathy plot is a helpful guide for epitope selection. If a hydrophilic region is selected, then peptide solubility should not be a problem. If there is a choice, one should select an antigenic peptide with as few hydrophobic residues (e.g., trytophan, valine, leucine, isoleucine, and phenylalanine) as possible. Glutamines may also cause insolubility, as it can form hydrogen bonds between peptide chains, so multiple glutamines in epitope should be avoided. The introduction of proline or tyrosine can induce structural motifs, thereby enhancing the immunogenic potential of the peptide. In addition, antibodies tend to bind with higher affinity to those epitopes that are flexible enough to move into accessible positions. However, unlike the C- or N-terminal ends of the channel, the transmembrane regions may have less flexibility.

7.2.2.3 Length of Epitope

The length of the epitope is also important. Longer antigenic peptides may have a greater conformational similarity to the native protein and are therefore more likely to induce antibodies that recognize the natural protein. However, if an epitope is too long, it may lose its specificity, especially when targeting the extracellular pore region. Data suggest that a single antigenic determinant, i.e., the smallest immunogenic peptide, is between five and eight amino acids, and therefore, a length of 15–20 amino acids is preferable for designing an antigenic peptide, as it should contain at least one epitope and adopt a limited number of conformations. Like other peptide-based antibody generations, a cysteine should be added to the end of the selected epitope for antibody affinity purification purpose if there is no internal cysteine in it.

7.2.3　Antigenic Peptide Synthesis and Animal Immunization

The selected epitope should be confirmed to be unique by BLAST searching against the GenBank protein database before sending the sequence for peptide synthesis. The integrity and purity of the peptide should be assayed with high-performance liquid chromatography. If the purity is less than 80%, it should not be used because low-purity peptides will likely yield poor-quality antibodies for pore-blocking functional studies.

Animal immunization is laborious work. However, many companies provide an antibody production service, such as the standard 77-day protocol for rabbit polyclonal antibody generation (Table 7.1). Briefly, rabbits approximately 2–3 kg are usually used for immunization. The synthetic peptide should be completely dissolved in a sterile saline at a concentration of 100–500 g/mL. About 1 mL of antigen solution is made into an emulsion with an equal amount of complete Freund's adjuvant (CFA)

TABLE 7.1
Schedule for Immunization and Bleeding[a]

Day	Procedure	Day	Procedure
0	Pre-bleed, Injection-CFA	49	Second bleed
14	Injection-IFA	56	Injection-IFA
28	Injection-IFA	63	Third bleed
35	First bleed	70	Injection-IFA
42	Injection-IFA	77	Final bleed

[a] Injection-CFA: injection with emulsion made with peptide and complete Freund's adjuvant; injection-IFA: injection with emulsion made with peptide and incomplete Freund's adjuvant.

(1 mg of dried Mycobacterium tuberculosis H37Ra, 0.85 mL of paraffin oil, and 0.15 mL of mannide monooleate). Then 1 mL of the emulsion is injected into the popliteal lymph nodes, and the remaining emulsion (~1 mL) can be subcutaneously injected into numerous small depots arranged along the spine.

The booster and bleeding protocol is also listed in Table 7.1. The incomplete Freund's adjuvant (IFA) (0.85 mL of paraffin oil and 0.15 mL of mannide monooleate) is used after the first CFA injection. The test bleeds will start at the fifth week of the protocol and terminate at day 77 or a later day depending on the antibody titration tested by an enzyme-linked immunosorbent assay.

7.2.4 PORE-BLOCKING ANTIBODY SCREENING

7.2.4.1 Enzyme-Linked Immunosorbent Assay

The immune serum should be screened by ELISA. The successful antibodies usually have a reasonable or high titration, for example, the anti-TRPC1 antibody (T1E3) had a titration higher than 1:50,000 dilution. The detailed procedure for ELISA is given as follows.

Coating antigen: Antigenic peptide dissolved in 50 mM Na_2HCO_3 (pH 9.6) at the concentration of 4 g/mL is used for plate coating. For example, for the 96-well Nunc-Immuno Plate, 50 L of the peptide solution is added to each well, and the plate is covered with a piece of paraffin membrane and kept at 4°C overnight.

Blocking nonspecific binding: The coated plate is washed three times with phosphate buffered saline (PBS) containing 0.2% Tween20 (PBS/Tween20) followed by an incubation with 300 L per well 1% dry milk in PBS at 37°C for 1 hour to block the nonspecific binding sites.

Incubation with antibody: The plate is then washed three times with PBS/Tween20 before the addition of 50 L of serial diluted antiserum (1:50 to 1:50,000 dilutions) to the wells. The plate is then incubated at 37°C for 2 hours and washed three times with PBS/Tween20 again. The secondary antibody (goat anti-rabbit IgG conjugated with horseradish peroxidase at 1:5,000 dilution) is added at 50 L/well and the plate incubated at 37°C for 1 hour.

Color development: After washing the plate with PBS/Tween20 three times, 50 L of the color development solution ((2,2′-azino-bis(3-ethylbenzthiazoline-6-sulfonic acid) diammonium salt 5.5 mg dissolved in a 10-mL phosphate citrate buffer plus 1 μL H_2O_2) is added into each well. The plate is gently shaken on a rotator and incubated at room temperature for 30 min in dark for color development, followed by absorbance reading at a wavelength of 405 nm using a plate reader.

7.2.4.2 Western Blotting and Immunostaining

Western blotting is useful for examining the specificity of antibody binding to the target protein. Cell lysates prepared from both native tissues and cultured cells transfected with the corresponding cDNA for the target protein should be used for validation of the pore-blocking antibody. In addition, binding to the extracellular region can be confirmed by immunocytochemistry performed on both permeabilized and nonpermeabilized cells. The extracellular binding antibody is expected to give positive staining regardless of whether the cells are permeabilized.

7.2.4.3   Fluorescence-Activated Cell Sorting (FACS)

Fluorescence-activated cell sorting is a powerful tool for screening the pore-blocking antibodies. The extracellular binding of E3-targeting antibodies can be examined by FACS using transfected cells or native cells if the channel is well expressed. For example, E3-targeting TRPC antibodies can be tested using the TRPC-transfected HEK-293 cells through a secondary antibody conjugated with fluorescent probes.

7.2.5   ANTIBODY PURIFICATION

Pore-blocking antibody should be purified. The protein A column and the immobile peptide column are used for purification. This step is essential, especially for these lower titration antisera. The solvent in the purified antibodies should be completely removed by dialysis in PBS. The purified antibodies are stored in a −80°C freezer in aliquots.

7.2.5.1   Affinity Purification of IgG Using Protein A

The protein A column can be used for purification of rabbit polyclonal antibodies. For example, the HiTrap affinity column (Amersham Pharmacia Biotech) has a high binding capacity for IgG. Briefly, the column is prewashed with 10 column volumes of the binding buffer (20 mM sodium phosphate, pH 7.4) and then loaded with rabbit serum samples diluted at 1:1 dilution in the binding buffer. The column is washed with 10 column volumes of the binding buffer or until no material appears in the effluent or the absorbance of the effluent at 280 nm approaches the background level. The bound IgG is then eluted from the column with four column volumes of the elution buffer (0.1 M citric acid). The eluted IgG is collected in tubes with the neutralization buffer (1 M Tris-HCl, pH 9.0). The eluted immunoglobulin is measured by a spectrophotometer at the wavelength of 280 nm. For mouse IgG purification, the protein G column should be used instead.

7.2.5.2   Affinity Purification Using Immobile Peptide Column

The immobile peptide column is prepared by conjugating the resin (SulfoLink coupling gel, Pierce Chemical Company, Rockford, Illinois, USA) with the antigenic peptide. The resin is prewashed with a coupling buffer (50 mM Tris and 5 mM EDTA, pH 8.5) and then mixed with an equal volume of 0.1% antigenic peptide in the coupling buffer. The mixture is rotated on a platform for 1 hour at room temperature and washed three times with the coupling buffer, each followed by centrifugation. The resin is then saturated by incubation with 50 mM cysteine, which is followed with washing in a wash buffer (1 M NaCl in PBS).

For antibody purification, an equal volume of PBS is mixed with the antiserum or the crude antibody and centrifuged at 10,000 rpm for 10 min at 4°C. The supernatant is then mixed with peptide-conjugated resin and incubated at 4°C overnight. The resin is loaded onto a small column, washed with the wash buffer, and then eluted using 200 mM glycine-HCl. The flow-throughs before and during the elution are collected into a series of test tubes containing the neutralization buffer. Aliquots of the eluted samples are measured at 280 nm and further tested by ELISA for the presence of proteins and desired antibodies, respectively.

7.2.6   FUNCTIONAL TEST OF PORE-BLOCKING ANTIBODIES

7.2.6.1   Patch-Clamp Recording

The blocking effect of the antibody should be tested by patch clamp recordings using either native or transfected cells that express the channel of interest, following examples for the TRPC pore-blocking antibodies[4,17] and the potassium channel antibodies.[23] Proper controls should be set in parallel, such as the antibody pre-incubated with the antigenic peptide to block the binding site, pre-immune IgG, and peptide alone. The dose-response curve may be useful for determining the antibody concentration needed for functional study. New patch-clamp recording systems may also be useful, for example, the planar patch;[17] however, the detached cells should be maintained at a good condition for high-quality recordings.

7.2.6.2   Calcium Imaging

For calcium-permeable channels, the pore-blocking effect can be tested using Ca^{2+} imaging, as exemplified for the TRPC1 and TRPC5 blocking antibodies.[4-6] Again, proper controls should be carried out in parallel or in alternating orders with the test samples in order to minimize the influence of experiment-to-experiment variations of Ca^{2+} imaging assay.

The FlexStation™ calcium assay system provides a fast, simple, and reliable fluorescence-based assay for detecting changes of intracellular Ca^{2+} concentrations. (See Chapter 1 for more details.) This high-throughput system should be effective for screening pore-blocking antibodies and/or new chemical ligands of ion channels.

7.3   ASSESSING TRPC CHANNEL FUNCTION USING PORE-BLOCKING ANTIBODIES

7.3.1   ASSESSING CA^{2+} INFLUX

7.3.1.1   TRPC1 Forms Store-Operated Channel Subunit

The pore-blocking antibody for TRPC1 (T1E3) was first applied to the freshly isolated rabbit cerebral arterioles to investigate its role in store-operated Ca^{2+} influx in the smooth muscle cells.[5] Using a Ca^{2+}-refilling protocol following store depletion by pretreatment of cells with thapsigargin, the store-operated Ca^{2+} influx was shown to be significantly inhibited by incubation with T1E3 (Figure 7.2), an effect that was prevented by pre-absorption of the antibody using the antigenic peptide absorbed. This experiment gave the first direct evidence that TRPC1 confers store-operated Ca^{2+} influx in vascular smooth muscle.[5] We suggested that TRPC1 could be a subunit of the store-operated channels or it could constitute a subpopulation of these channels because the inhibition of the store-operated Ca^{2+} influx by T1E3 was only around 25%. The larger part of the remaining T1E3-insentive store-operated Ca^{2+} entry in the smooth muscle cells could be mediated by other channels. As we have postulated, several store-operated Ca^{2+} entry pathways have recently been identified in vascular smooth muscle cells. These include heteromultimeric TRPC1/5 channels with store-operated properties[6,24] and Ca^{2+} entry mediated through STIM1[17,25] and Orai1.[26] The contribution of TRPC1 to the store-operated Ca^{2+} influx has also been

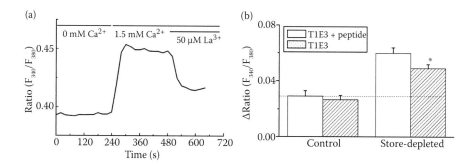

FIGURE 7.2   TRPC1 pore-blocking antibody T1E3 inhibits store-operated Ca^{2+} entry in freshly isolated arterial smooth muscle cells. (a) Ca^{2+} influx was measured by switching the external solution from 0 to 1.5 mM Ca^{2+} to arterial cells treated with and without thapsigargin. (b) Mean data for the T1E3 group and the antigenic peptide absorbed group (T1E3 + peptide). The Ca^{2+} influx in the store-depleted group is much higher than in the nondepleted group (see dotted line). In the store-depleted group, T1E3 significantly decreased the store-operated component ($n = 95$ each). Without store depletion, there was no effect of T1E3 ($n = 57$ each). (Adapted from Xu, S.Z., and D.J. Beech, *Circulation Research*, 88, 84, 2001. With permission.)

demonstrated in TRPC1 knockout mice[27] and in cells using the siRNA approach to knock down TRPC1 expression.[28–30]

We have not tested the T1E3 antibody in TRPC1-transfected cells because the TRPC1 current in the overexpression system is relatively small and very difficult to distinguish from the endogenous activity, despite the fact that the TRPC1 protein expression and subcellular localization have been confirmed in HEK293 cells and tsA 201 cells by the FLAG- and EYFP-tagged proteins, respectively, and by T1E3 antibody labeling (S. Z. Xu, unpublished data). It remains a mystery why the overexpressed TRPC1 cannot produce a remarkable current. The potential explanation could be due to its relatively high endogenous expression in many cell types and/or its lack of ability to traffic to the plasma membrane and assemble into store-operated channels when expressed alone.

7.3.1.2   Assessment of TRPC5 Function

The overexpressed TRPC5 channel has a typical outward–inward–outward rectification ("N" shaped) current–voltage (I–V) relationship[31] and can be activated by multiple mechanisms including a G-protein coupled receptor pathway and internal Ca^{2+} store depletion.[32] The TRPC5 channel can also be directly activated by lanthanides (Gd^{3+} and La^{3+})[22] and some reducing agents (thioredoxin, Tris[2-carboxyethyl] phosphine (TCEP), and dithiothreitol (DTT))[8] via modification of glutamate residues and a disulfide bridge in the S5–S6 region, respectively. The human TRPC5 current can be abolished by 2-APB.[31] Application of TRPC5 pore-blocking antibody (T5E3) also significantly inhibited the TRPC5 current (Figure 7.3). The effect is specific for TRPC5 because the pore-blocking antibody T5E3 has no effect on cells expressed with TRPC6, and the antigenic peptide-absorbed antibody did not show any effect.[6]

The store-operated Ca^{2+} entry has been assessed in the rabbit arteriole using T5E3. The store-operated Ca^{2+} entry is significantly inhibited by T5E3, but no effect

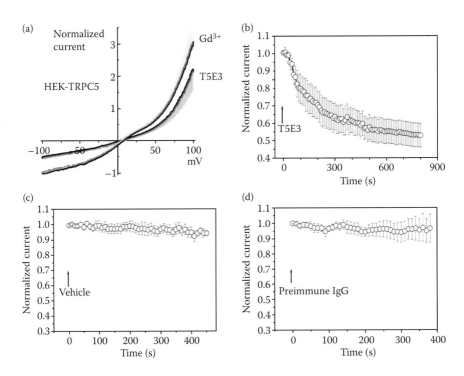

FIGURE 7.3 Pore-blocking antibody T5E3 inhibits the TRPC5 current. (a) Whole-cell current evoked by 10 μM gadolinium (Gd^{3+}) in HEK293 cells with inducible expression of human TRPC5 by tetracycline (HEK-TRPC5). Current–voltage relationship for TRPC5 activated by Gd^{3+} is similar to a previous report.[31] (b) The time course for acute application of T5E3 ($n = 7$). The current was measured at –80 mV. (c) Antibody vehicle (PBS) control ($n = 8$). (d) Preimmune IgG control ($n = 4$). (Adapted from Xu, S.Z. et al. *Nature Biotechnology*, 23, 1289, 2005. With permission.)

was observed in the arterioles without store depletion, suggesting that TRPC5 contributes to store-operated Ca^{2+} entry in the native cells.[6]

The blocking effect of T5E3 has also been examined in cerebral arterioles treated with La^{3+} because lanthanides have a unique stimulating effect on TRPC5 channels. After perfusion with 50 μM La^{3+}, the response to La^{3+} in the store-depleted cerebral arteriole displayed two phases, i.e., an initial inhibitory phase and a secondary stimulatory phase as the intracellular Ca^{2+} concentration gradually rose above the baseline.[4] T5E3 enhanced the inhibition of La^{3+} on the initial phase, which is presumed to be a store-operated component, and inhibited the stimulating phase of La^{3+}, which could be mediated by the homomeric TRPC5 channel.[4] These data suggest that native TRPC5 may also constitute a store-operated channel or contribute to the store-operated Ca^{2+} influx in smooth muscle cells.

7.3.1.3 Store-Operated TRPC1/5 Heteromultermeric Channel

The endogenous store-operated channel in cerebral arteriolar smooth muscle cells shows a tendency toward outward rectification,[6] which is quite different from the inward rectification or the "N"-shaped I–V curve seen for the overexpressed human TRPC5[31,32] and mouse TRPC5 channels.[33–38] Therefore, it is unlikely that the native store-operated channel is formed by homomeric TRPC1 or TRPC5 channels. Because overexpressed TRPC5 forms complexes with TRPC1 and heterologous co-expression of TRPC1 and TRPC5 gives rise to currents with outward rectification,[24,34] we tested the idea that the native store-operated channel in vascular smooth muscle could be a complex of TRPC5 and TRPC1. We found that TRPC1 immunoprecipitated with TRPC5 from cell lysates prepared from human saphenous vein and from tsA 201 cells co-expressing TRPC5 and TRPC1. The I–V curve of store-operated currents in the TRPC1/TRPC5 co-expressing cells is similar to that recorded from arteriolar smooth muscle cells. Therefore, we suggested that TRPC1/5 heteromultimeric channels could be an important component for the SOC in native smooth muscle cells.

7.3.1.4 Other Store-Operated Channels

STIM1 and ORAI1 have been reported to contribute to store-operated Ca^{2+} entry in vascular smooth muscle cells.[26,39] Recently, functional interactions among STIM, ORAI, and TRPC proteins have been demonstrated in several studies,[40–45] although independent signaling and localization of TRPCs and STIM1/ORAI1 have also been described.[46] An extracellular anti-STIM1 antibody significantly inhibited the store-operated current in the smooth muscle cells,[17] suggesting that plasma membrane STIM1 protein contributes to store-operated Ca^{2+} entry. This further demonstrates the utility of functional antibodies as tools for exploring ion channel functions in native cells.

7.3.2 Assessing Native Cell Function

7.3.2.1 Smooth Muscle Cell Proliferation and Migration

Proliferation and migration are two important non-contractile properties of smooth muscle cells. They are involved in neo-intimal growth, which is a key pathological process for coronary artery re-stenosis that usually happens after cardiac angioplasty surgery. The pore-blocking antibody T1E3 has been used as a specific tool for studying smooth muscle cell proliferation. Incubation with T1E3 significantly inhibited neo-intimal growth in human saphenous vein, and the effect was confirmed by transfection of TRPC1 antisense oligos.[47] In addition, proliferation of the smooth muscle cell line A7r5 was also inhibited by T1E3, while the control pre-immune serum had no effect on cell proliferation.[47]

To study cell migration, the wound injury model is used for the primary cultured smooth muscle cells.[24] Sphingsine-1 phosphate (S1P) can significantly induce smooth muscle migration. This effect is mediated by the TRPC5 channel because the incubation with the TRPC5 blocking antibody (T5E3) inhibited the S1P-induced migration, an effect also confirmed by transfection with a dominant negative TRPC5 mutant (DN-TRPC5).[24]

7.3.2.2 TRPCs Regulate Cell Secretory Function

Ca^{2+} is important in the regulation of cell secretion. The involvement of TRPCs in the cell secretory function has been described in salivary gland epithelial cells for fluid secretion,[27,48] GnRH neurons for gonadotropin-releasing hormone (GnRH) secretion,[49] and rat pituitary cells for adrenocorticotropin release.[50] The basal secretion of alkaline phosphatase is enhanced in COS-7 cells transfected with the cDNA for TRPC3 or TRPC7, but not that for TRPC1.[51]

Using TRPC1 and TRPC5 pore-blocking antibodies, it was shown that the levels of metalloproteinases MMP2 and MMP9 secreted by fibroblast-like synoviocytes (FLS cells) were significantly increased by the treatment of T1E3 and T5E3, suggesting that TRPC1 and TRPC5 channels are involved in the cell secretory function. This increasing effect by TRPC channel blockade has been confirmed by transfection with TRPC1 and TRPC5 siRNAs, suggesting that the constitutive and endogenous threodoxin-induced TRPC5 channel activity is essential for regulating cell secretion.[8]

7.3.2.3 TRPC Channel and Smooth Muscle Contraction

The role of TRPCs in smooth muscle contraction is still unclear, especially owing to the lack of in vivo experimental data. Application of the TRPC1 pore-blocking antibody for investigating smooth muscle contractility was reported in 2003.[7,52] Interestingly, the endothelin-1 induced contraction in rat tail artery was significantly inhibited by T1E3, but no effect was observed in the basilar arteries.[7] The reason is unclear, probably owing to the interaction with lipid raft or the differential expression of TRPCs in these tissues.[7]

TRPC6 is an important and highly expressed TRP isoform in vascular smooth muscle cells.[53,54] TRPC6 antisense oligodeoxynucleotides inhibited contraction in the organ cultured cerebral arteries.[55] On the contrary, elevated blood pressure was reported in TRPC6 knockout mice, which has been explained as due to compensation by TRPC3.[56]

The upregulation of the TRPC gene or protein expression has been reported in hypertension, such as TRPC6 in hypertensive rats,[57] TRPC3 and TRPC6 in idiopathic pulmonary hypertensive patients,[58,59] and TRPC3 and TRPC5 in essential hypertension.[60] These data suggest that TRPCs are important for regulating vascular tone, providing a clue that TRPCs could be good new molecular targets for antihypertensive therapy.

7.4 ADVANTAGES AND DISADVANTAGES

There are several advantages for application of pore-blocking antibodies: (1) The successful antibody can be used as a pharmacological tool for acute experiments in vivo. (2) Unlike synthetic chemicals, the pore-blocking antibodies are usually nontoxic and could be introduced for in vivo experiments. (3) The procedures for antibody generation are well established, so it should be a simple and quick way to make an antibody tool. (4) The pore-blocking antibody also can be used as an ordinary antibody for Western blotting, immunolabeling, immunoprecipitation, and ELISA if successful.

The disadvantages include the following: (1) Antibodies are unstable, and therefore, the proper storage and transportation are important. (2) Large-scale production of the antibody could be a problem, resulting in the limited use. (3) Batch-to-batch variations exist, calling for antibody titration for each batch. (4) Generation of a monoclonal antibody is still costly, and the affinities of monoclonal antibodies are generally lower than polyclonal antibodies. (5) Antibodies sometimes display unexpected cross reactions with unrelated antigens. (6) The ability for tissue penetration by the antibody should be considered if used for in vivo studies.

7.5　ENDOGENOUS FUNCTIONAL ANTIBODIES AND CLINICAL IMPLICATIONS

Many endogenous autoantibodies targeting ion channels have been identified; some of them are related to disease development. For example, antibodies to voltage-gated calcium, potassium channels, and glutamate receptors have been detected in the sera and cerebrospinal fluids of patients with ataxia, limbic encephalitis, and certain forms of epilepsy.[61–63] Some autoantibodies could be stimulatory, such as GluR3 autoantibodies.[64] Generally, the autoantibodies are regarded as pathogenic factors. Recently, a TRPC3 autoantibody has been reported in patients with myasthenia gravis, and this could contribute to the contractile abnormalities of the skeletal muscle.[65]

Therapeutic antibodies targeting endogenous proteins have been reported to treat a range of noninfectious diseases, such as cancer,[66] Alzheimer's disease,[67] and stroke.[68] Some targets are membrane proteins, such as the VEGF receptor,[69] suggesting that antibodies targeting specific endogenous protein or protein segment are of potential therapeutic significance. Although the pore-blocking antibodies for TRPC channels have not been tested in in vivo studies, there is a great potential for them in therapeutic development if TRPCs are linked to human diseases.

7.6　CONCLUSION

The pore-blocking antibodies can be used as powerful pharmacological tools to explore the ion channel function, especially for studying the native cell function via acute application. Unlike the procedures using siRNA or gene knockout, the blocking antibody is thought to directly interact with the channel pore and stop the ion flow. The E3-targeting methodology provides a foundation for pore-blocking antibody generation targeting ion channel subunits with six transmembrane segments. However, this strategy can be adopted for other ion channels, for instance, in the case of channels with one or two extracellular loops or even no loop, but an extracellular N-terminus. Although there is no report of a therapeutic antibody specifically developed for targeting ion channels, the successful generation and application of the pore-blocking antibodies may pave the way for further development in this field of potential immunotherapeutics for ion-channel-related diseases.

ACKNOWLEDGMENT

This work was supported by the British Heart Foundation and HYMS prime pump award.

REFERENCES

1. International Human Genome Sequencing Consortium. 2004. Finishing the euchromatic sequence of the human genome. *Nature* 431: 931–945.
2. Lander, E. S., L. M. Linton, B. Birren et al. 2001. Initial sequencing and analysis of the human genome. *Nature* 409: 860–921.
3. James, L. C., P. Roversi, and D. S. Tawfik. 2003. Antibody multispecificity mediated by conformational diversity. *Science* 299: 1362–1367.
4. Xu, S. Z., F. Zeng, M. Lei et al. 2005. Generation of functional ion-channel tools by E3 targeting. *Nature Biotechnology* 23: 1289–1293.
5. Xu, S. Z., and D. J. Beech. 2001. TrpC1 is a membrane-spanning subunit of store-operated Ca^{2+} channels in native vascular smooth muscle cells. *Circulation Research* 88: 84–87.
6. Xu, S. Z., G. Boulay, R. Flemming et al. 2006. E3-targeted anti-TRPC5 antibody inhibits store-operated calcium entry in freshly isolated pial arterioles. *American Journal of Physiology Heart and Circulatory Physiology* 291: H2653–H2659.
7. Bergdahl, A., M. F. Gomez, K. Dreja et al. 2003. Cholesterol depletion impairs vascular reactivity to endothelin-1 by reducing store-operated Ca^{2+} entry dependent on TRPC1. *Circulation Research* 93: 839–847.
8. Xu, S. Z., P. Sukumar, F. Zeng et al. 2008. TRPC channel activation by extracellular thioredoxin. *Nature* 451: 69–72.
9. Kwan, H. Y., B. Shen, X. Ma et al. 2009. TRPC1 associates with BK_{Ca} channel to form a signal complex in vascular smooth muscle cells. *Circulation Research* 104: 670–678.
10. Rosado, J. A., S. L. Brownlow, and S. O. Sage. 2002. Endogenously expressed Trp1 is involved in store-mediated Ca^{2+} entry by conformational coupling in human platelets. *Journal of Biological Chemistry* 277: 42157–42163.
11. Gomez-Varela, D., E. Zwick-Wallasch, H. Knotgen et al. 2007. Monoclonal antibody blockade of the human Eag1 potassium channel function exerts antitumor activity. *Cancer Research* 67: 7343–7349.
12. Klionsky, L., R. Tamir, B. Holzinger et al. 2006. A polyclonal antibody to the prepore loop of transient receptor potential vanilloid type 1 blocks channel activation. *Journal of Pharmacology Experimental Therapy* 319: 192–198.
13. Naylor, J., C. J. Milligan, F. Zeng et al. 2008. Production of a specific extracellular inhibitor of TRPM3 channels. *British Journal of Pharmacology* 155: 567–573.
14. Watschinger, K., S. B. Horak, K. Schulze et al. 2008. Functional properties and modulation of extracellular epitope-tagged Ca(V)2.1 voltage-gated calcium channels. *Channels (Austin)* 2: 461–473.
15. Conforti, L., I. Bodi, J. W. Nisbet et al. 2000. O2-sensitive K^+ channels: role of the Kv1.2-subunit in mediating the hypoxic response, *Journal of Physiology* 524 Pt 3: 783–793.
16. Miyazaki, S., M. Yuzaki, K. Nakada et al. 1992. Block of Ca^{2+} wave and Ca^{2+} oscillation by antibody to the inositol 1,4,5-trisphosphate receptor in fertilized hamster eggs. *Science* 257: 251–255.
17. Li, J., P. Sukumar, C. J. Milligan et al. 2008. Interactions, functions, and independence of plasma membrane STIM1 and TRPC1 in vascular smooth muscle cells. *Circulation Research* 103: e97–e104.
18. Yatani, A., H. Hamm, J. Codina et al. 1988. A monoclonal antibody to the alpha subunit of Gk blocks muscarinic activation of atrial K^+ channels. *Science* 241: 828–831.
19. Malouf, N. N., R. Coronado, D. McMahon et al. 1987. Monoclonal antibody specific for the transverse tubular membrane of skeletal muscle activates the dihydropyridine-sensitive Ca^{2+} channel. *Proceedings of the National Academy of Sciences of the United States of America* 84: 5019–5023.

20. Benham, C. D. 2005. Simple recipe for blocking ion channel. *Nature Biotechnology* 23: 1234–1235.

21. Feske, S. 2009. ORAI1 and STIM1 deficiency in human and mice: roles of store-operated Ca^{2+} entry in the immune system and beyond. *Immunology Review* 231: 189–209.

22. Jung, S., A. Muhle, M. Schaefer et al. 2003. Lanthanides potentiate TRPC5 currents by an action at extracellular sites close to the pore mouth. *Journal of Biological Chemistry* 278: 3562–3571.

23. Zhou, B. Y., W. Ma, and X. Y. Huang. 1998. Specific antibodies to the external vestibule of voltage-gated potassium channels block current. *Journal of General Physiology* 111: 555–563.

24. Xu, S. Z., K. Muraki, F. Zeng et al. 2006. A sphingosine-1-phosphate-activated calcium channel controlling vascular smooth muscle cell motility. *Circulation Research* 98: 1381–1389.

25. Ng, L. C., M. D. McCormack, J. A. Airey et al. 2009. TRPC1 and STIM1 mediate capacitative Ca^{2+} entry in mouse pulmonary arterial smooth muscle cells. *Journal of Physiology* 587: 2429–2442.

26. Baryshnikov, S. G., M. V. Pulina, A. Zulian et al. 2009. Orai1, a critical component of store-operated Ca^{2+} entry, is functionally associated with Na^+/Ca^{2+} exchanger and plasma membrane Ca^{2+} pump in proliferating human arterial myocytes. *American Journal of Physiology Cell Physiology* 297: C1103–C1112.

27. Liu, X., K. T. Cheng, B. C. Bandyopadhyay et al. 2007. Attenuation of store-operated Ca^{2+} current impairs salivary gland fluid secretion in TRPC1$^{-/-}$ mice. *Proceedings of the National Academy of Sciences of the United States of America* 104: 17542–17547.

28. Rao, J. N., O. Platoshyn, V. A. Golovina et al. 2006. TRPC1 functions as a store-operated Ca^{2+} channel in intestinal epithelial cells and regulates early mucosal restitution after wounding. *American Journal of Physiology Gastrointestinal and Liver Physiology* 290: G782–G792.

29. Beech, D. J., S. Z. Xu, D. McHugh et al. 2003. TRPC1 store-operated cationic channel subunit. *Cell Calcium* 33: 433–440.

30. Kim, M. S., J. H. Hong, Q. Li et al. 2009. Deletion of TRPC3 in mice reduces store-operated Ca^{2+} influx and the severity of acute pancreatitis. *Gastroenterology* 137: 1509–1517.

31. Xu, S. Z., F. Zeng, G. Boulay et al. 2005. Block of TRPC5 channels by 2-amino-ethoxydiphenyl borate: a differential, extracellular and voltage-dependent effect. *British Journal of Pharmacology* 145: 405–414.

32. Zeng, F., S. Z. Xu, P. K. Jackson et al. 2004. Human TRPC5 channel activated by a multiplicity of signals in a single cell. *Journal of Physiology* 559: 739–750.

33. Schaefer, M., T. D. Plant, A. G. Obukhov et al. 2000. Receptor-mediated regulation of the nonselective cation channels TRPC4 and TRPC5. *Journal of Biological Chemistry* 275: 17517–17526.

34. Strubing, C., G. Krapivinsky, L. Krapivinsky et al. 2001. TRPC1 and TRPC5 form a novel cation channel in mammalian brain. *Neuron* 29: 645–655.

35. Lee, Y. M., B. J. Kim, H. J. Kim et al. 2003. TRPC5 as a candidate for the nonselective cation channel activated by muscarinic stimulation in murine stomach. *American Journal of Physiology Gastrointestinal and Liver Physiology* 284: G604–G616.

36. Obukhov, A. G., and M. C. Nowycky. 2005. A cytosolic residue mediates Mg^{2+} block and regulates inward current amplitude of a transient receptor potential channel. *Journal of Neuroscience* 25: 1234–1239.

37. Okada, T., S. Shimizu, M. Wakamori et al. 1998. Molecular cloning and functional characterization of a novel receptor-activated TRP Ca^{2+} channel from mouse brain. *Journal of Biological Chemistry* 273: 10279–10287.

38. Philipp, S., J. Hambrecht, L. Braslavski et al. 1998. A novel capacitative calcium entry channel expressed in excitable cells. *EMBO Journal* 17: 4274–4282.

39. Potier, M., J. C. Gonzalez, R. K. Motiani et al. 2009. Evidence for STIM1- and Orai1-dependent store-operated calcium influx through ICRAC in vascular smooth muscle cells: role in proliferation and migration. *FASEB Journal* 23: 2425–2437.

40. Pani, B., H. L. Ong, S. C. W. Brazer et al. 2009. Activation of TRPC1 by STIM1 in ER-PM microdomains involves release of the channel from its scaffold caveolin-1. *Proceedings of the National Academy of Sciences of the United States of America* 106: 20087–20092.

41. Liao, Y. Y., N. W. Plummer, M. D. George et al. 2009. A role for Orai in TRPC-mediated Ca^{2+} entry suggests that a TRPC: Orai complex may mediate store and receptor operated Ca^{2+} entry. *Proceedings of the National Academy of Sciences of the United States of America* 106: 3202–3206.

42. Huang, G. N., W. Z. Zeng, J. Y. Kim et al. 2006. STIM1 carboxyl-terminus activates native SOC, I-crac and TRPC1 channels. *Nature Cell Biology* 8: 1003–1010.

43. Worley, P. F., W. Zeng, G. N. Huang et al. 2007. TRPC channels as STIM1 - regulated store-operated channels. *Cell Calcium* 42: 205–211.

44. Kim, M. S., W. Zeng, J. P. Yuan et al. 2009. Native store-operated Ca^{2+} influx requires the channel function of Orai1 and TRPC1. *Journal of Biological Chemistry* 284: 9733–9741.

45. Zeng, W. Z., J. P. Yuan, M. S. Kim et al. 2008. STIM1 gates TRPC channels, but not Orai1rai1, by electrostatic interaction. *Molecular Cell* 32: 439–448.

46. DeHaven, W. I., B. F. Jones, J. G. Petranka et al. 2009. TRPC channels function independently of STIM1 and Orai1. *Journal of Physiology* 587: 2275–2298.

47. Kumar, B., K. Dreja, S. S. Shah et al. 2006. Upregulated TRPC1 channel in vascular injury in vivo and its role in human neointimal hyperplasia. *Circulation Research* 98: 557–563.

48. Bandyopadhyay, B. C., W. D. Swaim, X. B. Liu et al. 2005. Apical localization of a functional TRPC3/TRPC6-Ca^{2+}-signaling complex in polarized epithelial cells—role in apical Ca^{2+} influx. *Journal of Biological Chemistry* 280: 12908–12916.

49. Zhang, C. G., T. A. Roepke, M. J. Kelly et al. 2008. Kisspeptin depolarizes gonadotropin-releasing hormone neurons through activation of TRPC-like cationic channels. *Journal of Neuroscience* 28: 4423–4434.

50. Yamashita, M., Y. Oki, K. Lino et al. 2009. The role of store-operated Ca^{2+} channels in adrenocorticotropin release by rat pituitary cells. *Regulatory Peptides* 156: 57–64.

51. Lavender, V., S. Chong, K. Ralphs et al. 2008. Increasing the expression of calcium-permeable TRPC3 and TRPC7 channels enhances constitutive secretion. *Biochemical Journal* 413: 437–446.

52. Bergdahl, A., M. F. Gomez, A. K. Wihlborg et al. 2005. Plasticity of TRPC expression in arterial smooth muscle: correlation with store-operated Ca^{2+} entry. *American Journal of Physiology Cell Physiology* 288: C872–C880.

53. Inoue, R., T. Okada, H. Onoue et al. 2001. The transient receptor potential protein homologue TRP6 is the essential component of vascular alpha(1)-adrenoceptor-activated Ca^{2+}-permeable cation channel. *Circulation Research* 88: 325–332.

54. Flemming, R., S. Z. Xu, and J. D. Beech. 2003. Pharmacological profile of store-operated channels in cerebral arteriolar smooth muscle cells, *British Journal of Pharmacology* 139: 955–965.

55. Welsh, D. G., A. D. Morielli, M. T. Nelson et al. 2002. Transient receptor potential channels regulate myogenic tone of resistance arteries. *Circulation Research* 90: 248–250.

56. Dietrich, A., Y. S. M. Mederos, M. Gollasch et al. 2005. Increased vascular smooth muscle contractility in TRPC6$^{-/-}$ mice. *Molecular Cell Biology* 25: 6980–6989.

57. Bae, Y. M., A. Kim, Y. J. Lee et al. 2007. Enhancement of receptor-operated cation current and TRPC6 expression in arterial smooth muscle cells of deoxycorticosterone acetate-salt hypertensive rats. *Journal of Hypertension* 25: 809–817.

58. Golovina, V. A., O. Platoshyn, C. L. Bailey et al. 2001. Upregulated TRP and enhanced capacitative Ca^{2+} entry in human pulmonary artery myocytes during proliferation. *American Journal of Physiology Heart and Circulatory Physiology* 280: H746–H755.

59. Yu, Y., I. Fantozzi, C. V. Remillard et al. 2004. Enhanced expression of transient receptor potential channels in idiopathic pulmonary arterial hypertension. *Proceedings of the National Academy of Sciences of the United States of America* 101: 13861–13866.

60. Liu, D. Y., A. Scholze, Z. M. Zhu et al. 2006. Transient receptor potential channels in essential hypertension. *Journal of Hypertension* 24: 1105–1114.

61. Lang, B., R. C. Dale, and A. Vincent. 2003. New autoantibody mediated disorders of the central nervous system. *Current Opinion in Neurology* 16: 351–357.

62. Lang, B., C. I. Newbold, G. Williams et al. 2005. Antibodies to voltage-gated calcium channels in children with falciparum malaria. *Journal of Infectious Diseases* 191: 117–121.

63. Hart, I. K., C. Waters, A. Vincent et al. 1997. Autoantibodies detected to expressed K^+ channels are implicated in neuromyotonia. *Annals of Neurology* 41: 238–246.

64. Levite, M., Y. Ganor, H. Goldberg-Stern et al. 2004. Autoimmune epilepsy: patients with epilepsy harbour autoantibodies to glutamate receptors and dsDNA on both sides of the blood-brain which drop after hemispherotomy. *Epilepsia* 45: 77–78.

65. Takamori, M. 2008. Autoantibodies against TRPC3 and ryanodine receptor in myasthenia gravis. *Journal of Neuroimmunology* 200: 142–144.

66. Plotkin, S. A. 2005. Vaccines: past, present and future. *Nature Medicine* 11: S5–S11.

67. Agadjanyan, M. G., A. Ghochikyan, I. Petrushina et al. 2005. Prototype Alzheimer's disease vaccine using the immunodominant B cell epitope from beta-amyloid and promiscuous T cell epitope pan HLA DR-binding peptide. *Journal of Immunology* 174: 1580–1586.

68. Takeda, H., M. Spatz, C. Ruetzler et al. 2002. Induction of mucosal tolerance to E-selectin prevents ischemic and hemorrhagic stroke in spontaneously hypertensive genetically stroke-prone rats. *Stroke* 33: 2156–2163.

69. Ferrara, N., K. J. Hillan, H. P. Gerber et al. 2004. Discovery and development of bevacizumab, an anti-VEGF antibody for treating cancer. *Nature Reviews Drug Discovery* 3: 391–400.

8 Strategies and Protocols to Generate Mouse Models with Targeted Mutations to Analyze TRP Channel Functions

Marc Freichel, Ulrich Kriebs, Dominik Vogt, Stefanie Mannebach, and Petra Weißgerber

CONTENTS

8.1 APPROACHES TO STUDYING THE FUNCTION OF TRP PROTEINS IN NATIVE SYSTEMS

Transient receptor potential (TRP) proteins constitute a superfamily of cation-permeable channels that show a broad range of activation, regulation, and selectivity mechanisms.[1,2] Four TRP proteins are assumed to form homo-oligomeric and hetero-oligomeric channels, and the biological roles of TRP channels appear to be highly diverse. They essentially contribute to thermosensation and pain perception, Ca^{2+} and Mg^{2+} absorption, endothelial permeability, smooth muscle proliferation, and mast cell degranulation in mice and other mammals.[3–5] The majority of our knowledge of the properties of TRP channels arises from diseases linked to mutations in TRP channels[6,7] and from recordings of the activity of ectopically expressed TRP channel proteins in cell lines such as HEK 293 cells. In this environment, they do not necessarily act in accordance with the native cellular context as in primary cells.

There are many obstacles that hinder the analysis of endogenous TRP channels, but studying TRP proteins in cell systems that most perfectly resemble the environment of their native occurrence will certainly improve our understanding of the molecular makeup, the properties, the physiological functions, and the regulatory mechanisms of TRP channels. Because of the lack of specific pharmacology for TRPC and most other TRP channels, genetic approaches are required to advance a causal understanding of their physiological functions in primary cells, in organs, for systemic functions of organisms, and for disease states. These approaches include overexpression of dominant negative variants, antisense oligonucleotides, and RNA interference (RNAi), as well as targeted deletion of the gene of interest using homologous recombination.

The use of overexpressing dominant negative variants, including domains of these proteins that interact with other TRP proteins[8,9] or pore mutants,[10–12] was shown to suppress TRP channel functions to various degrees. This approach can perturb homomeric and heteromeric channel complexes, but the effectiveness of neutralization by dominant negative variants may be limited in the analysis of proteins with a long turnover time, which is unknown for most native TRP proteins.

The application of antisense RNA to downregulate TRP gene expression was used initially after cloning of the first mammalian trp homologues[13] but was soon replaced by RNAi technologies that became popular, not just in the TRP field, after the demonstration that this principle is not only effective in nematodes,[14] but can also be used to knock down gene expression in cultured mammalian cells when small interfering RNA (siRNA) duplexes are used.[15] Despite the merits of this technology, it needs to be noted that off-target effects are not unusual[16] and may lead to unwanted toxicity and false-positive results. For example, Aarts et al. achieved complete suppression of TRPM7 mRNA 6–8 days after transfection of a TRPM7-specific siRNA in neurons.[17] However, this siRNA specifically targeting TRPM7 also significantly reduced TRPM2 expression levels. Also, specific antibodies are rare for most TRP proteins (see Chapter 6) and, accordingly, it is difficult to control the effectiveness of RNAi approaches. As a result, the situation arises in many cases that at the same time, a poorly characterized RNAi experiment is controlled by nonvalidated antibodies and vice versa. Another drawback of the use of RNAi is that effective knockdown of a TRP protein may take many days after the introduction of RNAi in cells until the assembled functional proteins at the cell surface are degraded. This is a major disadvantage for the analysis of the function of TRP proteins especially in those primary cells that cannot be cultured in vivo over such long periods without obvious morphological and functional changes, like cardiomyocytes from adult animals. RNAi-based approaches have been used in mice[18] but may activate the innate immune response.[19-21] These limitations can be overcome by the generation of transgenic animals in which the genetic modification is stably introduced into germ cells.

Genetically modified mice can be generated by either direct pronuclear injection of exogenous DNA into fertilized zygotes or injection of murine embryonic stem (ES) cells, with defined genetic mutations introduced by gene targeting into a blastocyst. Direct pronuclear injection results in random integration of the injected DNA into the genome, and the phenotypes rely on the extent of overexpression of the transgene if the functionality of endogenous gene loci is not affected. In this way, several TRP proteins were overexpressed, for example, under cardiomyocyte-specific promoters.[22-24] This approach demonstrates what these proteins are able to accomplish in cardiomyocytes residing in their native environment but cannot answer the question as to whether the corresponding TRP protein or channel is required for physiological functions in cardiomyocytes. It also remains to be determined whether such upregulation and gain of function is actually relevant under pathophysiological conditions.

A causal relation between the presence of a TRP protein and cellular or systemic functions in vivo can be established by using gene targeting approaches that are illustrated in this contribution. By this means, either deletions of essential parts of a TRP-encoding gene or subtle modification, such as pore mutations resulting in functionally inactive TRP proteins, can be evoked. Potentially, the latter strategy may have the advantage of creating functionally inactive TRP channel complexes without disturbing the channel architecture and subsequent alterations of signaling cascades that rely on direct interaction with TRP protein domains. Although gene targeting is time-consuming and laborious, the obtained animals, if they are viable and lack severe developmental alterations, can serve as an infinite source of primary cells and tissues in which defined TRP proteins are effectively deleted with sufficient specificity. This approach cannot exclude that inactivation of a given TRP protein may be compensated by upregulation or downregulation of other genes, including functionally related TRP genes. For example, upregulation of TRPC3,[25] but not of TRPC4,[26] has been shown in TRPC6-deficient mice. Conditional inactivation of the encoding gene in a time- or cell-type–dependent manner may be the solution in cases when compensation occurs. Alternatively, mouse models in which all redundant TRP proteins are inactivated simultaneously (compound knockout mice) can be used.

8.2 GENERATION OF TRANSGENIC ANIMALS BY GENE TARGETING STRATEGIES

ES cells have the advantage that they can be genetically modified by means of homologous recombination, a process by which a fragment of genomic DNA introduced into

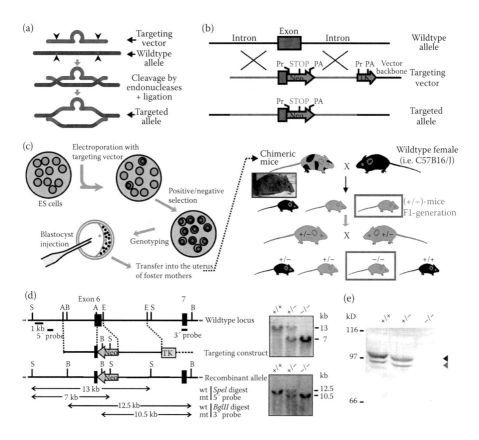

FIGURE 8.1 **(See color insert.)** Generation of knockout mice using homologous recombination in ES cells and classical gene targeting strategies. (a) Recombination can result in an exchange of homologous DNA sequences (blue) and the nonhomologous sequences in between (red). The nonhomologous sequences often encode a positive selection marker gene such as the neomycin resistance gene. (b) Targeted disruption of an endogenous gene using a classical strategy. Targeting vector sequences contain the Neomycin resistance (*Neo*) gene that is inserted in an exon of the targeted gene and a linked herpes virus thymidine kinase (*TK*) gene at one end. Each selection marker contains its own promoter (Pr) and polyadenylation (PA) signal. The vector is shown pairing with a chromosomal copy of the gene. (c) Generation of mouse germ line chimeras and mice homozygous for a targeted allele from embryonic stem cells. The first step involves the isolation of a clonal ES cell line that contains the desired mutation. Positive-negative selection can be used to enrich ES cell lines that contain the desired modified gene. The second step is that these ES cells are used to generate chimeric mice that are able to transmit the mutant gene to their progeny by injecting ES cells containing the desired targeted mutation into a blastocyst. These embryos are then surgically transferred to a recipient pseudopregnant foster mother. To facilitate isolation of the desired progeny, the ES cells and recipient blastocysts are derived from mice with distinguishable coat-color alleles (e.g., ES cells such as R1 from agouti brown mice and blastocysts from black mice such as C57Bl6/J). In this case, the extent of the contribution of ES cells to the formation of the chimeric mouse can be evaluated by visual assessment of coat color. Using R1 ES cells, breeding of chimeric mice with C57Bl6/J mice may lead to progeny with mice that are heterozygous for the mutation and that carry the agouti gene and have brown coat color.

a mammalian cell can recombine with the endogenous homologous sequence.[27,28] This process is known as "gene targeting." When such genetically modified ES cells are introduced into a pre-implantation embryo, they can contribute, even after extensive in vitro manipulation, to all cell lineages of the resulting chimeric animal. This contribution also comprises germ cells, so that the breeding of germ line chimeras, which transmit an ES cell-derived mutant chromosome to their progeny, allows the establishment of an animal heterozygous for the genetic alteration. A homozygous mutant mouse strain can be obtained by further breeding.

The gene targeting strategy takes full advantage of all resources provided by the known sequences of the mouse and human genomes (NCBI, http://www.ncbi.nlm.nih.gov/; Sanger Institute, http://www.sanger.ac.uk/). Compared to other procedures for introducing mutations into mice, the investigator can choose which genetic locus to mutate and has complete control on how to modulate the chosen genetic locus. Some loci are located very close to their neighboring genes. Depending on the targeting strategy, inactivation of the gene of interest may also affect and inactivate a closely adjacent neighbor gene. For instance, Bianco et al. have described a mouse model in which the TRPV6 knockout affects the closely adjacent EphB6 gene.[29]

8.2.1 "CLASSICAL" GENE TARGETING

A targeting vector is designed to recombine with and mutate a specific chromosomal locus of interest (Figure 8.1a). Because both the transfection efficiency and the targeting frequency of such a vector can be low, positive and negative selection markers such as a neomycin resistance gene and a thymidine kinase, which are both flanked by promoter sequences active in ES cells and polyadenylation sites, should be included to provide a strong selection for cells with the targeted recombination event.

In classical knockout strategies, the targeted exon(s) is (are) replaced by a positive selection cassette. This cassette often includes a stop codon, so that translation of transcripts from the targeted allele is interrupted. Homologous recombination between the targeting vector and the cognate chromosomal gene results in the disruption of one allele and the loss of the negative selection gene. Accordingly, cells in which this event has occurred will be heterozygous (+/−) for the targeted gene and will be resistant to both G418 and nucleoside analogs such as ganciclovir, owing to insertion of the neomycin resistance gene and the absence of the thymidine kinase (Figure 8.1b). However, the targeting vector will be most frequently integrated into the host cell genome at a random site through nonhomologous recombination. Because nonhomologous insertion of exogenous DNA into the host cell chromosome occurs

FIGURE 8.1 (Continued) Mice homozygous for the mutation are obtained by heterozygous intercrosses. (d) The genotype of the mice can be assessed by Southern blot analysis. Using a 5′ probe (3′ probe), a specific 13-kb (12.5-kb) band proves the wild-type allele, whereas a specific 7-kb (10.5-kb) band indicates the targeted allele. (e) Lack of expression of the gene is usually verified by Western blot analysis. Here, the two isoforms of the TRPC4 gene product (102 and 94 kDa) are not detectable in TRPC4$^{-/-}$ mice. (Modified from Freichel, M. et al., *Nature Cell Biology* 3, 121, 2001. With permission.)

through the ends of the linearized targeting vector, the negative selection gene will also be inserted into the genome and stay functional in the majority of cases. Cells derived from this type of recombination event are wild type for the targeted gene and contain both the positive and negative selection cassette. Therefore, these cells are resistant toward positive selection but are killed by negative selection.

This approach was frequently used for the generation of TRP-deficient mice[3] (Table 8.1). However, the analysis of gene functions by classical knockout strategies is hampered by several drawbacks of this approach, including the possibility of embryonic lethality when the gene is essential for early development. Furthermore,

---

TABLE 8.1

TRP-Deficient Mouse Lines Generated by a Classical Gene Targeting Approach

Gene	Strategy	Reference
TRPC1	Replacement of exon 8 by neo cassette	58
TRPC2	Replacement of exons 7–10 by neo cassette	59
TRPC2	Replacement of exons 6–11 by neo cassette	60
TRPC4	Replacement of exon 6 by neo cassette	53
TRPC6	Replacement of exon 7 by neo cassette	25
TRPC7	No mouse line reported	
TRPV1	Replacement of sequences encoding transmembrane segments 5 and 6 by neo cassette	61
TRPV1	Replacement of sequences encoding amino acids 460–555 of mTRPV1 by neo cassette	62
TRPV2	No mouse line reported	
TRPV3	Replacement of exons 14+15 by neo cassette	63
TRPV4	Replacement of exon 12 by neo cassette	64
TRPV6	Replacement of exons 9–15 by neo cassette	29
TRPM1	Replacement of exons 3–5 by a selection cassette	65, 66
TRPM1	Replacement of exons 4–6 by neo cassette	67
TRPM2	Replacement of the exon encoding the transmembrane segment 5 and part of the linker between segments 5 and 6 by neo cassette	68
TRPM3	No mouse line reported	
TRPM5	Replacement of exons 15–19 by neo cassette	69
TRPM5	Replacement of promoter and exons 1–4 by neo cassette	70
TRPM6	Replacement of exons 5–7 by neo cassette	71
TRPM8	Partial replacement of exons 13 and 14 by self-excising ACN cassette	37
TRPP1 (PKD1)	Replacement of exon 34 by neo cassette	72
TRPP2 (PKD2)	Insertion of neo cassette into exon 1	73
TRPP2 (PKD2)	Replacement of exon 1 by lacZ-neo cassette	74
TRPML1 (MCOLN1)	Replacement of exons 3, 4, and 5 by neo cassette	75
TRPA1	Replacement of parts of exon 23 by self-excising ACN cassette	38

the positive selection marker, such as the neomycin resistance cassette, remains in the genome and may cause position effects that can affect proper expression of neighboring genes.

8.2.2 CONDITIONAL GENE TARGETING USING SITE-SPECIFIC RECOMBINASE SYSTEMS

The application of site-specific recombinase systems, along with gene targeting techniques in ES cells, has now made it possible to modify the mouse genome in almost any desired manner, from creating specific point mutations to achieving large site-specific chromosomal rearrangements.[30–32] Site-specific recombinase systems consist of two basic elements: the recombinase enzyme and a target DNA sequence that is specifically recognized by the particular recombinase that catalyzes recombination between two of its recognition sites. This results in the modification of the associated DNA, such as deletion, insertion, inversion, or translocation, depending on the orientation and location of the recognition sites. Currently, two recombinase systems are frequently used in mouse gene targeting experiments: the Cre-*loxP* system from the bacteriophage P1 and the Flp-*FRT* system from the budding yeast *Saccharomyces cerevisiae*. Each of these recombinases (Cre or Flp) recognizes a 34-bp consensus sequence consisting of two 13-bp inverted repeats flanking an 8-bp nonpalindromic core that define the orientation of the overall sequence of the recognition site (*loxP* or *FRT*). This minimal 34-bp target site is very unlikely to occur at random in the mouse genome yet is small enough to be considered a "neutral" sequence when integrated into intronic chromosomal DNA sequences. Recombination can occur over large distances (up to several mega base pairs) and in a wide range of cell types both in vitro and in vivo. The use of site-specific recombinases allows for deletion of defined exons of a gene and/or the selection cassette after successful homologous recombination. If these sequences are flanked with *loxP* or *FRT* sites, they can be removed by Cre or Flp expression (Figure 8.2a, b). This was achieved in vitro by transient transfection of recombinant ES cells with a Cre- or Flp-encoding plasmid for generating a TRPC5 null allele,[33] but this modification requires additional passages of the ES cell clones and may therefore compromise germ line competency. Alternatively, recombination can be achieved in vivo by crossing mice containing the *loxP*-flanked marker with a mild Cre deleter mouse line such as EIIa-Cre[34,35] to overcome this problem. As a third possibility, the selection marker may be directly linked to a Cre transgene driven by a promoter that initiates transcription during spermatogenesis, with both transgenes being flanked by a common *loxP* element.[36] As a consequence, self-excision of the complete cassette occurs in chimeric mice in all sperm derived from the recombinant ES cells. This approach enables the excision of selection markers without additional breeding using mice expressing a Cre transgene and thus saves 3–6 months in comparison to conventional approaches. Such an approach was used to generate TRPM8-[37] and TRPA1-deficient mice.[38]

For a conditional (i.e., cell type restricted or time dependent) inactivation of a gene, two separate mouse strains are typically generated and intercrossed (Figure 8.2c). One mouse line expresses the recombinase in selected lineages or tissues, while the other mouse line carries a gene segment flanked by recognition target

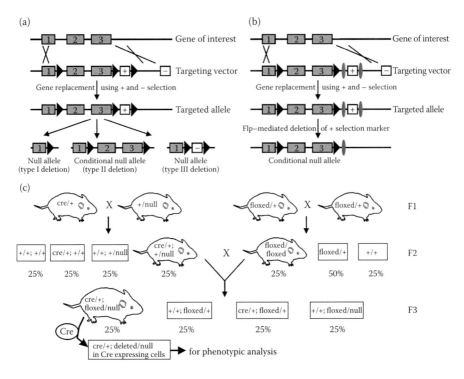

FIGURE 8.2 Strategies of conditional gene targeting. The gray boxes represent exons of the gene of interest; the box with "+" represents a positive selectable marker such as the neomycin resistance gene (neo); box with "−" depicts a negative selectable marker such as herpes simplex virus thymidine kinase (HSV-TK); black triangles, *loxP* sites; gray ovals, *FRT* sites: (a) Scheme of a 3-*loxP* strategy. After the gene replacement step, transient expression of Cre will generate three types of ES cell subclones due to partial recombination: one harboring a "floxed" allele with exons 2 and 3 flanked by *loxP*-sites (type II deletion) and two null alleles (type I and III deletions). The three different deletion events can be distinguished by PCR and Southern blot analyses. The negative selection marker is placed at one end of the targeting vector. (b) Scheme of a 2-*loxP*-2-*FRT* strategy: a two-step strategy using both the Cre-*loxP* and Flp-*FRT* recombination systems. The positive selection marker flanked by *FRT* sites is removed by Flp-mediated deletion either in vitro or in vivo after the gene replacement step. (c) Basic breeding scheme for conditional gene inactivation. Mice harboring the tissue-specific Cre transgene and heterozygous for the floxed and null alleles (*Cre/+; floxed/ null*) can be generated after two generations at a frequency of 25%, which will be used for phenotypic analysis. "+" represents the wild-type allele. (Modified from K.M. Kwan, *Genesis* 32, 49, 2002.)

sites ("floxed allele"). The standard breeding scheme to achieve conditional gene inactivation in mice in vivo is shown in Figure 8.2c. Basically, mice harboring a tissue-specific Cre transgene and that are heterozygous for the floxed and null alleles (*Cre/+; floxed/null*) can be generated after two generations at a 25% frequency. This frequency assumes that the floxed locus and the Cre transgene are not linked. In this

F3 generation, the target gene segment is deleted in cells that express the recombinase, whereas the target gene remains functional in cells of all other tissues where the recombinase is not expressed. The null allele can come from a previous knockout mouse line or the deleted conditional allele (type I deletion in Figure 8.2a) as a by-product of the excision of the selection marker. It is also possible to use *floxed/+* mice instead of *null/+* to generate (*Cre/+*; *floxed/+*) males in case there is no heterozygous null mouse strain available. However, this will impose a greater demand for the recombinase activity, as two floxed target alleles need to be excised to achieve conditional gene inactivation in the F3 offspring. However, many variations in these breeding schemes can be used, depending on genetic background requirements, fertility, and viability of each genotype.

These techniques have proven to be very useful tools for activating and inactivating specific genes in a "conditional" manner in spatially and temporally restricted patterns. This helps to circumvent the limitations of conventional gene inactivation, in particular, for a given gene that has multiple roles at different times and in different tissues. These methodologies are particularly useful for elucidating a complete picture of gene function in which the conventional knockout leads to an early lethal phenotype preventing the study of its later roles. The conditional technique is also very advantageous if the phenotype of the global knockout affects multiple tissues, preventing the detailed study of its function in a particular cell lineage. For example, the deletion of mouse Trpm7 revealed that it is essential for embryonic development, so that TRPM7$^{-/-}$ embryos died prenatally.[39,40] Using lck-Cre mice, Jin and coworkers were able to selectively delete Trpm7 in the T cell lineage.[39] Inactivation of TRPM7 abrogated Mg^{2+}-inhibited currents (I_{MIC}) in thymocytes, impaired developmental block in transition of CD4– CD8– thymocytes to CD4+ CD8+ cells, and caused a progressive depletion of thymic medullary cells. Additional mouse models that allow conditional inactivation of TRP genes are described in Table 8.2.

TABLE 8.2
TRP-Deficient Mouse Lines That Allow Conditional Inactivation

Gene	Strategy	Reference
TRPC3	Excision of exon 7 using Mox2-Cre transgenic mice	76
TRPC5	Excision of exon 5 using Cre mediated recombination in ES cells	33
TRPV4	Deletion of exon 4 using EIIa-cre transgenic mice	77
TRPV5	Deletion of exon 13 using EIIa-cre transgenic mice	78
TRPM4	Deletion of exons 15+16 using CMV-cre transgenic mice	79
TRPM7	Ubiquitous deletion of exon 17 using Cre deleter mice; generation of the floxed allele by Flp-recombinase-expressing transgenic mice; conditional disruption of Trpm7 in the T-cell-lineage using Lck-Cre transgenic mice	39
TRPP1 (PKD1)	Ubiquitous deletion of exons 2, 3, and 4 using Meox Cre mice; conditional disruption of PKD1 in various tissues including kidney and liver using MMTV-Cre transgenic mice	80

8.2.3 DETECTION OF THE INTRODUCED MUTATIONS USING PCR AND SOUTHERN BLOTTING

One of the most important aspects of any gene targeting experiment is to confirm that the desired genetic change has occurred. It is important to detect the different classes of integration events and particularly to distinguish correct targeting events from random integration of the entire vector. Therefore, the design of Southern blot strategies and testing of Southern probes should be done before constructing a gene targeting vector. Appropriate probes should be tested using genomic DNA isolated from wild-type ES cells and mouse tissue (e.g., tail biopsies).

For Southern blot analysis, an appropriate combination of restriction digests and 5′ as well as 3′ probes that bind to genomic sequences external to the gene targeting vector (external 5′ and/or 3′ probe) must be identified, to make it possible to distinguish the wild-type from the predicted homologous targeted allele. Usually, the restriction digest for the 5′ probe should be performed with an enzyme that cuts 5′ of the 5′ homologous region and in the inserted region. For the 3′ probe, a digestion should be used that cuts 3′ of the 3′ homologous region and in the inserted region (Figure 8.1d). There are restriction enzymes that are well suitable for digesting genomic DNA isolated from ES cells and from mouse biopsies, such as *BamHI*, *BglII*, *EcoRV*, *HindIII*, *KpnI*, *NheI*, *AflII*, *ScaI*, and *StuI*, whereas some other restriction enzymes do not perform well on DNA prepared from mouse biopsies, such as *NsiI*, *XhoI*, *SmaI*, *AclI*, and *SalI*.[41–43]

If a targeted allele is inherited in the expected Mendelian ratio, intercrosses of heterozygous mice produce wild-type (+/+), heterozygous (+/mutant), or homozygous (mutant/mutant) offspring, which now have to be defined by a PCR-based genotyping strategy. A commonly used strategy that distinguishes between a wild-type and a null allele and is based on two allele-specific primer pairs is shown in Figure 8.3.

8.2.4 KNOCK-IN OF REPORTER GENES

In addition to gene deletion and the introduction of defined mutations, gene targeting approaches have also been used to incorporate gene-based reporters for tagging

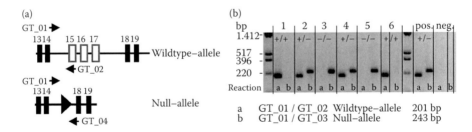

FIGURE 8.3 PCR-based genotyping of mutant mice. (a) Schematic overview of wild-type allele and null allele. Open boxes, targeted exons; closed boxes, exons; black triangle, *loxP* site; arrows indicate primers. (b) Genotyping of offspring from breedings of heterozygous mice with the detection of wild-type (wt/wt), heterozygous (+/−), and homozygous (−/−) mutant genotypes. Pos., positive reaction with DNA of heterozygous genotype; neg., control reaction lacking DNA template (H_2O_{PCR}).

TABLE 8.3
TRP-Deficient Mouse Lines with TRP-Dependent Reporter Expression

Gene	Strategy	Reference
TRPM8	Knock In of EGFP-SV40polyA-cassette into TRPM8 locus, 27 residues following the start codon in exon 5	48
TRPM8	Replacement of exon 1+2 by lacZ-Neo-cassette	44
TRPA1	Replacement of exons 22–24 by KDEL-IRES-PLAP-pA-FRT-Neo-FRT-cassette	45
TRPP1 (PKD1)	Replacement of exons 2–3 by lacZ-cassette cloned in frame to the 5′ end of exon 2 of Pkd1	81

different cellular populations. The bacterial beta-galactosidase gene (lacZ) has been the marker of choice for most studies in the mouse, such as the mouse line described by Colburn,[44] where exons 1 and 2 were replaced by a LacZ/neo cassette to create a TRPM8 null allele in which lacZ expression should be controlled under the TRPM8 promoter. A similar approach was used to generate a TRPP1 reporter allele (Table 8.3). Kwan et al. described mice in which the S5 and S6 transmembrane domains and the pore loop that contains the selectivity filter of TRPA1 were replaced with a cassette containing an internal ribosome entry site (IRES) with a human placental alkaline phosphatase (PLAP) gene and a polyadenylation sequence.[45] PLAP, which is a GPI-linked membrane protein located on the extracellular surface, was introduced into the TRPA1 gene locus with the intention that antibody staining or chromogenic development of cells expressing PLAP allows for molecular identification of TRPA1-expressing cells, and as a result, TRPA1 was recently identified using this approach in DRG neurons, as well as in epidermal and hair follicle keratinocytes.[46]

Over the past few years, fluorescent proteins have emerged as a unique alternative to other gene-based reporters in that their visualization is noninvasive and can be monitored in real time in vitro or in vivo. Several mutants of the original wild-type GFP gene with improved thermostability and fluorescence have been engineered,[47] and various GFP spectral variants have also been developed. Two of these novel color variants, enhanced yellow fluorescent protein (EYFP) and enhanced cyan fluorescent protein (ECFP), as well as fluorescent proteins with red chromophores such as mRFP and tandem dimer tomato (tdTomato), can also be used in mice. For example, the farnesylated enhanced green fluorescent protein (EGFPf) was fused to the start codon of the TRPM8 gene[48] to analyze the anatomical and functional properties of TRPM8-expressing nerve fibers. Using this strategy, the cellular expression of TRPM8 could be tracked to a defined population of DRG neurons.[49]

8.2.5 KNOCK-ADD-ON MOUSE MODELS

Knock-in approaches as described above in which the fluorescent protein is expressed under the control of a promoter region of a TRP gene allow the analysis of the cellular expression pattern of this TRP gene. Alternatively, the cDNA of a fluorescent protein, e.g., EGFP, can be integrated in-frame between the end of the coding sequence of a TRP gene and the beginning of the 3′ untranslated region, by the removal of the stop codon by means of homologous recombination without deleting any endogenous coding or regulatory sequences. In this case, a TRP-EGFP C-terminal fusion protein should be synthesized in physiological amounts in cells in which the TRP protein is found endogenously. In cells of these so-called "Knock-Add-On mice," one should be able to study not only cellular expression patterns but also the trafficking and movements of TRP-EGFP in real time and in live cells in the native environment under basal conditions as well as during stimulation. So far, there are no Knock-Add-On mouse models reported for TRP proteins, but the proof of principle was already reported for membrane proteins. For instance, Scherrer et al. created Knock-Add-On mice where the δ-opioid receptor (DOR) is replaced by an active DOR-EGFP fusion protein.[50] The authors claim that the DOR-EGFP mouse provides a unique approach to explore receptor localization and function in vivo. Confocal microscopy revealed that exposing primary neurons to different DOR agonists triggered clustering of the receptors along the plasma membrane. DOR-EGFP clusters were then progressively internalized, indicating a desensitization mechanism.

8.2.6 GENERATION OF GENETICALLY MODIFIED MOUSE MODELS USING GENE TRAPPING

Another possibility for generating genetically modified mouse models is to use the gene trap technique, which is based on the random integration of a selection marker into the ES cell genome to mutate and identify the trapped locus. The most widely used vectors contain a splice acceptor and polyadenylation signal flanking a βgeo (LacZ/neo) reporter gene, such that the reporter is activated upon insertion into introns of genes. Gene trap ES cell lines reliably contribute to the germ line, producing mutant mouse strains for the functional characterization of genes, and such mice are available for numerous genes from the International Gene Trap Consortium (www.genetrap.org). Although the insertion of the vector construct in a genomic region can result in complete inactivation of the "trapped" gene (a null allele), this is not guaranteed. In some cases, vector insertion creates a hypomorphic allele, which results in a reduction but not the elimination of wild-type levels of gene product or activity, often causing a less severe phenotype than a loss-of-function (or null) allele. Mouse models with mutated TRP genes obtained by this strategy have not been reported, but Vig et al. generated mice with an ES cell line from a gene-trapping source bearing a βgeo-based vector insertion in the first intron of the Orai1 gene.[51] These mice exhibited deficits in mast cell functions similar to Orai1-deficient mice produced via conditional gene targeting approaches[52] but had no perceptible defect in store-operated Ca^{2+} entry in T cells or apparent hair loss. The less pronounced phenotype in Orai1 gene trap mice could be explained if the gene trap allele corresponds to a hypomorphic rather than a null allele, from which still low levels of ORAI1 can be expressed.[52]

8.3 PROTOCOLS FOR GENE TARGETING EXPERIMENTS

In the following, several step-by-step protocols used for gene targeting experiments are described. More detailed information can be obtained from laboratory manuals that deal with gene targeting approaches in detail.[41–43]

8.3.1 MOUSE EMBRYONIC FIBROBLASTS (MEFS)

8.3.1.1 Preparation of MEFs

To maintain their pluripotency and undifferentiated state, ES cells are frequently cultured on a layer of mitotically inactive feeder cells derived from embryonic fibroblasts. These mouse embryonic feeder cells (EF cells) are prepared from day-14.5 embryos and are mitotically inactivated by γ irradiation or alternatively by treatment with mitomycin C. Preparation of EF cells obviously needs to be done in advance of any ES cell work. EF cells must be G418-resistant (or resistant to any other selection drug you wish to use in isolating homologous recombinants) and therefore must be prepared from mice harboring the neomycin resistance gene. We prepare our EF cells from embryos heterozygous for a mutated TRPC4 allele harboring a neomycin resistance gene driven by a phosphoglycerate-kinase (pgk) promoter.[53]

Reagents

- Trypsin: 0.5% Trypsin, 0.2% EDTA (Sigma T4174)
- PBS (Ca^{2+}, Mg^{2+} free): 500 mL   PBS w/o Ca^{2+}/Mg^{2+} (Invitrogen; 14190–094)
       add 200 μL   EDTA 0.5 M, pH = 8 (final conc. 0.2 mM)
- EF medium: 450 mL DMEM with Glutamax (Gibco; 31966 – 021)
        50 mL   fetal calf serum (FCS; 10% final)
        5 mL   nonessential amino acids (Gibco; 11140 – 035)
                    (0.1 mM final)
- 2× Freezing EF medium: EF medium containing 20% DMSO

Equipment

Sterile 10-cm plastic dishes; sterile 175-cm^2 flasks; cell culture centrifuge; conical tubes (50-mL screw cap); cryovials; sterilized surgical instruments for dissection; water bath at 37°C; 70% ethanol; binocular scope with a lamp, mask, hair cover, and gloves; laminar flow hood; 1 × 500-mL beaker; and 2 × 200-mL beaker.

Step-by-Step Protocol

- Prepare embryonic feeder cells on embryonic day (E) 14.5 from embryos carrying an allele with a neomycin resistance gene driven by a phospho-glycerate-kinase (pgk) promoter (see above).
- Transfer sterile surgical instruments (several fine and tough forceps and scissors) to the laminar flow hood and prepare several dishes of sterile PBS without Ca^{2+} and Mg^{2+} for washing of embryos. Sacrifice pregnant mouse

and rinse well with 70% EtOH in a beaker. Bring the mouse under the laminar flow and pin down.

- Open the peritoneal wall, dissect the uterus, and place it into sterile dish containing PBS without Ca^{2+} and Mg^{2+}. Wash the uterus several times by transferring to a new dish with PBS without Ca^{2+} and Mg^{2+}.
- Use two fine forceps to disrupt the uterine wall and isolate embryos. Cut away the head and place the rest of the embryo in a new dish in a minimal amount of PBS without Ca^{2+} and Mg^{2+}. Remove the fetal liver and heart ("red" parts) and place the embryo in a new dish with PBS w/o Ca^{2+} and Mg^{2+}. Finally, cut it into small pieces.
- Transfer each embryo together with 1 mL 1× trypsin/EDTA into a 50-mL screw cap tube (maximal five embryos per tube) and incubate at 37°C for 5–10 min in a water bath.
- Homogenize the tissue by pipetting up and down consecutively with a 25-mL pipette, 10-mL pipette, 5-mL pipette, 15–30 times for each pipette.
- Incubate again at 37°C for 5–10 min in a water bath.
- Repeat homogenizing by pipetting with 10-mL and 5-mL pipettes.
- Add EF medium to the tube (final volume 25-mL) and homogenize again with a 5-mL pipette for 5–10 min. Wait until the cell debris settles (about 5 min).
- Remove the supernatant carefully and disseminate the cells to two 175-cm^2 flasks per embryo: e.g., 5 embryos $\Rightarrow$ 10 flasks of 175 cm^2. Distribute the cells equally in an appropriate volume of EF medium.
- Incubate the cells at 37°C and 5% CO_2.

The medium is usually changed once after 24 hours. EF cells are then normally cultured until they reach confluency (6 days) without changing the medium and harvested after two additional days of culture (passage 0).

8.3.1.2   Harvesting EF Cells and γ-Irradiation

1. Wash the cells twice with PBS w/o Ca^{2+} and Mg^{2+}, trypsinize the cells with 1× trypsin/ EDTA (4 mL per flask), incubate for 5 min at 37°C (shorter is even better, certainly not too long), and neutralize with 6 mL EF medium. Collect 9 mL of the cells from the flask and leave 1 mL in each flask for continuation of the culture. Add 19 mL EF medium and culture these cells again until confluency (~8 days, passage P1). Then, proceed in the same way as described above. Ideally, use EF cells only from passages P0 and P1 for co-culture with ES cells.
2. Spin down the trypsinized cells for 5 min at 1000 rpm (~140 × g, cell culture centrifuge). Remove the supernatant and resuspend the cells again in 50 mL EF medium. Pellet the cells again by centrifugation (~140 × g, 5 min) and irradiate them with 60 Gray (e.g., using a [137]Caesium energy source).
3. Spin down the irradiated cells for 5 min at ~140 × g. Remove the supernatant but leave a little bit of medium to resuspend the cells with a polished Pasteur pipette. Resuspend the cells in an appropriate volume of EF

medium and add an equal volume of freshly prepared and cold EF freezing medium (containing 20% DMSO, final concentration of DMSO should be 10%). Quickly aliquot 1 mL of this cell suspension into cryovials and place the vials in a styrofoam box filled with dry ice, which allows them to cool down gradually for 15 min. Store the cells at −80°C or in liquid nitrogen for long-term storage. Distribute the cells to three cryovials per 175-cm² culture flask.

8.3.1.3 Thawing and Replating EF Cells

1. Quickly thaw an aliquot of irradiated frozen EF cells in a water bath at 37°C, and clean the outside of the vial(s) with 70% (v/v) ethanol before transferring to the laminar flow hood.
2. Transfer the EF cells into 10 mL of warm EF medium and pellet by centrifugation (5 min, 1000 rpm). This is to remove the DMSO contained in the freezing medium.
3. Aspirate the supernatant, resuspend cells in an appropriate volume of EF medium, and plate in a sterile tissue culture dish. Examine the feeder monolayer under the microscope to decide the right density for later ES cell culture.

8.3.2 MOUSE EMBRYONIC STEM CELLS

8.3.2.1 ES Cells—General Aspects of Handling and Use

Pluripotent embryonic stem (ES) cells were first established directly from murine blastocysts in 1981[54,55] after unsuccessful early attempts to obtain germ line chimerism with teratocarcinoma (EC) cells. In 1989, it was shown for the first time that ES cells efficiently produced germ line chimeras even after extensive in vitro manipulation, allowing for the generation of the first genetically modified mouse line using homologous recombination in ES cells.[56] Since then, a fairly large number of ES cell lines have been generated. The lines that are often used for gene targeting experiments include D3, E14, AB1, or R1, which are all derived from 129 substrains or 129 hybrids. There is a smaller number of less frequently used ES lines derived from other strains like ES623 and B6-III, which were established from C57BL/6 mice. The choice of which ES cell line to use may be dictated by factors such as the genetic background necessary for the experiments that are supposed to be performed with the resulting mouse model or the source of DNA used in the targeting construct. The appropriate genetic background can always be obtained by backcrossing (although often at a heavy price in terms of time), but the frequency of homologous recombinants might be low with nonisogenic DNA.

At the beginning of the work with ES cells, it is important to keep in mind that these cells are capable of participating in the development of a mouse. The highest risk is that mistreatment of ES cells will easily lead to differentiation from the pluripotent developmental state. Consequently, they may be unable to participate in development, and therefore, chimerism and/or germ line transmission will not occur. Hence, the primary goal in handling ES cells is to preserve their pluripotency.

Fortunately, this can easily be accomplished by careful tissue culture technique. ES cell maintenance cannot be postponed and requires daily feeding. If culturing high concentrations of ES cells on a single dish is required for a particular reason (e.g., prior to transfection), it would not be unusual to feed these cells early in the morning and again in the evening. ES cell density is very important, and cultures should be split frequently (every 2–3 days) so that individual colonies do not become too large (<1000 cells). Similarly, when trypsinizing ES cells, it is important to fully resuspend the cell clumps by gentle pipetting with a polished Pasteur pipette so that no large aggregates are transferred to the new plates. The reason for these precautions is that large colonies easily differentiate and likely lose their pluripotency. Also, the surface area of the tissue culture dish covered by ES cell colonies should not exceed ≈50%, because ES cultures will completely differentiate into endoderm-like cells within a few days if the cultures have ever been confluent. If ES cells are cultured appropriately, the fraction of differentiated cells, which can be recognized under a microscope by their different (e.g., fibroblast-like) morphology and size, is usually very low. This kind of overt differentiation is not problematic, because such differentiated cells have a limited lifespan and are continuously diluted out by splitting the cultures.

Considering that gene targeting experiments are a major laboratory effort, it is recommended to first establish (wild-type) subclones of an ES cell line and to test these subclones for germ line competence. Although this is a significant investment of energy and time, it will most likely circumvent the problem of inefficient ES cell germ line transmission of a targeted mutation. If a subclone with a good germ line record is identified, this cell population probably contains almost 100% "good" ES cells, and the majority of the derived transfectants should be germ line competent at least for the next 5–10 passages of the subclone.

ES cells are usually grown in the presence of "leukaemia inhibitory factor" (LIF) either on gelatinized culture plates or on feeder layers of primary embryonic fibroblasts (EF). LIF can be purchased (currently from Millipore/Chemicon), but cells transfected with an LIF expression vector are also available. We prefer to culture ES cells on a layer of (G418-resistant) EF cells in the presence of LIF when they are used for the generation of chimeric mice. G418-resistant feeder cells can be obtained from embryos carrying one allele with a neomycin resistance cassette containing a pgk promoter and polyadenylation signal (see above). If ES cells are just expanded for the preparation of genomic DNA, it is sufficient, and even beneficial, to culture them directly on plastic (gelatin-coated) plates in the presence of LIF without EF cells because the contamination of ES cell DNA with DNA from feeder cells is to be avoided.

The fetal calf serum (FCS) used for ES cell culture must be tested for support of ES cell growth and, upon identification of a good lot, a substantial number of bottles should be reserved as a stock. Different batches of FCS with low endotoxin levels should be tested by determining the plating efficiency and colony morphology after ES cells (grown before with previously tested FCS) are plated at a low density on EF layers in DMEM-ES (with and without LIF) prepared with 15% of the test FCS. ES cells are cultured for one passage in the media prepared with the various FCS lots (including the original one), and then 200–500 cells are plated on 6-cm dishes (in duplicate), and colonies are counted after one week. Plating efficiency (number of colonies/number of cells plated) is always well below 100% because some ES cells in suspension, obtained by trypsinization, form small aggregates.

Equipment for the Following Protocols Concerning ES Cell Culture
Tissue culture facility, preferably used only for ES cells, including a laminar flow cabinet; humidified incubator (37°C, 5% CO_2); inverted phase-contrast microscope with 4×, 10×, 20–25× objectives for routine observations; stereo microscope with transmitted light base for picking colonies, tabletop centrifuge; −70°C freezer; and liquid nitrogen tank. Electroporation apparatus with capacitance extender (for transfection), sterile disposable tissue culture grade plasticware (100-, 60-, 35-mm dishes; 6-, 24-, 48-, 96-well plates; centrifuge tubes; and cryovials), and multichannel pipettor.

ESL medium		Final concentration
400 mL	DMEM with Glutamax (Gibco; 31966 – 021)	
100 L	FCS (ES grade)	20%
5 mL	Nonessential amino acids (Gibco; 11140 – 035)	1 × (0.1 mM)
1 mL	LIF (5×10^5 U/mL) (ESGRO; Chemicon)	1000 U/mL
3.5 µL	ß-Mercaptoethanol (Sigma; M7522)	0.1 mM
G418-ESL medium		
500 mL	ESL-medium	
2.5 mL	G418 (#10131027; Invitrogen; 50 mg/mL)	250 µg/mL
Ganciclovir-G418-ESL medium		
500 mL	G418-ESL medium	
100 µL	Ganciclovir-solution (10 mM)	2 µM
PBS (Ca^{2+}, Mg^{2+} free)		
500 mL	PBS w/o Ca^{2+}/Mg^{2+} (Invitrogen;14190–094)	
200 µL	EDTA (0.5 M, pH = 8)	0.2 mM
Trypsin		
	0.5% Trypsin, 0.2% EDTA (Sigma T4174)	

8.3.2.2 Passage of ES Cells

Usually, ES cells are split at 1:4 to 1:8 ratios depending on their growth rate, ideally every other day (about 1×10^6 cells are seeded onto a 60-mm^2 plate). Cells should be trypsinized to a single cell suspension at every passage as large clumps differentiate. The passage number of ES cells must be kept as low as possible by keeping a stock of frozen vials in liquid nitrogen.

Step-by-Step Protocol

1. Replace the medium on the appropriate number of prepared feeder plates to ESL medium.
2. Aspirate the growth medium of the ES cell culture plate.
3. Rinse twice with PBS (w/o Ca^{2+}/Mg^{2+}).

4. Add trypsin (pre-incubated at 37°C) corresponding to the area of the plate.
5. Place the plate in the incubator for 5 min.
6. Check that the trypsin treatment has reached the desired level by rocking the plate to detach clumps from the bottom of the plate (or under microscope).
7. Add an equal volume of ESL to neutralize the trypsin.
8. Pipette up and down several times using a polished (small-diameter) Pasteur pipette.
9. Transfer the suspension into a 15-mL tube, and add ESL to a final volume of 12 mL.
10. Pellet the cells by low-speed centrifugation (~140 × g, cell culture centrifuge ~1000 rpm) for 5 min at RT (room temperature).
11. Aspirate the supernatant and add 1 drop of ESL to the pellet. Gently knock the tube on the bottom of the bench to resuspend the cells before adding more ESL.
12. Add ESL to a final volume of 12 mL to the tube and pipette gently to mix well.
13. Pellet the cells by low-speed centrifugation (~140 × g) for 5 min at RT and aspirate the supernatant.
14. Resuspend the cells with ~1 mL of ESL using a polished Pasteur pipette.
15. Add ESL to an appropriate volume and split the contents into the new plates containing a sufficient volume of medium. For future culturing, the medium has to be changed every day.

8.3.3 ELECTROPORATION

The technique of electroporation is a highly efficient way to transfect ES cells with a targeting vector. Usually ~1 × 10^7 cells are transfected with 20–50 µg of the targeting vector. The DNA to be electroporated has to be of high purity. It is advisable to remove endotoxins (e.g., using an EndoFree Plasmid Kit and Qiagen) that could affect transfection efficiency and viability of the transfected cells. The targeting vector has to be linearized to facilitate homologous recombination. To achieve this, a large-scale overnight digest with ~100 µg of the targeting vector is performed. It is advisable to clean the digested DNA to remove salts and enzymes before preparing the ES cells for electroporation.

Step-by-Step Protocol
Preparation of targeting vector for electroporation

1. Digest ~100 µg of the targeting vector with 100 units of an appropriate restriction enzyme overnight; perform phenol/chloroform/diethylether extraction.
2. Precipitate the DNA by adding sodium acetate (final concentration 0.3M) and 2× volume of 100% ethanol (v/v). DNA will be visible as a white precipitate that can be fished out with a sterile pipette tip and transferred into a new reaction tube.

3. After drying for about 20 min, resuspend the DNA with an appropriate volume of PBS. Measure DNA concentration using gel electrophoresis or spectral photometric methods.
4. Finally, the amount of targeting vector DNA to be electroporated (20–50 µg) has to be in a volume of 600 µL PBS (PBS/DNA solution).

Preparation of ES cells for electroporation

1. ~1 × 10^7 ES cells can be harvested from ~6 × 75-cm^2 culture flasks (~60% confluent).
2. Feed the ES cells with ESL medium 2 hours before harvesting.
3. Wash the cells twice with PBS w/o Ca^{2+}/Mg^{2+}.
4. Trypsinize the cells and collect them in 50 mL ESL medium.
5. Spin down 5 min (~140 × g) and resuspend the cell pellet with 14 mL ESL medium.
6. After counting the cells, centrifuge a volume containing ~1 × 10^7 cells (~140 × g), and resuspend cells in 600 µL PBS/DNA solution (ES cell/PBS/DNA solution).

Electroporation

1. Add 600 µL ES cell/PBS/DNA solution into an electroporation cuvette and incubate for 5 min at room temperature.
2. Perform electroporation (capacity, 3 µF; voltage, 0.8 kV; time constant, 0.1 s; GenePulser, BioRad, Munich) and incubate again for 5 min at room temperature.
3. Transfer the cells into 60 mL ESL medium and disperse into 6 × 75 cm^2 cell culture dishes with feeder cells.

Selection

1. 24 hours after electroporation: change medium using G418-ESL medium.
2. 48 hours after electroporation: change medium using G418-Ganciclovir ESL medium. Change medium twice daily if there are many dead cells in suspension due to selection.

8.3.4 Picking ES Cell Clones

The number of clones that must be recovered depends on a number of factors such as the type of selection (e.g., selection for homologous recombinants ± negative selection against random integrants), knowledge about the targeting frequency for the locus of interest, etc. For obvious reasons, however, the more clones that are isolated, the higher is the probability of recovering a homologous recombinant. As a very general number, 250–300 colonies are usually isolated for examination of a homologous recombination event.

Once the selection is completed (≈10 days after transfection and start of G418 and ganciclovir selection), resistant clones must be isolated. Colonies at this point should be fairly large in size (≈4000 cells) but still maintaining a discrete border or edge. Colonies may even have a dark "cap" indicative of a high density of cells at the top of the colony. Colonies which have begun to differentiate (and spread out) around the edges should not be isolated. Concerning the size of resistant colonies at a given time, there is a considerable variation among the population of resistant clones obtained after transfection. Thus, it is probably best to pick every day, starting at ≈day 10 after transfection. In each case, the colonies that have grown to the proper size are picked. After one week (≈day 17 after transfection), however, almost all colonies are overgrown, and the plates should be discarded. One should be cautious when picking colonies too early after selection. These colonies, while clearly visible, are often small, making it difficult to evaluate their state of differentiation.

Step-by-Step Protocol

1. One day before picking, 96-well plates with irradiated feeder cells (EF) are prepared to accommodate all clones. The medium of feeder dishes is changed before picking using 240 μL ESL per well of the master plate.
2. Add 1× trypsin/EDTA (50 μL/well) to a 96-well plate and 200 μL ESL medium to a set of 96-well plates for DNA analysis and an equivalent set of 96-well plates with irradiated feeder cells for master plates. U-bottom (or V-bottom) 96-well plates are preferable to flat bottom because this will make recovery much easier.
3. With a dissecting microscope in a tissue culture hood (with laminar flow) and using a 200-μL pipette, gently dislodge a colony from the feeder layer. Transfer it immediately to 96-well plate containing trypsin.
4. Pipette each well ~5 times up and down. After picking 12 colonies (one row), transfer the trypsinized cells to the wells of a 96-well master plate using a multichannel pipettor (Figure 8.4b). Resuspend cells in wells by pipetting up and down. (Air bubbles can help to memorize the position in each plate.) Now transfer 50 μL of the cell suspension to the equivalent wells of a DNA plate (Figure 8.4b). Once master clones have grown sufficiently, they can be frozen down in the 96-well plate. The cells in the DNA plates can be grown to confluency; genomic DNA can be prepared according to protocols described below.
5. When gelatin-treated plates are used for DNA plates, tissue culture dishes are first incubated with sterile, autoclaved 0.1% gelatin (in PBS or H_2O; Sigma G-1850, from porcine skin) in enough volume to cover the plate for at least 0.5–1 hour at room temperature in the hood.

8.3.5 FREEZING ES CELL CLONES

We usually freeze ES cell clones in 96-well plates. As a rule of thumb, there should be enough cells within each well such that there will still be enough cells to recover the clone even if only 10% are viable after thawing. On the other hand, too high a

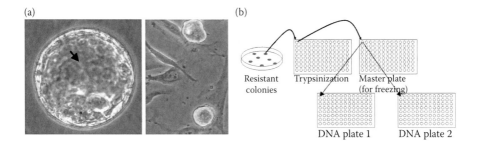

FIGURE 8.4 Picking ES cell clones. (a) Left: Blastocyst isolated on day 3.5 following conception. Asterisk indicates the inner cell mass, arrow the blastocoel. Right: Two colonies of R1 embryonic stem cells cultured on irradiated embryonic fibroblasts. (b) Following isolation from a 100-mm plate, the ES colonies are trypsinized to a single cell suspension and then passaged to a master plate and two independent DNA plates. Usually, two replica plates are used for DNA isolation. One is processed for Southern blot analysis, leaving the second plate as a backup.

density of cells often leads to differentiation. If the colonies show large differences in size and density between wells of the same 96-well plate, it is advisable to selectively freeze down individual clones in cryovials separately. Freezing medium is prepared freshly prior to use and kept on ice. We commonly use ~20% FCS and 10% DMSO as a final concentration. It is possible to increase the concentration of FCS in freezing medium up to 40% for better recovery of small amount of cells in 96-well plates. Freezing and thawing are usually counted as one passage.

Equipment and Preparations

1. 2× Freezing medium: ESL containing 20% DMSO (Sigma; D2650) for freezing in cryovials.
2. ESL medium containing 12.2% DMSO for freezing in 96-well plates.
3. Cryovials (1.8 mL; Nunc).
4. Change growth medium 2–3 hours before freezing the cells.
5. Freshly prepare freezing medium (ESL medium containing 12.2% DMSO); keep on ice.

Step-by-Step Protocol (96-Well Plate)

1. Aspirate the culture medium.
2. Wash twice with 250 μL PBS without Ca^{2+}/Mg^{2+}.
3. Add 50 μL 1× trypsin per well; incubate 4 min at 37°C.
4. Stop reaction by adding 230 μL ESL/12.2% DMSO (cold!) per well (final concentration of DMSO is 10%).
5. Resuspend cells using a 200-μL filter tip.
6. Seal the 96-well plate using cold-resistant sealing film (e.g., Sealview™).

Step-by-Step Protocol (Cryovials)

1. Harvest the cells in a 12-mL tube containing ESL by trypsinization as described in the protocol "Passage of ES cells." Pellet the cells at ~140 × g for 5 min at RT.
2. Aspirate the supernatant and add 1 drop of ESL to the pellet. Gently knock the tube on the bottom of the bench to resuspend the cells before adding more ESL.
3. Add ESL to a final volume of 12 mL to the tube and pipette gently to mix well.
4. Pellet the cells by low-speed centrifugation (~140 × g) for 5 min at RT.
5. Aspirate the supernatant and add 1 drop of ESL to the pellet.
6. Resuspend cells using a polished Pasteur pipette and add ESL to half of the final volume required. Then gradually add an equal volume of cold 2× freezing medium while shaking the tube and mix by pipetting up and down several times. Alternatively, gently resuspend the pellet in cold 1× freezing medium.
7. Quickly aliquot 1 mL of the cell suspension in freezing medium into labeled cryovials and put them on ice.

Final Steps for Both Protocols

1. Place the 96-well plates or cryovials in a styrofoam box filled with dry ice for ~15 min, which will allow them to cool down gradually.
2. Cells can be kept at −80°C for up to three months, but cryovials are better transferred into liquid nitrogen (−196°C) for long-term storage.

8.3.6 PREPARATION OF GENOMIC DNA FROM ES CELLS

Step-by-Step Protocol

1. Cultivate ES cells in "DNA 96-well plates" until complete confluency.
2. Aspirate the medium, and add 150 μL lysis buffer and incubate for 16 hours at 37°C.
3. Add 200 μL isopropanol and incubate overnight at 4°C using a rotation shaker.
4. Isolate precipitated DNA by winding on a 20-gauge needle and transfer it into a sterile 1.5-mL microcentrifuge tube.
5. After air-drying, dissolve DNA by rocking motion in 80–150 μL double distilled H_2O.
6. Use ~20– 25 μL of the DNA for digestion in Southern blot analysis and ~1 μL for PCR genotyping.

Lysis buffer	Tris, pH 8.5	100 mM
	EDTA	5 mM
	NaCl	200 mM
	SDS	0.2% (w/v)

After sterile filtration, add proteinase K (0.1 mg/mL; added freshly before use).

ACKNOWLEDGMENT

This work was supported by the Deutsche Forschungsgemeinschaft, the "HOMFOR" program, and Forschungsausschuss der Universität des Saarlandes. We thank A. Pfeifer for initial introduction of ES cell culture; S. Buchholz, J. Olausson, and R. Vennekens for their help in optimizing gene targeting protocols; F. Zimmermann and S. Dlugosch (Heidelberg) for blastocyst injections; C. Matka, K. Fischer, S. Hünecke, T. Volz, and S. Wagner for help with establishing protocols for Cre- and Flp-mediated recombination in vivo; all other members of the SPF facility for their help with breedings; and P. Wollenberg for the help with figures. Finally, the continuous support of V. Flockerzi over the years is very much appreciated; without it the establishment of a transgenic unit would not have been possible.

REFERENCES

1. Ramsey, I. S., M. Delling, and D. E. Clapham. 2006. An introduction to TRP channels. *Annual Review of Physiology* 68: 619.
2. Venkatachalam, K., and C. Montell. 2007. TRP channels. *Annual Review of Biochemistry* 76: 387.
3. Desai, B. N., and D. E. Clapham. 2005. TRP channels and mice deficient in TRP channels. *Pflügers Archiv* 451: 11.
4. Flockerzi, V. 2007. An introduction on TRP channels. *Handbook of Experimental Pharmacology* 1.
5. Nilius, B., and T. Voets. 2005. TRP channels: A TR(I)P through a world of multifunctional cation channels. *Pflügers Archiv* 451: 1.
6. Freichel, M., and V. Flockerzi. 2007. Biological functions of TRPs unravelled by spontaneous mutations and transgenic animals. *Biochemical Society Transactions* 35: 120.
7. Nilius, B., and G. Owsianik. 2010. Transient receptor potential channelopathies. *Pflügers Archiv* 460: 437.
8. Groschner, K., C. Rosker, and M. Lukas. 2004. Role of TRP channels in oxidative stress. *Novartis Foundation Symposium* 258: 222.
9. Xu, X. Z., H. S. Li, W. B. Guggino, and C. Montell. 1997. Coassembly of TRP and TRPL produces a distinct store-operated conductance. *Cell* 89: 1155.
10. Greka, A., B. Navarro, E. Oancea, A. Duggan, and D. E. Clapham. 2003. TRPC5 is a regulator of hippocampal neurite length and growth cone morphology. *Nature Neuroscience* 6: 837.
11. Kuzhikandathil, E. V., H. Wang, T. Szabo, N. Morozova, P. M. Blumberg, and G. S. Oxford. 2001. Functional analysis of capsaicin receptor (vanilloid receptor subtype 1) multimerization and agonist responsiveness using a dominant negative mutation. *Journal of Neuroscience* 21: 8697.
12. Hofmann, T., M. Schaefer, G. Schultz, and T. Gudermann. 2002. Subunit composition of mammalian transient receptor potential channels in living cells. *Proceedings of the National Academy of Sciences of the United States of America* 99: 7461.
13. Philipp, S., C. Trost, J. Warnat et al. 2000. TRP4 (CCE1) protein is part of native calcium release-activated Ca^{2+}-like channels in adrenal cells. *Journal of the Biological Chemistry* 275: 23965.
14. Fire, A., S. Xu, M. K. Montgomery, S. A. Kostas, S. E. Driver, and C. C. Mello. 1998. Potent and specific genetic interference by double-stranded RNA in Caenorhabditis elegans. *Nature* 391: 806.

15. Elbashir, S. M., J. Harborth, W. Lendeckel, A. Yalcin, K. Weber, and T. Tuschl. 2001. Duplexes of 21-nucleotide RNAs mediate RNA interference in cultured mammalian cells. *Nature* 411: 494.

16. Jackson, A. L., and P. S. Linsley. 2010. Recognizing and avoiding siRNA off-target effects for target identification and therapeutic application. *Nature Reviews. Drug Discovery* 9: 57.

17. Aarts, M., K. Iihara, W. L. Wei et al. 2003. A key role for TRPM7 channels in anoxic neuronal death. *Cell* 115: 863.

18. Kuhn, R., S. Streif, and W. Wurst. 2007. RNA interference in mice. *Handbook of Experimental Pharmacology* 149.

19. Bridge, A. J., S. Pebernard, A. Ducraux, A. L. Nicoulaz, and R. Iggo. 2003. Induction of an interferon response by RNAi vectors in mammalian cells. *Nature Genetics* 34: 263.

20. Robbins, M., A. Judge, and I. MacLachlan. 2009. siRNA and innate immunity. *Oligonucleotides* 19: 89.

21. Sledz, C. A., M. Holko, M. J. de Veer, R. H. Silverman, and B. R. Williams. 2003. Activation of the interferon system by short-interfering RNAs. *Nature Cell Biology* 5: 834.

22. Iwata, Y., Y. Katanosaka, Y. Arai, K. Komamura, K. Miyatake, and M. Shigekawa. 2003. A novel mechanism of myocyte degeneration involving the Ca^{2+}-permeable growth factor-regulated channel. *Journal of Cell Biology* 161: 957.

23. Kuwahara, K., Y. Wang, J. McAnally et al. 2006. TRPC6 fulfills a calcineurin signaling circuit during pathologic cardiac remodeling. *Journal of Clinical Investigations* 116: 3114.

24. Nakayama, H., B. J. Wilkin, I. Bodi, and J. D. Molkentin. 2006. Calcineurin-dependent cardiomyopathy is activated by TRPC in the adult mouse heart. *FASEB Journal* 20: 1660.

25. Dietrich, A., Y. S. M. Mederos, M. Gollasch et al. 2005. Increased vascular smooth muscle contractility in TRPC6$^{-/-}$ mice. *Molecular and Cell Biology* 25: 6980.

26. Tsvilovskyy, V. V., A. V. Zholos, T. Aberle et al. 2009. Deletion of TRPC4 and TRPC6 in mice impairs smooth muscle contraction and intestinal motility in vivo. *Gastroenterology* 137: 1415.

27. Smithies, O., R. G. Gregg, S. S. Boggs, M. A. Koralewski, and R. S. Kucherlapati. 1985. Insertion of DNA sequences into the human chromosomal beta-globin locus by homologous recombination. *Nature* 317: 230.

28. Thomas, K. R., and M. R. Capecchi. 1987. Site-directed mutagenesis by gene targeting in mouse embryo-derived stem cells. *Cell* 51: 503.

29. Bianco, S. D., J. B. Peng, H. Takanaga et al. 2007. Marked disturbance of calcium homeostasis in mice with targeted disruption of the Trpv6 calcium channel gene. *Journal of Bone and Mineral Research* 22: 274.

30. Branda, C. S., and S. M. Dymecki. 2004. Talking about a revolution: The impact of site-specific recombinases on genetic analyses in mice. *Developmental Cell* 6: 7.

31. Feil, R. 2007. Conditional somatic mutagenesis in the mouse using site-specific recombinases. *Handbook of Experimental Pharmacology* 3.

32. Lewandoski, M. 2001. Conditional control of gene expression in the mouse. *Nature Review of Genetics* 2: 743.

33. Riccio, A., Y. Li, J. Moon et al. 2009. Essential role for TRPC5 in amygdala function and fear-related behavior. *Cell* 137: 761.

34. Holzenberger, M., C. Lenzner, P. Leneuve, R. Zaoui, G. Hamard, S. Vaulont, and Y. L. Bouc. 2000. Cre-mediated germline mosaicism: a method allowing rapid generation of several alleles of a target gene. *Nucleic Acids Research* 28: E92.

35. Lakso, M., J. G. Pichel, J. R. Gorman et al. 1996. Efficient in vivo manipulation of mouse genomic sequences at the zygote stage. *Proceedings of the National Academy of Science of the United States of America* 93: 5860.

36. Bunting, M., K. E. Bernstein, J. M. Greer, M. R. Capecchi, and K. R. Thomas. 1999. Targeting genes for self-excision in the germ line. *Genes Development* 13: 1524.
37. Bautista, D. M., J. Siemens, J. M. Glazer et al. 2007. The menthol receptor TRPM8 is the principal detector of environmental cold. *Nature* 448: 204.
38. Bautista, D. M., S. E. Jordt, T. Nikai et al. 2006. TRPA1 mediates the inflammatory actions of environmental irritants and proalgesic agents. *Cell* 124: 1269.
39. Jin, J., B. N. Desai, B. Navarro, A. Donovan, N. C. Andrews, and D. E. Clapham. 2008. Deletion of Trpm7 disrupts embryonic development and thymopoiesis without altering Mg^{2+} homeostasis. *Science* 322: 756.
40. Weissgerber, P., A. Cavalié, S. Buchholz, V. Flockerzi, and M. Freichel. 2008. Early embryonic lethality following inactivation of the TRPM7 gene in mice. *Naunyn-Schmiedeberg's Archives of Pharmacology* 377 (Suppl 1): 1.
41. Joyner, A. L. 2000. *Gene Targeting—A Practical Approach*. New York: Oxford University Press.
42. Nagy, A. 2003. *Manipulating the Mouse Embryo: A Laboratory Manual*. Cold Spring Harbor, NY: Cold Spring Harbor Laboratory Press.
43. Torres, R., and R. Kühn. 1997. *Laboratory Protocols for Conditional Gene Targeting*. New York: Oxford University Press.
44. Colburn, R. W., M. L. Lubin, D. J. Stone, Jr. et al. 2007. Attenuated cold sensitivity in TRPM8 null mice. *Neuron* 54: 379.
45. Kwan, K. Y., A. J. Allchorne, M. A. Vollrath et al. 2006. TRPA1 contributes to cold, mechanical, and chemical nociception but is not essential for hair-cell transduction. *Neuron* 50: 277.
46. Kwan, K. Y., J. M. Glazer, D. P. Corey, F. L. Rice, and C. L. Stucky. 2009. TRPA1 modulates mechanotransduction in cutaneous sensory neurons. *Journal of Neuroscience* 29: 4808.
47. Shaner, N. C., G. H. Patterson, and M. W. Davidson. 2007. Advances in fluorescent protein technology. *Journal of Cell Science* 120: 4247.
48. Dhaka, A., A. N. Murray, J. Mathur, T. J. Earley, M. J. Petrus, and A. Patapoutian. 2007. TRPM8 is required for cold sensation in mice. *Neuron* 54: 371.
49. Dhaka, A., T. J. Earley, J. Watson, and A. Patapoutian. 2008. Visualizing cold spots: TRPM8-expressing sensory neurons and their projections. *Journal of Neuroscience* 28: 566.
50. Scherrer, G., P. Tryoen-Toth, D. Filliol et al. 2006. Knockin mice expressing fluorescent delta-opioid receptors uncover G protein-coupled receptor dynamics in vivo. *Proceedings of the National Academy of Science of the United States of America* 103: 9691.
51. Vig, M., W. I. DeHaven, G. S. Bird et al. 2008. Defective mast cell effector functions in mice lacking the CRACM1 pore subunit of store-operated calcium release-activated calcium channels. *Nature Immunology* 9: 89.
52. Gwack, Y., S. Srikanth, M. Oh-Hora et al. 2008. Hair loss and defective T- and B-cell function in mice lacking ORAI1. *Molecular Cell Biology* 28: 5209.
53. Freichel, M., S. H. Suh, A. Pfeifer et al. 2001. Lack of an endothelial store-operated Ca^{2+} current impairs agonist-dependent vasorelaxation in TRP4$^{-/-}$ mice. *Nature Cell Biology* 3: 121.
54. Evans, M. J., and M. H. Kaufman. 1981. Establishment in culture of pluripotential cells from mouse embryos. *Nature* 292: 154.
55. Martin, G. R. 1981. Isolation of a pluripotent cell line from early mouse embryos cultured in medium conditioned by teratocarcinoma stem cells. *Proceedings of the National Academy of Science of the United States of America* 78: 7634.
56. Thompson, S., A. R. Clarke, A. M. Pow, M. L. Hooper, and D. W. Melton. 1989. Germ line transmission and expression of a corrected HPRT gene produced by gene targeting in embryonic stem cells. *Cell* 56: 313.

57. Kwan, K. M. 2002. Conditional alleles in mice: practical considerations for tissue-specific knockouts. *Genesis* 32: 49.

58. Dietrich, A., H. Kalwa, U. Storch et al. 2007. Pressure-induced and store-operated cation influx in vascular smooth muscle cells is independent of TRPC1. *Pflügers Archiv* 455: 465.

59. Stowers, L., T. E. Holy, M. Meister, C. Dulac, and G. Koentges. 2002. Loss of sex discrimination and male-male aggression in mice deficient for TRP2. *Science* 295: 1493.

60. Leypold, B. G., C. R. Yu, T. Leinders-Zufall, M. M. Kim, F. Zufall, and R. Axel. 2002. Altered sexual and social behaviors in trp2 mutant mice. *Proceedings of the National Academy of Science of the United States of America* 99: 6376.

61. Caterina, M. J., A. Leffler, A. B. Malmberg et al. 2000. Impaired nociception and pain sensation in mice lacking the capsaicin receptor. *Science* 288: 306.

62. Davis, J. B., J. Gray, M. J. Gunthorpe et al. 2000. Vanilloid receptor-1 is essential for inflammatory thermal hyperalgesia. *Nature* 405: 183.

63. Moqrich, A., S. W. Hwang, T. J. Earley et al. 2005. Impaired thermosensation in mice lacking TRPV3, a heat and camphor sensor in the skin. *Science* 307: 1468.

64. Suzuki, M., A. Mizuno, K. Kodaira, and M. Imai. 2003. Impaired pressure sensation in mice lacking TRPV4. *Journal of the Biological Chemistry* 278: 22664.

65. Morgans, C. W., J. Zhang, B. G. Jeffrey et al. 2009. TRPM1 is required for the depolarizing light response in retinal ON-bipolar cells. *Proceedings of the National Academy of Science of the United States of America* 106: 19174.

66. Shen, Y., J. A. Heimel, M. Kamermans, N. S. Peachey, R. G. Gregg, and S. Nawy. 2009. A transient receptor potential-like channel mediates synaptic transmission in rod bipolar cells. *Journal of Neuroscience* 29: 6088.

67. Koike, C., T. Obara, Y. Uriu et al. 2010. TRPM1 is a component of the retinal ON bipolar cell transduction channel in the mGluR6 cascade. *Proceedings of the National Academy of Science of the United States of America* 107: 332.

68. Yamamoto, S., S. Shimizu, S. Kiyonaka et al. 2008. TRPM2-mediated Ca^{2+} influx induces chemokine production in monocytes that aggravates inflammatory neutrophil infiltration. *Nature Medicine* 14: 738.

69. Zhang, Y., M. A. Hoon, J. Chandrashekar et al. 2003. Coding of sweet, bitter, and umami tastes: different receptor cells sharing similar signaling pathways. *Cell* 112: 293.

70. Damak, S., M. Rong, K. Yasumatsu et al. 2006. Trpm5 null mice respond to bitter, sweet, and umami compounds. *Chemical Senses* 31: 253.

71. Walder, R. Y., B. Yang, J. B. Stokes et al. 2009. Mice defective in Trpm6 show embryonic mortality and neural tube defects. *Human Molecular Genetics* 18: 4367.

72. Lu, W., B. Peissel, H. Babakhanlou et al. 1997. Perinatal lethality with kidney and pancreas defects in mice with a targetted Pkd1 mutation. *Nature Genetics* 17: 179.

73. Wu, G., V. D'Agati, Y. Cai et al. 1998. Somatic inactivation of Pkd2 results in polycystic kidney disease. *Cell* 93: 177.

74. Pennekamp, P., C. Karcher, A. Fischer et al. 2002. The ion channel polycystin-2 is required for left-right axis determination in mice. *Current Biology* 12: 938.

75. Venugopal, B., M. F. Browning, C. Curcio-Morelli et al. 2007. Neurologic, gastric, and ophthalmologic pathologies in a murine model of mucolipidosis type IV. *American Journal of Human Genetics* 81: 1070.

76. Hartmann, J., E. Dragicevic, H. Adelsberger et al. 2008. TRPC3 channels are required for synaptic transmission and motor coordination. *Neuron* 59: 392.

77. Liedtke, W., and J. M. Friedman. 2003. Abnormal osmotic regulation in trpv4$^{-/-}$ mice. *Proceedings of the National Academy of Science of the United States of America* 100: 13698.

78. Hoenderop, J. G., J. P. van Leeuwen, B. C. van der Eerden et al. 2003. Renal Ca^{2+} wasting, hyperabsorption, and reduced bone thickness in mice lacking TRPV5. *Journal of Clinical Investigations* 112: 1906.

79. Vennekens, R., J. Olausson, M. Meissner et al. 2007. Increased IgE-dependent mast cell activation and anaphylactic responses in mice lacking the calcium-activated nonselective cation channel TRPM4. *Nature Immunology* 8: 312.

80. Piontek, K. B., D. L. Huso, A. Grinberg et al. 2004. A functional floxed allele of Pkd1 that can be conditionally inactivated in vivo. *Journal of American Society of Nephrology* 5: 3035.

81. Bhunia, A. K., K. Piontek, A. Boletta et al. 2002. PKD1 induces p21(waf1) and regulation of the cell cycle via direct activation of the JAK-STAT signaling pathway in a process requiring PKD2. *Cell* 109: 157.

9 Studying Endogenous TRP Channels in Visceral and Vascular Smooth Muscles

Alexander V. Zholos

CONTENTS

9.1 INTRODUCTION

9.1.1 Nonselective Cation Channels in Smooth Muscles

As in all other types of muscles, calcium is the critical signal for initiating and maintaining contractions of smooth muscle (SM) cells. However, in comparison to skeletal and cardiac muscles, SM appears to be especially enriched with a multitude of ion channels that function in concert to regulate Ca^{2+} influx and thus the intracellular free calcium concentration ($[Ca^{2+}]_i$). Under physiological conditions, changes in SM tension (either contraction or reduction of existing tone) are usually induced by, or at least associated with, changes in the membrane potential. Thus, SM excitation in the form of action potentials in those types of SM capable of generating them or in the form of more slowly developing potential changes in less excitable or nonexcitable SM (e.g., slow waves) causes Ca^{2+} entry via voltage-gated Ca^{2+} channels (VGCCs). This, in turn, can trigger Ca^{2+}-induced Ca^{2+} release (CICR), bringing an additional source of Ca^{2+} into action. Similarly, membrane hyperpolarization causes SM relaxation by inhibiting action potential discharge and reducing activity of VGCCs.

Although VGCCs and CICR provide the bulk of the activator Ca^{2+} for SM contraction, it is another group of ion channels, termed nonselective cation channels (NSCCs), that appears to play the most essential role in SM responsiveness to various stimuli. Indeed, various NSCCs expressed in SM mediate the primary response to numerous neurotransmitters and hormones, stretch, lipid messengers, and various metabolic/environmental factors. Moreover, recent evidence suggests that these NSCCs are critical not only for the regulation of SM contractile state but also for the regulation of slow phenotypic changes, SM cell proliferation, and migration processes that are especially important in vascular function and disease.[1,2]

NSCCs present in SM cells provide a route for Na^+, Ca^{2+}, and Mg^{2+} entry, which not only directly contributes to calcium signaling but also causes membrane depolarization and, hence, opening of VGCCs, increased electrical activity, and eventually, SM contraction. The importance of NSCCs for SM function has been appreciated for more than 40 years. Numerous early studies have documented changes in ion permeability of SM cell membrane in response to various naturally occurring ligands (e.g., acetylcholine, catecholamines, angiotensin II, histamine, bradykinin, oxytocin, prostaglandins) and revealed the critically important link between these changes and SM contractility.[3] The resistance of these responses to the action of the "classical" Ca^{2+} channel blockers, such as verapamil, D600, or nifedipine, and the non-voltage-dependent nature of their initiation implied the involvement of a different, non-VGCC, class of ion channels.

Further studies revealed that the properties of these channels were far more complex than initially thought.[4–8] Thus, currently, there is convincing evidence indicating that in SM, there are at least three distinct Ca^{2+} entry pathways for contraction initiation and maintenance. These are (1) VGCCs, (2) receptor-operated Ca^{2+}-permeable cation channels (ROCs), and (3) store depletion-induced Ca^{2+}-permeable cation channels (SOCs). Moreover, their properties (e.g., pharmacology, biophysical "signatures," regulation) and relative contributions show considerable variations in different types of SM, likely contributing to the significant diversity of SM electrical and contractile activity patterns observed in different organs (e.g., "phasic" and "tonic" SM).

The molecular basis of the complexity of SM NSCCs remained unclear until recently, when important clues started to rapidly emerge, indicating a connection between these channels and mammalian homologues of the *Drosophila* transient receptor potential (TRP) protein.[9–12]

9.1.2 TRP CHANNEL PROTEINS AS MOLECULAR COUNTERPARTS OF SMOOTH MUSCLE NONSELECTIVE CATION CHANNELS

Various TRPs are currently considered as the leading candidate proteins mediating diverse nonvoltage-activated Ca^{2+} entry pathways in visceral and vascular SM.[7] These channels have been implicated in various physiological responses, some of which are briefly outlined below.

9.1.2.1 TRP Channels in Visceral Smooth Muscles

In visceral SM, NSCCs have long been recognized as important targets for the action of excitatory neurotransmitters, signaling via G proteins.[3–6] The first conclusive evidence for the presence of a receptor-operated cation conductance was obtained by patch-clamp recording of responses to acetylcholine, the major excitatory neurotransmitter in various visceral SM, including rabbit jejunal SM.[13] This study showed, in particular, that acetylcholine induces an inward cation current that was characterized by an unusual voltage dependence (U-shaped current–voltage relationship at negative potentials). The underlying channels (conductance 20–25 pS) permeable to both Na^+ and K^+ have been later characterized in guinea pig ileal myocytes.[14] This muscarinic cation current (mIcat) remains the most extensively studied example of a G protein coupled receptor (GPCR) regulated NSCC current in visceral SM. Similar cation currents have been reported in gastric, colonic, and airway SM,[15–17] and thus, this current appears to be of widespread importance for visceral SM function.

Already, in early studies of mIcat, its complex regulation by intracellular Ca^{2+} and membrane potential, as well as the involvement of G proteins in its activation, became evident.[18–22] Further studies elucidated synergistic roles of the M_2 and M_3 acetylcholine receptor subtypes and complex signal transduction pathways involving both phospholipase C (PLC) activation (the $M_3/G_{q/11}/$PLC pathway) and $G_{i/o}$ protein activation (the M_2 effect) involved in mIcat generation.[4–6,23–25] At the single-channel level, at least three different NSCCs mediate mIcat, and there is evidence that M_2 and M_3 receptor subtypes differentially regulate these channels via three distinct cholinergic signaling pathways.[26,27]

As already described, for a receptor-operated conductance, the mIcat shows an unusual current–voltage (I-V) relationship, which is U-shaped at negative potentials and, in addition, shows pronounced double rectification around the reversal potential. These are typical "biophysical signature" elements of heterologously expressed TRPC4 and TRPC5 channels, which also show other features (most importantly, activation by G proteins and intracellular Ca^{2+} dependence) that are consistent with the mIcat properties.[28] Moreover, TRPC4 associates with PLC via a PDZ domain-containing protein, NHERF.[29]

Consistently, surveys of TRP expression in SM of the gastrointestinal (GI) tract by RT-PCR showed expression of TRPC4/6/7 genes with the highest expression of TRPC4.[30] Finally, recent studies using receptor and channel knock-out (KO) mice have characterized the specific roles of M_2 and M_3 receptors in mIcat regulation[27] and provided conclusive evidence for TRPC4 (>80%) and TRPC6 (residual current) involvement in mIcat generation.[31] Thus, cholinergic excitation-contraction coupling in visceral SM includes signaling leading to an increase in $[Ca^{2+}]_i$ as a result of Ca^{2+} release induced by InsP$_3$ (the M_3/G$_{q/11}$/PLC system), as well as activation of Ca^{2+} entry via VGCCs due to TRPC4/6-mediated membrane depolarization (evoked in synergy between the M_3/G$_{q/11}$/PLC system and the M_2/G$_{i/o}$ systems). Because VGCCs are well expressed in GI muscles, TRPC4/6 channels probably play a relatively minor direct role in Ca^{2+} influx.

9.1.2.2 Vascular TRPs

The expression patterns and functional roles of TRP proteins in vascular SM have been studied even more extensively. In the vasculature, a number of neurotransmitters and hormones exert their action through activation of NSCCs in a receptor- or store-operated mode. Similarly to the above-described action of muscarinic agonists on visceral SM, activation of α-adrenoceptors (AR) by norepinephrine or epinephrine in vascular SM has long been known to increase membrane permeability to sodium, potassium, and chloride.[3]

The α_1-AR is widely present in vascular SM, where it plays a central role in the sympathetic control of blood pressure. The signaling is largely similar to that already described for muscarinic receptors in visceral SM and involves both stimulation of PLC, followed by InsP$_3$-induced Ca^{2+} release, and activation of an inward cation current (termed α_1-AR-NSCC). Again, these channels can admit Ca^{2+} both directly and indirectly via membrane depolarization and VGCC opening.[32] The native current is activated via the α_1-AR/G$_{q/11}$/PLC/diacylglycerol (DAG) system, and it shows characteristic S-shaped voltage dependence.[33] Interestingly, external Ca^{2+} exerts both stimulatory and inhibitory effects on α_1-AR-NSCC.[32] This feature, namely, consistency in voltage-dependent and pharmacological properties between native α_1-AR-NSCC and heterologously expressed TRPC6 and selective suppression of native current with TRPC6 antisense oligonucleotides, indicated that TRPC6 was an essential component of α_1-AR-gated native NSCC.[34] It should be noted that other PLC-stimulating receptors, including vasopressin and platelet-derived growth factor receptor, can also induce TRPC6 activation, implying that this channel can serve as a common pathway for the action of different receptor agonists causing vasoconstriction.

In addition, other studies implicated TRPC6 as stretch-[35] and store-operated channels,[36] the latter study also suggesting a role for TRPC6 in pulmonary vascular SM cell proliferation. There is also strong evidence for the involvement of TRPC1/4/5 in SOC in vascular SM, while other TRP channels (notably TRPV2 and TRPM4) were implicated in SM stretch-activated currents (SAC).[8] There is also emerging evidence for functional expression of TRPM8 and TRPV4 in blood vessels.[37,38]

In addition to playing roles in SOC, ROC, and SAC, vascular TRPs have been described for numerous other functions. These include their roles in Mg^{2+} homeostasis, oxidative stress, cell proliferation and migration, and vascular injury.[7,8,39–42]

9.2 ELECTROPHYSIOLOGICAL ANALYSIS OF NATIVE TRP CURRENTS

There is currently overwhelming evidence that heterologously overexpressed TRP proteins form NSCCs.[9–12] Therefore, once plasmalemmal presence of a certain TRP isoform is established, it is a reasonable expectation that the corresponding native TRP channel would function as an NSCC. Further confirmation ultimately needs selective TRP pharmacological tools (e.g., channel agonists and/or blockers), but these are generally lacking. Therefore, current strategies for studying native TRPs include not only molecular identification of their expression at the gene and protein level and their localization study by immunocytochemistry but also the use of molecular biology tools (e.g., gene knock-out or gene knock-down with the use of siRNA or antisense oligonucleotides) and blocking antibodies for functional channel identification. Finally, biophysical characterization of native currents, including analysis of single-channel properties, and appropriate comparisons with heterologously expressed TRP channels are performed.

Except for the initial stage of TRP expression analysis with molecular biology tools, these strategies require patch-clamp recordings, membrane potential assays, and/or calcium imaging techniques for native channel functional characterization. Although intracellular calcium recordings are much more productive and straightforward, direct patch-clamp recordings of channel activity in the whole-cell and, especially, in the single-channel mode remain the gold standard. This chapter will consider the electrophysiological approach in detail. There are also many added features, as patch-clamp analysis can be combined with other techniques (e.g., rapid drug application, flash-photolysis of "caged" "compounds," intracellular calcium recordings) that aid the study of endogenous TRP channels; these approaches will also be outlined.

9.2.1 PATCH-CLAMP SYSTEM FOR LOW NOISE CURRENT RECORDINGS

Whether or not single-channel recordings are planned, it is always desirable to achieve the minimal current noise when constructing a patch-clamp system. Exceptionally, native TRP currents in SM can be as large as 2–3 nA (e.g., m*I*cat in guinea pig ileal myocytes), but in most cases, because of a low level of endogenous TRP expression or low single-channel conductance, the recorded currents are in the order of 5–50 pA (e.g., α_1-AR-NSCC, SOC currents). Thus, technical refinements of the system in terms of noise, electrical hum, vibration, etc. should approach the standard of single-channel recording.

9.2.1.1 System Configuration and Noise Minimization

For more detailed descriptions, The Axon Guide can be consulted (http://www.moleculardevices.com/pages/instruments/axon_guide.html). This laboratory guide expertly explains many practical aspects of high-quality patch-clamp recordings.[43] Here, we will only briefly illustrate and explain the patch-clamp system configuration used in the author's laboratory (Figure 9.1). The system is based around the Nikon TE2000S inverted microscope (1) which is positioned on an Intracel Isolate system 2000 air table (2). *Critical:* mechanical vibrations can not only damage the cell during recordings but also cause low-frequency electrical noise that can be mistaken for line electrical interference. For the purpose of shielding, the patch-clamp system is typically placed inside a large Faraday cage. Apart from a larger footprint, problems with electrical interference may occur when additional pieces of equipment are installed inside the cage (e.g., solution changers, temperature controllers). Therefore, we use a small custom-made cage, which is attached directly to the stainless steel plate (3) bolted just above the microscope stage. A roll of foil covered with adhesive film (4) is used instead of front doors, as lowering the foil does not produce any mechanical disturbance to the system. *Critical:* it is essential that all components of the system such as the microscope stage, steel plate, cage panels, and parts of the micromanipulator are securely connected by copper wires. To avoid earth loops (i.e., two different earths connected to the same system), the only grounding is provided by connecting the steel plate with the ground pin on the back of the amplifier head

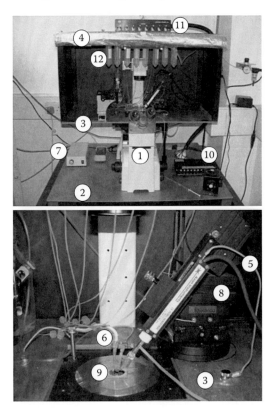

FIGURE 9.1 Patch-clamp system for low noise current recordings. Various parts of the system are indicated by the numbers. See text for details.

stage (5) (Figure 9.1, bottom). Note that the indifferent Ag/AgCl electrode is also connected to the same high-quality signal ground (6). There is no need to ground the antivibration table in this configuration; in fact, it is better to place the microscope on an insulating plate to avoid contact between the microscope chassis and the table, thus avoiding any unnecessary load on the signal ground.

This configuration is similar to the suggested use of a large coffee can for shielding the Axopatch 200B headstage during the functional checkout of the amplifier (http://www.axon.com/pages/mdc_kb_article.cfm?ArticleID=685). In fact, root-mean-squared (RMS) noise values, which are a good indicator of proper system configuration, are as described in the typical Axopatch 200B initial functional checkout with the amplifier disconnected from any other equipment (e.g., 0.05 pA in PATCH mode and 0.55 pA in whole-cell $\beta = 1$ mode). The only possible source of electrical noise is the power supply for the microscope light (7), and if this occurs, the radical solution is to use a car battery.

A motorized or piezo-driven micromanipulator is essential for the precise and smooth positioning of the patch pipette. We use the Scientifica PatchStar motorized micromanipulator (8), which allows 20 nm resolution and has a drift of less than 1 micron over 2 hours. Its "home in/home out" feature (memorizing the position to automatically bring the pipette to the fixed place) allows convenient and fast exchange of the pipettes, but these should be standard in length. There is an offset function that makes this requirement less stringent. *Critical:* the micromanipulator is bolted directly to the steel plate at the shortest possible distance from the bath to ensure minimal thermal drift of the pipette.

For data acquisition and analysis, we use the DigiData 1322 digitizer and the pClamp 9 software (Molecular Devices, Sunnyvale, California, USA).

Patch pipettes are fabricated from borosilicate glass capillaries with 1.5 mm OD and 0.86 mm ID (Harvard Apparatus, Edenbridge, UK) using a Narishige PC-10 puller. After heat polishing of the pipette tip (*critical:* use platinum wire and spread some melted pipette glass on the new filament to prevent pipette contamination), they have a resistance of 1–3 MOhm when filled with the pipette solution. It is more difficult to patch SM myocytes with larger pipettes, but smaller pipettes compromise the efficiency of cell perfusion with internal solutions. *Critical:* efficient intracellular perfusion is especially important if cation currents are activated by infusion of GTPγS or when studying the effects of TRP blocking antibodies in the whole-cell configuration.

For single-channel recordings, pipettes are coated with one-part Dow Corning R-6101 elastomer. With gigaseals >10 GOhm, this system configuration allows patch-clamp recordings with current noise <0.25 pA (RMS, 5 kHz bandwidth).

9.2.1.2 Solution Exchange

Rapid noise-free solution exchange is an essential requirement for the study of pharmacological properties of ion currents. Cells are placed in a small CoverWell perfusion chamber (Grace Bio-Labs, 20 mm diameter; Sigma catalog number Z379050) pressed-to-seal to cover slips (9). The volume of the solution in the chamber is about 0.5 mL. Routinely, solutions are replaced by perfusing the whole chamber through one of the eight separate polyethylene tubes glued together and positioned just above the surface of the solution (Figure 9.2a, top). This arrangement excludes the possibility of solution cross-contamination. The solution is removed via a suction tube positioned at the opposite side of the chamber into a bottle that should be placed inside (*critical*) the cage. Any cables or tubes containing salt solutions entering the Faraday cage compromise its purpose. ValveLink 8.2 solution changer from AutoMate Scientific is used to switch the solution on and off by activating pinch solenoid valves. *Critical:* both the controller (10) and the valves (11) are kept outside the cage to avoid any electrical interference, while pressurized solution reservoirs (12) are placed inside the cage. The pressure-regulated system allows fine control of solution flow and ensures steady flow rate. If recordings are made in 35-mm Petri dishes, a perfusion insert chamber (AutoMate Scientific) can be used to minimize the volume and thus accelerate solution exchange.

For fast application and washout of receptor agonists, we use a double-barreled application pipette fabricated from theta glass tubing as described in detail elsewhere[44] (Figure 9.2b). The pipette is attached to a piezo-driven micromanipulator that allows its precise positioning and rapid controlled movement (e.g., TTL pulse can be used to control the moment of solution application, and this can even be synchronized with the voltage protocol used for current recordings). This system allowed the study of m*I*cat on- and off-kinetics while agonists were applied and removed within few milliseconds.[45]

The rate of solution exchange can be evaluated simply by applying deionized water to the bath (Figure 9.2a, bottom) or directly to the pipette tip (Figure 9.2b, bottom). This reduces pipette current (produced by a small negative holding potential) almost to zero level (dotted line). Complete solution exchange in the above-described 0.5-mL chamber can be achieved in about 0.5 s, while local drug application with a piezoelectric-driven positioning system can be performed on a millisecond time scale. Another advantage of the latter system is that it ensures highly reproducible drug applications (e.g., five traces are superimposed in Figure 9.2b, bottom).

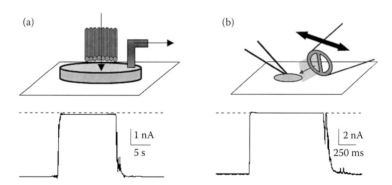

FIGURE 9.2 Two approaches to drug application during patch-clamp recordings: (a) External solutions are applied via individual polyethylene tubes positioned just above the experimental chambers; and (b) "concentration clamp" technique. Theta-style pipette with tip of about 50 μm is mounted using piezoelectric-driven positioning system driven by a voltage pulse. Current traces at the bottom were recorded during application of deionized water (a) to the bath or (b) locally to the patch pipette tip.

9.2.2 PROCEDURES FOR CELL ISOLATION

To bring about cell dispersal, the longitudinal layer of the guinea pig distal ileum is dissected and placed in a physiological salt solution (PSS) of the following composition (in mM): NaCl, 120; KCl, 6; $CaCl_2$, 2.5; $MgCl_2$, 1.2; glucose, 12; HEPES 10; pH adjusted to 7.4 with NaOH. The tissue is cut into small pieces, which are immediately transferred to divalent cation-free PSS for about 5 min. The pieces are incubated in 2 mL of this solution containing collagenase (type 1A, Sigma), soybean trypsin inhibitor, and bovine serum albumin (all at 1 mg/mL) at 36°C for 20 min.

After tissue digestion, the pieces are carefully washed in enzyme-free divalent cation-free PSS and agitated in 1 mL of this solution using a small-bore glass pipette until it becomes very cloudy, indicating successful tissue dispersal. Nondispersed SM pieces and myenteric plexuses are discarded by cell suspension filtering through 80-μm nylon net filters (NY80 from Millipore), and the purified cell suspension is diluted using normal PSS to obtain approximately 0.8 mM $CaCl_2$ and 0.4 mM $MgCl_2$ final concentration. Cells are then kept in this solution in the above-described experimental chambers under a saturated atmosphere at 4°C until use. Mouse ileal myocytes are isolated similarly, but enzyme treatment time is reduced to 15–17 min.

Vascular myocytes (e.g., from rat tail arteries) can be isolated in a similar way using a combination of collagenase (type XI, 1 mg/mL), papain (1 mg/mL), and dithiothreitol (0.8 mM).[38] Other enzymes are often used for vascular SM cell isolation, such as protease Type I or Type XIV, dispase, and elastase, as it is generally more difficult to isolate vascular myocytes compared to visceral myocytes.

9.2.3 TRP CURRENT ISOLATION AND IDENTIFICATION

SM cells express numerous voltage- and Ca^{2+}-activated channels permeable to Ca^{2+}, Na^+, and Cl^-. Pharmacological inhibition of these non-TRP channels would require a combination of various blockers, raising concerns that some of these blockers may affect TRP channels under study. Thus, studying SM TRP channels with quasi-physiological ion gradients presents significant problems. These can be at least partially resolved by using appropriate voltage protocols (e.g., by applying slow negatively going voltage ramps following a voltage step to a very positive potential of duration sufficient to inactivate VGCCs). A more productive approach to this problem is to use internal and external solutions designed to minimize non-TRP currents while maximizing TRP currents. The latter is also important for improving signal-to-noise ratio, which is essential for biophysical analysis of TRP conductances (current kinetics, activation curves).

9.2.3.1 Composition of Internal and External Solutions

Because all TRP channels are permeable to monovalent cations, both Na^+ and Cs^+ can be used as the main cation for the study of endogenous SM TRP channels. K^+ is unsuitable for these studies since SM cells express large conductance BK_{Ca} and many other types of K^+ channels. Cs^+ is the preferred cation because it eliminates K^+ currents as it permeates most types of K^+ channels poorly and in fact even blocks many of them. If Na^+ is used, one needs to consider that some SM express non-TRP

Na$^+$-permeable channels (I$_f$ or I$_h$ current, metabotropic and ionotropic purinocep-
tors), and some SM even express voltage-gated Na$^+$ channels.[46] Chloride currents
can be minimized by dialyzing cells with large impermeable anions such as methanesulfonate or aspartate as the major anion. *Critical:* the commonly used Ag/AgCl
interface for transferring flow of ions into electron flow performs adequately only
when at least a few millimoles of Cl$^-$ are present.

In addition, Ca^{2+}-activated conductances (e.g., BK$_{Ca}$ and Cl$_{Ca}$ channels) can be
easily suppressed by including a Ca^{2+} chelator, such as EGTA or BAPTA, in the
pipette solution at a high concentration, typically 5 or 10 mM. However, some TRP
channels (e.g., TRPC4/5 and TRPM4) also need intracellular Ca^{2+} for their activity,
and this maneuver will also suppress Ca^{2+}-dependent TRP currents. In such cases, it
may be appropriate to "clamp" intracellular [Ca^{2+}]$_i$ at a certain level (e.g., 100 nM)
using Ca^{2+}/EGTA or Ca^{2+}/BAPTA mixtures. BAPTA performs better for these purposes since it is a faster Ca^{2+} chelator, and it is also less pH-sensitive.

External Ca^{2+} is another major factor affecting activity of endogenous SM TRP
channels. For example, α_1-AR-NSCC is facilitated by micromolar concentrations of
Ca^{2+}$_o$ but inhibited by millimolar Ca^{2+}$_o$,[32,47] and m/cat is also inhibited by millimolar
Ca^{2+}$_o$ and Mg^{2+}$_o$.[48]

After optimal ion conditions for recording of endogenous SM TRP have been
established, it is important that comparisons with the corresponding recombinant
TRP channel overexpressed in HEK293 or CHO cells are carried out using the same
ionic gradients.

9.2.3.2 Channel Identification and Single-Channel Recordings

The gold standard of patch-clamp techniques is the measurement of single-channel
activity in cell-attached or cell-free (inside-out or outside-out) patches. Using patch
pipettes with small tips (e.g., resistance of about 10 MOhm), it is often possible to
record the activity of only one channel protein. This approach completely eliminates
the above-described problems of TRP current isolation from other contaminating
currents in whole-cell recordings. In multichannel patches, any interfering channels
can be easily identified by differences in their kinetics and unitary current amplitudes. However, with this approach, there is also an issue of how well one single-
channel protein represents TRP ensemble behavior that can only be evaluated by
recording whole-cell currents.

Because highly selective and potent TRP pharmacological tools are generally
lacking, the current emphasis is on the use of molecular biology approaches for
native TRP isoform identification. Comparative studies of single-channel activity in wild-type and gene knock-out (KO) myocytes may provide ultimate proof of
TRP involvement in endogenous current. Such proof has been recently provided
for TRPC4 forming a 55-pS cation channel in mouse ileal myocytes that underlies
>80% of the integral current.[31] Activity of this channel was frequently observed in
wild type, but it was completely absent in TRPC4 KO myocytes. However, it should
be noted that data from transgenic mice often do not confirm data obtained using
other approaches, and such apparent contradictions are especially evident for the
role of native TRPs in in vivo functions such as vasoconstriction and blood pressure
control.

In the vasculature, the most conclusive evidence based on TRPC4 KO exists for the role of this channel as an important component of SOC in endothelial signaling.[49] Consistently, TRPC4$^{-/-}$ mice showed impaired agonist-induced vasorelaxation. In contrast, studies from many laboratories showing a role for TRPC1 in SOC using the "knock-down" approach are at variance with the KO data.[50] One likely explanation for resolving this paradox is that other TRP channels (TRPC4/5) can compensate for the loss of TRPC1 function. Such compensation through overexpression of TRPC3 proteins was indeed found to take place in TRPC6$^{-/-}$ mice that showed increased vascular contractility and higher blood pressure, which is contrary to expectations because of the involvement of TRPC6 in α_1-AR-NSCC.[51]

Studies utilizing antisense oligonucleotides or siRNA technologies have provided a more consistent picture for the critical roles of various vascular TRP subtypes (notably TRPC1/3/6 and TRPM4) in diverse vascular SM functions, including membrane potential control, Ca^{2+} influx via SOC, ROC, and SAC, myogenic response, and cell proliferation and migration. This vast area of research has been recently reviewed.[7,8,39–42,52]

9.2.3.3 TRP Antibodies

The molecular biology approaches to TRP identification using knock-down technologies necessitate cell culture because a certain period of time (1–3 days depending on the rate of protein turnover) is needed to bring down the level of channel protein expression. However, cell culture conditions can alter the expression of other TRP and non-TRP channels. For example, organ culture of rat cerebral arteries causes a more than fivefold increase in TRPC1/6 mRNA levels, whereas TRPC3 expression is decreased.[53] In addition, culturing of vascular myocytes causes transition from a contractile to a proliferative cell phenotype.

An alternative approach to channel identification is to employ anti-TRP antibodies, in which case freshly dispersed cells that represent their physiological phenotype most closely can be used. In its most straightforward form, the technique requires isolation of inside-out patches, allowing the antibodies that are raised against intracellular TRP epitopes to be directly applied to the cytoplasmic side of the membrane. The advantage of this approach is that channel activity can be investigated in the same patch before and after application of the blocking antibody. This facilitates data analysis as the patch serves as its own control, and thus, paired tests are possible.

There are, however, important concerns regarding selectivity of TRP antibodies[54] (also see Chapter 6). Nevertheless, there are also remarkable examples of high potency and selectivity of TRPC antibodies when they act directly on excised membrane patches.[41] Thus, a spontaneously active Ca^{2+}-permeable cation channel, which contributes to resting membrane conductance and basal Ca^{2+} influx in rabbit ear artery myocytes, was identified as TRPC3 as its activity in inside-out patches was selectively inhibited by anti-TRPC3 but not by anti-TRPC1/4/5/6/7 antibodies.[55] In addition, the channel showed a pharmacological profile similar to that of the expressed TRPC3 channels, while immunocytochemical analysis confirmed its predominant presence at, or close to, the plasma membrane. In mesenteric artery myocytes, angiotensin II activates two different cation conductances. The signal transduction involves a common 1,2 diacylglycerol (DAG) pathway, which acts either in a protein kinase C (PKC)-independent (I_{cat1}) or -dependent (I_{cat2}) manner. Application of anti-bodies to inside-out patches revealed that, in this case, TRPC6 and TRPC1 proteins are important components of I_{cat1} and I_{cat2}, respectively.[56] In rabbit coronary artery myocytes, endothelin activates Ca^{2+}-permeable NSCC to which both TRPC3 and TRPC7 contribute, as was shown with the use of TRP antibodies.[57] Again, their action was highly selective as anti-TRPC1/4/5/6 antibodies had no effect. Finally, using TRPC antibodies, it proved possible to characterize the extremely complex makeup of native SOC channels present in different blood vessels (coronary and mesenteric arteries, portal vein) whereby TRPC1, TRPC5, TRPC6, and TRPC7 make distinct contributions.[58]

A new strategy for generating ion channel inhibitors taking advantage of antibody specificity has been recently proposed (see Chapter 7 for details). These so-called E3-antibodies target the third extracellular region of a channel, with the obvious advantage of allowing their efficient use in whole-cell patch-clamp and intracellular calcium recordings. The T1E3 and T5E3 proved to be efficient tools in the identification of TRPC1 and TRPC5 currents in vascular SM.[53,59]

9.3 ANALYSIS OF SIGNAL TRANSDUCTION PATHWAYS USING PATCH-CLAMP TECHNIQUES

Patch-clamp recording techniques can be further combined with other approaches and tools (pharmacological analysis, agents for G protein research, antibodies, and calcium measurements) that greatly enhance their analytical power. In this section, we briefly outline these additional approaches while highlighting their usefulness for TRP research in SM.

9.3.1 RECEPTOR AND G PROTEIN SIGNALING

In contrast to rather nonselective agents available for TRP identification, receptor agonists are highly selective molecules, and agonists and/or antagonists for different receptor subtypes are often available. Thus numerous receptor agonists (e.g., car-bachol, histamine, endothelin, angiotensin II, and bradykinin) have been used to characterize GPCR-mediated regulation of native SM TRPs. They are often used at a single submaximal concentration capable of producing a robust current response. Selective receptor antagonists can then be applied in the presence of an agonist, or cells can be pretreated with the antagonist to ensure that equilibrium at the receptor site is reached before the agonist is applied for the second time. In the latter approach, it is important to evaluate the extent of receptor desensitization during repeated agonist applications. Strong desensitization may preclude the use of this protocol.

For mIcat, we found that desensitization of current responses can be minimized if 1 mM GTP is added to the pipette solution.[60] This allowed quantitative pharmacological analysis (known as Schild analysis[61]) to be performed.[62] In this approach, concentration-effect curves are generated by applying the agonist at ascending concentrations, while ensuring that the current reaches a steady state level at each agonist dose. The experiment is then repeated, on the same cell, in the presence of a fixed antagonist concentration. Data analysis is performed by fitting concentration-effect data points with a logistic function:

$$G/G_{max} = \{1 + ([EC_{50}]/[A])^b\}^{-1}$$

where G/G_{max} is the normalized cationic conductance (G_{max} is achieved at a maximally effective agonist concentration), EC_{50} is the agonist concentration ([A]) when $G/G_{max} = 0.5$, and b is the slope factor of the agonist curve. The dose-ratio (DR) is then obtained by dividing the EC_{50} value in the presence of the antagonist by that in control. Finally, apparent antagonist affinities (pA_2) can be obtained by Schild regression analysis (plot of log(DR-1) versus log antagonist concentration), providing experiments with three to four antagonist concentrations are performed. This work showed that both the M_2 and the M_3 acetylcholine receptors synergize in mIcat activation.

Many excellent tools also exist for investigation of roles of G proteins in TRP regulation. A significant latency between agonist application and current response is an early indication that a G protein link may be involved in channel regulation. Indeed, an appreciable latency of the order of 100 ms between stimulation of the cholinergic nerves (or iontophoretic application of muscarinic agonists) and membrane depolarization has been measured in various visceral SM preparations.[3] This is in contrast to the opening of fast ligand-gated channels, where an agonist-binding site is part of the channel molecule. Similarly, a significant latency of about 1 s has been described for norepinephrine-induced cation current in rabbit portal vein myocytes.[63]

To further access whether G-protein activation is involved in current responses, GTPγS, a hydrolysis-resistant analogue of GTP, is commonly used. Inclusion of GTPγS in the pipette solution (typically, 0.2–0.5 mM) causes a slowly developing inward cation current both in visceral and vascular myocytes. These currents are characterized by voltage-dependent properties identical to agonist-induced mIcat and α_1-AR-NSCC currents,[21,26,33] respectively. Alternatively, GDPβS, which blocks G protein activation, can be included in the pipette solution (1–5 mM) resulting in the inhibition of these currents.

Pertussis toxin (PTX) is another useful tool that allows highly selective disruption of $G_{i/o}$ signaling. Incubation time and PTX concentration are important factors. For example, we found that relatively high concentration (6 µg/mL) and incubation at 36°C for 4–9 hours are needed to abolish mIcat and muscarinic suppression of VGCC activity.[64] In addition, we noted differences between the potency of PTX supplied in the form of lyophilized powder and reconstituted in water and that supplied in 50% glycerol solution.[64]

Finally, antibodies specific for different G protein α-subunits can be infused via patch pipette in the attempt to pinpoint specific G protein(s) involved in the signal transduction. The antibodies are diluted with the intracellular solution to 1:200 or 1:100 v/v, which is about 5–10 higher a concentration than that used for Western blotting. This allows faster access of the antibodies to the target proteins, but even at such high concentration, it is important to allow at least 20–25 min for sufficient cell dialysis before the agonist is applied. As a control, the same antibodies can be heated to inactivate them (e.g., at 80°–90°C for 5–10 min). These studies revealed that $G\alpha_o$ protein couples muscarinic receptors to NSCC mediating m/cat, while $G\beta\gamma$ is not involved.[65]

It should be noted that proteins included in the patch pipette have a significant negative impact on the success of gigaseal formation. However, the strategies that are used, for example, in perforated patch technologies (e.g., filling the pipette tip with normal solution and back-filling the pipette with solution containing membrane pore-forming agents) are not acceptable for antibody infusion because of very slow diffusion rates of these large molecules.

Direct application of the antibody to the internal side of the membrane in inside-out configuration avoids these problems. Using this approach,[66] it was shown that constitutively active cation channel currents in rabbit ear artery myocytes (these were later identified as TRPC3 channels[55]) were potentiated by anti-$G\alpha_q$/$G\alpha_{11}$ antibodies, whereas a combination of anti-$G\alpha_{i1-3}$/$G\alpha_o$ antibodies rapidly and reversibly inhibited channel activity.

9.3.2 PHOSPHOLIPASE C SIGNALING

Activation of various $G_{q/11}$ coupled receptors expressed in SM cells causes, via PLC activation and InsP$_3$ formation, Ca^{2+} release and contraction. PLC activation also commonly gates TRPC channels, although the molecular scenarios are diverse and may involve all relevant lipids such as DAG in the case of TRPC2/3/6/7, InsP$_3$ formation, and/or Ca^{2+} store depletion in the case of TRPC1/3.[12] In recent years, patch-clamp techniques that allow direct applications of relevant signaling molecules to both the external and internal side of the membrane have been extremely instrumental in addressing questions on the specific roles of PLC products in TRP regulation.[67]

In SM research, several studies documented critical roles of the PLC substrate, phosphatidylinositol 4,5-bisphosphate (PI(4,5)P$_2$), in the regulation of native TRP channels. In these experiments, diC8-PI(4,5)P$_2$, the more water-soluble short form of PI(4,5)P$_2$, was included in the patch pipette (20–100 µM) in whole-cell recordings or added to the bath during recordings from inside-out patches. This caused significant suppression of m/cat (as well as heterologously expressed TRPC4α isoform) in guinea pig and mouse ileal myocytes,[31,68] as well as inhibition of angiotensin II-induced NSCC and native TRPC6 activity in rabbit mesenteric artery myocytes.[69] The inhibitory action of PI(4,5)P$_2$ on these native conductances is highly selective as, for example, infusion of PI(3,4,5)P$_3$, PI(3,4)P$_2$, and PI(4)P did not affect TRPC4 activity, while InsP$_6$ and PI(3,5)P$_2$ potentiated the current.[68]

Additional techniques can be used for altering the level of PI(4,5)P$_2$ and investigating its functional role in TRP regulation. PI(4,5)P$_2$ levels can be reduced by cell treatment with wortmannin (20–30 µM for 10–60 min) or LY294002 (100 µM) to inhibit PI-4 kinase or by applying anti-PI(4,5)P$_2$ antibodies (1:200 v/v) or PI(4,5)P$_2$ scavenger poly-L-lysine (50 µg/mL applied via pipette or to inside-out patches).[68,70]

It is now becoming increasingly evident that ROCs and SOCs have very diverse properties in different blood vessels, which is likely due to variable contribution of diverse TRPC isoforms to their formation. It is thus perhaps not surprising that the action of signaling molecules occurring downstream of PLC activation on other SM ROCs and SOCs is diverse and complex. For example, PI(4,5)P$_2$ has contrasting activating effects on native TRPC1-like SOC, and both $PI(4,5)P_2$ and $PI(3,4,5)P_3$ synergize to activate TRPC1/5/6 channels during the action of endothelin on rabbit coronary artery myocytes.[70,71] To reveal the roles of $PI(3,4,5)P_3$, technical approaches similar to the above described for $PI(4,5)P_2$ can be used, including application of diC8-$PI(3,4,5)P_3$, wortmannin (50 nM) or PI-828 (3 μM) (to inhibit PI-3 kinase in this case), and anti-$PI(3,4,5)P_3$ antibodies.[71]

The products of PLC enzymatic activity, $InsP_3$ and DAG, are not required for mIcat (TRPC4) activation,[72] but both molecules synergize for optimal $α_1$-AR-NSCC activation.[73] Moreover, DAG activates vascular TRPC-mediated ROCs in a PKC-independent manner but inhibits them via a PKC-dependent mechanism, whereas both $PI(4,5)P_2$ and PKC are of critical importance for vascular SOC stimulation.[74]

9.3.3 CALCIUM DEPENDENCE

Activity of some TRP channels (e.g., already mentioned TRPC4/5 and TRPM4) is strongly influenced by $[Ca^{2+}]_i$ level. It is still unclear whether these effects involve any direct binding of Ca^{2+} to the channel protein. It is likely that they are mediated by calmodulin and other Ca^{2+}-binding proteins.[75] At least for mIcat, the role of calmodulin in its regulation is well established.[76]

Several technical approaches can be used for the investigation of $[Ca^{2+}]_i$ regulation of native SM TRP currents. The most straightforward approach uses cell perfusion with solutions of known $[Ca^{2+}]_i$. These can be prepared by adding variable amounts of Ca^{2+} to solutions containing 10 mM BAPTA or EGTA. Equilibrium solution composition can be calculated using software packages such as EqCal (Biosoft) or MaxChelator (http://maxchelator.stanford.edu). There are, however, caveats with this approach, as there is some disagreement on the stability constants and their dependence on many factors such as ionic strength of solution, temperature, and pH.

Even more importantly, prolonged maintenance of $[Ca^{2+}]_i$ at levels significantly different from a resting level of about 100 nM can disrupt physiological channel regulation. For example, prolonged infusion of 500 nM $[Ca^{2+}]_i$ inhibits mIcat,[77] although it is well established that under more physiological conditions, rapid Ca^{2+} release events in the form of Ca^{2+} oscillations strongly potentiate the current.[78–80]

Using this approach, Inoue and Isenberg[19] showed that buffering $[Ca^{2+}]_i$ with 40-mM Ca^{2+}–EGTA mixtures modulates mIcat in a manner such that half-maximal and maximal activation occur at about 200 nM and 1 μM, respectively. One advantage of this approach is that voltage-dependent properties of the current at different steady $[Ca^{2+}]_i$ levels can be easily evaluated using voltage step or ramp protocols (see below). In contrast, fairly rapid oscillations of $[Ca^{2+}]_i$ and mIcat when $[Ca^{2+}]_i$ is not buffered practically preclude measurements of a I-V relationship, even by a single-voltage ramp. Indeed, as the voltage ramp progresses, activation of channels is altered by constantly oscillating $[Ca^{2+}]_i$, which makes it impossible to measure a steady state I-V curve.

An ultimate solution to this problem lies in the combination of patch-clamp recordings, calcium imaging, and flash photolysis of "caged" Ca^{2+}, a technically challenging approach. It was used to characterize mIcat regulation by $[Ca^{2+}]_i$ using simultaneous patch-clamp recordings and confocal Ca^{2+} imaging. Myocytes are loaded with 0.1 mM fluo-3 via patch pipette. In addition, the pipette solution contains 5 mM photolabile NP-EGTA and 3.8 mM $CaCl_2$. Assuming K_D values for fluo-3 and NP-EGTA of 526 and 80 nM, respectively, one can calculate the initial $[Ca^{2+}]_i$ being about 100 nM. Flash photolysis of NP-EGTA, brought about by a brief flash of UV light (emitted from a xenon arc lamp and filtered at 300–380 nm), causes an extremely rapid, uniform, and sustained $[Ca^{2+}]_i$ increase throughout the myocyte, as can be verified by fluo-3 fluorescence intensity measurement. Correspondingly, there are step-like increases in mIcat size.[81] In addition, $InsP_3$ can be applied intracellularly by flash photolysis of "caged" $InsP_3$. This is a very useful technique for rapid and controlled drug application, especially intracellularly, as changing composition of the pipette solution can be problematic. In case of mIcat regulation, we found that Ca^{2+} release via $InsP_3$ receptors, but not ryanodine receptors, plays a central role in mIcat potentiation.[81] Many compounds for signal transduction research, including receptor agonists, are available as biologically inert "caged" precursors.

9.4   DATA ANALYSIS AND INTERPRETATION

Analysis of patch-clamp data can be as simple as deriving a single I-V relationship measured by voltage ramp or as complicated as discovering a single-channel mechanism by analysis of the number of open and closed states, connections between them, and kinetics of transitions between various channel conformations. It is worthwhile to consider here at least some relevant issues as, in the end, imperfect voltage protocols and time-consuming data analysis may be one of the most significant caveats/bottlenecks in ion channel research.

9.4.1   VOLTAGE PROTOCOLS

Voltage-gated ion channels are mainly studied with the use of voltage step protocols, which is dictated by the very nature of these channels showing rapid activation, inactivation, and deactivation kinetics that can only be studied with this approach. However, non- or weakly-voltage-dependent TRP channels are typically investigated using voltage ramps. These have the advantage of rapid evaluation of the I-V relationship in a wide range of potentials. Typically, the ramps are applied at a regular interval (e.g., 1–30 s), allowing a convenient plot of the time course of current changes at certain chosen potentials. This is very useful for monitoring the action of TRP pharmacological modulators, as well as the impact of various signal transduction molecules leading to TRP activation or inhibition. Another advantage of voltage ramps over voltage steps is the more precise measurement of the current reversal potential. *Important:* voltage ramps generate sustained current defined as $I = C_M\, dV/dt$, where C_M is membrane capacitance and dV/dt is the rate of voltage change. This current should be subtracted to obtain correct reversal potential value, and this is especially important for small endogenous TRP currents measured by fast ramps in relatively large cells, such as SM cells.

It is also important to make an informed choice of the voltage range and consider whether the ramp should go in the negative or positive direction. SM cells do not tolerate extreme voltages, and negative potentials seem especially detrimental. For TRP currents that activate slowly but deactivate rapidly, it is better to use a voltage step to a positive value, maintain that value for the time needed for full current activation, and only then ramp to a negative value. This facilitates the achievement of the steady state activation in the voltage range studied. Even if a positively going voltage ramp is applied, it is important to clamp the membrane potential at the starting value, at least for the duration of the capacitive transient.

Ramp duration is another critical factor to be considered. In the TRP literature, various protocols are used, and sometimes, this makes it difficult to compare the results obtained in different laboratories. Only two types of ramps have biophysical meaning: (1) very fast ramps that produce the so-called "instantaneous" I-V relationship and that characterize the open-channel I-V relationship (usually linear), and (2) ramps that are slow enough to approximate steady state channel activation. In case of weakly voltage-dependent TRP channels, the difference between these two can be negligible, but some TRPs show prominent voltage-dependent behavior and, in addition, show shifts of the activation curve on the voltage axis that depend on the extent of G protein activation, temperature, or PI(4,5)P$_2$.

Figure 9.3 illustrates currents recorded in HEK293 cells expressing TRPC4 channels. The current was induced by intracellular infusion of GTPγS. These responses develop within several minutes and afterward remain very stable, allowing various voltage protocols to be tested. Figure 9.3a shows the voltage step protocol and corresponding superimposed current traces, while Figure 9.3b shows the voltage ramp protocol and TRPC4 current trace measured with this protocol, which spans the same range of potentials as in Figure 9.3a. The gray trace shows the current measured immediately after breakthrough. We do not subtract "leak" currents, as it is not clear to what extent spontaneous activity of TRP channels may contribute to these. Figure 9.3c shows three I-V relationships. Open circles and a solid line represent steady state currents measured at the end of each voltage step and current measured by voltage ramp, respectively. In this case, it is evident that the 6-s voltage ramp from 80 to −120 mV is sufficiently long to approximate steady state activation of TRPC4. If voltage steps are not practical (e.g., current response is not sustained), one can simply evaluate whether ramp I-V curve represents steady state channel activation by increasing ramp duration severalfold. The third I-V curve shown by open triangles represents "instantaneous" I-V relationship for currents measured at the very beginning of each voltage step in Figure 9.3a. A similar curve represents an open-channel I-V relationship in single-channel recordings, which is nonlinear for TRPC4 as there is a channel "flickering" block around its reversal potential.[26]

Activation curves can be constructed in several different ways. The full conductance curve can be conveniently calculated and plotted by dividing current amplitude at each test potential V_T by the driving force ($V_T - V_{REV}$) at that potential, where V_{REV} is the reversal potential. Alternatively, relative activation can be evaluated by plotting tail current amplitude measured upon the return from V_T to the holding potential (Figure 9.3a) and normalized by the current amplitude before voltage step (Figure 9.3c, inset). The activation curves are fitted with the Boltzmann equation to obtain the potential of half-maximal activation and slope factor. These are important elements of the "biophysical signature" of TRP channels.

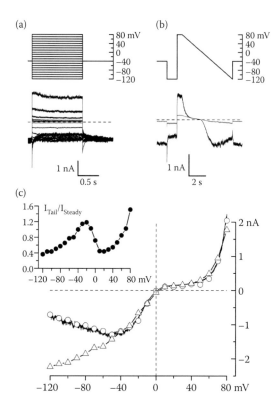

FIGURE 9.3 Analysis of TRPC4 voltage dependence using voltage step and voltage ramp protocols. Recording conditions were as described.[68] (a, b) GTPγS (200 µM)-induced currents were recorded at the holding potential of −40 mV, and their voltage dependence was investigated by applying a series of voltage steps or slow (6 s duration) voltage ramps from 80 to −120 mV. In the latter case, additional voltage steps to −120 and 80 mV provide useful information on current deactivation and activation kinetics, respectively. (c) Instantaneous (triangles) and steady state I-V curves measured by voltage steps (circles) or ramp (solid line). The inset shows the TRPC4 activation curve obtained by the analysis of tail currents in panel a. (A. V. Zholos and M. X. Zhu, unpublished data.)

9.4.2 ANALYSIS OF SINGLE-CHANNEL ACTIVITY

Other characteristic biophysical TRP properties can be measured using the single-channel approach. Among them, the knowledge of single-channel conductance can be of significant discriminative value for TRP identification. Unitary amplitudes are obtained from the amplitude histograms, constructed either in the form of all-point or fitted-levels amplitudes. The latter is the preferred method for rapidly gating channels as the numerous transitional values causing smearing of the peaks in the all-point approach are excluded from analysis. After executing the single-channel search algorithm, pClamp automatically calculates the channel open probability.

Single-channel kinetics (e.g., the number of open and closed states, mean open and closed times, and burst activity) can also provide important clues on TRP identity, if similar data are available for heterologously expressed counterparts. Ultimately, detailed analysis of channel kinetics can reveal channel gating mechanism and explain the nature of channel activation or ligand interaction with the channel. The interested reader can find detailed descriptions of protocols in the recently published *Nature Protocols* article[82] and our recent paper on TRPM8 single-channel mechanism.[83]

9.4.3 AUTOMATION OF DATA ANALYSIS

Finally, we would like to highlight the benefit of automation of patch-clamp data analysis. Some voltage protocols, such as those designed to investigate channel activation and inactivation properties, recovery from inactivation, etc. can be very complex (e.g., double-pulse protocols and variable pulse durations). The data then need to be extracted in an error-free, efficient, and consistent manner and further analyzed by performing various fitting operations using Boltzmann, exponential, logistic, and other functions. All these routine steps can be efficiently automated, allowing the investigator to concentrate on the interpretation of the findings. In our laboratory, we use the combination of pClamp and Origin for this purpose. In the case study published on the Origin Web site, we show, as an example, how the VGCC "window" current can be computed in about 30 s compared to 20–25 min required for the same task in the manual mode (http://www.originlab.com/index.aspx?s=9&lm=68&pid=1540).

Even automation of simple, but frequently repeated, tasks can greatly increase the efficiency of patch-clamp data analysis. In the example below, several steps are programmed using the Origin LabTalk™ scripting language to extract multiple I-V relationships from the data file obtained with the ramp protocol illustrated in Figure 9.3b. The following script is executed (by creating a user button on the toolbar or simply from the script window) after importing the pClamp file into the Origin worksheet:

```
get %H_time -e last;           //finds the last row
                               containing data
mark -d %H_time -b 9163 -e last; //deletes data after the
                               end of the ramp
mark -d %H_time -b 1 -e 3161;  //deletes data before
                               the start of the ramp
for (i=1;i<6002;i+1) {%H_time[i]=(i-1)*(-200/6000)+80};
//converts time values into the corresponding voltage
values. In this example, the ramp starts at 80 mV, ends
at -120 mV (hence 200 in the formula), and contains 6000
data points from row 3162 to row 9162.
```

9.5 CONCLUSION

During the past decade, experiments combining various experimental approaches have provided overwhelming evidence for expression and function of TRP channels in visceral and vascular SM. The strongest evidence exists for the physiological functions of TRPC isoforms, which combine apparently in a different, tissue-specific manner to form diverse ROC, SOC, and SAC channels. Some of these channels are also constitutively active and make important contributions to the resting membrane potential and basal $[Ca^{2+}]_i$. More recently, members of the TRPV and TRPM families have also begun to emerge as important determinants of SM function. Patch-clamp recording techniques, in combination with molecular biology methods and other approaches, have played a leading role in these discoveries, guided, to a large extent, by the knowledge of structure, activation mechanisms, and function of heterologously expressed TRP channels.

ACKNOWLEDGMENTS

The author would like to thank M. X. Zhu for critical reading of the manuscript. Research in the author's laboratory is currently funded by the British Heart Foundation (PG/09/063), U.S. National Institutes of Health (DK081654), and Northern Ireland Chest Heart and Stroke (NICHS).

REFERENCES

1. Beech, D. J. 2007. Ion channel switching and activation in smooth-muscle cells of occlusive vascular diseases. *Biochemical Society Transactions* 35: 890–894.
2. House, S. J., M. Potier, J. Bisaillon, H. A. Singer, and M. Trebak. 2008. The non-excitable smooth muscle: Calcium signaling and phenotypic switching during vascular disease. *Pflügers Archiv* 456: 769–785.
3. Bolton, T. B. 1979. Mechanisms of action of transmitters and other substances on smooth muscle. *Physiological Reviews* 59: 606–718.
4. Kuriyama, H., K. Kitamura, T. Itoh, and R. Inoue. 1998. Physiological features of visceral smooth muscle cells, with special reference to receptors and ion channels. *Physiology Reviews* 78: 811–920.
5. Bolton, T. B., S. A. Prestwich, A. V. Zholos, and D. V. Gordienko. 1999. Excitation-contraction coupling in gastrointestinal and other smooth muscles. *Annual Review of Physiological* 61: 85–115.
6. Sanders, K. M. 1998. G protein-coupled receptors in gastrointestinal physiology. IV. Neural regulation of gastrointestinal smooth muscle. *American Journal of Physiology* 275: G1–G7.
7. Beech, D. J., K. Muraki, and R. Flemming. 2004. Non-selective cationic channels of smooth muscle and the mammalian homologues of *Drosophila* TRP. *Journal of Physiology* 559: 685–706.
8. Albert, A. P., and W. A. Large. 2006. Signal transduction pathways and gating mechanisms of native TRP-like cation channels in vascular myocytes. *Journal of Physiology* 570: 45–51.
9. Venkatachalam, K., and C. Montell. 2007. TRP Channels. *Annual Review of Biochemistry* 76: 387–417.
10. Clapham, D. E., D. Julius, C. Montell, and G. Schultz. 2005. International Union of Pharmacology. XLIX. Nomenclature and structure-function relationships of Transient Receptor Potential channels. *Pharmacological Reviews* 57: 427–450.
11. Montell, C., L. Birnbaumer, V. Flockerzi et al. 2002. A unified nomenclature for the superfamily of TRP cation channels. *Molecular Cell* 9: 229–231.
12. Ramsey, I. S., M. Delling, and D. E. Clapham. 2006. An introduction to TRP channels. *Annual Review of Physiology* 68: 619–647.

13. Benham, C. D., T. B. Bolton, and R. J. Lang. 1985. Acetylcholine activates an inward current in single mammalian smooth muscle cells. *Nature* 316: 345–347.
14. Inoue, R., K. Kitamura, and H. Kuriyama. 1987. Acetylcholine activates single sodium channels in smooth muscle cells. *Pflügers Archiv* 410: 69–74.
15. Janssen, L. J., and S. M. Sims. 1992. Acetylcholine activates non-selective cation and chloride conductances in canine and guinea-pig tracheal myocytes. *Journal of Physiology* 453: 197–218.
16. Vogalis, F., and K. M. Sanders. 1990. Cholinergic stimulation activates a non-selective cation current in canine pyloric circular muscle cells. *Journal of Physiology* 429: 223–236.
17. Lee, H. K., O. Bayguinov, and K. M. Sanders. 1993. Role of nonselective cation current in muscarinic responses of canine colonic muscle. *American Journal of Physiology* 265: C1463–C1471.
18. Inoue, R., and G. Isenberg. 1990. Acetylcholine activates nonselective cation channels in guinea pig ileum through a G protein. *American Journal of Physiology* 258: C1173–C1178.
19. Inoue, R., and G. Isenberg. 1990. Intracellular calcium ions modulate acetylcholine-induced inward current in guinea-pig ileum. *Journal of Physiology* 424: 73–92.
20. Inoue, R., and G. Isenberg. 1990. Effect of membrane potential on acetylcholine-induced inward current in guinea-pig ileum. *Journal of Physiology* 424: 57–71.
21. Komori, S., M. Kawai, T. Takewaki, and H. Ohashi. 1992. GTP-binding protein involvement in membrane currents evoked by carbachol and histamine in guinea-pig ileal muscle. *Journal of Physiology* 450: 105–126.
22. Komori, S., and T. B. Bolton. 1990. Role of G-proteins in muscarinic receptor inward and outward currents in rabbit jejunal smooth muscle. *Journal of Physiology* 427: 395–419.
23. Zholos, A. V., T. B. Bolton, A. V. Dresvyannikov et al. 2004. Cholinergic excitation of smooth muscles: multiple signaling pathways linking M_2 and M_3 muscarinic receptors to cationic channels. *Neurophysiology* 36: 398–406.
24. Zholos, A. V. 2006. Regulation of TRP-like muscarinic cation current in gastrointestinal smooth muscle with special reference to PLC/InsP$_3$/Ca^{2+} system. *Acta Pharmacologica Sinica* 27: 833–842.
25. Unno, T., H. Matsuyama, H. Okamoto et al. 2006. Muscarinic cationic current in gastrointestinal smooth muscles: Signal transduction and role in contraction. *Autonomic Autacoid Pharmacology* 26: 203–217.
26. Zholos, A. V., A. A. Zholos, and T. B. Bolton. 2004. G-protein-gated TRP-like cationic channel activated by muscarinic receptors: Effect of potential on single-channel gating. *Journal of General Physiology* 123: 581–598.
27. Sakamoto, T., T. Unno, T. Kitazawa et al. 2007. Three distinct muscarinic signalling pathways for cationic channel activation in mouse gut smooth muscle cells. *Journal of Physiology* 582: 41–61.
28. Plant, T. D., and M. Schaefer. 2005. Receptor-operated cation channels formed by TRPC4 and TRPC5. *Naunyn-Schmiedeberg's Archives of Pharmacology* 371: 266–276.
29. Tang, Y., J. Tang, Z. Chen et al. 2000. Association of mammalian Trp4 and phospholipase C isozymes with a PDZ domain-containing protein, NHERF. *Journal of Biological Chemistry* 275: 37559–37564.
30. Walker, R. L., J. R. Hume, and B. Horowitz. 2001. Differential expression and alternative splicing of TRP channel genes in smooth muscles. *American Journal of Physiology* 280: C1184–C1192.
31. Tsvilovskyy, V. V., A. V. Zholos, T. Aberle et al. 2009. Deletion of TRPC4 and TRPC6 in mice impairs smooth muscle contraction and intestinal motility in vivo. *Gastroenterology* 137: 1415–1424.

32. Helliwell, R. M., and W. A. Large. 1996. Dual effect of external Ca^{2+} on noradrenaline-activated cation current in rabbit portal vein smooth muscle cells. *Journal of Physiology* 492: 75–88.

33. Helliwell, R. M., and W. A. Large. 1997. α_1-adrenoceptor activation of a non-selective cation current in rabbit portal vein by 1,2-diacyl-sn-glycerol. *Journal of Physiology* 499: 417–428.

34. Inoue, R., T. Okada, H. Onoue et al. 2001. The transient receptor potential protein homologue TRP6 is the essential component of vascular α_1-adrenoceptor-activated Ca^{2+}-permeable cation channel. *Circulation Research* 88: 325–332.

35. Welsh, D. G., A. D. Morielli, M. T. Nelson, and J. E. Brayden. 2002. Transient receptor potential channels regulate myogenic tone of resistance arteries. *Circulation Research* 90: 248–250.

36. Yu, Y., M. Sweeney, S. Zhang et al. 2003. PDGF stimulates pulmonary vascular smooth muscle cell proliferation by upregulating TRPC6 expression. *American Journal of Physiology* 284: C316–C330.

37. Yang, X. R., M. J. Lin, L. S. McIntosh, and J. S. K. Sham. 2006. Functional expression of transient receptor potential melastatin- and vanilloid-related channels in pulmonary arterial and aortic smooth muscle. *American Journal of Physiology* 290: L1267–L1276.

38. Johnson, C. D., D. Melanaphy, A. Purse et al. 2009. Transient receptor potential melastatin 8 channel involvement in the regulation of vascular tone. *American Journal of Physiology* 296: H1868–H1877.

39. Beech, D. J. 2005. Emerging functions of 10 types of TRP cationic channel in vascular smooth muscle. *Clinical and Experimental Pharmacology and Physiology* 32: 597–603.

40. Albert, A. P., S. N. Saleh, C. M. Peppiatt-Wildman, and W. A. Large. 2007. Multiple activation mechanisms of store-operated TRPC channels in smooth muscle cells. *Journal of Physiology* 583: 25–36.

41. Albert, A. P., S. N. Saleh, and W. A. Large. 2009. Identification of canonical transient receptor potential (TRPC) channel proteins in native vascular smooth muscle cells. *Current Medical Chemistry* 16: 1158–1165.

42. Inoue, R., L. J. Jensen, J. Shi et al. 2006. Transient receptor potential channels in cardiovascular function and disease. *Circulation Research* 99: 119–131.

43. Hamill, O. P., A. Marty, E. Neher, B. Sakmann, and F. J. Sigworth. 1981. Improved patch-clamp techniques for high-resolution current recording from cells and cell-free membrane patches. *Pflügers Archiv* 391: 85–100.

44. Jonas, P. 1995. Fast applications of agonists to isolated membrane patches. In *Single-channel recording*, ed. B. Sakmann, and E. Neher, 231–243. New York: Plenum Press.

45. Bolton, T. B., and A. V. Zholos. 2003. Potential synergy: Voltage-driven steps in receptor-G protein coupling and beyond. *Science STKE* pe52.

46. Smirnov, S. V., A. V. Zholos, and M. F. Shuba. 1992. Potential-dependent inward currents in single isolated smooth muscle cells of the rat ileum. *Journal of Physiology* 454: 549–571.

47. Helliwell, R. M., and W. A. Large. 1998. Facilitatory effect of Ca^{2+} on the noradrenaline-evoked cation current in rabbit portal vein smooth muscle cells. *Journal of Physiology* 512: 731–741.

48. Zholos, A. V., and T. B. Bolton. 1995. Effects of divalent cations on muscarinic receptor cationic current in smooth muscle from guinea-pig small intestine. *Journal of Physiology* 486: 67–82.

49. Freichel, M., S. H. Suh, A. Pfeifer et al. 2001. Lack of an endothelial store-operated Ca^{2+} current impairs agonist-dependent vasorelaxation in TRP4$^{-/-}$ mice. *Nature Cell Biology* 3: 121–127.

50. Dietrich, A., H. Kalwa, U. Storch et al. 2007. Pressure-induced and store-operated cation influx in vascular smooth muscle cells is independent of TRPC1. *Pflügers Archiv* 455: 465–477.
51. Dietrich, A., M. Schnitzler, M. Gollasch et al. 2005. Increased vascular smooth muscle contractility in TRPC6$^{-/-}$ mice. *Molecular Cell Biology* 25: 6980–6989.
52. Watanabe, H., M. Murakami, T. Ohba, Y. Takahashi, and H. Ito. 2008. TRP channel and cardiovascular disease. *Pharmacology Therapy* 118: 337–351.
53. Bergdahl, A., M. F. Gomez, A. K. Wihlborg et al. 2005. Plasticity of TRPC expression in arterial smooth muscle: correlation with store-operated Ca^{2+} entry. *American Journal of Physiology* 288: C872–C880.
54. Flockerzi, V., C. Jung, T. Aberle et al. 2005. Specific detection and semi-quantitative analysis of TRPC4 protein expression by antibodies. *Pflügers Archiv* 451: 81–86.
55. Albert, A. P., V. Pucovsky, S. A. Prestwich, and W. A. Large. 2006. TRPC3 properties of a native constitutively active Ca^{2+}-permeable cation channel in rabbit ear artery myocytes. *Journal of Physiology* 571: 361–369.
56. Saleh, S. N., A. P. Albert, C. M. Peppiatt, and W. A. Large. 2006. Angiotensin II activates two cation conductances with distinct TRPC1 and TRPC6 channel properties in rabbit mesenteric artery myocytes. *Journal of Physiology* 577: 479–495.
57. Peppiatt-Wildman, C. M., A. P. Albert, S. N. Saleh, and W. A. Large. 2007. Endothelin-1 activates a Ca^{2+}-permeable cation channel with TRPC3 and TRPC7 properties in rabbit coronary artery myocytes. *Journal of Physiology* 580: 755–764.
58. Saleh, S. N., A. P. Albert, C. M. Peppiatt-Wildman, and W. A. Large. 2008. Diverse properties of store-operated TRPC channels activated by protein kinase C in vascular myocytes. *Journal of Physiology* 586: 2463–2476.
59. Xu, S. Z., G. Boulay, R. Flemming, and D. J. Beech. 2006. E3-targeted anti-TRPC5 antibody inhibits store-operated calcium entry in freshly isolated pial arterioles. *American Journal of Physiology* 291: H2653–H2659.
60. Zholos, A. V., and T. B. Bolton. 1996. A novel GTP-dependent mechanism of ileal muscarinic metabotropic channel desensitization. *British Journal of Pharmacology* 119: 997–1012.
61. Arunlakshana, O., and H. O. Schild. 1959. Some quantitative uses of drug antagonists. *British Journal of Pharmacology* 14: 48–58.
62. Zholos, A. V., and T. B. Bolton. 1997. Muscarinic receptor subtypes controlling the cationic current in guinea-pig ileal smooth muscle. *British Journal of Pharmacology* 122: 885–893.
63. Byrne, N. G., and W. A. Large. 1988. Membrane ionic mechanisms activated by noradrenaline in cells isolated from the rabbit portal vein. *Journal of Physiology* 404: 557–573.
64. Pucovsky, V., A. V. Zholos, and T. B. Bolton. 1998. Muscarinic cation current and suppression of Ca^{2+} current in guinea pig ileal smooth muscle cells. *European Journal of Pharmacology* 346: 323–330.
65. Yan, H. D., H. Okamoto, T. Unno et al. 2003. Effects of G-protein-specific antibodies and Gβγ subunits on the muscarinic receptor-operated cation current in guinea-pig ileal smooth muscle cells. *British Journal of Pharmacology* 139: 605–615.
66. Albert, A. P., and W. A. Large. 2004. Inhibitory regulation of constitutive transient receptor potential-like cation channels in rabbit ear artery myocytes. *Journal of Physiology* 560: 169–180.
67. Rohacs, T., and B. Nilius. 2007. Regulation of transient receptor potential (TRP) channels by phosphoinositides. *Pflügers Archiv* 455: 157–168.
68. Otsuguro, K. I., J. Tang, Y. Tang et al. 2008. Isoform-specific inhibition of TRPC4 channel by phosphatidylinositol 4,5-bisphosphate. *Journal of Biological Chemistry* 283: 10026–10036.

69. Albert, A. P., S. N. Saleh, and W. A. Large. 2008. Inhibition of native TRPC6 channel activity by phosphatidylinositol 4, 5-bisphosphate in mesenteric artery myocytes. *Journal of Physiology* 586: 3087–3095.

70. Saleh, S. N., A. P. Albert, and W. A. Large. 2009. Obligatory role for phosphatidylinositol 4,5-bisphosphate in activation of native TRPC1 store-operated channels in vascular myocytes. *Journal of Physiology* 587: 531–540.

71. Saleh, S. N., A. P. Albert, and W. A. Large. 2009. Activation of native TRPC1/C5/C6 channels by endothelin-1 is mediated by both PIP$_3$ and PIP$_2$ in rabbit coronary artery myocytes. *Journal of Physiology* 587: 5361–5375.

72. Zholos, A. V., Y. D. Tsytsyura, D. V. Gordienko, V. V. Tsvilovskyy, and T. B. Bolton. 2004. Phospholipase C, but not InsP$_3$ or DAG, -dependent activation of the muscarinic receptor-operated cation current in guinea-pig ileal smooth muscle cells. *British Journal of Pharmacology* 141: 23–36.

73. Albert, A. P., and W. A. Large. 2003. Synergism between inositol phosphates and diacylglycerol on native TRPC6-like channels in rabbit portal vein myocytes. *Journal of Physiology* 552: 789–795.

74. Large, W. A., S. N. Saleh, and A. P. Albert. 2009. Role of phosphoinositol 4,5-bisphosphate and diacylglycerol in regulating native TRPC channel proteins in vascular smooth muscle. *Cell Calcium* 45: 574–582.

75. Zhu, M. X. 2005. Multiple roles of calmodulin and other Ca^{2+}-binding proteins in the functional regulation of TRP channels. *Pflügers Archiv* 451: 105–115.

76. Kim, S. J., S. C. Ahn, I. So, and K. W. Kim. 1995. Role of calmodulin in the activation of carbachol-activated cationic current in guinea-pig gastric antral myocytes. *Pflugers Arch* 430: 757–762.

77. Tsytsyura, Y., A. V. Zholos, M. F. Shuba, and T. B. Bolton. 2000. Effect of intracellular Ca^{2+} on muscarinic cationic current in guinea pig ileal smooth muscle cells. *Neurophysiology* 32: 198–199.

78. Komori, S., M. Kawai, P. Pacaud, H. Ohashi, and T. B. Bolton. 1993. Oscillations of receptor-operated cationic current and internal calcium in single guinea-pig ileal smooth muscle cells. *Pflügers Archiv* 424: 431–438.

79. Pacaud, P., and T. B. Bolton. 1991. Relation between muscarinic receptor cationic current and internal calcium in guinea-pig jejunal smooth muscle cells. *Journal of Physiology* 441: 477–499.

80. Zholos, A. V., S. Komori, H. Ohashi, and T. B. Bolton. 1994. Ca^{2+} inhibition of inositol trisphosphate-induced Ca^{2+} release in single smooth muscle cells of guinea-pig small intestine. *Journal of Physiology* 481: 97–109.

81. Gordienko, D. V., and A. V. Zholos. 2004. Regulation of muscarinic cationic current in myocytes from guinea-pig ileum by intracellular Ca^{2+} release: A central role of inositol 1,4,5-trisphosphate receptors. *Cell Calcium* 36: 367–386.

82. Mortensen, M., and T. G. Smart. 2007. Single-channel recording of ligand-gated ion channels. *Nature Protocols* 2: 2826–2841.

83. Fernández, J. A., R. Skryma, G. Bidaux et al. 2011. Voltage and cold dependent gating of single TRPM8 ion channels. *Journal of General Physiology* (in Press).

10 TRPC Channels in Neuronal Survival

Junbo Huang, Wanlu Du, Hailan Yao,
and Yizheng Wang

CONTENTS

10.1 INTRODUCTION

The family of transient receptor potential canonical (TRPC) proteins was the first group of TRP homologs cloned in mammals[1] after discovery of the TRP protein in *Drosophila*.[2] There are seven members in this family (TRPC1–7). The TRPC channels formed by homo- or heteromeric TRPC proteins are Ca^{2+}-permeable nonselective cation channels. According to the similarity in amino acid sequence, the mammalian TRPCs can be classified into four subgroups: TRPC1, TRPC2, TRPC3/6/7, and TRPC4/5,[3] while the TRPC2 is a pseudogene in primates.[4] The TRPC proteins have six transmembrane domains, and both N- and C-termini of these proteins are intracellular, suggesting that these channels can be regulated by intracellular signaling molecules. These channels can be activated in various cell types by G-protein-coupled receptors (GPCR) and receptor tyrosine kinases (RTK) through a phospholipase C (PLC)-dependent mechanism. Therefore, TRPC channels may act as a sensor for environmental cues.

Neuronal survival is important for brain development and for pathogenesis of certain diseases in the central nervous system (CNS). During development, groups of neurons are chosen to keep alive and to establish elaborated networks. Under pathologic conditions, such as ischemic injury and degenerative diseases, neuronal survival is compromised, leading to brain injury and psychological disorders. To promote neuronal survival, complicated processes are involved in which both neurotrophins and Ca^{2+} play critical roles. Because Ca^{2+} is known to mediate cell survival, it is possible that Ca^{2+} influx via TRPC channels is required for neuronal survival promoted by neurotrophins, the receptors of which are RTKs. All TRPC channels have been found in mammalian CNS. As possible modulators or integrators of various cellular signals, TRPC channels can upregulate a series of pathways for neuronal survival. Studies about the roles of TRPC channels in neuronal survival could help us to understand how neurons endure and survive in the complex situations during maturation and under insults. This review will focus on the possible roles of TRPC channels in neuronal survival.

10.2   PROPERTIES OF TRPC CHANNELS

10.2.1   STRUCTURE OF TRPC PROTEINS

As a subfamily of TRP channels, each of the seven members of TRPCs has six transmembrane domains predicted by the amino acid sequence. There is as yet no crystal structure of any TRP channel. It is believed that their three-dimensional structures could be roughly similar to those of K^+ channels.[5] TRP proteins form tetrameric structures when they assemble as a channel, just like K^+ channels, and recent studies of electron cryomicroscopy of TRPV1 confirm the prediction.[6,7] Figure 10.1 shows

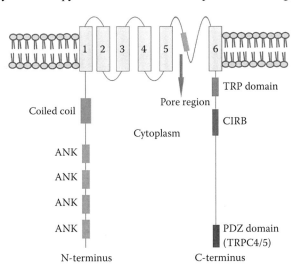

FIGURE 10.1   Schematic structure of TRPC proteins. 1-6: transmembrane domains of TRPC proteins. ANK: ankyrin-like repeat. CIRB: calmodulin/IP$_3$ receptor binding region. The PDZ binding domain is specific for TRPC4 and TRPC5.

the schema of a general structure of a TRPC channel protein on the plasma membrane according to the work of Vannier et al.[8] Both the N- and C-termini are cytoplasmic. At the N-terminal, there are three to four ankyrin-like repeats, followed by a coiled coil region. Ankyrin-like repeats are a common protein–protein interaction domain, although there is no conserved binding partner. Several proteins have been reported to interact with this region of different TRPCs. MxA, an interferon-induced GTPase that can inhibit the multiplication of RNA viruses, has been shown to interact with the second ankyrin-like repeat domain in mammalian TRPCs, and it can regulate TRPC channel activities.[9] RNF24, a membrane RING-H2 protein, was found to interact with the ankyrin-like repeat domain of TRPC6 and colocalize with TRPC3.[10] This protein is thought to affect trafficking of TRPC proteins in the Golgi apparatus.[10] Moreover, these repeat domains have been recently found necessary for the assembly of TRPC4 and TRPC5,[11] while only the coiled coil region was shown to contribute to the homo- and heteromeric TRPC channel formation previously.[12] This function of the coiled coil region is plausible because this protein motif is commonly used to control oligomerization.[13,14] Notably, the N-terminal coiled coil regions of TRPC4 and TRPC5, but not other TRPCs, interact with stathmin, a microtubule destabilizing phosphoprotein. This interaction with stathmin may play a role in TRPC4- and TRPC5-mediated inhibition of neurite outgrowth during development.[15]

The putative pore region of TRPCs is located between the fifth and the sixth transmembrane domains. The conserved hydrophobic α-helix in this region has been confirmed by mutagenesis analysis to be important for the opening and selectivity of TRPC channels.[16,17] Compared with the N-terminus, the C-terminus of TRPC proteins has been reported to interact with regulatory proteins. The EWKFAR motif, together with the proline-rich motif (LPXPFXXXPSPK) that follows it, is highly conserved in the TRPC family. This 25-amino-acid region is also termed TRP domain.[3] In TRPCs, this domain has been reported to be necessary for interaction with Homer (only for TRPC1[18] and possibly for TRPC3[19]) and immunophilin.[20] After the TRP domain, a calmodulin (CaM)/inositol 1,4,5-trisphosphate (IP_3) receptor binding (CIRB) region has been identified in all TRPC proteins,[21] while TRPC1 and TRPC5 have an additional CaM binding site downstream of CIRB.[22,23] An extended region, including a PDZ binding domain, is found in TRPC4 and TRPC5 at the end of the C-terminus. This domain is responsible for TRPC interaction with the Na^+/H^+ exchanger regulatory factor (NHERF). This interaction likely enables TRPC4 and TRPC5 to form signaling complexes with PLCβ and to link with the actin cytoskeleton.[24,25]

10.2.2 Activation and Regulation

It has been proposed that TRPC channels are involved in the processes of store-operated Ca^{2+} entry (SOCE) and receptor-operated Ca^{2+} entry (ROCE).[26,27] SOCE, also termed store depletion-activated Ca^{2+} entry, describes Ca^{2+} influx following the depletion of intracellular Ca^{2+} stores. ROCE refers to Ca^{2+} influx triggered by PLC activation when GPCR or RTK pathways are stimulated. Activated PLCs cleave phosphatidylinositol 4, 5-bisphosphate (PIP_2) to generate diacylglycerol (DAG) and IP_3, and ROCE can be enhanced by IP_3-induced intracellular Ca^{2+} store depletion.

Therefore, ROCE and SOCE are different pathways with reciprocal regulation.[27] All TRPC channels can be activated by stimulating PLCs,[28] with either Gq-coupled pathways stimulating PLCβ or RTK signaling pathways activating PLCγ. These two signaling pathways are the typical modes for TRPC channel activation through the ROCE mechanism. There are two proposals about how TRPC channels are activated by PLCs. In one, IP_3 causes Ca^{2+} depletion from the intracellular stores by interacting with IP_3 receptors, and the Ca^{2+} store depletion can lead to activation of TRPC channels. In the other, DAG directly activates TRPC channels.[3,28]

In heterologous expression systems, TRPC channels have been reported to be activated by Ca^{2+} store depletion, except for TRPC6.[29–33] In native conditions, TRPC3 can be activated by Ca^{2+} store depletion.[34] In rat primary pontine neurons, the native TRPC3 channel conductance (I_{BDNF}) can be initiated by brain-derived neurotrophic factor (BDNF). The activation of I_{BDNF} requires PLC. Also, if the Ca^{2+} level is elevated in the cytoplasm or IP_3 receptors are blocked, I_{BDNF} cannot be induced. Furthermore, I_{BDNF} is not initiated by DAG,[34] suggesting that the activation of the endogenous TRPC3 likely depends on Ca^{2+} store depletion.

DAG and its analogs can also trigger TRPC conductance in native conditions. In cultured B lymphocytes and smooth muscle cells, TRPC6 and TRPC7 can be activated by DAG, and such activation is abolished when TRPC6 and TRPC7 proteins are eliminated.[35,36] TRPC3 and TRPC6 can be activated by DAG independent of Ca^{2+} store depletion when they are overexpressed in CHO-K1 cells.[37] However, DAG can also activate PKC, which inhibits TRPC channels.[38] For this reason, in DT40 B cells, TRPC7 can be activated by DAG only when PKC activity is suppressed.[39] TRPC4 and TRPC5 cannot be activated by DAG.[37,38]

Additionally, altering the expression of TRPC proteins in the plasma membrane may also affect TRPC activation. Several in vitro studies support this proposal.[40–43] In HEK293 cells stably transfected with TRPC6, the level of the TRPC6 protein in the plasma membrane increased within the first 30 s of GPCR stimulation or store depletion. PLCs may also be involved in this process to regulate Ca^{2+} influx via TRPC channels.[42] In hippocampal neurons, TRPC5 is translocated to the plasma membrane when RAC1 is activated. This process involves synthesis of PIP_2 and is thought to be important in Ca^{2+}-dependent neurite growth repression.[44]

10.3   DISTRIBUTION OF TRPCs IN THE NERVOUS SYSTEM

The mRNA expression pattern of TRPCs in different tissues from rodents to human shows that TRPCs are highly expressed in all regions of the CNS.[45–47] Results of immunohistochemistry studies also confirm the expression of TRPC proteins in the CNS. In the hippocampus, TRPC1, TRPC3, TRPC4, and TRPC5 are found in the pyramidal cell bodies of CA1 or CA3 regions, as well as in the granule cell bodies of the dentate gyrus, while TRPC6 is dispersedly stained only at the molecular layer of the dentate gyrus.[48] In rat substantia nigra, TRPC6 is found at the proximal dendrites and axon hillock of tyrosine hydroxylase (TH)-positive neurons, well colocalized with mGluR1, but with little signal in nuclei and presynaptic regions.[49] TRPC3 has been reported to have preferential distribution in oligodendrocytes.[50] TRPC4 and TRPC5 are the predominant TRPC subtypes in the adult rat brain because both are expressed highly in the frontal cortex, pyramidal cell layer of the hippocampus, and dentate gyrus.[47] Comparatively, TRPC4 is specifically detected throughout layers 2–6 of the prefrontal cortex, motor cortex, and somatosensory cortex, while TRPC5 seems to be present in layers 2, 3, 5, and 6 of the prefrontal cortex and anterior cingulated.[47]

In brain development, the mRNA for TRPCs, except TRPC4, can be detected in the E13 mouse brain.[51,52] During this period, the expression of TRPC1 is specifically detected in some post-mitotic neurons distinct from Cajal-Retzius cells at the preplate, while it overlaps with 80% of the proliferative neural stem cells in the embryonic telencephalon.[51,53] TRPC3, together with TRPC6, localize to BrdU-positive cells, while TRPC6 is also found in neuronal cells.[51] TRPC4 mRNA is not expressed until E14.5. It is then found prominently in the cortex, septal area, pyramidal cells of the hippocampus, granule cells of the dentate gyrus, and cerebellum. During development to adulthood, TRPC4 mRNA levels decline significantly in these brain regions except for the hippocampus.[52] TRPC3 and TRPC6 are also found to have a peak of postnatal expression in the cerebellum, and their expression patterns are consistent with their important roles in neuronal survival during brain development.[54]

The presence of TRPC mRNA is also detected in the mouse dorsal root ganglion and nodose ganglion by in situ hybridization. TRPC1, TRPC3, and TRPC6 are found as the primary subsets from E12 on and then increase gradually to adult levels. TRPC2 has a reverse trend in development, while TRPC4, TRPC5, and TRPC7 expression starts from E12 and reaches a peak level at E18. Specifically, TRPC3 is detected exclusively in isolectin B4-positive cells, which are TRPV1-negative, while TRPC1 and TRPC2 localize in the neurofilament 200-positive large-size subclass of neurons.[55]

The omnipresent distribution of TRPCs in the nervous system suggests their importance to the development and function of the system. Because these proteins can form heterotetramers with each other, the heterogeneity of their spatiotemporal expression may be far more complex than what has been found so far.

10.4 CA²⁺ AND NEURONAL SURVIVAL

During development, many neurons are generated, while about 70% of them are lost later in life because of natural cell death. This phenomenon was first studied in the twentieth century and has been found to occur during development of both the CNS and peripheral nervous system (PNS) in various species.[56] The programmed neuronal cell death and survival of limited neuronal populations are believed to be essential for formation of appropriate neural networks and to delete incorrect connections.[57] Neurotrophic factors (NTF) play important roles in neuronal survival. In fact, studies of neuronal survival led directly to the discovery and exploration of NTFs.[56,58] Moreover, studies of CNS development have established that neuronal activity is also crucial for neurons to survive.[59,60] In both NTF-induced and activity-dependent survival, Ca²⁺ acts as a pivotal ion.

In the CNS, neuronal survival in the visual, olfactory, and auditory sensory systems was found to be afferent dependent.[60] It is believed that activity of the target neurons is important for their survival. Depolarization promotes various neurons in the

CNS to survive in a cultured system. When cultured in a nerve growth factor (NGF)-deprived medium, rat sympathetic neurons die within 3 days, while depolarization by elevating extracellular K^+ prevents the death and supports neuronal survival.[61] Activation of L-type voltage-gated Ca^{2+} channels (VGCC) and sustained elevation of Ca^{2+} in cytosol are required in depolarization-induced neuronal survival.[62,63] L-type VGCC inhibitors suppress depolarization and neuronal survival,[62,63] while the agonists of L-type VGCCs, as well as thapsigargin, which releases Ca^{2+} from the internal stores, can mimic the survival-promoting effect of depolarization by elevating cytosolic Ca^{2+} levels.[61,64]

Classic NTF-dependent neuronal survival has been thought to share a similar mechanism as activity-dependent survival. Stimulating neurons with NGF did not induce apparent cytosolic Ca^{2+} elevation,[65,66] indicating that sustained Ca^{2+} elevation might not be needed for NTF-induced survival promotion. However, if intracellular Ca^{2+} concentration is kept at extremely low levels, NTFs cannot promote neuronal survival.[61] Therefore, proper Ca^{2+} levels in the cytosol are needed for NTF function. Moreover, the protection by IGF-1 of granule cells exposed to low K^+ medium depends on the activity of L-type VGCCs.[67] Subsequent studies revealed that BDNF stimulation induces intracellular Ca^{2+} elevation in neurons and that this elevation is important for neuronal survival.[54] BDNF binding to TrkB induces dimerization of TrkB, which in turn activates PLCγ, thereby stimulating TRPC channels to elevate the cytosolic Ca^{2+} level.

It is known that moderate intracellular Ca^{2+} elevation can promote neuronal survival through various pathways, including the PI_3K-AKT pathway. Several NTFs (NGF, BDNF, GDNF, and IGF) can activate the PI_3K-AKT cascade[68–70] to promote neuronal survival. In cerebellar granule neurons, phosphorylated AKT can inactivate FKHRL1. Inactivation of this Forkhead transcriptional regulator suppresses the expression of apoptotic genes.[71] Active AKT can also phosphorylate BAD[72] and caspase-9[73] directly to induce apoptotic machinery dysfunction. The AKT pathway can upregulate the activity of CREB and NF-κB, which are transcriptional factors known to promote neuronal survival.[74,75] Ca^{2+} elevation can lead to phosphorylation of AKT independent of PI_3K or MAP kinases.[76] Moreover, Ca^{2+} elevation can be regulated by activation of AKT. Phosphorylation of L-type VGCCs by AKT is essential for IGF-1-induced L-type VGCC potentiation.[67] The Ca^{2+}/CaM-dependent protein kinase II can promote activity-dependent neuronal survival by inhibiting histone deacetylase-5 (HDAC5C), which is the repressor of MEF2, a transcriptional factor involved in neuronal survival.[77,78]

The production of NTFs can also be affected by intracellular Ca^{2+} elevation. In hippocampal neurons, neuronal activity can induce postsynaptic secretion of BDNF and NT3. The secretion is dependent on Ca^{2+} influx and release.[79,80] Non-neuronal cells can also secrete NTFs in response to intracellular Ca^{2+} elevation. For example, astrocytes can release BDNF when stimulated with glutamate.[81] Brain endothelial cells release BDNF in response to hypoxia.[82] Vascular smooth muscle cells secrete NGF when stimulated with thrombin.[83] These processes are Ca^{2+} dependent.[81–83] Together, these results suggest that during development or under pathological conditions, NTFs can be secreted in Ca^{2+}-dependent manners to promote neuronal survival in non-cell autonomous ways.

It has also been known that improper Ca^{2+} elevation induces neuronal cell death. Under pathological conditions such as hypoxia,[84] ischemia,[85] trauma,[86] and neurodegenerative disease,[87] Ca^{2+} elevation is a key factor for neuronal death. Increases in the Ca^{2+} level can initialize apoptosis in many systems.[88,89] Although it is not clear why changes in Ca^{2+} level lead to different outcomes, the Ca^{2+} set point hypothesis suggests that the Ca^{2+} concentration is the key.[61] Another possibility is that Ca^{2+} influx from different routes can induce different fates of neurons, which is consistent with the fact that neurons have many routes for Ca^{2+} influx.[90,91] Activation of NMDA receptors readily evokes neurotoxicity, although the activation induces an increase in the Ca^{2+} level equal to that induced by depolarization via elevating extracellular K^+.[91] The spatial and temporal patterns of the Ca^{2+} signal due to Ca^{2+} entry from different routes vary, and this may lead to activation of different downstream pathways[91,92] that affect neuronal fate.

10.5 TRPC CHANNELS IN NEURONAL SURVIVAL

Members of the TRPC subfamily have roles in multiple processes, including neuronal development, survival, and proliferation of neural stem cells. TRPC1 is involved in mGluR1-mediated slow excitatory postsynaptic conductance (EPSC) in cerebellar Purkinje cells,[93] in hippocampal glutamate-induced cell death,[94] and in embryonic neural stem cell proliferation.[53] TRPC2 may play a role in pheromone transduction in the vomeronasal system.[95] TRPC3 is involved in BDNF-induced dendritic spine formation,[96] cerebellar granule neuron (CGN) survival,[54] and motor coordination.[97] TRPC4 may have a role in neurite extension in post-mitotic neurons.[98] TRPC5 is involved in controlling neurite extension and growth cone morphology[15] in hippocampal neurons[99] and fear-related behavior.[100] TRPC6 plays essential roles in BDNF-mediated survival of CGNs during development[54] and in dendritic growth,[101] as well as synapse formation by hippocampal neurons.[102] Recent studies also indicate that TRPC channels mediate muscarinic receptor-induced slow afterdepolarization in pyramidal cells of the cerebral cortex[103] and neuroprotection promoted by platelet-derived growth factor.[104] TRPC channels also play a role in chemokine (C-C motif) ligand 2 (CCL2)-mediated neuroprotection in rat primary midbrain neurons.[105] Among a variety of physiological functions of TRPC channels, promoting neuron survival makes them a promising target for future therapeutic strategies in diseases.

10.5.1 SIGNALING PATHWAYS AND NEURONAL SURVIVAL

It has been known that neuronal survival can be regulated by intrinsic death pathways and extrinsic trophic signaling processes, among which neurotropin signaling is critical for neuron survival both in development and in adulthood.[70,106–109] The NTFs are a pleiotropic group of secreted growth factors that regulate multiple aspects of neuronal development, including neuronal survival. In the mammalian CNS, there are four NTFs, which are composed of a family of structurally related proteins: NGF, BDNF, neurotrophin 3 (NT-3), and NT-4. Neurotrophin binding to its receptor, usually together with intrinsic tyrosine kinase activity, triggers one or more intracellular signaling pathways responsible for neuronal survival.[110,111] These pathways inhibit intrinsic apoptosis machinery from carrying out a cellular suicide program.[112–114] Two intracellular signaling pathways that are activated by RTKs and are crucial in promoting neuronal survival include the extracellular signal-regulated kinase (ERK)/cAMP response element-binding protein (CREB) pathway and the PI$_3$K/AKT pathway.[115–117] These two pathways can activate both intracellular mechanisms and nuclear transcriptional mechanisms that inhibit cell death.

During CNS development, restriction of neuronal death by survival signaling is important for proper formation of neuronal networks,[118] while in adulthood, neuronal death is mainly related to pathological conditions, including stroke, neurodegenerative diseases, and trauma.[119,120] To protect against neuronal death, survival signaling is enhanced and neuroprotective mechanisms are activated to preserve neuronal survival. Loss or reduction in these intrinsic europrotections might contribute to or accelerate neuronal damage, whereas their activation might rescue neurons from brain injuries.

10.5.2 TRPCs as a Transducer of BDNF-Mediated Survival

TRPC3 and TRPC6 play key roles in neuronal survival via transmitting BDNF-mediated signals (Figure 10.2). It has been found that RTKs can stimulate PLCγ to activate TRPC channels.[28,121] BDNF is essential for survival of a variety of neurons,[122] including CGNs[74,123,124] and striatal neurons. In pontine neurons, BDNF triggers a nonselective inward current that resembles that of TRPC channels.[34] In both CGNs and *Xenopus laevis* spinal neurons, TRPC channels are essential for BDNF-triggered growth cone turning.[125–127] On the basis of these findings, Jia et al. proposed that TRPC channels may transmit BDNF survival signals and promote CGN survival.[54] They have examined the role of TRPCs in neuronal survival in both CGN cultures and that neonatal cerebellum. They found that TRPC3 and TRPC6 are required for BDNF-mediated neuronal protection, BDNF-triggered intracellular Ca^{2+} elevation, and BDNF-induced CREB activation. Overexpressing TRPC3 or TRPC6 markedly enhanced CREB phosphorylation and increased CREB-dependent transcription. Furthermore, overexpressing these channels protected CGNs against serum deprivation-induced cell death via CREB activation. In contrast, knocking down the expression level of TRPC3 or TRPC6 led to increased cell apoptosis in the neonatal cerebellum. Taken together, these findings provide in vitro and in vivo evidence that TRPC channels play a critical role in promoting neuronal survival and indicate that activation of CREB is a key downstream event for the neuronal protective effect of TRPC channels.

A recent study using mice with a mutation in the TRPC3 gene revealed the importance of TRPC3 channels for Purkinje cell survival during development. The mutation causes an alteration in TRPC3 phosphorylation and abnormal gating of the channel, leading to cell death and ataxia in mice.[128] These results suggest that the normal gating of TRPC3 channels is critical to initiate the downstream events for neuronal survival, while the aberrant opening of the channel could lead to neuronal death.

10.5.3 TRPCs in Neuroprotection against Neuronal Injury

TRPC1 has been reported to protect human SH-SY5Y neuroblastoma cells against salsolinol-mediated cytotoxicity[129] and 1-methyl-4-phenylpyridinium ion

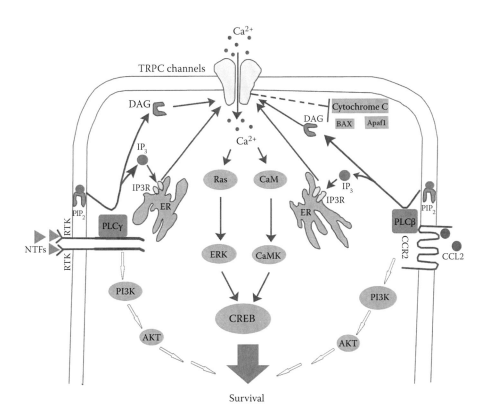

FIGURE 10.2 Possible mechanisms for TRPC channels in neuronal survival. NTFs and CCL2 activate PLC to cleave PIP_2 into DAG and IP_3. DAG and IP_3 activate TRPC channels, and Ca^{2+} influx via TRPC channels promotes neuronal survival by the Ras/ERK- and CaM/CaMK-CREB pathway. TRPC channels can also inhibit the release of cytochrome C, Bax, and Apaf-1 to protect neurons.

(MPP[+])-mediated neurotoxicity[130] likely by inhibiting apoptosis. Both TRPC1 activation and TRPC1 over-expression protect SH-SY5Y cells from apoptosis. Similarly, the increase in TRPC1 protein expression decreased the level of proteins required for carrying out apoptosis, such as cytochrome C, Bax, and Apaf1, by unknown mechanisms. It is thus suggested that TRPC1 may inhibit apoptotic signaling to provide neuroprotection against agents that induce Parkinson's disease.

It has been recently reported that inhibition of TRPC channels suppresses the CCL2-mediated neuroprotection against HIV-1 transactivator protein (Tat) toxicity in rat neurons.[105] TRPC channels are involved in the process of CCL2 protection by elevating intracellular Ca^{2+} and activating the ERK/CREB pathway in a manner similar to CGN survival promoted by TRPC3/TRPC6[105] (Figure 10.2). Further research on functions of TRPC channels under pathologic conditions may provide new avenues for treatment of brain damage and neurodegenerative diseases.

10.6 TOOLS TO STUDY TRPC CHANNELS

Several approaches have been employed to study TRPCs in mammalian cells, including expression of the wild-type or dominant-negative form of TRPCs in exogenous expression systems, use of specific antibodies, and pharmacological inhibitors or agonists. Moreover, generation of transgenic mice and genetic ablation in knockout mice are also powerful approaches to investigate TRPCs in vivo. Perhaps one of the most convenient ways to specifically manipulate TRPCs at the present time is to use wild-type and dominant-negative constructs of TRPCs. As the investigation of TRPCs evolves, several pharmacological inhibitors against certain subtypes of TRPCs have been developed, such as Pyr3 for TRPC3. It is important to realize the limitations of these approaches. For instance, TRPC channels formed by overexpressed TRPC proteins in heterologous expression systems might be different from the endogenous channels because overexpression might lead to formation of homomeric channels. Moreover, the specificities of pharmacological inhibitors and antibodies commonly used are not entirely clear. For example, SKF96365, which is widely used to inhibit TRPCs,[126,131] can inhibit both receptor-mediated and voltage-gated Ca^{2+} entry.[131] La^{3+}, which is used to inhibit several TRPCs,[15,132–134] can inhibit nonspecific Ca^{2+} channels and mitochondrial cationic channel(s).[135] At present, one of the biggest hurdles to the study of possible roles of TRPCs in vivo is the lack of specific antagonists for TRPCs. In addition to genetic interference and pharmacological inhibition, generation of specific antibodies against the pore region of TRPCs to block the ion flow could greatly help the study of TRPC functions (see Chapters 6 and 7). Moreover, TRPC proteins can form homo- and heteromeric channels. Therefore, to investigate the contribution of an individual channel protein to channel functions is a challenge. Additionally, the interpretation of the data obtained through experimental modulations of TRPC channels may be hampered by the fact that the inhibition or activation may affect other channels, receptors, or intracellular molecules. Studying the regulation of TRPC channels in both heterologous expression systems and native physiological systems would be important for a better understanding of these channels. For the latter, the generation of conditional TRPC knock-out mice and/or the use of RNAi gene silencing could be powerful tools (see Chapter 8).

10.7 PERSPECTIVES

TRPC channels are activated by GPCRs and RTKs through a PLC-dependent mechanism. Because both GPCRs and RTKs are important for neuronal survival, TRPC channels could serve as a sensor of the extracellular signals that activate these receptors in the CNS. Accumulating evidence so far supports a critical role of TRPCs in neuronal survival under physiological conditions. These findings have laid a foundation for further investigating their possible roles in pathological conditions. The broad expression profiles of TRPCs in the CNS point to the possibility that they might also participate in the pathogenesis of neurodegenerative disorders, such as Alzheimer's disease and Parkinson's disease. Future studies will show whether and how these channels contribute to neuronal survival in pathological conditions.

It should be pointed out that TRPM7 channels are involved in ischemic neuronal death.[136] Suppression of TRPM7 protein expression in hippocampal neurons prevents delayed neuronal death in global cerebral ischemia.[137] This raises a question as to why different members of the TRP family, almost all of which mediate Ca^{2+} influx, have different roles in neuronal fate. Hopefully, mediators underlying the pro-survival role of TRPC channels and the pro-death role of TRPM7 in neurons will be uncovered in the near future. Factors that determine whether TRP channels promote or inhibit neuronal survival also warrant future exploration. Identification of these factors may be critical for the development of strategies to target the TRP cascade in combating neurological diseases.

ACKNOWLEDGMENT

This work was supported by a grant from the 973 Program (2011CB809000) of China.

REFERENCES

1. Wes, P. D., J. Chevesich, A. Jeromin, C. Rosenberg, G. Stetten, and C. Montell. 1995. TRPC1, a human homolog of a Drosophila store-operated channel. *Proceedings of the National Academy of Sciences of the United States of America* 92: 9652–9656.
2. Montell, C., and G. M. Rubin. 1989. Molecular characterization of the Drosophila trp locus: a putative integral membrane protein required for phototransduction. *Neuron* 2: 1313–1323.
3. Venkatachalam, K., and C. Montell. 2007. TRP channels. *Annual Review of Biochemistry* 76: 387–417.
4. Liman, E. R., and H. Innan. 2003. Relaxed selective pressure on an essential component of pheromone transduction in primate evolution. *Proceedings of the National Academy of Sciences of the United States of America* 100: 3328–3332.
5. Long, S. B., X. Tao, E. B. Campbell et al. 2007. Atomic structure of a voltage-dependent K^+ channel in a lipid membrane-like environment. *Nature* 450: 376–382.
6. Moiseenkova-Bell, V. Y., L. A. Stanciu, I. I. Serysheva, B. J. Tobe, and T. G. Wensel. 2008. Structure of TRPV1 channel revealed by electron cryomicroscopy. *Proceedings of the National Academy of Sciences of the United States of America* 105: 7451–7455.
7. Myers, B. R., Y. Saimi, D. Julius et al. 2008. Multiple unbiased prospective screens identify TRP channels and their conserved gating elements. *Journal of General Physiology* 132: 481–486.
8. Vannier, B., M. X. Zhu, D. Brown et al. 1998. The membrane topology of human transient receptor potential 3 as inferred from glycosylation-scanning mutagenesis and epitope immunocytochemistry. *Journal of Biological Chemistry* 273: 8675–8679.
9. Lussier, M. P., S. Cayouette, P. K. Lepage et al. 2005. MxA, a member of the dynamin superfamily, interacts with the ankyrin-like repeat domain of TRPC. *Journal of Biological Chemistry* 280: 19393–19400.
10. Lussier, M. P., P. K. Lepage, S. M. Bousquet et al. 2008. RNF24, a new TRPC interacting protein, causes the intracellular retention of TRPC. *Cell Calcium* 43: 432–443.
11. Schindl, R., I. Frischauf, H. Kahr et al. 2008. The first ankyrin-like repeat is the minimum indispensable key structure for functional assembly of homo- and heteromeric TRPC4/TRPC5 channels. *Cell Calcium* 43: 260–269.

12. Engelke, M., O. Friedrich, P. Budde et al. 2002. Structural domains required for channel function of the mouse transient receptor potential protein homologue TRP1beta. *FEBS Letters* 523: 193–199.

13. Lupas, A. N., and M. Gruber. 2005. The structure of alpha-helical coiled coils. *Advances in Protein Chemistry* 70: 37–78.

14. Walter, J. K., C. Rueckert, M. Voss et al. 2009. The oligomerization of the coiled coil-domain of occludin is redox sensitive. *Annals of the New York Academy of Sciences* 1165: 19–27.

15. Greka, A., B. Navarro, E. Oancea et al. 2003. TRPC5 is a regulator of hippocampal neurite length and growth cone morphology. *Nature Neuroscience* 6: 837–845.

16. Liu, X., B. B. Singh, and I. S. Ambudkar. 2003. TRPC1 is required for functional store-operated Ca^{2+} channels. Role of acidic amino acid residues in the S5-S6 region. *Journal of Biological Chemistry* 278: 11337–11343.

17. Owsianik, G., K. Talavera, T. Voets et al. 2006. Permeation and selectivity of TRP channels. *Annual Review of Physiology* 68: 685–717.

18. Yuan, J. P., K. Kiselyov, D. M. Shin et al. 2003. Homer binds TRPC family channels and is required for gating of TRPC1 by IP3 receptors. *Cell* 114: 777–789.

19. Woo, J.S., H. Kim do, P. D. Allen et al. 2008. TRPC3-interacting triadic proteins in skeletal muscle. *Biochemical Journal* 411: 399–405.

20. Sinkins, W. G., M. Goel, M. Estacion et al. 2004. Association of immunophilins with mammalian TRPC channels. *Journal of Biological Chemistry* 279: 34521–34529.

21. Tang, J., Y. Lin, Z. Zhang, S. Tikunova, L. Birnbaumer, and M. X. Zhu. 2001. Identification of common binding sites for calmodulin and inositol 1,4,5-trisphosphate receptors on the carboxyl termini of trp channels. *Journal of Biological Chemistry* 276: 21303–21310.

22. Ordaz, B., J. Tang, R. Xiao et al. 2005. Calmodulin and calcium interplay in the modulation of TRPC5 channel activity. Identification of a novel C-terminal domain for calcium/calmodulin-mediated facilitation. *Journal of Biological Chemistry* 280: 30788–30796.

23. Singh, B. B., X. Liu, J. Tang et al. 2002. Calmodulin regulates Ca^{2+}-dependent feedback inhibition of store-operated Ca^{2+} influx by interaction with a site in the C terminus of TrpC1. *Molecular Cell* 9: 739–750.

24. Tang, Y., J. Tang, Z. Chen et al. 2000. Association of mammalian trp4 and phospholipase C isozymes with a PDZ domain-containing protein, NHERF. *Journal of Biological Chemistry* 275: 37559–37564.

25. Mery, L., B. Strauss, J. F. Dufour et al. 2002. The PDZ-interacting domain of TRPC4 controls its localization and surface expression in HEK293 cells. *Journal of Cell Science* 115: 3497–3508.

26. Salido, G. M., S. O. Sage, and J. A. Rosado. 2009. TRPC channels and store-operated Ca^{2+} entry. *Biochimica et Biophysica Acta* 1793: 223–230.

27. Birnbaumer, L. 2009. The TRPC class of ion channels: a critical review of their roles in slow, sustained increases in intracellular Ca^{2+} concentrations. *Annual Review of Pharmacology and Toxicology* 49: 395–426.

28. Montell, C. 2005. The TRP superfamily of cation channels. *Science STKE* 2005: re3.

29. Philipp, S., A. Cavalié, M. Freichel et al. 1996. A mammalian capacitative calcium entry channel homologous to *Drosophila* TRP and TRPL. *EMBO Journal* 15: 6166–6171.

30. Zhu, M. X., M. Jiang, M. Peyton et al. 1996. trp, a novel mammalian gene family essential for agonist-activated capacitative Ca^{2+} entry. *Cell* 85: 661–671.

31. Zitt, C., A. Zobel, A. G. Obukhov et al. 1996. Cloning and functional expression of a human Ca^{2+}-permeable cation channel activated by calcium store depletion. *Neuron* 16: 1189–1196.

32. Philipp, S., J. Hambrecht, L. Braslavski et al. 1998. A novel capacitative calcium entry channel expressed in excitable cells. *EMBO Journal* 17: 4274–4282.

33. Vannier, B., M. Peyton, G. Boulay et al. 1999. Mouse trp2, the homologue of the human trpc2 pseudogene, encodes mTrp2, a store depletion-activated capacitative Ca^{2+} entry channel. *Proceedings of the National Academy of Sciences of the United States of America* 96: 2060–2064.

34. Li, H. S., X. Z. Xu, and C. Montell. 1999. Activation of a TRPC3-dependent cation current through the neurotrophin BDNF. *Neuron* 24: 261–273.

35. Lievremont, J. P., T. Numaga, G. Vazquez et al. 2005. The role of canonical transient receptor potential 7 in B-cell receptor-activated channels. *Journal of Biological Chemistry* 280: 35346–35351.

36. Soboloff, J., M. Spassova, W. Xu, L. P. He, N. Cuesta, and D. L. Gill. 2005. Role of endogenous TRPC6 channels in Ca^{2+} signal generation in A7r5 smooth muscle cells. *Journal of Biological Chemistry* 280: 39786–39794.

37. Hofmann, T., A. G. Obukhov, M. Schaefer, C. Harteneck, T. Gudermann, and G. Schultz. 1999. Direct activation of human TRPC6 and TRPC3 channels by diacylglycerol. *Nature* 397: 259–263.

38. Venkatachalam, K., F. Zheng, and D. L. Gill. 2003. Regulation of canonical transient receptor potential (TRPC) channel function by diacylglycerol and protein kinase C. *Journal of Biological Chemistry* 278: 29031–29040.

39. Vazquez, G., G. S. Bird, Y. Mori et al. 2006. Native TRPC7 channel activation by an inositol trisphosphate receptor-dependent mechanism. *Journal of Biological Chemistry* 281: 25250–25258.

40. Lockwich, T., B. B. Singh, X. Liu et al. 2001. Stabilization of cortical actin induces internalization of transient receptor potential 3 (Trp3)-associated caveolar Ca^{2+} signaling complex and loss of Ca^{2+} influx without disruption of Trp3-inositol trisphosphate receptor association. *Journal of Biological Chemistry* 276: 42401–42408.

41. Mehta, D., G. U. Ahmmed, B. C. Paria et al. 2003. RhoA interaction with inositol 1,4,5-trisphosphate receptor and transient receptor potential channel-1 regulates Ca^{2+} entry. Role in signaling increased endothelial permeability. *Journal of Biological Chemistry* 278: 33492–33500.

42. Cayouette, S., M. P. Lussier, E. L. Mathieu et al. 2004. Exocytotic insertion of TRPC6 channel into the plasma membrane upon Gq protein-coupled receptor activation. *Journal of Biological Chemistry* 279: 7241–7246.

43. Itagaki, K., K. B. Kannan, B. B. Singh et al. 2004. Cytoskeletal reorganization internalizes multiple transient receptor potential channels and blocks calcium entry into human neutrophils. *Journal of Immunology* 172: 601–607.

44. Bezzerides, V. J., I. S. Ramsey, S. Kotecha et al. 2004. Rapid vesicular translocation and insertion of TRP channels. *Nature Cell Biology* 6: 709–720.

45. Riccio, A., A. D. Medhurst, C. Mattei et al. 2002. mRNA distribution analysis of human TRPC family in CNS and peripheral tissues. *Brain Research* 109: 95–104.

46. Kunert-Keil, C., F. Bisping, J. Kruger et al. 2006. Tissue-specific expression of TRP channel genes in the mouse and its variation in three different mouse strains. *BMC Genomics* 7: 159.

47. Fowler, M. A., K. Sidiropoulou, E. D. Ozkan et al. 2007. Corticolimbic expression of TRPC4 and TRPC5 channels in the rodent brain. *PloS one* 2: e573.

48. Chung, Y. H., H. Sun Ahn, D. Kim et al. 2006. Immunohistochemical study on the distribution of TRPC channels in the rat hippocampus. *Brain Research* 1085: 132–137.

49. Giampa, C., Z. DeMarch, S. Patassini et al. 2007. Immunohistochemical localization of TRPC6 in the rat substantia nigra. *Neuroscience Letters* 424: 170–174.

50. Fusco, F. R., A. Martorana, C. Giampà et al. 2004. Cellular localization of TRPC3 channel in rat brain: preferential distribution to oligodendrocytes. *Neuroscience Letters* 365: 137–142.

51. Boisseau, S., C. Kunert-Keil, S. Lucke et al. 2009. Heterogeneous distribution of TRPC proteins in the embryonic cortex. *Histochemistry and Cell Biology* 131: 355–363.
52. Zechel, S., S. Werner, and O. von Bohlen Und Halbach. 2007. Distribution of TRPC4 in developing and adult murine brain. *Cell and Tissue Research* 328: 651–656.
53. Fiorio Pla, A., D. Maric, S. C. Brazer et al. 2005. Canonical transient receptor potential 1 plays a role in basic fibroblast growth factor (bFGF)/FGF receptor-1-induced Ca^{2+} entry and embryonic rat neural stem cell proliferation. *Journal of Neuroscience* 25: 2687–2701.
54. Jia, Y., J. Zhou, Y. Tai et al. 2007. TRPC channels promote cerebellar granule neuron survival. *Nature Neuroscience* 10: 559–567.
55. Elg, S., F. Marmigere, J. P. Mattsson et al. 2007. Cellular subtype distribution and developmental regulation of TRPC channel members in the mouse dorsal root ganglion. *Journal of Comparative Neurology* 503: 35–46.
56. Bennet, M. R., W. G. Gibson, and G. Lemon. 2002. Neuronal cell death, nerve growth factor and neurotrophic models: 50 years on. *Autonomic Neuroscience* 95: 1–23.
57. Cohen, S., and M. E. Greenberg. 2008. Communication between the synapse and the nucleus in neuronal development, plasticity, and disease. *Annual Review of Cell and Developmental Biology* 24: 183–209.
58. Yuen, E. C., C. L. Howe, Y. Li et al. 1996. Nerve growth factor and the neurotrophic factor hypothesis. *Brain and Development* 18: 362–368.
59. Catsicas, M., Y. Pequignot, and P. G. Clarke. 1992. Rapid onset of neuronal death induced by blockade of either axoplasmic transport or action potentials in afferent fibers during brain development. *Journal of Neuroscience* 12: 4642–4650.
60. Linden, R. 1994. The survival of developing neurons: a review of afferent control. *Neuroscience* 58: 671–682.
61. Franklin, J. L., and E. M. Johnson, Jr. 1992. Suppression of programmed neuronal death by sustained elevation of cytoplasmic calcium. *Trends in Neurosciences* 15: 501–508.
62. Collins, F., and J. D. Lile. 1989. The role of dihydropyridine-sensitive voltage-gated calcium channels in potassium-mediated neuronal survival. *Brain Research* 502: 99–108.
63. Koike, T., D. P. Martin, and E. M. Johnson, Jr. 1989. Role of Ca^{2+} channels in the ability of membrane depolarization to prevent neuronal death induced by trophic-factor deprivation: evidence that levels of internal Ca^{2+} determine nerve growth factor dependence of sympathetic ganglion cells. *Proceedings of the National Academy of Sciences of the United States of America* 86: 6421–6425.
64. Lampe, P. A., E. B. Cornbrooks, A. Juhasz et al. 1995. Suppression of programmed neuronal death by a thapsigargin-induced Ca^{2+} influx. *Journal of Neurobiology* 26: 205–212.
65. Franklin, J. L., C. Sanz-Rodriguez, A. Juhasz et al. 1995. Chronic depolarization prevents programmed death of sympathetic neurons in vitro but does not support growth: requirement for Ca^{2+} influx but not Trk activation. *Journal of Neuroscience* 15: 643–664.
66. Tolkovsky, A. M., A. E. Walker, R. D. Murrell et al. 1990. Ca^{2+} transients are not required as signals for long-term neurite outgrowth from cultured sympathetic neurons. *Journal of Cell Biology* 110: 1295–1306.
67. Blair, L. A., K. K. Bence-Hanulec, S. Mehta, T. Franke, D. Kaplan, and J. Marshall. 1999. Akt-dependent potentiation of L channels by insulin-like growth factor-1 is required for neuronal survival. *Journal of Neuroscience* 19: 1940–1951.
68. Crowder, R. J., and R. S. Freeman. 1998. Phosphatidylinositol 3-kinase and Akt protein kinase are necessary and sufficient for the survival of nerve growth factor-dependent sympathetic neurons. *Journal of Neuroscience* 18: 2933–2943.
69. Hetman, M., K. Kanning, J. E. Cavanaugh et al. 1999. Neuroprotection by brain-derived neurotrophic factor is mediated by extracellular signal-regulated kinase and phosphatidylinositol 3-kinase. *Journal of Biological Chemistry* 274: 22569–22580.

70. Segal, R. A., and M. E. Greenberg. 1996. Intracellular signaling pathways activated by neurotrophic factors. *Annual Review of Neuroscience* 19: 463–489.

71. Brunet, A., A. Bonni, M. J. Zigmond et al. 1999. Akt promotes cell survival by phosphorylating and inhibiting a Forkhead transcription factor. *Cell* 96: 857–868.

72. Datta, S. R., H. Dudek, X. Tao et al. 1997. Akt phosphorylation of BAD couples survival signals to the cell-intrinsic death machinery. *Cell* 91: 231–241.

73. Cardone, M. H., N. Roy, H. R. Stennicke et al. 1998. Regulation of cell death protease caspase-9 by phosphorylation. *Science* 282: 1318–1321.

74. Bonni, A., A. Brunet, A. E. West, S. R. Datta, M. A. Takasu, and M. E. Greenberg. 1999. Cell survival promoted by the Ras-MAPK signaling pathway by transcription-dependent and -independent mechanisms. *Science* 286: 1358–1362.

75. Maggirwar, S. B., P. D. Sarmiere, S. Dewhurst et al. 1998. Nerve growth factor-dependent activation of NF-kappaB contributes to survival of sympathetic neurons. *Journal of Neuroscience* 18: 10356–10365.

76. Yano, S., H. Tokumitsu, and T. R. Soderling. 1998. Calcium promotes cell survival through CaM-K kinase activation of the protein-kinase-B pathway. *Nature* 396: 584–587.

77. Mao, Z., A. Bonni, F. Xia et al. 1999. Neuronal activity-dependent cell survival mediated by transcription factor MEF2. *Science* 286: 785–790.

78. Linseman, D. A., C. M. Bartley, S. S. Le et al. 2003. Inactivation of the myocyte enhancer factor-2 repressor histone deacetylase-5 by endogenous Ca^{2+}/calmodulin-dependent kinase II promotes depolarization-mediated cerebellar granule neuron survival. *Journal of Biological Chemistry* 278: 41472–41481.

79. Kolarow, R., T. Brigadski, and V. Lessmann. 2007. Postsynaptic secretion of BDNF and NT-3 from hippocampal neurons depends on calcium calmodulin kinase II signaling and proceeds via delayed fusion pore opening. *Journal of Neuroscience* 27: 10350–10364.

80. Lessmann, V., and T. Brigadski. 2009. Mechanisms, locations, and kinetics of synaptic BDNF secretion: an update. *Neuroscience Research* 65: 11–22.

81. Jean, Y. Y., L. D. Lercher, and C. F. Dreyfus. 2008. Glutamate elicits release of BDNF from basal forebrain astrocytes in a process dependent on metabotropic receptors and the PLC pathway. *Neuron Glia Biology* 4: 35–42.

82. Wang, H., N. Ward, M. Boswell et al. 2006. Secretion of brain-derived neurotrophic factor from brain microvascular endothelial cells. *European Journal of Neuroscience* 23: 1665–1670.

83. Sherer, T. B., D. B. Clemow, and J. B. Tuttle. 2000. Calcium homeostasis and nerve growth factor secretion from vascular and bladder smooth muscle cells. *Cell and Tissue Research* 299: 201–211.

84. Siesjo, B. K. 1989. Calcium and cell death. *Magnesium* 8: 223–227.

85. Mody, I., and J. F. MacDonald. 1995. NMDA receptor-dependent excitotoxicity: the role of intracellular Ca^{2+} release. *Trends in Pharmacological Sciences* 16: 356–359.

86. Tymianski, M., and C. H. Tator. 1996. Normal and abnormal calcium homeostasis in neurons: a basis for the pathophysiology of traumatic and ischemic central nervous system injury. *Neurosurgery* 38: 1176–1195.

87. Kawahara, M., M. Negishi-Kato, and Y. Sadakane. 2009. Calcium dyshomeostasis and neurotoxicity of Alzheimer's beta-amyloid protein. *Expert Review of Neurotherapeutics* 9: 681–693.

88. McConkey, D. J., and S. Orrenius. 1997. The role of calcium in the regulation of apoptosis. *Biochemical and Biophysical Research Communications* 239: 357–366.

89. Wang, H. G., N. Pathan, I. M. Ethell et al. 1999. Ca^{2+}-induced apoptosis through calcineurin dephosphorylation of BAD. *Science* 284: 339–343.

90. Choi, D. W. 1988. Calcium-mediated neurotoxicity: relationship to specific channel types and role in ischemic damage. *Trends in Neurosciences* 11: 465–469.

91. Tymianski, M., M. P. Charlton, P. L. Carlen et al. Source specificity of early calcium neurotoxicity in cultured embryonic spinal neurons. *Journal of Neuroscience* 13: 2085–2104.

92. Ghosh, A., and M. E. Greenberg. 1995. Calcium signaling in neurons: molecular mechanisms and cellular consequences. *Science* 268: 239–247.

93. Kim, S. J., Y. S. Kim, J. P. Yuan, R. S. Petralia, P. F. Worley, and D. J. Linden. 2003. Activation of the TRPC1 cation channel by metabotropic glutamate receptor mGluR1. *Nature* 426: 285–291.

94. Narayanan, K. L., K. Irmady, S. Subramaniam et al. 2008. Evidence that TRPC1 is involved in hippocampal glutamate-induced cell death. *Neuroscience Letters* 446: 117–122.

95. Lucas, P., K. Ukhanov, T. Leinders-Zufall et al. 2003. A diacylglycerol-gated cation channel in vomeronasal neuron dendrites is impaired in TRPC2 mutant mice: mechanism of pheromone transduction. *Neuron* 40: 551–561.

96. Amaral, M. D., and L. Pozzo-Miller. 2007. TRPC3 channels are necessary for brain-derived neurotrophic factor to activate a nonselective cationic current and to induce dendritic spine formation. *Journal of Neuroscience* 27: 5179–5189.

97. Hartmann, J., E. Dragicevic, H. Adelsberger et al. 2008. TRPC3 channels are required for synaptic transmission and motor coordination. *Neuron* 59: 392–398.

98. Weick, J. P., M. A. Johnson, and S. C. Zhang. 2009. Developmental regulation of human embryonic stem cell-derived neurons by calcium entry via transient receptor potential channels. *Stem Cells* 27: 2906–2916.

99. Davare, M. A., D. A. Fortin, T. Saneyoshi et al. 2009. Transient receptor potential canonical 5 channels activate Ca^{2+}/calmodulin kinase Igamma to promote axon formation in hippocampal neurons. *Journal of Neuroscience* 29: 9794–9808.

100. Riccio, A., Y. Li, J. Moon et al. 2009. Essential role for TRPC5 in amygdala function and fear-related behavior. *Cell* 137: 761–772.

101. Tai, Y., S. Feng, R. Ge et al. 2008. TRPC6 channels promote dendritic growth via the CaMKIV-CREB pathway. *Journal of Cell Science* 121: 2301–2307.

102. Zhou, J., W. Du, K. Zhou et al. 2008. Critical role of TRPC6 channels in the formation of excitatory synapses. *Nature Neuroscience* 11: 741–743.

103. Yan, H. D., C. Villalobos, and R. Andrade. 2009. TRPC channels mediate a muscarinic receptor-induced afterdepolarization in cerebral cortex. *Journal of Neuroscience* 29: 10038–10046.

104. Yao, H., F. Peng, Y. Fan, M. X. Zhu, G. Hu, and S. J. Buch. 2009. TRPC channel-mediated neuroprotection by PDGF involves Pyk2/ERK/CREB pathway. *Cell Death Differ* 16: 1681–1693.

105. Yao, H., F. Peng, N. Dhillon et al. 2009. Involvement of TRPC channels in CCL2-mediated neuroprotection against tat toxicity. *Journal of Neuroscience* 29: 1657–1669.

106. Bibel, M., and Y. A. Barde. 2000. Neurotrophins: key regulators of cell fate and cell shape in the vertebrate nervous system. *Genes Development* 14: 2919–2937.

107. Pettmann, B., and C. E. Henderson. 1998. Neuronal cell death. *Neuron* 20: 633–647.

108. Huang, E. J., and L. F. Reichardt. 2003. Trk receptors: roles in neuronal signal transduction. *Annual Review of Biochemistry* 72: 609–642.

109. Sofroniew, M. V., C. L. Howe, and W. C. Mobley. 2001. Nerve growth factor signaling, neuroprotection, and neural repair. *Annual Review of Neuroscience* 24: 1217–1281.

110. Reichardt, L. F. 2006. Neurotrophin-regulated signalling pathways. Philosophical Transactions of the Royal Society of London. Series B, Biological Sciences 361: 1545–1564.

111. Chao, M. V., and B. L. Hempstead. 1995. p75 and Trk: a two-receptor system. *Trends in Neuroscience* 18: 321–326.

112. Raff, M. C. 1992. Social controls on cell survival and cell death. *Nature* 356: 397–400.

113. Raff, M. C., B. A. Barres, J. F. Burne, H. S. Coles, Y. Ishizaki, and M. D. Jacobson. 1993. Programmed cell death and the control of cell survival: lessons from the nervous system. *Science* 262: 695–700.

114. Burek, M. J., and R. W. Oppenheim. 1996. Programmed cell death in the developing nervous system. *Brain Pathology* 6: 427–446.

115. Nunez, G., and L. del Peso. 1998. Linking extracellular survival signals and the apoptotic machinery. *Current Opinion in Neurobiology* 8: 613–618.

116. Datta, S. R., A. Brunet, and M. E. Greenberg. 1999. Cellular survival: a play in three Akts. *Genes Development* 13: 2905–2927.

117. Grewal, S. S., R. D. York, and P. J. Stork. 1999. Extracellular-signal-regulated kinase signalling in neurons. *Current Opinion in Neurobiology* 9: 544–553.

118. Yuan, J., and B. A. Yankner. 2000. Apoptosis in the nervous system. *Nature* 407: 802–809.

119. Kermer, P., N. Klocker, and M. Bahr. 1999. Neuronal death after brain injury. Models, mechanisms, and therapeutic strategies in vivo. *Cell Tissue Research* 298: 383–395.

120. Gorman, A. M. 2008. Neuronal cell death in neurodegenerative diseases: recurring themes around protein handling. *Journal of Cellular and Molecular Medicine* 12: 2263–2280.

121. Clapham, D. E. 2003. TRP channels as cellular sensors. *Nature* 426: 517–524.

122. Davies, A. M. 1994. The role of neurotrophins during successive stages of sensory neuron development. *Progress in Growth Factor Research* 5: 263–289.

123. Minichiello, L., and R. Klein. 1996. TrkB and TrkC neurotrophin receptors cooperate in promoting survival of hippocampal and cerebellar granule neurons. *Genes Development* 10: 2849–2858.

124. Schwartz, P. M., P. R. Borghesani, R. L. Levy et al. 1997. Abnormal cerebellar development and foliation in BDNF⁻/⁻ mice reveals a role for neurotrophins in CNS patterning. *Neuron* 19: 269–281.

125. Shim, S., E. L. Goh, S. Ge et al. 2005. XTRPC1-dependent chemotropic guidance of neuronal growth cones. *Nature Neuroscience* 8: 730–735.

126. Li, Y., Y. C. Jia, K. Cui et al. 2005. Essential role of TRPC channels in the guidance of nerve growth cones by brain-derived neurotrophic factor. *Nature* 434: 894–898.

127. Wang, G. X., and M. M. Poo. 2005. Requirement of TRPC channels in netrin-1-induced chemotropic turning of nerve growth cones. *Nature* 434: 898–904.

128. Becker, E. B., P. L. Oliver, M. D. Glitsch et al. 2009. A point mutation in TRPC3 causes abnormal Purkinje cell development and cerebellar ataxia in moonwalker mice. *Proceedings of the National Academy of Sciences of the United States of America* 106: 6706–6711.

129. Bollimuntha, S., M. Ebadi, and B. B. Singh. 2006. TRPC1 protects human SH-SY5Y cells against salsolinol-induced cytotoxicity by inhibiting apoptosis. *Brain Research* 1099: 141–149.

130. Bollimuntha, S., B. B. Singh, S. Shavali et al. 2005. TRPC1-mediated inhibition of 1-methyl-4-phenylpyridinium ion neurotoxicity in human SH-SY5Y neuroblastoma cells. *Journal of Biological Chemistry* 280: 2132–2140.

131. Merritt, J. E., W. P. Armstrong, C. D. Benham et al. 1990. SKandF 96365, a novel inhibitor of receptor-mediated calcium entry. *Biochemistry Journal* 271: 515–522.

132. Hochstrate, P. 1989. Lanthanum mimicks the trp photoreceptor mutant of Drosophila in the blowfly Calliphora. *Journal of Comparative Physiology* 166: 179–187.

133. Strubing, C., G. Krapivinsky, L. Krapivinsky et al. 2001. TRPC1 and TRPC5 form a novel cation channel in mammalian brain. *Neuron* 29: 645–655.

134. Jung, S., A. Mühle, M. Schaefer, R. Strotmann, G. Schultz, and T. D. Plant. 2003. Lanthanides potentiate TRPC5 currents by an action at extracellular sites close to the pore mouth. *Journal of Biological Chemistry* 278: 3562–3571.

135. Chinopoulos, C., A. A. Starkov, S. Grigoriev et al. 2005. Diacylglycerols activate mito-chondrial cationic channel(s) and release sequestered Ca^{2+}. *Journal of Bioenergetics and Biomembranes* 37: 237–247.
136. Aarts, M., K. Iihara, W. L. Wei et al. 2003. A key role for TRPM7 channels in anoxic neuronal death. *Cell* 115: 863–877.
137. Sun, H. S., M. F. Jackson, L. J. Martin et al. 2009. Suppression of hippocampal TRPM7 protein prevents delayed neuronal death in brain ischemia. *Nature Neuroscience* 12: 1300–1307.

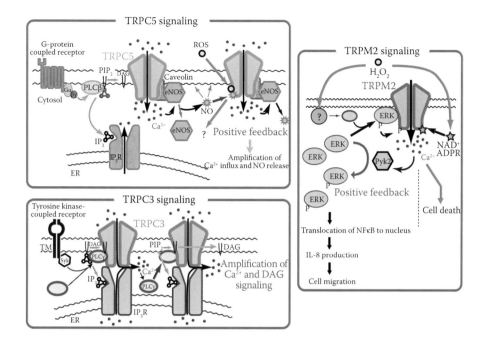

FIGURE 3.1

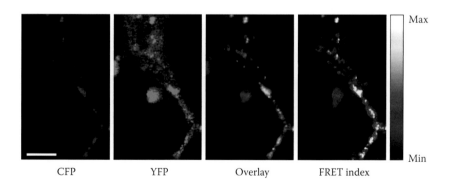

| CFP | YFP | Overlay | FRET index |

FIGURE 4.1

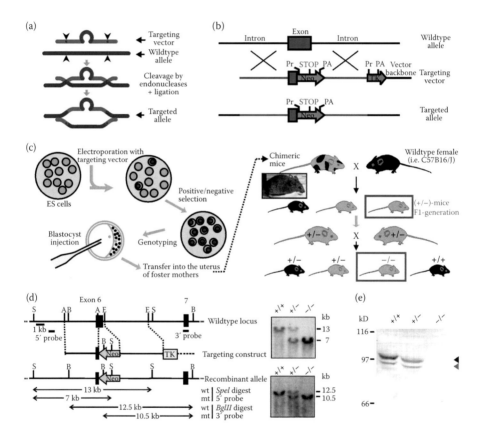

FIGURE 8.1

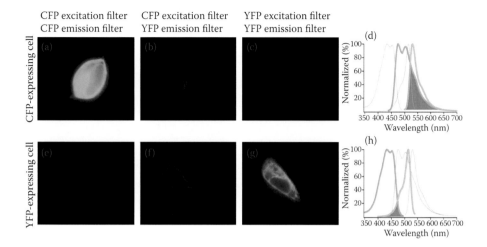

FIGRUE 15.1

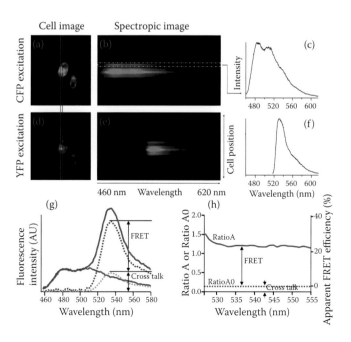

FIGURE 15.2

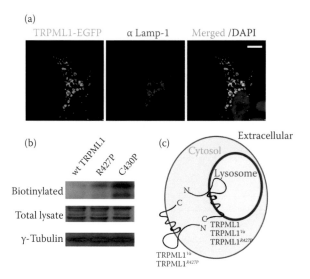

(a)

TRPML1-EGFP α Lamp-1 Merged /DAPI

(b)

wt TRPML1 R427P C430P

Biotinylated

Total lysate

γ-Tubulin

(c)

Extracellular

Cytosol

Lysosome

N

C

C

N TRPML1
   TRPML1^{Va}
   TRPML1^{R427P}

TRPML1^{Va}
TRPML1^{R427P}

FIGURE 19.1

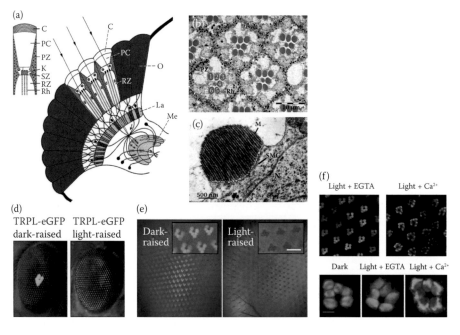

(a)

C

PC

PZ

K

SZ

RZ

Rh

C

PC

O

RZ

La

Me

(b)

PZ

Rh

10 μm

(c)

M.

SMC

N

500 nm

(d)

TRPL-eGFP
dark-raised

TRPL-eGFP
light-raised

(e)

Dark-
raised

Light-
raised

(f)

Light + EGTA Light + Ca^{2+}

Dark Light + EGTA Light + Ca^{2+}

FIGURE 20.1

11 Electrophysiological Methods for the Study of TRP Channels

Karel Talavera and Bernd Nilius

CONTENTS

11.1 INTRODUCTION

Transient receptor potential (TRP) cation channels function as polymodal cellular sensors involved in the fine-tuning of many physiological processes.[1–8] Intriguingly, mutations in the encoding genes cause a plethora of diseases, and TRP channels are considered, to date, the most attractive channel targets by the pharmaceutical industry. This exploding interest in TRP channels makes it necessary to revisit some of the elementary methods and tools to describe these proteins as what they are, ion channels. Without any doubt, the most direct way to approach function of ion channels is the use of patch-clamp methods. Here we give a short description of how those methods can be used to evaluate gating properties, pore and permeation properties, selectivity sequences of TRP channel pores, and the measurement of fractional Ca^{2+} currents, i.e., the determination of the percentage of a current carried by Ca^{2+} as charge carrier. We also include some description of methods, which enable a researcher to obtain information about the pore structure, e.g., sizing of the pore, and discuss some modeling approaches used for determination of gating schemes. It is to be noted, however, that we do not intend by any means to supplant the role of more extensive treatises on ion channel and membrane biophysics,[9–12] to which we keenly refer the reader.

11.2 PATCH-CLAMP TECHNIQUE

Patch-clamp remains the most direct technique to study the properties of ion channels. However, some experimental protocols need to be adapted to meet the particularities of TRPs (see Section 11.4). Special attention should be taken when recording TRP currents in heterologous expression systems, where typical TRP currents are on the order of several nA. Voltage-clamp errors (V_{Error}) may be in the order of tens of mV even after substantial electronic compensation of the series resistance. This can be estimated by $V_{Error} = I \cdot R_{Series}$, where I is the measured current amplitude and R_{Series} is the series resistance. It must also be kept in mind that the possibility of high electronic compensation of R_{Series} is not a guarantee of good recording conditions. A high R_{Series} is also indicative of poor intracellular access and thus of compromised intracellular dialysis. Large TRP currents combined with poor intracellular perfusion may also raise the problem of depletion and/or accumulation of permeant ions, which result in time-dependent decrease of outward and inward currents, respectively.

The current–voltage relationship of several TRP channels becomes quasi-linear during application of strong stimuli. This unavoidably complicates the dissection of the TRP currents from leak and/or background currents. Notably, extreme temperatures and high chemical concentrations worsen recording conditions and tend to enhance leak currents. Unfortunately, dissection of TRP and leak components is uncommon in the TRP literature, making it likely that some studies have arrived at wrong conclusions about gating properties during TRP channel stimulation (see Section 11.4.1).

To determine if leak has affected the recordings, one can check the reversibility of the effect of temperature changes and compound application or apply specific channel inhibitors at high concentrations. We have employed two other methods to determine leak contribution. One is to apply a holding potential equal to the theoretical equilibrium potential of the permeant ion, which is, in turn, chosen to be different from zero by using asymmetrical ionic concentrations. This permits the measurement of leak currents at the level of the holding potential.[13,14] The other method is to replace the permeant ions by impermeant ions in one side of the membrane, usually the extracellular one.[15] This method has the shortcoming that it assumes that the permeability of the impermeant ion does not change during stimulation, which is not always valid for TRP channels.[16–18]

Special consideration should also be given to the choice of the patch-clamp configuration, as washout of intracellular modulators in whole-cell and inside-out recordings is prone to alter channel properties. It has been shown, for example, that TRPA1 activation requires intracellular polyphosphates,[19] PIP$_2$ or MgATP$_2$[20] in excised patches, and recordings in sensory neurons are more stable in perforated patches.[21] More intriguing are the cases of TRPM4 and TRPM5. TRPM4 allows for stable recordings in inside-out but not in whole-cell configuration, whereas the opposite occurs for TRPM5.[22]

11.3 STUDY OF PORE PROPERTIES

11.3.1 CONDUCTION PROPERTIES

11.3.1.1 Characterizing Ionic Permeability

Ionic permeation through biological channels has been described using a wide variety of formalisms, including the Goldman–Hodgkin–Katz (GHK) constant field theory,[10] Eyring's theory for absolute reaction rates,[23] the Poisson–Nernst–Planck theory,[24,25] and molecular dynamics.[26–28] Given its simplicity, the most commonly employed is undoubtedly the GHK theory, which reduces the permeation process to the partition of ions between the aqueous milieu and the lipid membrane. In this formalism, ions are considered to move independently from each other, in accordance with simple electrodiffusion under a constant transmembrane electric field. Under these premises, it is possible to derive the GHK current equation:

$$I_C = P_C z_C^2 \frac{VF^2}{RT} \frac{[C]_i - [C]_o \exp(-z_C FV/RT)}{1 - \exp(-z_C FV/RT)}$$

where P_C is the permeability and z_C is the valence of the permeating ion, the subscripts i and e denote intra- and extracellular concentrations, respectively, F is the Faraday constant, R is the gas constant, V is the membrane potential, and T is absolute temperature. P_C is the only parameter describing the ion-membrane interactions and is given by $\beta_C D_C/a$, where a is the membrane thickness and β_C and D_C are the ion's partition and diffusion coefficients, respectively. This oversimplification makes the absolute value of P_C actually of little use. In practice, it is more useful to determine the permeability of several different ions relative to a certain ion of reference, for it has been found that the determination of the ionic selectivity series based on relative permeabilities is very helpful, for instance, for relating recombinant channels expressed in heterologous systems to their native counterparts.[29] Relative permeabilities are determined using an equation derived from the equilibrium condition, which reflects when total inward and outward currents add up to zero at the equilibrium (or reversal) potential, V_{rev}. In the cases of mixture of a monovalent cation X$^+$ with the reference cation Na$^+$:

$$\frac{P_X}{P_{Na}} = \frac{[Na^+]_e - [Na^+]_i \exp(V_{rev}FV/RT)}{[X^+]_i \exp(V_{rev}F/RT) - [X^+]_e}$$

To determine the permeability of a divalent cation, such as Ca^{2+} relative to a monovalent cation X$^+$, the easiest approach is to use extracellular and intracellular solution containing only Ca^{2+} and the monovalent cation, respectively. In such a condition:

$$\frac{P_{Ca}}{P_{Na}} = \frac{[X^+]_i}{4[Ca^{2+}]_e} \exp(V_{rev}F/RT)(\exp(V_{rev}F/RT)+1)$$

The determination of relative permeabilities in other useful scenarios with different ionic compositions is described elsewhere.[30]

These calculations should take into account two important corrections. First, reversal potentials determined experimentally should be corrected for liquid junction potentials $V_{rev,corrected} = V_{rev,measured} - V_{LJ}$, where V_{LJ} is the liquid junction potential that can be calculated according to Refs. 31 and 32. Second, ionic concentrations should actually reflect ionic activities according to $a_i = \gamma_i \cdot [X_i]$, where the activity coefficient γ_i can be approximated by

$$\log \gamma_i = \frac{-0.51 \cdot z_i^2 \sqrt{I}}{1+\sqrt{I}}$$

where z_i is the valence of the ion and I is the ionic strength.

11.3.1.2 Selectivity Sequences (Eisenman Series)

An important feature of ion channel pores is the so-called selectivity sequence for ions of equal charge. By considering a rather simple electrostatic model for the equilibrium between dehydration and binding of ions to a charged site, Eisenman explained why, out of the 120 formal possibilities to write different selectivity sequences for 5 monovalent cations, only 11 are commonly found in ion channels.[10,30] In Eisenman's model, the energy of the transition from a hydrated cation to a dehydrated cation bound to a negatively charged site can be approximated by

$$\Delta G = \frac{z_{site} z_C q_e^2}{4\pi\varepsilon_0(r_{site}+r_C)} - \frac{z_C^2 q_e^2}{8\pi\varepsilon_0 r_C}\left(\frac{1}{\varepsilon} - \frac{1}{\varepsilon_0}\right)$$

where z_{site} and z_C are the valences of the charges of the binding site and the cation, respectively, q_e is the unitary charge, ε is the dielectric permittivity of water, ε_0 is the dielectric permittivity of vacuum, and r_{site} and r_C are the radius of the binding site and the cation, respectively. The first term corresponds to the energy of electrostatic interaction (U) between the ion and the binding site, whereas the second term represents the energy of dehydration, or Born energy. If the site is large, the dehydration term would have more weight than U, and therefore, larger ions are going to be favored for binding as they have less dehydration energy. Thus, a weak-field binding site would show the selectivity series I: $Cs^+ > Rb^+ > K^+ > Na^+ > Li^+$. On the other hand, very small sites favor the binding of smaller cations, as U would be the dominating component, especially for smaller cations, which can get closer to the interaction site. In the extreme, a very strong-field binding site would show a selectivity series XI: $Li^+ > Na^+ > K^+ > Rb^+ > Cs^+$. Binding sites with intermediate strengths allow for nine other selectivity sequences.[10,30]

In practice, the selectivity sequence for a particular channel can be obtained by comparing the relative permeabilities for different cations of equal charge. In a typical experiment, the cation species of the extracellular solution are consecutively exchanged, while keeping constant an intracellular solution containing only one permeant ion. The reversal potential is measured for each condition, which allows determination of the permeability of each cation species relative to the cation species present in the intracellular solution.

11.3.1.3 Pore Block

The passage of ions through open channels is generally affected by the presence of other chemical species, often resulting in pore block. This phenomenon not only is relevant for the pathophysiological roles of the channels but also serves to compare the properties of recombinant channels to those of their native counterparts.[29,30]

Among the effects of pore blockers we can expect: (1) changes in the current rectification pattern resulting from voltage-dependent block, (2) time dependence of currents that reflect the movement of blocker ions in and out of the pore, (3) competition between blockers and permeating ions, (4) flickering behavior of single-channel currents, and (5) reduction of the single-channel conductance. These features can be described in terms of kinetic models, which have been developed from the seminal work of Woodhull on the voltage-dependent inhibition of Na^+ permeability by protons at nodes of Ranvier.[33] This model is based on the idea that binding of the blocker to a high-affinity site in the permeation pathway obstructs the passage of permeant ions. Blocking and unblocking events are described in terms of a simple kinetic scheme ruling channel transitions between blocked and unblocked states. Under the assumption of equilibration of the blocking and unblocking events, the probability of finding the channel in the unblocked state is given by

$$P_{Unblocked} = \frac{k_{-1} + k_2}{k_{-1} + k_2 + k_1[B]_o + k_{-2}[B]_i},$$

where k_1 and k_2 are the rate constants governing the exit of the blocker to the extra- and intracellular moieties, respectively, and $k_1[B]_o$ and $k_{-2}[B]_i$ are the rate constants governing the entry of the blocker from the extra- and intracellular moieties, respectively.

Blocker–channel interactions are accounted for by means of a Gibb's free energy profile that stretches along the conduction pore consisting of one well flanked by two barriers. The depth of the energy well reflects the affinity of the binding site for the blocker, whereas the heights of the flanking barriers reflect the difficulty of accession to the binding site from the extra- and intracellular moieties. The rates of blocking and unblocking events can be quantified using Eyring's theory for absolute reaction rate:[23]

$$k_1 = \kappa \frac{k_B T}{h} \exp\left(-\frac{G_o + ze\delta V/2}{k_B T}\right), k_{-1} = \kappa \frac{k_B T}{h} \exp\left(-\frac{G_o - G_w - ze\delta V/2}{k_B T}\right)$$

$$k_2 = \kappa \frac{k_B T}{h} \exp\left(-\frac{G_i - G_w + ze(1-\delta)V/2}{k_B T}\right)$$

$$k_{-2} = \kappa \frac{k_B T}{h} \exp\left(-\frac{G_i - ze(1-\delta)V/2}{k_B T}\right)$$

where T is the absolute temperature, k_B and h are the Boltzmann's and Planck's constants, respectively, e is the unitary charge, and z is the valence of the blocker. The prefactor κ is the maximal probability of the transition and, given that is unknown, is usually taken as 1. G_w is the energy at the binding site, G_o and G_i are the energies of the outer and inner barriers, respectively, and the electric distance δ is the fraction of the transmembrane electric field dropped at the position of the binding site. The parameters G_w, G_o, G_i, and δ can be determined by fitting the voltage dependence of finding the channel in the unblocked conformation at different blocker concentrations.

Woodhull's model assumes that blocker binding to the pore is not affected by permeating ions, which is a reasonable approximation only if the affinity of the blocker is much higher than the affinity of the permeating ions. An example of the breakdown of this assumption is the blocking effect of divalent cations on currents carried by monovalent cations. In order to account for this complexity, one must consider the interactions of all ionic species with the pore (i.e., the individual energy profiles) and the interionic interactions in the permeation pathway (electrostatic forces), in the framework of more complex kinetic schemes.[10] Such quantitative analyses have never been performed on TRP channels. Nevertheless, the paper of Vennekens et al.[34] discusses a model of selectivity and permeation in TRPV5 based on Eyring's rate theory, whereas the paper of Voets et al.[35] on the block of TRPV6 by Mg^{2+} provides for a particularly good example of a thorough characterization of voltage- and time-dependent pore block and the competition of the blocker with ions of large permeability.

Importantly, pore block should be distinguished from other types of effects that lead to current reduction, such as allosteric inhibition of channel activation and shift of voltage dependence of channel activation by screening of surface charges (see below).

The structural determinants of pore block can be directly assessed by mutation of residues in putative pore regions. Being cationic channels, and by analogy to the original work done on voltage-gated Na^+ and Ca^{2+} channels,[10] one is first poised to test the contribution of acidic residues. Typically, the residues that determine pore block by ions are the same as those responsible for the selectivity and permeation properties, i.e., divalent/monovalent cation relative permeabilities, Eisenman selectivity sequence, and single-channel conductance.[30]

11.3.1.4 Fractional Ca²⁺ Currents

With the exception of TRPM4 and TRPM5, all TRP channels are permeable to Ca2+, and therefore constitute Ca^{2+} entry pathways in multiple cell types. Provided the key roles of Ca^{2+} as intracellular messenger, it is important to know the fraction of the total current that is carried by this ion. The so-called "fractional Ca^{2+} current," or $P_f\%$,[36] can be determined by measuring the increase in intracellular Ca^{2+} relative to the total charge permeating the channels, with a combination of whole-cell patch-clamp and photometry experiments. Using a high concentration of the Ca^{2+}-sensitive dye Fura-2 in the intracellular solution (~2 mM) ensures that this Ca^{2+} chelator overwhelms the buffering capacity of endogenous Ca^{2+}-binding molecules, so that the Ca^{2+} entry can be directly related to the decrease of fluorescence of Fura-2 at 510 nm when excited with 380 nm, F^{380}.[37] To translate the decrease of F^{380} into increase of intracellular Ca^{2+} concentration, it is sufficient to determine the time integral of the current ($\int I_{Ca} dt$) recorded in the sole presence of Ca^{2+} as extracellular permeating ion.

Then, $P_f\%$ can be determined by $P_f\% = 100 \times \dfrac{\int I_{Ca} dt}{\int I_{Ca,X} dt} \times \dfrac{\Delta F_{Ca,X}^{380}}{\Delta F_{Ca}^{380}}$, where $\int I_{Ca,X} dt$ is the time integral of the current in the presence of Ca^{2+} and another extracellular cation X, and $\Delta F_{Ca,X}^{380}$ and ΔF_{Ca}^{380} are the corresponding changes in fluorescence. Typically, test currents are evoked at physiological membrane potentials, from a "resting" condition that ensures no Ca^{2+} influx, for example, at –60 mV in the absence of channel agonists, like that for TRPV1[38] or, at a very positive potential, like that for TRPA1.[18] Surprisingly, $P_f\%$ has only been determined for TRPV1 and TRPA1, but the results are so striking that they raise the interest to study other TRP channels. First, $P_f\%$ depends upon the stimulus, so that for TRPV1, the value is larger when the currents are stimulated by capsaicin than by low extracellular pH,[38] and for TRPA1, $P_f\%$ is larger in the presence of the agonist allyl isothiocyanate.[18] Second, the values of $P_f\%$ measured experimentally do not fit the predictions of the GHK theory, indicating that Ca^{2+} impairs the permeation of other ions.[18] Third, the larger $P_f\%$ of TRPA1, in comparison to that of TRPV1, may result in differential contribution of these channels to Ca^{2+}-dependent processes in nociceptive neurons, where these channels are co-expressed.[18]

11.3.2 Pore Structure

11.3.2.1 Pore Diameter and Pore Dilation (Stokes, Excluded Volume)

A hydrodynamic approach to ionic permeation through ion channels allows the derivation of an equation relating the relative ionic permeabilities to the diameter of the permeant ions and the pore diameter.[10,39,40] Thus, one can assume that a permeating ion X^+ is not subjected to friction with the pore:

$$\frac{P_X}{P_{Na}} = k \left(1 - \frac{d_X}{d_{Pore}} \right)^2$$

where P_X/P_{Na} is the relative permeability of X^+ with respect to Na^+, k is a constant factor, and d_X and d_{Pore} are the diameters of the ion X^+ and the pore, respectively. When considering friction, the equation takes the form

$$\frac{P_X}{P_{Na}} = \frac{k}{d_X}\left(1 - \frac{d_X}{d_{Pore}}\right)^2$$

In a typical experiment, the relative permeabilities of several ions of increasing size are plotted versus their respective diameter, and k and d_{Pore} are determined from the fit of the data with one of the equations given above. Notably, it has been shown that TRPV1[16] and TRPA1,[17,18] but not TRPM8,[17] undergo a significant increase in pore diameter during agonist-induced stimulation.

11.3.2.2 Pore Architecture

The architecture of the pore can be assessed using the substituted cysteine accessibility method (SCAM), whereby pore-lining residues are identified according to the accessibility of hydrophilic agents, such as Ag^+ and methanethiosulfonate (MTS) reagents to substituted cysteines.[10,41] In the first of such studies on TRP channels, Voets et al. showed that extracellular application Ag^+ did not modify currents through wild-type TRPV6 channels, while it inhibited a number of cysteine mutants.[35] Positions in a model helical wheel structure were then classified according to the rate of Ag^+-induced current inhibition. Remarkably, all rapidly reacting positions were located on the same side of the helix, whereas the nonreacting positions clustered on the opposite side. Furthermore, these two groups were separated by two sets of slowly reacting positions. These results clearly indicated that residues Pro526 to Thr538 form a pore α-helix. Cationic MTS reagents showed accessibilities that were consistent with the α-helical structure and, given their larger size with respect to Ag^+, helped to determine the residues located beyond the narrowest point of the pore. Similar experiments confirmed an analogous structural model for TRPV5,[42] and a more recent study indicates that the side chains on amino acids in the selectivity filter point toward the permeation pathway, in contrast to the structure of K^+ channels, where backbone carbonyls project toward pore.[43]

11.4 GATING PROPERTIES

11.4.1 Determining the Intrinsic Voltage Dependence of Channel Gating

For several years, the gating of TRP channels was thought to be intrinsically voltage independent because these channels were shown to be either constitutively active or were believed to need other specific stimuli to be activated (chemical and thermal). It was 3 years after the cloning of the capsaicin receptor TRPV1 that, using voltage step protocols, Gunthorp et al. found that this channel has rectification properties consistent with voltage-dependent gating.[44] Since then, other TRP channels have been shown to be intrinsically voltage gated: TRPM4,[45] TRPM5,[13,46] TRPM8,[47,48] TRPV3,[49,50] TRPM3,[51] TRPA1,[52] and TRPP3.[53] Of note, TRPM4 and TRPM5 seem to be the only voltage-gated TRP channels shown so far that require the presence of a specific activator (intracellular Ca^{2+}) to manifest voltage-dependent gating.[22]

To expose the intrinsic voltage dependence of the gating of TRP channels, it is necessary to avoid voltage-dependent effects of modulatory ions and to use appropriate voltage step protocols. The first requirement is met by removing all ions that may produce an open pore block, such as Ca^{2+}, Mg^{2+},[44,45,47] and intracellular polyamines (inside-out recordings).[54]

TRP channels are commonly studied using protocols with fixed membrane negative potentials or voltage ramps. These allow basic characterization of TRP channel modulation but do not lead to significant mechanistic insight into the relationship between voltage-dependent gating and the action of gating modulators. As for any other voltage-gated ion channel, the properties of voltage-dependent gating of TRPs can be studied using voltage-step protocols. Typically, one can apply a series of increasing depolarizing pulses, followed by an invariant pulse to a potential at which it is possible to reliably record tail currents (see below). Of note, the low voltage sensitivity of channel gating requires the application of step protocols over ample voltage ranges.[13,45–48,55] Importantly, voltage pulses should be of enough duration, typically several hundred milliseconds, to allow for reaching the steady state of current amplitudes at each potential. Another important aspect is to set the holding potential at the reversal potential (typically 0 mV), in order to avoid ionic depletion and/or accumulation of the permeant ions.

A priori, it is expected that at the holding potential, the open probability is different from zero. Thus, when membrane is hyperpolarized from the holding potential, TRP currents decay as channels transit from open to closed state(s) (deactivation). Conversely, when membrane is depolarized, the open probability $P_{\text{Open}}(V)$ will increase. In general, the steady-state open probability can be described using a Boltzmann function of the form:

$$P_{\text{Open}}(V) = P_{-\infty} + \frac{P_{+\infty} - P_{-\infty}}{1 + \exp(-(V - V_{\text{act}})/s_{\text{act}})}$$

where $P_{-\infty}$ and $P_{+\infty}$ are the open probabilities at very negative and very positive potentials, respectively. V_{act} is the voltage for half-maximal voltage-dependent activation, and s_{act} is the so-called slope factor, which indicates how shallow the voltage dependence is. Special attention should be given to the value of $P_{-\infty}$, for when it is different from zero, this indicates that there is a voltage-independent component of channel activation. It should be noticed, however, that a nonzero value of $P_{-\infty}$ can result from the presence of leak currents at negative potentials. In our experience, $P_{-\infty}$ is not significantly different from zero for TRPM8, TRPV1,[47] TRPM4, and TRPM5.[13,14] On the other hand, $P_{+\infty} < 1$ has been taken as an indication that voltage is only a partial activator of the channel,[56] but this is a rather common property of voltage-gated channels.

In practice, $P_{\text{Open}}(V)$ can be determined using two equivalent methods. One is to fit the voltage dependence of steady-state currents to an equation of the form

$$I_{\text{SS}}(V) = G_{\text{Max}} \cdot (V - V_r) P_{\text{Open}}(V) = G_{\text{Max}} \cdot (V - V_r) \cdot \left(P_{-\infty} + \frac{P_{+\infty} - P_{-\infty}}{1 + \exp(-(V - V_{\text{act}})/s_{\text{act}})} \right)$$

where G_{Max} is the maximal conductance and V_r is the reversal potential. Note, however, that this equation assumes that the single-channel current has an Ohmic behavior. In the general case,

$$I_{SS}(V) = G \cdot i(V) \cdot P_{\text{Open}}(V) = G \cdot i(V) \cdot \left(P_{-\infty} + \frac{P_{+\infty} - P_{-\infty}}{1 + \exp(-(V - V_{\text{act}})/s_{\text{act}})} \right)$$

where G is a scaling factor and $i(V)$ is the voltage dependence of the single-channel current. In general, such an equation involves too many parameters, which cannot be fitted independently. Such cases require the independent determination of $i(V)$, i.e., the rectification pattern of the currents (see, e.g., Refs. 57 and 58).

The other way to determine $P_{\text{Open}}(V)$ is using the voltage dependence of tail currents. In the simplest alternative of this method, one determines the amplitude of tail currents evoked by an invariant pulse to a potential at which the steady-state open probability is very close to zero, i.e., a very negative potential. Then

$$P_{\text{Open}}(V) = \frac{I_{\text{Tail}}(V)}{I_{\text{Tail},+\infty}},$$

where $I(V)$ is the tail current amplitude as a function of the pre-pulse potential and $I_{\text{Tail},+\infty}$ is the tail current amplitude corresponding to pre-pulses to very positive potentials. Given the very shallow character of $P_{\text{Open}}(V)$ in TRP channels, $I_{\text{Tail},+\infty}$ is usually determined from the projection of the fit of $I_{\text{Tail}}(V)$ with a Boltzmann-type function at very positive potential. This procedure is, of course, valid only when sufficient saturation of the open probability is achieved at the most positive pulses of the protocol. Importantly, tail currents may be too fast to accurately determine their amplitude at very negative potentials. In such cases, one can determine tail current amplitudes at a positive potential, where the rate of current relaxation is much slower.[59] To be rigorous, tail amplitudes should be determined from the fit of the deactivation time course with one (or the sum of) exponential function(s). Importantly, one should try to avoid including in the fit the initial part of the current decay in order to exclude the artifact caused by the capacitive transient and, in some cases, such as in TRPA1, the recovery from inactivation after pre-pulses to very positive voltages.[15]

Interestingly, there is a rather common scenario in which the voltage dependence of TRP channels can be masked. When gating modulators produce a very strong shift of the voltage dependence to negative potentials, currents become linear owing to saturated open probability. Typical examples of this are TRPV1 and TRPM8 when exposed to high concentrations of capsaicin and menthol, respectively.[47]

Another typical characteristic of voltage-gated TRPs is the shallow voltage dependence of the time constants of current relaxation. This has been explicitly determined for TRPM4 and TRPM5,[13] for which, considering a simple close-open gating model, the time constant of current relaxation can be written as

$$\tau(V,T) = \frac{h/kT}{\exp\left(\dfrac{-\Delta G_O + zF\delta V}{RT}\right) + \exp\left(\dfrac{-\Delta G_C + zF(1-\delta)V}{RT}\right)},$$

where h, k, F, and R are the Planck's, Boltzmann's, Faraday's, and gas constants, respectively; T is the absolute temperature; ΔG_O and ΔG_C are the molar Gibb's free energies associated with channel opening and closing, respectively; z is the valence of the apparent gating charge of the channel; and δ is the fraction of the transmembrane electric field across which the gating charges move.

11.4.2 LIMITING SLOPE METHOD

Fitting the voltage dependence of channel activation with a Boltzmann type of function allows the determination of the valence of the apparent gating charge:

$$z_{App} = \frac{RT}{s_{act}F}$$

However, this is valid only for a simple closed-open two-state gating model. A more rigorous estimate of gating charge can be done with the limiting slope method.[60,61] It can be derived that the total gating charge z_T is the limit of displaced charge $q_a(V)$ when $P_{Open}(V)$ tends to zero at very negative potentials. Thus

$$z_T = kT \lim_{P_{Open} \to 0} q_a(V) = kT \lim_{P_{Open} \to 0} \frac{d\ln(P_{Open}(V))}{dV}$$

$P_{Open}(V)$ can be determined by dividing the measured whole-cell ionic current by the driving force $(V - V_r)$ and by the maximal conductance. In turn, currents can be measured during application of a slow voltage ramp in a voltage range at which $P_{Open}(V)$ is expected to be very low (typically from 10^{-5} to 10^{-1}). Importantly, this procedure requires the subtraction of leak currents, which can be estimated from the projection of a linear fit of the current trace at very negative potentials (e.g., between -250 and -220 mV). So far, this method has only been applied for wild-type and mutant TRPM8 channels,[62] giving an estimate of 0.9 elementary charges for wild-type TRPM8 and between 0.6 and 0.7 elementary charges for the mutants R842A and K856A. Notably, these estimates are very close to the values determined from the Boltzmann fit of $P_{Open}(V)$, suggesting that the closed-open model is a rather good approximation for the voltage-dependent gating of TRPM8.

11.5 GATING MODIFIERS

11.5.1 GATING MODULATORS

Contrasting with the fundamental role of TRPs as molecular integrators of physical and chemical stimuli,[63–69] there is a remarkable lack of studies addressing the mechanisms of modulation from a biophysical point of view. It is only very recently that comprehensive models of ion channel gating have been put forward to explain the modulation by temperature changes and channel agonists. Central to these advances was the discovery that several TRP channels are voltage gated,[55] for it is now established that various gating modulators act via direct[5,13,47,55,62,65,70–72] or allosteric[48,73–76] interactions with the voltage-sensing and/or gating machineries. The relationship between voltage-dependent gating and the action of a TRP agonist was first modeled for TRPM4.[70] Using a simple sequential model of activation in which Ca^{2+} binding acts as a permissive step for voltage-dependent gating, Nilius et al. were able to reproduce the effects of intracellular Ca^{2+} on the voltage dependence of channel activation and kinetics. However, this initial model rapidly evolved to accommodate the interaction of Ca^{2+} with the open state of the channel and modulation by PIP_2.[77]

Thermal modulation of TRP channels was first analyzed in terms of a gating model for TRPM8 and TRPV1 by Voets et al. who demonstrated that changes in temperature result in graded shifts of their voltage dependence of activation and proposed that thermal and voltage-dependent gating are tightly linked.[47] In this model, the cold- or heat-dependent activation is determined by whether the voltage-dependent closing or opening transition is the one with the highest temperature dependence, respectively. Almost simultaneously, Brauchi et al. proposed an alternative model for TRPM8, suggesting that voltage and temperature act allosterically.[48] Both models have been criticized, the first for neglecting the known complexities of single-channel properties,[78–80] and the second for having been originally based on a rather incorrect analysis of the voltage dependence determined from tail current measurements.[62] Regardless of their weaknesses, these models represent the very important first steps in the understanding of TRP function as molecular integrators.[81] For example, TRPM8 modulation by voltage, temperature, and menthol can be very closely described in terms of a Monod–Wyman–Changeux (MWC) type of model, in which each subunit of the tetrameric channel can independently bind a single menthol molecule and the four subunits undergo a concerted voltage-dependent transition between the closed and open states.[62,66] Furthermore, the MWC model suggests that, in agreement with binding experiments, mutations of voltage-sensing residues produce a decrease in menthol binding affinity without altering its efficacy.[62] The relationship between voltage-dependent gating and chemical modulation has served for the analysis of other phenomena, such as the stimulatory action of intracellular Ca^{2+},[52] menthol,[71] and nicotine[15] on TRPA1; the inhibition of TRPM8 by BCTC, SKF96365, and phenanthroline;[82] and the inhibition of TRPM5 by quinine.[14]

11.5.2 Formal Considerations on Thermosensing in TRP Channels

The thermosensory properties of TRPs have been traditionally characterized by their so-called "temperature threshold for activation." Depending on whether the temperature threshold is low or high, each thermoTRP has been attributed with a possible role in the perception of pleasant or unpleasant temperature changes. However, we would like to argue that the concept of channel thermal threshold is not formally justified and actually can be misleading when trying to understand the thermosensory role of TRP channels.

As mentioned above, temperature changes result in graded shifts of the voltage dependence for activation of TRPM8, TRPV1,[47] TRPM4, TRPM5,[13] and TRPA1.[72] Thus, thermally induced activation is not due to an abrupt phenomenon, which is sufficient to conceptually invalidate the use of thermal threshold. In addition, the practical methods to estimate thermal thresholds are rather arbitrary and in most cases have no relevance to physiological settings. For example, the most popular method in patch-clamp experiments is determining the value of temperature at which the curves fitting the temperature dependence of background and TRP currents intercept each other. However, such interception points depend on the relative size of the TRP and background and leak currents, which are different for each expression system and recording condition. For these reasons, we believe that temperature thresholds of channel activation should be considered with extreme care when seeking the understanding of the thermosensory role of TRPs.

11.5.3 SCREENING OF SURFACE CHARGES

The function of ion channels is generally affected by the presence of membrane surface charges arising from negative membrane phospholipid head groups, sialic acid residues at glycosylation sites, and charged amino acids of the channel. By producing a local disturbance of the electric potential, surface charges alter the local ionic concentration at the pore entrances and modify the electric field acting on the channel's voltage sensors.[10] Specifically, the presence of negative surface charges produces a decrease in the local electric potential that, in turn, reduces the effective electric potential drop across the membrane. This reduction in local electric potential can be compensated by the presence of cations, especially divalent cations and protons. This effect, called screening of surface charges, increases the drop of electric potential across the membrane (hyperpolarization) and thus reduces membrane excitability.[10]

The Gouy–Chapman model assumes that the surface charge is homogeneously distributed and that ions do not bind to the membrane. Thus, the surface charge density σ can be determined by

$$\sigma = \sqrt{2\varepsilon RT \sum_i C_i \left(\exp\left(\frac{-z_i F\Phi}{RT} \right) - 1 \right)}$$

where ε is the dielectric permittivity of the solution, z_i and C_i are the valence and concentration of the ith ion, and Φ is the surface potential.[83] To account for possible ion binding to specific sites at the membrane surface, the Gouy–Chapman–Stern theory yields

$$\sigma = \frac{\sqrt{2\varepsilon RT \sum_i C_i \left(\exp\left(\frac{-z_i F\Phi}{RT} \right) - 1 \right)}}{1 + \sum_i \frac{C_i}{K_i} \exp\left(\frac{-z_i F\Phi}{RT} \right)}$$

where K_i is the binding constant of the ith ion.[84]

The screening of surface charges is expected to influence all voltage-dependent processes at the plasma membrane, so voltage-dependent gating of TRP channels is no exception. However, so far this has been studied only in TRPM8 channels.[85] It was shown that increase of extracellular concentration of Mg^{2+}, Ca^{2+}, Ba^{2+}, and H^+ produced a decrease in TRPM8 currents, which was consistent with shifts of the activation curve to more positive potentials. Notably, these effects were distinguishable from open pore block, for the shifts were not accompanied by changes in the slope factor of the activation curve. The experimental data were fitted by a model with a surface charge density of around 0.01 charges per $\mathring{A}^2$, a Ca^{2+} binding constant between 2.4 and 7.7 M, H^+ binding with a pK_a of 5.7, and very weak or no binding for Mg^{2+} ($K_{Mg} > 27$–2000 M).

REFERENCES

1. Nilius, B., G. Owsianik, T. Voets, and J. A. Peters. 2007. Transient receptor potential channels in disease. *Physiological Reviews* 87: 165.
2. Pedersen, S. F., and B. Nilius. 2007. Transient receptor potential channels in mechanosensing and cell volume regulation. *Methods in Enzymology* 428: 183.
3. Pedersen, S. F., G. Owsianik, and B. Nilius. 2005. TRP channels: An overview. *Cell Calcium* 38: 233.
4. Nilius, B., and T. Voets. 2005. Trp channels: A TR(I)P through a world of multifunctional cation channels. *Pflügers Archiv* 451: 1.
5. Voets, T., K. Talavera, G. Owsianik, and B. Nilius. 2005. Sensing with TRP channels. *Nature Chemical Biology* 1: 85.
6. Ramsey, I. S., M. Delling, and D. E. Clapham, D. E. 2006. An introduction to TRP channels. *Annual Review of Physiology* 68: 619.
7. Venkatachalam, K., and C. Montell. 2007. TRP Channels. *Annual Review of Biochemistry* 76: 387.
8. Hoenderop, J. G., B. Nilius, and R. J. Bindels. 2005. Calcium absorption across epithelia. *Physiological Review* 85: 373.
9. DeFelice, L. J. 1997. *Electrical Properties of Cells. Patch Clamp for Biologists.* New York: Plenum Press.
10. Hille, B. 2001. *Ionic Channels of Excitable Membranes.* 3rd ed. Sunderland: Sinauer Associates.
11. Molleman, A. 2003. *Patch Clamping. An Introductory Guide to Patch Clamp Electrophysiology.* New York: John Wiley.
12. Sakmann, B., and E. Neher. 2009. *Single-Channel Recording.* 2nd ed. New York: Springer.
13. Talavera, K., K. Yasumatsu, T. Voets et al. 2005. Heat activation of TRPM5 underlies thermal sensitivity of sweet taste. *Nature* 438: 1022.
14. Talavera, K., K. Yasumatsu, R. Yoshida et al. 2008. The taste transduction channel TRPM5 is a locus for bitter-sweet taste interactions. *FASEB Journal* 22: 1343.
15. Talavera, K., M. Gees, Y. Karashima et al. 2009. Nicotine activates the chemosensory cation channel TRPA1. *Nature Neuroscience* 12: 1293.
16. Chung, M. K., A. D. Guler, and M. J. Caterina. 2008. TRPV1 shows dynamic ionic selectivity during agonist stimulation. *Nature Neuroscience* 11: 555.
17. Chen, J., D. Kim, B. R. Bianchi et al. 2009. Pore dilation occurs in TRPA1 but not in TRPM8 channels. *Molecular Pain* 5: 3.
18. Karashima, Y., J. Prenen, K. Talavera et al. 2010. Agonist-induced changes in Ca^{2+} permeation through the nociceptor cation channel TRPA1. *Biophysics Journal* 98: 773.

19. Kim, D., and E. J. Cavanaugh. 2007. Requirement of a soluble intracellular factor for activation of transient receptor potential A1 by pungent chemicals: role of inorganic polyphosphates. *Journal of Neuroscience* 27: 6500.
20. Karashima, Y., J. Prenen, V. Meseguer et al. 2008. Modulation of the transient receptor potential channel TRPA1 by phosphatidylinositol 4, 5-biphosphate manipulators. *Pflügers Archiv* 457: 77.
21. Fajardo, O., V. Meseguer, C. Belmonte, and F. Viana. 2008. TRPA1 channels mediate cold temperature sensing in mammalian vagal sensory neurons: pharmacological and genetic evidence. *Journal of Neuroscience* 28: 7863.
22. Ullrich, N. D., T. Voets, J. Prenen et al. 2005. Comparison of functional properties of the Ca^{2+}-activated cation channels TRPM4 and TRPM5 from mice. *Cell Calcium* 37: 267.
23. Woodbury, J. W. 1972. Eyring rate theory model of the current-voltage relationships of ion channels in excitable membranes. In *Chemical Dynamics: Paper in Honor of Henry Eyring*, ed. J. Hirchenfelder and D. Henderson, 601–607. New York: John Wiley.
24. Nonner, W., and B. Eisenberg. 1998. Ion permeation and glutamate residues linked by Poisson-Nernst-Planck theory in L-type calcium channels. *Biophysical Journal* 75: 1287.
25. Nonner, W., D. P. Chen, and B. Eisenberg. 1998. Anomalous mole fraction effect, electrostatics, and binding in ionic channels. *Biophysical Journal* 74: 2327.
26. Corry, B., T. W. Allen, S. Kuyucak, and S. H. Chung. 2000. A model of calcium channels. *Biochimica et Biophysica Acta* 1509: 1.
27. Corry, B., T. W. Allen, S. Kuyucak, and S. H. Chung. 2001. Mechanisms of permeation and selectivity in calcium channels. *Biophysics Journal* 80: 195.
28. Barreiro, G., C. R. Guimaraes, and R. B. de Alencastro. 2002. A molecular dynamics study of an L-type calcium channel model. *Protein Engineering* 15: 109.
29. Voets, T., and B. Nilius. 2003. The pore of TRP channels: trivial or neglected? *Cell Calcium* 33: 299.
30. Owsianik, G., K. Talavera, T. Voets, and B. Nilius. 2006. Permeation and selectivity of TRP channels. *Annual Review of Physiology* 68: 685.
31. Neher, E. 1992. Correction for liquid junction potentials in patch clamp experiments. *Methods in Enzymology* 207: 123.
32. Barry, P. H. 1994. JPCalc, a software package for calculating liquid junction potential corrections in patch-clamp, intracellular, epithelial and bilayer measurements and for correcting junction potential measurements. *Journal of Neuroscience Methods* 51: 107.
33. Woodhull, A. M. 1973. Ionic blockage of sodium channels in nerve. *Journal of General Physiology* 61: 687.
34. Vennekens, R., J. Prenen, J. G. Hoenderop et al. 2001. Pore properties and ionic block of the rabbit epithelial calcium channel expressed in HEK 293 cells. *Journal of Physiology* 530: 183.
35. Voets, T., A. Janssens, J. Prenen, G. Droogmans, and B. Nilius. 2003. Mg^{2+}-dependent gating and strong inward rectification of the cation channel TRPV6. *Journal of General Physiology* 121: 245.
36. Schneggenburger, R., Z. Zhou, A. Konnerth, and E. Neher. 1993. Fractional contribution of calcium to the cation current through glutamate receptor channels. *Neuron* 11: 133.
37. Neher, E. 1995. The use of fura-2 for estimating Ca buffers and Ca fluxes. *Neuropharmacology* 34: 1423.
38. Samways, D. S., B. S. Khakh, and T. M. Egan. 2008. Tunable calcium current through TRPV1 receptor channels. *Journal of Biological Chemistry* 283: 31274.
39. Robinson, R. A., and R. H. Stokes. 1959. *Electrolyte Solution*. London: Butterworth.
40. Bormann, J., O. P. Hamill, and B. Sakmann. 1987. Mechanism of anion permeation through channels gated by glycine and gamma-aminobutyric acid in mouse cultured spinal neurones. *Journal of Physiology* 385: 243.

41. Karlin, A., and M. H. Akabas. 1998. Substituted-cysteine accessibility method. *Methods in Enzymology* 293: 123.

42. Dodier, Y., U. Banderali, H. Klein et al. 2004. Outer pore topology of the ECaC-TRPV5 channel by cysteine scan mutagenesis. *Journal of Biological Chemistry* 279: 6853.

43. Dodier, Y., F. Dionne, A. Raybaud, R. Sauve, and L. Parent. 2007. Topology of the selectivity filter of a TRPV channel: rapid accessibility of contiguous residues from the external medium. *American Journal of Physiology. Cell Physiology* 293: C1962.

44. Gunthorpe, M. J., M. H. Harries, R. K. Prinjha, J. B. Davis, and A. Randall. 2000. Voltage- and time-dependent properties of the recombinant rat vanilloid receptor (rVR1). *Journal of Physiology* 525(Part 3): 747.

45. Nilius, B., J. Prenen, G. Droogmans et al. 2003. Voltage dependence of the Ca^{2+}-activated cation channel TRPM4. *Journal of Biological Chemistry* 278: 30813.

46. Hofmann, T., V. Chubanov, T. Gudermann, and C. Montell. 2003. TRPM5 is a voltage-modulated and Ca^{2+}-activated monovalent selective cation channel. *Current Biology* 13: 1153.

47. Voets, T., G. Droogmans, U. Wissenbach et al. 2004. The principle of temperature-dependent gating in cold- and heat-sensitive TRP channels. *Nature* 430: 748.

48. Brauchi, S., P. Orio, and R. Latorre. 2004. Clues to understanding cold sensation: thermodynamics and electrophysiological analysis of the cold receptor TRPM8. *Proceedings of the National Academy of Sciences of the United States of America* 101: 15494.

49. Chung, M. K., H. Lee, A. Mizuno, M. Suzuki, and M. J. Caterina. 2004. 2-aminoethoxy-diphenyl borate activates and sensitizes the heat-gated ion channel TRPV3. *Journal of Neuroscience* 24: 5177.

50. Xu, H., I. S. Ramsey, S. A. Kotecha et al. 2002. TRPV3 is a calcium-permeable temperature-sensitive cation channel. *Nature* 418: 181.

51. Grimm, C., R. Kraft, G. Schultz, and C. Harteneck. 2005. Activation of the melastatin-related cation channel TRPM3 by D-erythro-sphingosine. *Molecular Pharmacology* 67: 798.

52. Zurborg, S., B. Yurgionas, J. A. Jira, O. Caspani, and P. A. Heppenstall. 2007. Direct activation of the ion channel TRPA1 by Ca^{2+}. *Nature Neuroscience* 10: 277.

53. Shimizu, T., A. Janssens, T. Voets, and B. Nilius. 2009. Regulation of the murine TRPP3 channel by voltage, pH, and changes in cell volume. *Pflügers Archiv* 457: 795.

54. Nilius, B., J. Prenen, T. Voets, and G. Droogmans. 2004. Intracellular nucleotides and polyamines inhibit the Ca^{2+}-activated cation channel TRPM4b. *Pflügers Archiv* 448: 70.

55. Nilius, B., K. Talavera, G. Owsianik et al. 2005. Gating of TRP channels: a voltage connection? *Journal of Physiology* 567: 35.

56. Matta, J. A., and G. P. Ahern. 2007. Voltage is a partial activator of rat thermosensitive TRP channels. *Journal of Physiology* 585: 469.

57. Talavera, K., A. Janssens, N. Klugbauer, G. Droogmans, and B. Nilius. 2003. Extracellular Ca^{2+} modulates the effects of protons on gating and conduction properties of the T-type Ca^{2+} channel α_{1G} (CaV3.1). *Journal of General Physiology* 121: 511.

58. Talavera, K., A. Janssens, N. Klugbauer, G. Droogmans, and B. Nilius. 2003. Pore structure influences gating properties of the T-type Ca^{2+} channel α_{1G}. *Journal of General Physiology* 121: 529.

59. Meseguer, V., Y. Karashima, K. Talavera et al. 2008. Transient receptor potential channels in sensory neurons are targets of the antimycotic agent clotrimazole. *Journal of Neuroscience* 28: 576.

60. Almers, W. 1978. Gating currents and charge movements in excitable membranes. *Reviews of Physiology, Biochemistry, and Pharmacology* 82: 96.

61. Sigworth, F. J. 1994. Voltage gating of ion channels. *Quarterly Review of Biophysics* 27: 1.

62. Voets, T., G. Owsianik, A. Janssens, K. Talavera, and B. Nilius. 2007. TRPM8 voltage sensor mutants reveal a mechanism for integrating thermal and chemical stimuli. *Nature Chemical Biology* 3: 174.

63. Huang, J., X. Zhang, and P. A. McNaughton. 2006. Modulation of temperature-sensitive TRP channels. *Seminars in Cell and Developmental Biology* 17: 638.

64. Caterina, M. J. 2007. Transient receptor potential ion channels as participants in thermosensation and thermoregulation. *American Journal of Physiology. Regulatory, Integrative, and Comparative Physiology* 292: R64.

65. Talavera, K., T. Voets, and B. Nilius. 2008. Mechanisms of thermosensation in TRP channels. In *Sensing with Ion Channels*, ed. B. Martinac, 101–20. Berlin: Springer Verlag.

66. Talavera, K., B. Nilius, and T. Voets. 2008. Neuronal TRP channels: thermometers, pathfinders and life-savers. *Trends in Neuroscience* 31: 287.

67. Bessac, B. F., and S. E. Jordt. 2008. Breathtaking TRP channels: TRPA1 and TRPV1 in airway chemosensation and reflex control. *Physiology* 23: 360.

68. Brooks, S. M. 2008. Irritant-induced chronic cough: irritant-induced TRPpathy. *Lung* 186(Suppl 1): S88.

69. Damann, N., T. Voets, and B. Nilius. 2008. TRPs in our senses. *Current Biology* 18: R880.

70. Nilius, B., J. Prenen, A. Janssens, T. Voets, and G. Droogmans. 2004. Decavanadate modulates gating of TRPM4 cation channels. *Journal of Physiology* 560: 753.

71. Karashima, Y., N. Damann, J. Prenen et al. 2007. Bimodal action of menthol on the transient receptor potential channel TRPA1. *Journal of Neuroscience* 27: 9874.

72. Karashima, Y., K. Talavera, W. Everaerts et al. 2009. TRPA1 acts as a cold sensor in vitro and in vivo. *Proceedings of the National Academy of Science of the United States of America* 106: 1273.

73. Brauchi, S., G. Orta, M. Salazar, E. Rosenmann, and R. Latorre. 2006. A hot-sensing cold receptor: C-terminal domain determines thermosensation in transient receptor potential channels. *Journal of Neuroscience* 26: 4835.

74. Brauchi, S., G. Orta, C. Mascayano et al. 2007. Dissection of the components for PIP2 activation and thermosensation in TRP channels. *Proceedings of the National Academy of Science of the United States of America* 104: 10246.

75. Latorre, R., S. Brauchi, G. Orta, C. Zaelzer, and G. Vargas. 2007. ThermoTRP channels as modular proteins with allosteric gating. *Cell Calcium* 42: 427.

76. Latorre, R., C. Zaelzer, and S. Brauchi. 2009. Structure-functional intimacies of transient receptor potential channels. *Quarterly Review of Biophysics* 42: 201.

77. Nilius, B., F. Mahieu, J. Prenen et al. 2006. The Ca^{2+}-activated cation channel TRPM4 is regulated by phosphatidylinositol 4,5-biphosphate. *EMBO Journal* 25: 467.

78. Premkumar, L. S., S. Agarwal, and D. Steffen. 2002. Single-channel properties of native and cloned rat vanilloid receptors. *Journal of Physiology* 545: 107.

79. Hui, K., B. Liu, and F. Qin. 2003. Capsaicin activation of the pain receptor, VR1: multiple open states from both partial and full binding. *Biophysics Journal* 84: 2957.

80. Liu, B., K. Hui, and F. Qin. 2003. Thermodynamics of heat activation of single capsaicin ion channels VR1. *Biophysics Journal* 85: 2988.

81. Zhu, M. X. 2007. Understanding the role of voltage gating of polymodal TRP channels. *Journal of Physiology* 585: 321.

82. Malkia, A., R. Madrid, V. Meseguer et al. 2007. Bidirectional shifts of TRPM8 channel gating by temperature and chemical agents modulate the cold sensitivity of mammalian thermoreceptors. *Journal of Physiology* 581: 155.

83. Grahame, D. C. 1947. The electrical double layer and the theory of electrocapillarity. *Chemical Reviews* 41: 441.

84. Gilbert, D. L., and G. Ehrenstein. 1969. Effect of divalent cations on potassium conductance of squid axons: determination of surface charge. *Biophysics Journal* 9: 447.
85. Mahieu, F., A. Janssens, M. Gees et al. 2010. Modulation of the cold-activated cation channel TRPM8 by surface charge screening. *Journal of Physiology* 588: 315.

12 High-Throughput Approaches to Studying Mechanisms of TRP Channel Activation

Hongzhen Hu, Michael Bandell,
Jorg Grandl, and Matt Petrus

CONTENTS

12.1 INTRODUCTION

12.1.1 "ThermoTRPs"

Transient receptor potential (TRP) channels are calcium-permeable nonselective cation channels with six transmembrane (TM) domains and a putative pore loop between TM5 and TM6. About 28 mammalian TRP channels have been identified so far, with different numbers of splicing variants for each channel gene. TRP channels have been classified into six different subgroups, including TRPV (1–6), TRPM (1–8), TRPC (1–7), TRPA1, TRPP (1–3), and TRPML (1–3), according to their sequence similarities.[1] In general, TRP channels are involved in calcium handling (e.g., intracellular calcium mobilization and calcium reabsorption) and a broad range of sensory modalities, including pain, temperature, taste, etc.[2,3] TRP channelopathies are part of important mechanisms in a variety of diseases such as neurodegenerative disorders, diabetes mellitus, inflammatory bowel diseases, epilepsy, cancer, etc.[4–10] Several members of the TRP family, TRPV1-4, TRPM8, and TRPA1, also called "ThermoTRPs," are involved in the detection of temperature changes, thus acting as the molecular thermometers of our body.[11–14] In addition, they are also polymodal nociceptors that integrate painful stimuli such as noxious temperatures and chemical insults. For example, TRPV1 channel mediates thermal hyperalgesia and pain induced by capsaicin and acid.[12,14–16] TRPA1 is a nociceptor that integrates many noxious environmental stimuli including oxidants and electrophilic agents.[14,17,18] Recently, gene deletion animals have been created to study the role of TRP channels in pain and nociception, and involvement of TRPV1, TRPV3, TRPV4, and TRPA1 in nociception has been confirmed.[19–24] The physiology and pathophysiology of TRP channels have been covered in some excellent recent reviews[13–15,25–29] and will not be discussed in this chapter.

12.1.2 Mutagenesis Approaches to Studying TRP Channel Function

Mutations in mammalian genomes are part of important disease mechanisms. Organisms surviving naturally occurring mutations are often associated with phenotypes that result from gain of function or loss of function of the mutated proteins. Mutations in TRP channels also have profound effects and cause diseases in mammals. For instance, point mutations of TRPV3 at Gly573 to Ser or Cys have recently been linked to autosomal dominant hairless phenotypes and spontaneous dermatitis in rodents, implicating an important role for TRPV3 in alopecia and skin diseases. Furthermore, the G573S mutation is a cause of pruritus and/or dermatitis associated with scratching when it is overexpressed in transgenic mice at the TRPV3 locus.[30–32] The gain of function phenotype has also been studied in heterologous systems where G573S and G573C mutations of murine TRPV3 are found to be constitutively active.[33]

G573S and G573C substitutions render the TRPV3 channel spontaneously active under normal physiological conditions, which in turn alters ion homeostasis and membrane potentials of skin keratinocytes, leading to hair loss and dermatitis-like skin diseases. These studies establish the TRPV3 channel as a potential target for therapeutic intervention in skin inflammation and pruritus, as well as cosmetic practices.[34]

Genetic screening has identified multiple human diseases associated with TRPV4 mutations. Point mutations I331F, D333G, R594H, R616Q, A716S, and P799L are associated with spondylometaphyseal dysplasia (SPSMA), Kozlowski-type, and Metatropic dysplasia. Point mutations R269C, R269H, R315W, and R316C cause Charcot–Marie–Tooth disease type 2C (CMT2C). Both SPSMA and CMT2C are genetically heterogeneous disorders caused by degeneration of peripheral nerves. Individuals associated with SPSMA show loss and progressive weakness of scapular and peroneal muscle tissue, bone abnormalities, and laryngeal palsy. CMT2C, the most common inherited neurological disease, leads to progressive weakness of distal limbs, vocal cords, diaphragm, and intercostal and laryngeal muscles. R616Q and V620I substitutions of TRPV4 lead to a gain-of-function phenotype associated with constitutive activity and cause brachyolmia in human carriers.[35] On the other hand, a loss-of-function nonsynonymous polymorphism (P19S) in TRPV4 is associated with human hyponatremia.[36–41]

Although spontaneous mutations have shown that specific amino acids in TRP channels are important for their normal function, systemic mutagenesis studies will be more informative of the overall picture of the ion channel by mapping important amino acids and specific protein domains. Therefore, structure–function mutagenesis studies are critical to understanding the underlying activation mechanisms of TRP channels, especially in the absence of the crystal structure of a complete TRP channel.[42] Site-directed and random mutagenesis approaches are commonly used methods. Site-directed mutagenesis usually involves knowledge about sequence and/or structure of the protein to be mutated. It is best suited for testing model-guided hypotheses by analyzing functions of amino acid residues at highly conserved positions. Site-directed mutagenesis has been extensively used to map important amino acid residues involved in ligand-channel interactions of TRP channels. For instance, it has been used to identify putative activation sites for capsaicin, protons, ATP, and protein kinases in TRPV1.[39,43–55] Site-directed mutagenesis approaches have also been applied to identify amino acid residues involved in chemical modification of TRPA1 channels.[56–64] Furthermore, side-directed mutagenesis has been used to reveal elements involved in TRP channel gating and trafficking.[65–70] Most of the above studies are based on sequence homology among TRP channels and structure homology of the pore loop between TRP channels and voltage-gated potassium channels, for some of which the crystal structures have been resolved and a large body of knowledge of channel gating has been accumulated.[71–76] Site-directed mutagenesis generally deals with single point mutations and possibly small insertions or deletions. However, it is also possible to swap large domains among different TRP channel proteins with high structure homology. The disadvantages of the site-directed mutagenesis are that the number of mutants is limited, the process of generating and analyzing individual mutant channels is laborious, and the design of these mutations is often driven by preconceptions focusing on particular regions of the ion channels.

Taking a high-throughput structure function approach

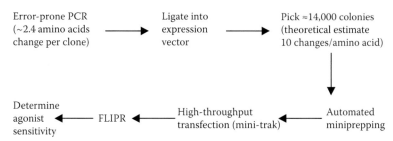

FIGURE 12.1 Taking a high-throughput structure–function approach. The schematic diagram illustrates the steps involved in the high-throughput approach to studying the TRP channel activation mechanism. Error-prone PCR is used to generate mutants carrying random mutations. The PCR products are ligated into an expression vector and transformed. About 14,000 colonies are picked, the followed by automated miniprep and high-throughput transfection. FLIPR calcium assay is used to determine agonist sensitivity of the wild-type and mutant channels.

These disadvantages can be overcome by random mutagenesis that requires less information about relevant positions. The beauty of random mutagenesis is that it is an unbiased approach and poised to understand the full spectrum of molecular processes that govern ion channel activity. Random mutagenesis often requires generation of mutant libraries and efficient screening systems. It is possible for a saturating mutagenesis analysis if the size of the protein is small. The disadvantage of random mutagenesis is that it usually demands highly advanced automation to accomplish mutant library generation and high-throughput functional assays. Taking advantage of high-throughput methods, Bandell et al.[77] have combined random mutagenesis and FLIPR calcium assay to study the menthol interaction sites in TRPM8 (Figure 12.1). This novel approach has also been used to study the activation mechanisms of TRPV1 and TRPV3 channels[78–80] and will be the focus of this chapter.

12.2 HIGH-THROUGHPUT MUTAGENESIS APPROACH

12.2.1 Mutant Library Construction

12.2.1.1 Materials

LB agar miller 3.5 L (35 g/L); Terrific Broth media 6 L (47.6 g/L) + glycerol (8 mL/L); round-tip toothpicks; Corning square bioassay dishes, 500 cm² surface area (need about 12–13 plates for picking up 4000 colonies); Costar aluminum sealing tape; Qiagen AirPore microporous tape sheets; VWR plate vortexer; RoboPrep 2500 (MWG); MWG 4204 liquid handling robot; BioTek PowerWave XS 96-well plate spectrometers; NucleoSpin Robot-96 Plasmid Core Kit (Macherey-Nagel); Costar 96-well UV plates; MJ Research PTC-200 Gradient Thermal Cycler; Diversify PCR Random Mutagenesis Kit (Clonetech); Minlute gel purification kit (Qiagen); LigaFast™ Rapid DNA Ligation System (Promega); One Shot® MAX Efficiency® DH5α™ (Invitrogen); Qfill Plate Dispenser (Genetix); Qpix (Genetix); Black/Clear, tissue culture treated sterile, poly-D-lysine coated 384 plates (Greiner Bio-one);

Fugene 6.0 (Roche); Dulbecco's modified Eagle's medium (DMEM) (Hyclone); Opti-MEM media (Invitrogen); 0.05% Trypsin/EDTA solution (Hyclone); Dulbecco's Phosphate Buffered Saline (DPBS) (Hyclone).

12.2.1.2 Preparing Agar Plates

Autoclave LB agar miller and fill each of the Corning square bioassay dishes with about 250 mL LB agar (cool down to warm temperature before adding antibiotics [100 µg/mL, Carbenicillin]). Use gas flame to remove bubbles formed during the process. Lift cover and allow to sit for 2 hours to dry the LB agar, or dry the plates thoroughly (1.5 hours) in a biosafety hood. Then cover the plate with plastic film and store at 4°C.

12.2.1.3 Error-Prone PCR, Ligation, and Transformation

We have randomly mutagenized mouse TRPM8, TRPV3, and TRPV1 fragments using a Diversify® PCR Random Mutagenesis Kit as per manufacturer's instructions. Two approaches have been used to generate mutant clones. The first approach involves error-prone PCR on the entire gene using forward and reverse primers to sequences outside the open reading frame (with a restriction site at each end of the PCR product). Reaction conditions are chosen according to the manufacturer's protocol so as to give a mutation frequency of 2.0 mutations per 1000 base pairs. The PCR mixture contains: water 40 µL, 10 × titanium Taq buffer 5 µL, dGTP 1 µL, 50 × diversify dNTP mix, primer mix (10 µM) 1 µL, template DNA (1 ng/ µL) 1 µl, and titanium Taq polymerase 1 µL. The PCR protocol is: 94°C, 30 s → 25 cycles (94°C, 30 s → 68°C, 3 min) → 68°C, 1 min → 4°C. We used 4–8 PCR runs and pooled them together to ensure enough materials. The pooled PCR products are column purified using the Qiagen PCR cleanup kit. The PCR product is then cut with respective restriction enzymes and gel purified using the Qiagen Minielute gel purification kit. This fragment is then ligated back into pcDNA5-FRT mammalian expression vector using specific restriction sites located upstream of the start and downstream of the stop codon with the Promega LigaFast™ Rapid DNA Ligation System (incubate 1 hour at RT), following manufacturer's instructions on the amounts of vector and insert to use. The ligation mix (3 µL) is added to one vial (50 µL) of One Shot MAX Efficiency DH5α (Invitrogen) for transformation, with 30 µL of the transformation mix used to plate. However, under this condition, the mutation rate is approximately 5 to 6 base changes (~4 amino acid substitutions) per clone. Therefore, the second approach can be used to decrease the mutation rate. Here, instead of cloning the full-length PCR product, the product is digested at a restriction site in the middle of the gene (NsiI for TRPM8 and AvrII for TRPV3), thereby generating two fragments of approximately equal size representing the two halves of the gene. These two fragments are each ligated individually into pcDNA5 containing the wild-type TRPM8 or TRPV3 gene, from which the corresponding fragment has been removed. This approach yields mutations at a rate of approximately two amino acid substitutions per clone.

12.2.1.4 Colony Picking

Antibiotics are added to autoclaved Terrific Broth solution, and 1.5–1.7 mL of the Terrific Broth + antibiotics was added to each well of 96-well deep bottom plates using Q-fill (included in the NucleoSpin Robot-96 Plasmid Core Kit). The plates are covered with Costar aluminum sealing tape to make sure each well is separated and clean. In our TRPM8, TRPV3, and TRPV1 libraries, we have picked about 14,000 colonies with Qpix (Genetix) or by hand with round-tip toothpicks. Make sure to write the plate number on the side of the plate in case plate covers get lost. Use Qiagen AirPore microporous tape sheets to seal the plates to avoid crossover to neighboring wells. Culture the plates in a shaker overnight (37°C, 270 rpm rotation).

12.2.1.5   Glycerol Backups

After about 24 hours, the plates are removed from the shaker and 70 μL of Terrific Broth culture mix are added to each well of a 96-well plate preloaded with 30 μL autoclaved 50% glycerol. The glycerol backup plates are sealed with Costar aluminum sealing tape and stored at −80°C. These backup plates are critical for follow-up studies or rescreening if needed.

12.2.1.6   High-Throughput Minipreps

The Terrific Broth cultures are centrifuged at 4°C, 3200 rpm, 5 min. Pour out Terrific Broth and tap on a coversheet to remove residual Terrific Broth before adding 250 μL A1 buffer (with RNase added, stored at 4°C) by Q-fill. Seal the plates with Costar aluminum sealing tape and resuspend the pellet by shaking for 2–5 min at maximum speed on a VWR plate vortexer. Store the plates at 4°C before miniprep. The minipreps are processed on a MWG Biotech RoboPrep 2500s, and plasmid DNA was eluted into Costar 96-well UV plates.

12.2.1.7   DNA Normalization

Because the well-to-well differences of DNA concentrations in the minipreps are large and DNA yields in some wells are typically too high to permit direct spotting on to 384-well assay plates, DNA concentrations of the raw minipreps are then normalized using a two-step dilution protocol, using the integrated MWG 4204 liquid handling robot. The diluent used is 5 mM Tris-HCl, pH 8.5. The integrated normalization suite is equipped with integrated BioTek PowerWave XS 96-well plate spectrometers that enable importation of UV data to the liquid handlers, which are equipped with database software that tracks DNA concentrations and calculates normalization volumes. Typically, cDNA collections are normalized to 40 ng/μL with a final volume of 80 μL. Round-bottom 96-well plates are employed as destination plates during normalization, which enables the spotting robot to maximally spot the DNA. DNA in the UV plates at lower concentrations will not be transferred to the round-bottom normalization plate, i.e., these wells contain no DNA in the normalized plates. Normalized DNA can be stored for extended periods in foiled plates at −20°C prior to spotting on to assay plates.

12.2.2   DNA Transfection

DNA transfection efficiency depends on the amount of DNA spotted per well, the quality of the minipreps, and the desired ratio of DNA to transfection reagents, which depends heavily on the type of transfection reagents used. The liposome-based transfection reagents such as Fugene 6.0 (Roche), Lipofectamine series (Invitrogen), and TransIT®-293 (Mirus) can all be used for transfection of HEK293 cells. For cDNA overexpression assays, Fugene 6.0 is the optimal choice of transfection reagent. DNA is spotted into the assay plates using a Perkin–Elmer MiniTrak fitted with disposable tips. Each well of the 384-well plate usually receives 75 ng DNA. One 96-well round-bottomed plate containing normalized plasmid DNA is placed on the desk, and a small aliquot of DNA (1.875 µL, 40 ng/µL) from each well is spotted into each 384-well assay plate as quadruplicates. Each spotted 384-well plate is covered with Costar aluminum sealing tape. Positive (wild-type TRP channels) and negative (mock transfection) assay controls are included on each plate. The 384-well assay plates can then be stored at −20° or −80°C for future use. A Fugene 6.0-DNA ratio of 1.76:1 for HEK293 cells is optimal for the 384-plate assay in our hands. For each plate, about 10 mL Fugene 6.0/Opti-MEM mix, i.e., 101.5 µL Fugene 6.0 in 10 mL Opti-MEM, is prepared, of which 20 µL is dispensed into each well of 384-well plates, prespotted with a mutant DNA from the library using µ-Fill (Bio-Tek). It is critical to prime µ-Fill with 25 mL 100% ethanol/50 mL PBS followed with 25 mL Fugene 6.0/Opti-MEM mix to make sure the equipment is sterile and clean. Incubate at room temperature for about 20–40 min. In the meantime, trypsinize HEK293T cells and dilute them in culture media containing 20% of the fetal bovine serum. Add 20 µL of HEK293T cells at a density of about 150,000 to 300,000 cells/mL to each well.

12.3 HIGH-THROUGHPUT FUNCTIONAL ASSAYS

12.3.1 FLIPR CALCIUM ASSAY

FLIPR refers to Fluorometric Imaging Plate Readers. The latest version is the FLIPR[TETRA] from Molecular Devices. FLIPR has been a popular tool in biotech and pharmaceutical companies for high-throughput screening and lead optimization. It has capabilities to include 96, 384, and 1536 simultaneous liquid transfers and multi-wavelength kinetic reading. We have used FLIPR calcium assay to screen interesting TRP channel mutants by taking advantage of the fact that most of the TRP channels, especially TRPV channels, are highly calcium permeable.

12.3.1.1 Materials

FLIPR[TETRA] (Molecular Devices) (we have also used FLIPR384 in primary screening); Fluo-3 AM (Invitrogen); pluronic F-127 (Invitrogen); probenecid (Sigma); DMSO (Sigma); microplate washers ELx405 (Biotek); uFill (Biotek); MiniTrak (Perkin–Elmer); compound plates (384-well format) (Genomics Institute of the Novartis Research Foundation); 10× Hank's Buffered Salt Solution (HBSS) (Hyclone); HEPES (1 M) (Hyclone); consumables from Fisher Scientific.

12.3.1.2 Buffer Preparation and Dye Loading

Two days after transfection, cells are washed three times with assay buffer (10 mM HEPES buffered HBSS, containing 2.5 mM probenecid, pH 7.0) using a microplate washer ELx405 (Bio-tek), set to leave 25 µL buffer per well. Fluo-3 imaging has been used to study the spatial dynamics of many elementary processes in Ca^{2+} signaling. A stock solution of Fluo-3 AM is made in DMSO at 2 mM (this stock can be stored at $-20°C$ for extended time). The stock solution is further diluted at a 1:1 ratio with pluronic F-127 (20% in DMSO) to 1 mM. The working solution of Fluo-3 AM is made in HEPES-buffered HBSS (containing 5 mM probenecid) at 8 µM, and 25 µL is added to each well (4 µM final); then, the plates are incubated for 1 hour in a 37°C CO_2 incubator. It should be noted that probenecid has a nonspecific effect on some TRP channels, including TRPV2, and special attention should be paid with data interpretations.[81] Cells are then washed three times with the assay buffer using a microplate washer ELx405 (Bio-tek), set to leave 25 µL buffer per well.

12.3.1.3   Preparation of Compound Plates

It is critical to decide which compounds are to be screened against the mutant library. These should include not only the compounds whose interaction sites are pursued in the research plans but also positive and negative controls. For instance, we used cold response as a negative control for TRPM8 mutagenesis screening of menthol interaction sites and camphor and heat as negative controls for TRPV3 mutagenesis screening of the 2APB interaction site(s).[77,78,80] For positive controls, ATP and ionomycin are commonly used in calcium assays, mainly to confirm the viability of the cells and to normalize responses of the test compounds. The concentrations of agonists used for primary screening have to be chosen carefully. The best agonist concentration should fit the following categories: (1) low or no background signals, i.e., the agonist does not evoke endogenous calcium response in the cell lines used (such as HEK293T cells) at the concentration employed for screening; (2) the agonist-evoked calcium response should be large enough for statistical analysis. The dilution series of the compounds are made on 384-well plates using a robotic liquid handling system (MiniTrak, Perkin–Elmer). Dilutions are made with the same buffer used for dye loading. "Boat" plate configuration is another great feature of FLIPR in case the whole library is to be screened against a specific test compound, such as menthol for TRPM8. In this case, a boat that holds a large volume of the test compound can be used to replace the compound plates, which will save time and decrease expense because the pipette tips can be reused. The lid of a FLIPR tip box fits well with the dimension of the head of the pipettor and can be used as a "boat." Compound plates should be made on the same day FLIPR experiments are performed. For compounds whose stock solutions are dissolved in DMSO, the final DMSO concentration in each well should be lower than 1%, since higher DMSO concentrations have a nonspecific effect on HEK293 cells. FLIPR liquid handling has some dead volume, and usually, an extra volume (~10 µL more than the desired amount) should be added to the compound plate in order to avoid shortage of the compound solution.

12.3.1.4   Temperature Activation Using FLIPR

In order to test activation of thermoTRPs by temperature, two approaches have proven to be useful: (1) Pipetting pre-chilled buffer into the test well.[77] This method requires adjusting the temperature of a buffer bath so that after mixing, the final temperature in the test well will be changed and cause activation of thermoTRPs. Buffer addition from a boat is best suited owing to the large volume. Temperature adjustment can be achieved by using ice (cold) or a microwave (heat), and the precise temperatures after mixing need to be monitored with a thermometer. Although this method is easy to use, the precision and temperature range are limited. (2) A custom-made device consisting of a metal 384-pin plate that is cooled or heated by electric Peltier elements can be used to rapidly alter the temperature of a 384-well plate over a wide range of temperatures. Using this method, we have achieved temperature changes of 2°C/s and a precision of <0.5°C from 4° to 70°C.[78] In addition, the temperature profile is user defined throughout the assay, and thus, a complete temperature-activation curve can be measured.[79]

12.3.1.5 Reading a FLIPR Plate

FLIPR[TETRA] utilizes LED modules with distinct ranges to excite cellular fluorescence. Filtered light is captured using a CCD camera and displayed in real time by the software. These features are controlled by the drag-and-drop user software that displays data in real-time kinetics. To operate FLIPR, turn on first the chiller, then FLIPR[TETRA] and the computer. Wait for about 10 min for FLIPR[TETRA] to autocheck and the camera temperature to cool down. Then open ScreenWorks 3, which is the software used to acquire data. Here we will discuss only the basics of the software. Readers are referred to the user's manual of FLIPR[TETRA] for more details. ScreenWorks 3 has four basic tasks to be addressed before reading a plate: settings, analysis, transfer fluid, and read with transfer fluid. The settings include setup read mode and assign plate position. Setup read mode defines excitation wavelength and emission wavelengths; for Fluo-3, the excitation wavelength should be set between 470 and 495 nm. The emission wavelength should be set between 517 and 575 nm. The setup read mode also sets camera gain, exposure time, excitation intensity, and gate opening. These parameters can be set using default values or modified according to signal test result. Assign plate position basically defines the positions of FLIPR pipette tips, compound plates, and read plate. Read plate is preset at plate position 3, and most of the time, pipette tip box is placed at plate position 1. Two compound plates can be placed at plate positions 2 and 4. For an assignment like this, it is possible to do two compound additions from two separate compound plates sequentially. TETRAcycler, the optional internal plate handler on FLIPR[TETRA], exchanges microplates between the system plate stage and an external third-party plate handler. TETRAcycler minimizes downtime between experiments, thus increasing throughput, by exchanging plates and tips while the experiment is in process. In the analysis window, data are displayed as relative fluorescence unit (RFU) over time. Two graphs are displayed: the main graph illustrates calcium responses in each well of the 384-well plates, while the graph detail window displays the overlapping curves from wells selected from the main graph. The windows will autoscale according to the magnitude of the minimal and maximal responses. Transfer fluid is a very important setting for fluid handling. It defines the source plate position, source plate format, volume to be transferred, transfer speed, and how low the pipette tips will go. It also controls the hold volume in the tips and tip up speed after transfer. Read with transfer fluid allows simultaneous data acquisition before, during, and after compound addition. It defines the read time interval, baseline duration (number of reads before dispense), and compound response (number of reads after dispense).

There is a second interval that can add more reads if long reads are required; however, the total number of reads cannot exceed 800 in two intervals. A signal test should be run in a couple of cell plates to determine the dye loading and background signal intensity. After acquiring baseline calcium levels, 25 μL assay buffer containing known concentrations of agonists is added to each well, yielding a 50 μL total volume and half the concentration from the original compound plate. We routinely measure fluorescence for 2 min after compound addition with about 20 s baseline reading. At the end of the experiments, close ScreenWorks, shut down the computer, and turn off FLIPR and the chiller.

12.3.1.6   FLIPR Data Analysis

Screening data are exported from ScreenWork3. ScreenWork3 has the capability for basic statistical analysis such as concentration–response curves, EC_{50} values, Z factors, etc. Data can also be exported using options such as "Time Sequence," "Statistics," and "Group Statistics" (see ScreenWork3 manual for other applications); here we discuss only "Statistics" commonly used to analyze screen data. In Statistics, baseline subtraction (Max–Min) is used to calculate the net response after a compound addition, where Max is the maximal response and Min is the baseline. However, in case of mutants associated with constitutive activity, caution should be excised when doing the baseline subtraction because the net response often is truncated by a high baseline response. Data are exported using Excel format as a default. Usually, data are averaged if the same experiments are executed in multiple wells. Averaged data can be analyzed with Excel or commercially available software such as IGORPro (Wavemetrics), Sigmaplot, Origin, etc. Time courses for compound-induced activation can be fitted for each single well by a mono-exponential function to obtain rate values. Histograms of maximal activations, their ratios, and rates of signal increases are produced for all mutants, wild-type TRP channels, and pcDNA controls. Histograms of controls are fitted with Gaussian distributions to obtain averages and standard deviations, which are used to calculate cut-off values for hit selection. Cut-off values for hit confirmation are obtained from calculating average values and standard deviations of pooled data from controls from all plates. For concentration–response curves, maximal responses are calculated after baseline subtraction. Averages and standard deviations are calculated from at least four wells per concentration; responses are normalized to unity and fitted by a Hill equation. The cDNA clones of the hits are repicked and transformed to DH5α, and single colonies are picked for DNA isolation. Another alternative is to use the glycerol backup to grow more bacterial cultures of the clones.

12.3.2   Other Alternative High-Throughput Approaches

12.3.2.1   Yeast Screening

Budding yeast allows convenient and efficient genetic and molecular manipulations. Zhou et al. have used a random mutagenesis approach and identified gain-of-function mutants at the C terminus of the predicted sixth transmembrane helix of yeast TRPY1, which is a mechanosensitive TRP channel.[82–84] These particular mutants respond more strongly to mild osmotic upshocks (osmolarity increases) and remain responsive to membrane stretch force and to Ca^{2+}, indicating primary defects in the gate region but not in the sensing of gating principles (Figure 12.2). Another advantage of using yeast as a screening tool is that its vacuoles released by hypotonic swelling can be used for patch-clamp recordings (whole-vacuole mode or excised cytoplasmic-side-out mode) in which membrane stretch evokes current influx. The electrophysiological recordings on yeast vacuoles provide an excellent approach to follow-up the hits from forward mutagenesis screens.

Yeast has also been used as an alternative to cultured mammalian cells to study ion channel structure and function. Indeed, rat TRPV1 and rat TRPV4 channels have been successfully expressed in yeast.[85] Yeast cells are first transformed with a plasmid that produces aequorin in order to monitor cytoplasmic Ca^{2+} by luminescence. They are then transformed with a CEN plasmid bearing the rat TRPV1 or TRPV4 gene. The transformed yeast culture is monitored with a luminometer and hypotonically shocked by dilution. By using the yeast expression system, the authors were able to show that hypotonic stress, but not heat, activates TRPV4 independent of polyunsaturated fatty acids. Myers et al. also generated TRPV1 mutant libraries by using error-prone PCR. The mutant cDNA clones were then transformed into yeast. The investigators identified amino acid residues at the pore area that affect the TRPV1 channel gating by following up with mutants associated with constitutive activity and increased cobalt uptake.[85–87]

12.3.2.2 $^{45}Ca^{2+}$ Assay

This is another high-throughput functional assay for TRP channels. Basically, cells are transfected with TRP channel cDNAs and plated in a 96-well plate at a density of about 20,000 cells/well. Cells that stably overexpress a specific TRP channel gene such as TRPV1 can be plated directly onto 96-well assay plates and maintained in 37°C CO_2 incubators. After 48 hours, the culture medium is aspirated, and a buffer consisting of 50 µL F-12 medium containing 15 mM HEPES, 0.1 mg/mL bovine serum albumin, 1 µM capsaicin, and 20 µCi $^{45}Ca^{2+}$ is added to the cells. After 2 min incubation at room temperature, the cells are washed with 100 µL PBS. After PBS

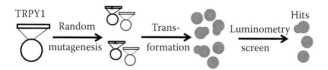

FIGURE 12.2 Forward mutagenesis screen using luminometry assay in yeast. The entire open-reading frame of TRPY1 is mutagenized and transformed into yeast cells (in TRPY1 knockout background, yvc1Δ) that were first transformed with a plasmid that produces aequorin to monitor cytoplasmic Ca^{2+} by luminescence. The transformed yeast culture is monitored with a luminometer and is hypotonically shocked by dilution. Mutants associated with "gain of function" or "loss of function" are picked as hits of the screen. (Modified from Zhou, M. X. et al., *Proceedings of the National Academy of Sciences of the United States of America* 104, 15555–15559, 2007.)

is aspirated, scintillation fluid is added to the well, and the counts are read using a microbeta or top count.[49,88]

12.4 FOLLOW-UP STUDIES

12.4.1 SITE-DIRECTED MUTAGENESIS

Identification of hits from the forward mutagenesis screen is only the first step toward understanding the role of specific amino acid residues in ligand interaction or channel gating of TRP channels. To dissect the role of individual amino acids, it is necessary to follow up with site-directed mutagenesis, which usually is done by using commercially available mutagenesis kits (such as the QuickChange Site-directed Mutagenesis kits from Stratagene). To examine the effect of site chain properties of individual amino acid residues (such as charges, hydrophobicity scales, and sizes), we usually engineer all 19 other substitutions. Another alternative approach is to design a pair of primers and assign "NNN" for the three nucleotides that code the specific amino acid, followed by PCR using the QuickChange kit. In this way, the targeted amino acid will be replaced with one of the other 19 amino acids randomly. After transformation, we routinely pick up about 50 colonies for bacterial culture, miniprep, and sequencing. Detailed analysis of concentration–response relationships of the mutants is critical for reaching conclusions about these point mutations.

12.4.2 CALCIUM IMAGING ASSAY

The ratiometric calcium imaging assay is a standard functional assay for calcium-permeable TRP channels. It measures calcium influx through TRP channels upon activation in individual live cells in real time. The other advantage of calcium imaging assay is that repeated applications of agonists to the cells can be achieved by a perfusion system. It is used to validate the hits from the FLIPR screen. Methods regarding calcium imaging assays have been covered by other chapters of this volume and will not be discussed here.

12.4.3 ELECTROPHYSIOLOGICAL RECORDINGS

FLIPR and calcium imaging are indirect measurements of TRP channel activity. Electrophysiological recordings are still the gold standard for studying ion channels by directly measuring ion flux across biological membrane. Whole-cell patch-clamp recordings on mammalian heterologous expression systems, such as HEK293 and CHO cells, and two-electrode voltage clamp recordings on a *Xenopus oocyte* expression system are most commonly used to characterize the electrophysiological properties of TRP channel mutants.[77–80,86] These methods are often used to examine current density, voltage modulation, ion selectivity, detailed concentration–response curves, pharmacological properties, etc. Sometimes, single-channel recordings are also used to compare the single-channel properties of the wild-type and mutant channels, such as unitary conductance and so on. Readers are referred to other chapters of this volume for detailed protocols on electrophysiological recordings.

12.5 CONCLUSIONS

Mutagenesis approaches are well-established methods that have been used to study structural and functional relationships of numerous proteins including enzymes, ion channels, G protein-coupled receptors (proteins involved in stimulus–response pathways), transporters, etc. In this chapter, we have described a novel high-throughput approach to studying TRP channel activation mechanisms. It includes generation of a large random mutagenesis library and a functional FLIPR calcium assay. Our new high-throughput screening methodology allowed us to analyze 14,000 mutants, out of which we have isolated mutants that specifically affected menthol sensitivity in TRPM8 and 2-APB sensitivity in TRPV3. Most importantly, we have identified specific amino acid residues in the pore loop that specifically dictate the temperature responses of TRPV3[78] and TRPV1.[79] Our experiments have yielded significant insights into the functional elements of "ThermoTRPs" that would have been difficult to obtain using other mutagenesis methods. Specifically, the new methodology could be used to identify amino acid residues in certain ion channel proteins and G protein-coupled receptors that are involved in interactions with small molecules that affect their function. Therefore, it provides a framework for high-throughput approaches for future studies on the structural and functional relationships of cellular proteins.

ACKNOWLEDGMENTS

The Experimental work and methodology development described here were carried out in the laboratory of Ardem Patapoutian. We thank Myleen Medina, Anthony Marelli, Jia Zhang, and Tony Orth for preparing miniprep DNA, and members of A.P. laboratory for helpful discussion. The research was supported by grants from the U.S. National Institutes of Health and the Novartis Research Foundation (to A.P.).

REFERENCES

1. Clapham, D. E. 2007. SnapShot: mammalian TRP channels. *Cell* 129: 220.
2. Clapham, D. E. 2003. TRP channels as cellular sensors. *Nature* 426: 517–524.
3. Nilius, B., K. Talavera, G. Owsianik et al. 2005. Gating of TRP channels: a voltage connection? *Journal of Physiology* 567: 35–44.
4. Baron, R. 2009. Neuropathic pain: a clinical perspective. *Handbook of Experimental Pharmacology* 194: 3–30.
5. Birder, L. A. 2007. TRPs in bladder diseases. *Biochimica et Biophysica Acta* 1772: 879–884.
6. Nilius, B. 2007. TRP channels in disease. *Biochimica Biophysica Acta* 1772: 805–812.
7. Sprenger, T., and P. J. Goadsby. 2009. Migraine pathogenesis and state of pharmacological treatment options. *BMC Medical* 7: 71.
8. Storr, M. 2007. TRPV1 in colitis: is it a good or a bad receptor?—a viewpoint. *Neurogastroenterology Motility* 19: 625–629.
9. Watanabe, H., M. Murakami, T. Ohba, K. Ono, and H. Ito. 2009. The pathological role of transient receptor potential channels in heart disease. *Circulation Journal* 73: 419–427.
10. Woudenberg-Vrenken, T. E., R. J. Bindels, and J. G. Hoenderop. 2009. The role of transient receptor potential channels in kidney disease. *Nature Reviews. Nephrology* 5: 441–449.

11. Patapoutian, A. 2005. TRP channels and thermosensation. *Chemical Senses* 30(Suppl. 1): i193–i194.

12. Dhaka, A., V. Viswanath, and A. Patapoutian. 2006. Trp ion channels and temperature sensation. *Annual Review of Neuroscience* 29: 135–161.

13. Caterina, M. J. 2007. Transient receptor potential ion channels as participants in thermosensation and thermoregulation. *American Journal of Physiology. Regulatory, Integrative and Compative Physiology* 292: R64–R76.

14. Bandell, M., L. J. Macpherson, and A. Patapoutian. 2007. From chills to chilis: mechanisms for thermosensation and chemesthesis via thermoTRPs. *Current Opinion in Neurobiology* 17: 490–497.

15. Basbaum, A. I., D. M. Bautista, G. Scherrer, and D. Julius. 2009. Cellular and molecular mechanisms of pain. *Cell* 139: 267–284.

16. Julius, D., and A. I. Basbaum. 2001. Molecular mechanisms of nociception. *Nature* 413: 203–210.

17. Patapoutian, A., and L. Macpherson. 2006. Channeling pain. *Nature Medicine* 12: 506–507.

18. Bessac, B. F., and S. E. Jordt. 2008. Breathtaking TRP channels: TRPA1 and TRPV1 in airway chemosensation and reflex control. *Physiology* 23: 360–370.

19. Caterina, M. J., A. Leffler, A. B. Malmberg et al. 2000. Impaired nociception and pain sensation in mice lacking the capsaicin receptor. *Science* 288: 306–313.

20. Davis, J. B., J. Gray, M. J. Gunthorpe et al. 2000. Vanilloid receptor-1 is essential for inflammatory thermal hyperalgesia. *Nature* 405: 183–187.

21. Moqrich, A., S. W. Hwang, T. J. Earley et al. 2005. Impaired thermosensation in mice lacking TRPV3, a heat and camphor sensor in the skin. *Science* 307: 1468–1472.

22. Alessandri-Haber, N., J. J. Yeh, A. E. Boyd et al. 2003. Hypotonicity induces TRPV4-mediated nociception in rat. *Neuron* 39: 497–511.

23. Bautista, D. M., S. E. Jordt, T. Nikai et al. 2006. TRPA1 mediates the inflammatory actions of environmental irritants and proalgesic agents. *Cell* 124: 1269–1282.

24. Kwan, K. Y., A. J. Allchorne, M. A. Vollrath et al. 2006. TRPA1 contributes to cold, mechanical, and chemical nociception but is not essential for hair-cell transduction. *Neuron* 50: 277–289.

25. Venkatachalam, K., and C. Montell. 2007. TRP channels. *Annual Review of Biochemistry* 76: 387–417.

26. Talavera, K., B. Nilius, and T. Voets. 2008. Neuronal TRP channels: thermometers, pathfinders and life-savers. *Trends in Neuroscience* 31: 287–295.

27. Damann, N., T. Voets, and B. Nilius. 2008. TRPs in our senses. *Current Biology* 18: R880–R889.

28. Abramowitz, J., and L. Birnbaumer. 2009. Physiology and pathophysiology of canonical transient receptor potential channels. *FASEB Journal* 23: 297–328.

29. Patapoutian, A., S. Tate, and C. J. Woolf. 2009. Transient receptor potential channels: targeting pain at the source. *Nature Reviews. Drug Discovery* 8: 55–68.

30. Asakawa, M., T. Yoshioka, T. Matsutani et al. 2006. Association of a mutation in TRPV3 with defective hair growth in rodents. *Journal of Investigative Dermatology* 126: 2664–2672.

31. Imura, K., T. Yoshioka, I. Hikita et al. 2007. Influence of TRPV3 mutation on hair growth cycle in mice. *Biochemical and Biophysical Research Communications* 363: 479–483.

32. Yoshioka, T., K. Imura, M. Asakawa et al. 2009. Impact of the Gly573Ser substitution in TRPV3 on the development of allergic and pruritic dermatitis in mice. *Journal of Investigative Dermatology* 129: 714–722.

33. Xiao, R., J. Tian, J. Tang, and M. X. Zhu. 2008. The TRPV3 mutation associated with the hairless phenotype in rodents is constitutively active. *Cell Calcium* 43: 334–343.

34. Steinhoff, M., and T. Biro. 2009. A TR(I)P to pruritus research: role of TRPV3 in inflammation and itch. *Journal of Investigative Dermatology* 129: 531–535.
35. Rock, M. J., J. Prenen, V. A. Funari et al. 2008. Gain-of-function mutations in TRPV4 cause autosomal dominant brachyolmia. *Nature Genetics* 40: 999–1003.
36. Nilius, B., and G. Owsianik. 2010. Channelopathies converge on TRPV4. *Nature Genetics* 42: 98–100.
37. Auer-Grumbach, M., A. Olschewski, L. Papic et al. 2010. Alterations in the ankyrin domain of TRPV4 cause congenital distal SMA, scapuloperoneal SMA and HMSN2C. *Nature Genetics* 42: 160–164.
38. Deng, H. X., C. J. Klein, J. Yan et al. 2010. Scapuloperoneal spinal muscular atrophy and CMT2C are allelic disorders caused by alterations in TRPV4. *Nature Genetics* 42: 165–169.
39. Landoure, G., A. A. Zdebik, T. L. Martinez et al. 2010. Mutations in TRPV4 cause Charcot-Marie-Tooth disease type 2C. *Nature Genetics* 42: 170–174.
40. Tian, W., Y. Fu, A. Garcia-Elias et al. 2009. A loss-of-function nonsynonymous polymorphism in the osmoregulatory TRPV4 gene is associated with human hyponatremia. *Proceedings of the National Academy of Sciences of the United States of America* 106: 14034–14039.
41. Krakow, D., J. Vriens, N. Camacho et al. 2009. Mutations in the gene encoding the calcium-permeable ion channel TRPV4 produce spondylometaphyseal dysplasia, Kozlowski type and metatropic dysplasia. *American Journal of Human Genetics* 84: 307–315.
42. Latorre, R., C. Zaelzer, and S. Brauchi. 2009. Structure-functional intimacies of transient receptor potential channels. *Quarterly Review of Biophysics* 42: 201–246.
43. Yao, J., and F. Qin. 2009. Interaction with phosphoinositides confers adaptation onto the TRPV1 pain receptor. *PLoS Biology* 7: e46.
44. Lishko, P. V., E. Procko, X. Jin, C. B. Phelps, and R. Gaudet. 2007. The ankyrin repeats of TRPV1 bind multiple ligands and modulate channel sensitivity. *Neuron* 54: 905–918.
45. Chuang, H. H., and S. Lin. 2009. Oxidative challenges sensitize the capsaicin receptor by covalent cysteine modification. *Proceedings of the National Academy of Sciences of the United States of America* 106: 20097–20102.
46. Jordt, S. E., and D. Julius. 2002. Molecular basis for species-specific sensitivity to "hot" chili peppers. *Cell* 108: 421–430.
47. Liu, B., W. Ma, S. Ryu, and F. Qin. 2004. Inhibitory modulation of distal C-terminal on protein kinase C-dependent phospho-regulation of rat TRPV1 receptors. *Journal of Physiology* 560: 627–638.
48. Chou, M. Z., T. Mtui, Y. D. Gao, M. Kohler, and R. E. Middleton. 2004. Resiniferatoxin binds to the capsaicin receptor (TRPV1) near the extracellular side of the S4 transmembrane domain. *Biochemistry* 43: 2501–2511.
49. Gavva, N. R., L. Klionsky, Y. Qu et al. 2004. Molecular determinants of vanilloid sensitivity in TRPV1. *Journal of Biological Chemistry* 279: 20283–20295.
50. Vlachova, V., J. Teisinger, K. Susankova et al. 2003. Functional role of C-terminal cytoplasmic tail of rat vanilloid receptor 1. *Journal of Neuroscience* 23: 1340–1350.
51. Bhave, G., H. J. Hu, K. S. Glauner et al. 2003. Protein kinase C phosphorylation sensitizes but does not activate the capsaicin receptor transient receptor potential vanilloid 1 (TRPV1). *Proceedings of the National Academy of Sciences of the United States of America* 100: 12480–12485.
52. Brauchi, S., G. Orta, M. Salazar, E. Rosenmann, and R. Latorre. 2006. A hot-sensing cold receptor: C-terminal domain determines thermosensation in transient receptor potential channels. *Journal of Neuroscience* 26: 4835–4840.
53. Jordt, S. E., M. Tominaga, and D. Julius. 2000. Acid potentiation of the capsaicin receptor determined by a key extracellular site. *Proceedings of the National Academy of Sciences of the United States of America* 97: 8134–8139.

54. Ryu, S., B. Liu, J. Yao, Q. Fu, and F. Qin. 2007. Uncoupling proton activation of vanilloid receptor TRPV1. *Journal of Neuroscience* 27: 12797–12807.

55. Dhaka, A., V. Uzzell, A. E. Dubin et al. 2009. TRPV1 is activated by both acidic and basic pH. *Journal of Neuroscience* 29: 153–158.

56. Hu, H., M. Bandell, M. J. Petrus, M. X. Zhu, and A. Patapoutian. 2009. Zinc activates damage-sensing TRPA1 ion channels. *Nature Chemical Biology* 5: 183–190.

57. Macpherson, L. J., A. E. Dubin, M. J. Evans et al. 2007. Noxious compounds activate TRPA1 ion channels through covalent modification of cysteines. *Nature* 445: 541–545.

58. Miyamoto, T., A. E. Dubin, M. J. Petrus, M. J., and A. Patapoutian. 2009. TRPV1 and TRPA1 mediate peripheral nitric oxide-induced nociception in mice. *PLoS One* 4: e7596.

59. Xiao, B., A. E. Dubin, B. Bursulaya et al. 2008. Identification of transmembrane domain 5 as a critical molecular determinant of menthol sensitivity in mammalian TRPA1 channels. *Journal of Neuroscience* 28: 9640–9651.

60. Hinman, A., H. H. Chuang, D. M. Bautista, and D. Julius. 2006. TRP channel activation by reversible covalent modification. *Proceedings of the National Academy of Sciences of the United States of America* 103: 19564–19568.

61. Macpherson, L. J., B. Xiao, K. Y. Kwan et al. 2007. An ion channel essential for sensing chemical damage. *Journal of Neuroscience* 27: 11412–11415.

62. Yoshida, T., R. Inoue, T. Morii et al. 2006. Nitric oxide activates TRP channels by cysteine S-nitrosylation. *Nature Chemical Biology* 2: 596–607.

63. Cruz-Orengo, L., A. Dhaka, R. J. Heuermann et al. 2008. Cutaneous nociception evoked by 15-delta PGJ2 via activation of ion channel TRPA1. *Molecular Pain* 4: 30.

64. Fujita, F., K. Uchida, T. Moriyama et al. 2008. Intracellular alkalization causes pain sensation through activation of TRPA1 in mice. *Journal of Clinical Investigations* 118: 4049–4057.

65. Voets, T., A. Janssens, G. Droogmans, and B. Nilius. 2004. Outer pore architecture of a Ca^{2+}-selective TRP channel. *Journal of Biological Chemistry* 279: 15223–15230.

66. Voets, T., G. Owsianik, A. Janssens, K. Talavera, and B. Nilius. 2007. TRPM8 voltage sensor mutants reveal a mechanism for integrating thermal and chemical stimuli. *Nature Chemical Biology* 3: 174–182.

67. Watanabe, H., J. Vriens, A. Janssens et al. 2003. Modulation of TRPV4 gating by intra- and extracellular Ca^{2+}. *Cell Calcium* 33: 489–495.

68. Nilius, B., H. Watanabe, and J. Vriens. 2003. The TRPV4 channel: structure-function relationship and promiscuous gating behaviour. *Pflügers Archiv* 446: 298–303.

69. Tsuruda, P. R., D. Julius, and D. L. Minor, Jr. 2006. Coiled coils direct assembly of a cold-activated TRP channel. *Neuron* 51: 201–212.

70. Garcia-Sanz, N., P. Valente, A. Gomis et al. 2007. A role of the transient receptor potential domain of vanilloid receptor I in channel gating. *Journal of Neuroscience* 27: 11641–11650.

71. Armstrong, C. M. 2003. Voltage-gated K channels. *Science STKE* 2003: re10.

72. Doyle, D. A., J. Morais Cabral, R. A. Pfuetzner et al. 1998. The structure of the potassium channel: molecular basis of K^+ conduction and selectivity. *Science* 280: 69–77.

73. Mackinnon, R. 2004. Structural biology. Voltage sensor meets lipid membrane. *Science* 306: 1304–1305.

74. Gutman, G. A., K. G. Chandy, J. P. Adelman et al. 2003. International Union of Pharmacology. XLI. Compendium of voltage-gated ion channels: potassium channels. *Pharmacological Reviews* 55: 583–586.

75. Jiang, Y., A. Lee, J. Chen et al. 2003. X-ray structure of a voltage-dependent K^+ channel. *Nature* 423: 33–41.

76. Long, S. B., X. Tao, E. B. Campbell, and R. MacKinnon. 2007. Atomic structure of a voltage-dependent K^+ channel in a lipid membrane-like environment. *Nature* 450: 376–382.

77. Bandell, M., A. E. Dubin, M. J. Petrus et al. 2006. High-throughput random mutagenesis screen reveals TRPM8 residues specifically required for activation by menthol. *Nature Neuroscience* 9: 493–500.

78. Grandl, J., H. Hu, M. Bandell et al. 2008. Pore region of TRPV3 ion channel is specifically required for heat activation. *Nature Neuroscience* 11: 1007–1013.

79. Grandl, J., S. E. Kim, V. Uzzell et al. 2010. Temperature-induced opening of TRPV1 ion channel is stabilized by the pore domain. *Nature Neuroscience* 13: 708–714.

80. Hu, H., J. Grandl, M. Bandell, M. Petrus, and A. Patapoutian. 2009. Two amino acid residues determine 2-APB sensitivity of the ion channels TRPV3 and TRPV4. *Proceedings of the National Academy of Sciences of the United States of America* 106: 1626–1631.

81. Bang, S., K. Y. Kim, S. Yoo, S. H. Lee, and S. W. Hwang. 2007. Transient receptor potential V2 expressed in sensory neurons is activated by probenecid. *Neuroscience Letters* 425: 120–125.

82. Zhou, X., Z. Su, A. Anishkin et al. 2007. Yeast screens show aromatic residues at the end of the sixth helix anchor transient receptor potential channel gate. *Proceedings of the National Academy of Sciences of the United States of America* 104: 15555–15559.

83. Su, Z., X. Zhou, W. J. Haynes et al. 2007. Yeast gain-of-function mutations reveal structure-function relationships conserved among different subfamilies of transient receptor potential channels. *Proceedings of the National Academy of Sciences of the United States of America* 104: 19607–19612.

84. Su, Z., X. Zhou, S. H. Loukin et al. 2009. The use of yeast to understand TRP-channel mechanosensitivity. *Pflügers Archiv* 458: 861–867.

85. Loukin, S. H., Z. Su, and C. Kung. 2009. Hypotonic shocks activate rat TRPV4 in yeast in the absence of polyunsaturated fatty acids. *FEBS Letters* 583: 754–758.

86. Myers, B. R., C. J. Bohlen, and D. Julius. 2008. A yeast genetic screen reveals a critical role for the pore helix domain in TRP channel gating. *Neuron* 58: 362–373.

87. Myers, B. R., Y. Saimi, D. Julius, and C. Kung. 2008. Multiple unbiased prospective screens identify TRP channels and their conserved gating elements. *Journal of General Physiology* 132: 481–486.

88. Gavva, N. R., R. Tamir, L. Klionsky et al. 2005. Proton activation does not alter antagonist interaction with the capsaicin-binding pocket of TRPV1. *Molecular Pharmacology* 68: 1524–1533.

13 Methods Used for Studying TRP Channel Functions in Sensory Neurons

Louis S. Premkumar

CONTENTS

13.1 INTRODUCTION

Over 30 members of the transient receptor potential (TRP) family of ion channels have been cloned. Several of these TRP channels are expressed in subpopulations of primary sensory afferent neurons. TRP channels are Ca^{2+} permeant nonselective cation channels and are activated by physical (temperature and mechanical force) and chemical stimuli.[1–8]

The cell bodies of sensory neurons innervating the head and neck are located in the trigeminal ganglia (TG), and the cell bodies innervating the rest of the body are located in the dorsal root ganglia (DRG). TG neurons form their synapses at the caudal spinal trigeminal nucleus (CSTN), and DRG neurons form synapses at the dorsal horn (DH) lamina I and lamina II of the spinal cord. TRP channels are distributed in the peripheral and central terminals of sensory neurons, and they play a role in nociceptive transmission by initiating action potentials at the nerve terminals and modulating neurotransmitter release at the first sensory synapse.

Sensory TRP channels are sensitized by pro-inflammatory agents and mediate heightened pain sensitivity.[3,9–12] In order to gain insight into the process of nociception, it is first necessary to understand the characteristics of TRP channels that respond to physical and chemical stimuli. Nociceptive TRP channels are targets for next-generation analgesics. In fact, some TRP channels are specifically involved following inflammation.[4,7,12] On the basis of their distribution in nociceptors, the targets include TRPVanilloid 1, TRPVanilloid 3, TRPVanilloid 4, TRPAnkyrin 1, and TRPMelastatin 8. The methods that can be adopted to study the expression and function of TRP channels in sensory neurons are described in this chapter.

13.2   IN VITRO NEURONAL PREPARATIONS TO STUDY SENSORY NEURONS

13.2.1   DRG AND TG NEURONAL CULTURES FROM EMBRYONIC RATS OR MICE

The advantages of using embryonic DRG/TG neurons in culture are that they can be used after four days and can survive for weeks. Only embryonic neurons grown in culture have been shown to readily form synaptic connections between sensory neurons and second-order neurons.

Pregnant rats/mice are deeply anesthetized with a lethal dose of Nembutal (80 mg kg^{-1}, I.P.), and embryonic day 18 (E18) embryos are removed. The embryos are decapitated in hypothermic anesthesia. The heads are used to collect TG, and the bodies are used for collecting DRG. Both TG and DRG are removed under a dissecting microscope.

For TG dissections, the head of the embryo is hemisected sagittally, and TG is exposed in the medial region. For DRG dissections, the skin on the back of the embryo is split with fine forceps, and the spinal cord is exposed. The spinal column is removed, and DRG can be located on either side of the vertebral column. TG/DRG are removed and placed into a 15-mL conical tube with a 5-mL cold L-15 medium (Sigma, St. Louis, MO, USA) and centrifuged for 5 min at 500 × g.

After centrifugation, the supernatant is discarded, and the pellet is resuspended in 5 mL Hank's buffered salt solution (Ca^{2+} and Mg^{2+}-free, HBSS) containing 5 mg collagenase D, Worthington type 2 (0.1%) (Roche Molecular Biochemicals, Indianapolis, IN, USA), and 5 mg trypsin, type 1 (0.1%) (Sigma). DRG are digested for 45 min in a 37°C water bath with rotation at 150 rpm. After digestion, the samples are centrifuged at 500 × g for 5 min. The supernatant is discarded, and the pellet is resuspended in an 8-mL Neurobasal medium (NBM) supplemented with B-27, a serum free supplement (2 mM), L-Glutamine (10 μL/mL) (Invitrogen, Carlsbad, CA, USA), nerve growth factor (NGF 2.5S, 100 ng/mL, Sigma), and penicillin–streptomycin (Invitrogen). DRG are triturated 30–50 times with a siliconized fire-polished 9-inch Pasteur pipette and plated on glass cover slips previously coated with poly-D-lysine (10 μg/mL, Sigma) in a 24-well plate containing 0.5 mL medium per well. Samples are incubated at 5% CO_2 and 37°C. On day 2, the medium is changed. The neurons are ready to use after 4 days in culture.

13.2.2   DRG AND TG DISSOCIATED NEURONS FROM ADULT RATS OR MICE

In order to address ontogenic changes and while studying disease models, it is necessary to use sensory neurons from adult animals. Rats/mice are deeply anesthetized with isoflurane (Abbott Labs, Chicago, IL, USA) and, when they no longer respond to the foot pinch, are decapitated for TG collection. The brain is removed from the head, and TG can be located in the trigeminal notch of the petrosal bone. For DRG collection, the spinal column is removed, trimmed, cut into two segments, and placed into an L-15 medium in a Petri dish. The dorsal side of the vertebral column is cut by a pair of fine scissors, and the spinal cord is exposed. DRG are located in the intervertebral foramen. TG and DRG are collected with fine forceps and placed in an ice-cold L-15 medium. The processing procedure is the same between embryonic and adult tissues. Neurons are maintained at 37°C in an incubator (humidified atmosphere of 5% CO_2) and are used within 24 hours.

13.2.3 DRG-DH Co-Cultures from Embryonic Rats or Mice

In order to study the role of TRP channels expressed in the central terminals of DRG/TG neurons, embryonic DRG/TG neurons are co-cultured with spinal DH or CSTN neurons. Embryonic neuronal cultures readily form synapses as compared to adult neurons.

Embryonic DRG/TG neurons are prepared as described above. The spinal cord is collected after the meninges are pulled away with two pairs of forceps. The spinal cord is placed on its ventral side, and the medial ventral horn part is pressed down. The DH, which is located on both lateral sides, is cut by a blade longitudinally. The DH of the spinal cord is transferred to a 15-mL tube containing HBSS with digestive enzymes, and neurons are dissociated as described above. The DH, together with the DRG neurons, is plated on poly-D-lysine coated glass cover slips. In some dishes, only DH or DRG neurons are plated to obtain DH or DRG neuronal monocultures, respectively. The co-cultures are ready to use after one week. DRG neurons and DH neurons are easily distinguished by their morphology. DRG neurons are rounded with long axons, whereas DH neurons are pyramidal shaped. In a similar manner, CSTN neurons can be isolated for TG-CSTN neuronal co-cultures.

13.2.4 Spinal Cord and CSTN Slice Preparations

Rats/mice between the ages of 2 and 10 weeks old are anesthetized with isoflurane and decapitated with a guillotine. The brain and the spinal cord are removed and placed in an ice-cold sucrose solution for 2 min containing (in mM): 209 sucrose, 2 KCl, 1.25 NaH_2PO_4, 5 $MgCl_2$, 0.5 $CaCl_2$, 26 $NaHCO_3$ and 10 D-glucose, titrated to pH 7.4 (290 mOSM). The solution/tissue is aerated with a 95% O_2–5% CO_2 gas mixture. Horizontal slices of the CSTN are obtained upon isolating a section of medulla trimmed caudally at the cervical spinal cord and rostrally at the obex.[13] Transverse spinal cord slices are obtained after hydraulic extrusion of the spinal cord. An OTS 3000 vibratome (Leica, Nussloch, Germany) or a Precisionary VF-200 vibratome (Greenville, NC, USA) is used to cut 250-μm slices of the desired tissue in ice-cold aerated sucrose. Slices are immediately harvested and placed in an incubation chamber containing oxygenated hibernate A (Brain Bits, Springfield, IL, USA) at 32°C. The temperature of the incubation chamber is allowed to fall to room temperature as slices incubate, and slices are allowed to recover for 1 hour after slicing before being placed in the recording chamber.

13.3 EXPRESSION OF TRP CHANNELS IN DRG/TG NEURONS

13.3.1 RT-PCR

The reverse transcription-polymerase chain reaction (RT-PCR) technique is used to identify the RNA of interest in a given tissue. Total RNA from DRG/TG neurons is extracted by Trizol reagent (Invitrogen) and reversely transcribed to cDNA by using a cDNA synthesis kit (Promega, Madison, WI, USA). PCR is performed by using different cDNAs as templates in 30 cycles with 30-s denaturation at 95°C, 30-s annealing at 58°C, and 30-s extension at 68°C using PCR green master mix (Promega). Appropriate primer pair sequences and orientation (F, forward; R, reverse) must be designed and obtained; for example, to identify the presence of TRPV1 mRNA, the primer pair is F: accacggctgcttactatcg, R: ctccagtgacacg gaaatagtcc. A housekeeping gene, hypoxanthine guanine phosphoribosyl transferase I (HPRT), is used as a control, and the primer pair is F: gcttcctcctcagaccgcttt, R: ctggttcatcatcgctaatcacg. Samples are treated with DNase (Ambion, Austin, TX, USA) followed by cDNA generation. Products are run in an agrose gel (1.5%) with ethidium bromide in the Tris/Borate/EDTA (TBE) buffer. The gel is imaged by the Versa Doc system (Bio-Rad, Hercules, CA, USA), and the band density is quantified by Quantity One (Bio-Rad).

Because DRG neurons consist of multiple-cell types, single-cell RT-PCR can be performed by harvesting cell contents in identified cells after recording currents using a patch-clamp technique as described by Monyer and Jonas.[14]

To quantify the amount of RNA, real-time RT-PCR is performed with SYBR Green fluorescent tag (New England Biolabs, Ipswich, MA, USA) using a BioRad Cycler. Real-time RT-PCR conditions are as follows: 95°C for 15 min, then 25 cycles of 95°C for 10 s, 62°C for 25 s, then 72°C for 30 s. Results are quantified using HPRT as a reference in the comparative C_T method. For this method, the amplification efficiencies of the target gene and the housekeeping gene have to be similar. The comparative C_T method is also known as the $2^{-\Delta\Delta C_T}$ method, where

$$\Delta\Delta C_T = \Delta C_{T,sample} - \Delta C_{T,reference}$$

Here, $\Delta C_{T,sample}$ is the C_T value for any sample normalized to the endogenous housekeeping gene, and $\Delta C_{T,reference}$ is the C_T value for the calibrator also normalized to the endogenous housekeeping gene.

13.3.2 IMMUNOHISTOCHEMISTRY

Rats/mice are anesthetized by ketamine and xylazine (85 and 5 mg/kg, respectively, I.P.) and are transcardially perfused with freshly made fixative consisting of 4% paraformaldehyde in Sörenson's K-Na phosphate buffer. To prepare the fixative, 1.78 g monobasic potassium phosphate and 14.38 g dibasic sodium phosphate are dissolved in 750 mL dH$_2$O, and 250 mL of 16% of paraformaldehyde stock solution is added and titrated to pH 7.4. Tissue samples of spinal cord segments, DRG, and paw skin are harvested and kept in the same fixative for 2 hours. After immersion in 15 and 30% sucrose for successive 24-hour periods at 4°C, tissues are then embedded with an embedding medium (Triangle Biomedical Sciences, Durham, NC, USA) and quickly frozen with liquid nitrogen. The spinal cord, DRG, and paw skin tissues are sectioned into 20-, 10-, and 20-μm sections, respectively, using a cryostat (Leica CM 1850, Nussloch, Germany). Immunostaining can be done either on slides or on free-floating tissues. After samples are permeabilized with phosphate buffered saline (PBS) containing 0.1% Triton X-100 for 20 min and blocked with 10% donkey serum for 30 min, the sections are incubated with polyclonal rabbit anti-TRP antibodies (Affinity BioReagents, Golden, CO, USA; Alomone Laboratories, Jerusalem, Israel; Abcam Ltd., Cambridge, UK) and monoclonal mouse anti-NeuN antibody (Millipore, Billerica, MA, USA) for 1 hour at room temperature. The sections are then washed with PBS and incubated with secondary antibodies (rhodamine red-X-conjugated donkey anti-goat IgG and FITC-conjugated donkey anti-rabbit IgG) (Jackson Immuno, West Grove, PA, USA) for 1 hour at room temperature. Sections are washed, mounted onto Superfrost/Plus slides, and covered with cover slips with the use of Vectashield (Vector Laboratories, Burlingame, CA, USA). Images are obtained using a confocal microscope (Olympus Fluoview, Tokyo, Japan). The intensity of staining is determined by measuring the gray value of the stained region using Image J (Research Service Branch, NIMH, Bethesda, MD, USA). Data from at least six sections of each animal and from at least three different animals are collected and averaged. The imaging system is calibrated with a density step tablet to assure that optical densities obtained are within the linear response range of the system. The identity of all sections can be concealed/blinded to insure unbiased quantification.[15] The specificity of antibodies can be determined by pre-incubating a peptide sequence against which the antibody has been generated to prevent binding or by using knockout animals. Also, see Chapter 6 for discussion on TRP antibodies.

13.3.3 IN SITU HYBRIDIZATION

In situ hybridization is a powerful technique that allows researchers to determine the distribution of DNA and RNA sequences in specific cell types.[16] Radioactive and nonradioactive probes can be used for in situ hybridization. Nonradioactive probes (sense and antisense) using digoxigenin (Roche Molecular Biochemicals) are often preferred because they have high sensitivity and are easy and safe to use. Tissues of rats/mice are fixed by transcardial perfusion of 4% paraformaldehyde in PBS. DRG are dissected, postfixed for an additional 2 hours in paraformaldehyde, and cryoprotected overnight in 30% sucrose. Sections (10 μm) are treated with 1 μg/mL proteinase K (Sigma) for 5 min, acetylated for 10 min with 0.25% (v/v) acetic anhydride in 0.1 M triethanolamine, pre-hybridized for 4 hours at 56°C, and hybridized with probes overnight at 56°C. Following post hybridization washes and blocking, sections are incubated for 30 min in anti-DIG antibody conjugated with horseradish peroxidase (Roche Molecular Biochemicals). The signal is visualized using tyramide signal amplification (Perkin Elmer, Waltham, MA, USA).

Radioactive probes can be used in the in situ hybridization study. Selective oligonucleotide probes (36- to 45-mer) are synthesized and purified by Sigma Genosys (Woodlands, TX, USA). The probes are 3′ end labeled (^{35}S) in the presence of terminal deoxynucleotidyl transferase (TdT) (Thermo Fisher Scientific, Waltham, MA, USA) and purified using micro BioSpin-30 columns (Bio-Rad) at $1000 \times$ g. To prepare sections for in situ hybridization, the sections are fixed in freshly prepared 4% paraformaldehyde (in 0.1M PBS, pH7.4), acetylated for 10 min at pH 8.0 with 0.25% acetic anhydride in 0.1 M triethanolamine and 0.9% sodium chloride, dehydrated using an ethanol series, and delipidated with 100% chloroform. The sections are incubated with the hybridization mixture, which consists of the labeled probe, 50% formamide, 4× sodium chloride-sodium citrate, 250 mg/mL yeast tRNA, 250 mg/mL sheared salmon sperm DNA, 10% dextran sulfate, 1× Denhardts, 25 mM sodium phosphate, 1 mM sodium pyrophosphate, and 10 mM dithiothreitol, overnight (17–22 hours) at 42°C in a humidified chamber. Posthybridization steps are performed as follows: slides are rinsed 2× 15 min in 1× SSC with 1 mM dithiothreitol at room temperature, rinsed in the same solution for 1 hour at 55°C, and then transferred through 1× SSC (Saline–Sodium Citrate), 0.1× SSC, 70% ethanol, and 95% ethanol (two dips in each). Tissues are air dried, dipped in Kodak NTB photographic emulsion (Eastman Chemical Co., Kingsport, TN, USA), and exposed in the dark for 3 weeks at 4°C. Slides are then developed at 15°C in D-19 (Eastman Chemical Co.) for 4 min, rinsed with distilled H$_2$O for 30 s, fixed for 5 min in fixer (Eastman Chemical Co.), and counterstained with thionin. Two different controls are used for specificity controls on adjacent sections. First, competitive blocking of the labeled oligonucleotide is performed using excess concentrations (50-fold) of unlabeled oligonucleotides. Second, sections are incubated with labeled sense oligonucleotides, which should not be hybridized with cellular mRNAs.

Images are captured using a CoolSnap high-resolution digital camera connected to an *Image-Pro4* system under 25× objective.[17] The accumulation of grains over cell bodies is interpreted as hybridization of the probe to its corresponding mRNAs in these neurons, and the number of grains over somata (grains/100 mm^2) is counted. Values are corrected by subtracting average background labeling obtained from five random off-tissue areas on each slide. Ten to 15 cells/section are selected and analyzed in a blinded fashion. Only grains within the perimeter of the cell are counted (number of grains/100 mm^2 of the cell area).

13.3.4 WESTERN BLOT

Rats/mice are anesthetized and sacrificed. DRG/TG are removed as described above and placed in a lysis buffer (0.1% SDS, 1% Triton X-100, 1% deoxycholate, protease and phosphatase inhibitor cocktail, 1:100) (Sigma) and are then homogenized and centrifuged. The protein concentration is measured by the bicinchoninic acid (BCA) assay. Protein is separated by 10% SDS-PAGE and transferred to a nitrocellulose membrane (Bio-Rad). Membranes are probed overnight with rabbit anti-TRP and anti-actin (Sigma) antibodies followed by incubation with horseradish peroxidase-conjugated (HRP) goat anti-rabbit IgG (Santa Cruz Biotechnology) for 1 hour. After incubation with enhanced chemiluminescence reagents (Santa Cruz Biotechnology), membranes are scanned using the Hitachi genetic systems (Hitachi Software Engineering, Tokyo, Japan), and blots are analyzed using GeneTools Analysis Software (SynGene, Frederick, MD, USA).

13.3.5 TRITIATED AGONIST AND ANTAGONIST BINDING

Binding studies with [³H]agonists and antagonists are carried out according to methods described by Szallasi and Blumberg.[18,19] Nonspecific ligand binding is reduced by adding bovine α1-acid glycoprotein (100 μg per tube) after the binding reaction has been terminated. Binding assay mixtures contain [³H]ligand, nonradioactive ligands, and 0.25 mg/mL bovine serum albumin (BSA, Cohn fraction V), along with at least 40 μg of DRG membrane proteins. The final volume is adjusted to 500 μL (competition binding assays) or 1000 μL (saturation binding assays). Nonspecific binding is defined as binding that occurs in the presence of 1-μM nonradioactive ligand. For saturation binding, [³H]ligands are added in increasing concentrations. Competition binding assays are performed in the presence of a fixed concentration of [³H]ligand and various concentrations of competing ligands. The binding reaction is initiated by placing the assay mixtures into a 37°C water bath and is terminated after a 15-min incubation period by cooling the tubes on ice. Membrane-bound ligand is separated from free and the α1-acid glycoprotein-bound ligands by pelleting the membranes in a Beckman 12 bench top centrifuge (15 min, maximal velocity). The radioactivity is determined by using a scintillation counter (Beckman Instruments, Fullerton, CA, USA).

13.3.6 RELEASE OF NEUROPEPTIDES (CGRP AND SP) FROM NERVE TERMINALS

DRG neurons are peptidergic or nonpeptidergic. Calcitonin gene-related peptide (CGRP) and substance P (SP) are stored in peptidergic nerve terminals (central and peripheral). To study the changes in peripheral and central nerve terminals, CGRP and SP release can be measured in response to TRP channel activation from paw skin and spinal cord tissues, respectively.[20] The tissues are transferred into a beaker containing 200–300 mL of Kreb's solution that contains (in mM): 119 NaCl, 25 $NaHCO_3$, 1.2 KH_2PO_4, $MgSO_4$, 2.5 $CaCl_2$, 4.7 KCl, and 11 D-glucose. To minimize peptide degradation, 0.1% BSA, 1 mM phosphoramidon, and 1 mM captopril are added to Kreb's solution. The solution is aerated with 5% CO_2 and 95% O_2, brought to 37°C, and left to aerate for 1 hour for stabilization. Each tissue piece is transferred into an Eppendorf tube containing 400 μL Kreb's or experimental solution. After 10 min, the tissue is transferred to tubes containing TRP channel agonists for 10 min. The samples are then dried to remove excess liquid and weighed within 0.1 mg. The test solutions are freeze-dried and stored at −80°C.[21] CGRP release is measured by using the ¹²⁵I CGRP radio immunoassay kit from Peninsula Labs (San Carlos, CA, USA) following the manufacturer's protocol. Similarly, freeze-dried samples are analyzed for SP concentration using an SP ELISA kit, following the protocol of the manufacturer (Cayman Chemical Company, Ann Arbor, MI, USA).

13.4   FUNCTIONAL STUDIES OF TRP CHANNELS IN SENSORY NEURONS

13.4.1   Function of TRP Channels in Isolated Sensory Neurons

13.4.1.1 Ca^{2+} Fluorescence Imaging

Relative Ca^{2+} Flux: To detect acute increases in intracellular Ca^{2+} levels by TRP channel activation, cultured or dissociated DRG neurons grown on cover slips are used. In order to obtain relative change in fluorescence representing Ca^{2+} influx, the neurons are incubated with 2 µM Fluo-4 AM (Invitrogen) for 20 min at 37°C. Fluo-4 is excited at 488 nm, emitted fluorescence is detected with a 535 ± 25 nm bandpass filter using a Leica microscope (DMIRE2), and data are read into a computer running Scanalytics software (Rockville, MD, USA). Changes in fluorescence are expressed as F/F_0, where F is the fluorescence at time t, and F_0 is the background fluorescence.

Ratiometric Ca^{2+} Imaging: To quantify the increase in intracellular Ca^{2+} concentrations, ratiometric Ca^{2+} imaging is performed with Fura-2 AM dye (Invitrogen) and analyzed using Scanalytics software. DRG/TG neurons are loaded with 3 µM Fura-2 AM and placed into a recording chamber containing (in mM): 140 NaCl, 4 KCl, 2 $CaCl_2$, 1 $MgCl_2$, 5 glucose, and 10 HEPES, titrated to pH 7.4 with NaOH. Pairs of images are collected every 2 s at alternating exposures of 340 and 380 nm using a Polychrome V monochromator (Lamda DG4, Sutter Instruments, Novato, CA, USA) and a CCD camera (Retiga EX, Leeds Precision Instruments Inc. Minneapolis, MN, USA). After the subtraction of background fluorescence, the ratio of fluorescence at 340 and 380 nm is calculated following application of TRP channel agonists. By constructing a standard curve of known concentrations of Ca^{2+}, the absolute concentration of intracellular Ca^{2+} can be determined.

Neurons under study are perfused continuously with the control solution from a 300-µm barrel positioned 50–100 µm away from the neuron (complete solution exchange is achieved in <100 ms). Solutions containing agonists and antagonists are applied by activating solenoid valves (ASCO, Florham Park, NJ, USA) that switch between different solutions. Thermal stimuli (heat/cold) are applied through a computer-controlled heating/cooling peltier device (Warner Instrument Corporation, Hamden, CT, USA). The temperature is measured at the mouth of the flow pipes using a thermocouple (Warner Instrument).

Ca^{2+} Permeability: Changes in the reversal potential (ΔE_{rev}) are used as an index of Ca^{2+} permeability relative to those for Cs^+. Change in the reversal potential (ΔE_{rev}) for TRP channel currents is measured after replacing a Cs^+-based reference solution (140 mM CsCl, 10 mM HEPES, titrated to pH 7.2 with CsOH) with a solution in which CsCl is substituted with a different concentration of Ca^{2+} and *N*-methyl-D-glucamine (NMDG) and titrated to pH 7.2 with HCl. The pipette solution consists of (in mM): 140 CsCl, 10 EGTA, and 10 HEPES, titrated to pH 7.2 with CsOH. Because

TRP channels exhibit strong rectification, a ramp protocol that changes the voltage from −100 to +100 mV in 1–2 s can be used. By subtracting the base line current, agonist-induced currents are plotted to determine the reverse potential. While perfusing the solutions of different compositions, it is necessary to adjust the junction potential[22] (http://web.med.unsw.edu.au/PHBSoft/). ΔE_{rev} values are converted to P_{Ca}/P_{Cs} using the Lewis equation, as described by Wollmuth and Sakmann.[23]

Fractional Ca^{2+} Currents (P_f): Ca^{2+} permeability estimated by changing the extracellular Ca^{2+} and determining the reversal potential requires the use of unphysiological concentrations of Ca^{2+}. In order to determine the Ca^{2+} flux through TRP channels at physiological concentrations of the ion, the fractional Ca^{2+} current (P_f) can be determined by simultaneous recording of Ca^{2+} fluorescence by imaging and membrane current using the patch-clamp technique. DRG neurons are patched with an intracellular solution that contains (in mM): 140 KCl, 10 HEPES and 1 K_5·fura-2 (Invitrogen), titrated to pH 7.2 with KOH. The extracellular bath solution contains (in mM): 140 NaCl, 2 $CaCl_2$, 1 $MgCl_2$, 10 glucose, and 10 HEPES, titrated to pH 7.4 with NaOH.[24–26]

Fura-2 fluorescence (380-nm excitation and 510-nm emission) is gathered using a photomultiplier (Photon Technology International, South Brunswick, NJ, USA). In order to standardize the sensitivity of the microscope and the photomultiplier tube, the fura-2 signal is normalized to a "bead unit" (BU). One BU equals the average fluorescence of seven Fluoresbrite carboxy BB 4.6-µm microspheres (Polysciences, Warrington, PA, USA). Fractional Ca^{2+} currents (P_f) are quantified using the following equation:

$$P_f (\%) = 100 \times Q_{Ca}/Q_T$$

where Q_T is the total charge and is equal to the integral of the agonist-induced transmembrane current. Q_{Ca} is the part of Q_T carried by Ca^{2+} and is equal to ΔF_{380} divided by the calibration factor F_{max}. F_{max} is calculated in a separate series of experiments under conditions in which Q_T is expected to equal Q_{Ca}. Data are analyzed in Clampfit 8.1 (Molecular Devices, Sunnyvale, CA, USA), and calculations are performed using Microcal Origin (Northampton, MA, USA).

13.4.1.2 Whole-Cell Patch-Clamp Recording (Voltage-Clamp and Current-Clamp)

Voltage-Clamp Recording: Patch-clamp techniques[27,28] are used to record membrane currents from cultured or dissociated DRG/TG neurons. Neurons for the study can be chosen according to their size, which largely represents the type of neuron (small diameter <25 µm are nonmyelinated C-fibers; medium diameter <40 µm are thinly myelinated Aδ fibers; >40 µm are Aβ fibers). The type of TRP channels expressed depends on the type of neuron. Patch pipettes are fabricated from borosilicate glass (World Precision Instruments, Sarasota, FL, USA) using a two-step electrode puller (Narishige, Tokyo, Japan). The electrode is lowered to the neuronal plane using a micromanipulator (EXFO Life Sciences Group, Mississauga, Ontario, Canada), and a gigaseal is obtained by gently pressing the electrode against the neuronal cell body and applying gentle suction. After forming a gigaseal, the amount of suction is gradually increased causing the membrane within the patch pipette to rupture without disrupting the gigaseal, resulting in a whole-cell configuration mode. In the whole-cell mode, currents are recorded in response to application of TRP channel agonists and blockade of agonist responses by antagonists. Currents are recorded at a holding potential of –60 and +60 mV, using voltage steps of 20-mV increments, or using a ramp protocol that changes the voltage from –100 to +100 mV in 1–2 s. Because most of the TRP channels exhibit outward rectification (the current flow into the cell is restricted), it is necessary to record and analyze currents at positive and negative potentials to quantify the extent of rectification. The currents are recorded using a patch-clamp amplifier (Axopatch 200B, Molecular Devices or EPC10, HEKA Elektronik, Lambrecht/Pfalz, Germany), and the data are collected with an open filter in the amplifier, digitized (VR-10B, Instrutech Corp., Great Neck, NY, USA), and stored either on videotapes or directly in the computer. For whole-cell recording, the bath solution contains (in mM): 140 Na-gluconate, 10 NaCl, 1 or 2 $MgCl_2$, 10 HEPES, and 10 EGTA, titrated to pH 7.3 with NaOH, and the pipette solution contains (in mM): 140 Na-gluconate/140 K-gluconate, 10 NaCl, 1 or 2 $MgCl_2$, 10 EGTA/BAPTA, 10 HEPES, 2 K_2ATP, and 0.25 GTP, titrated to pH 7.3 with NaOH/KOH. For whole-cell current analysis, the data are filtered at 2.5 kHz (–3 dB frequency with an eight-pole low-pass Bessel filter, LP10, Warner Instrument) and digitized at 5 kHz. Data analysis is done by pCLAMP software (Molecular Devices).

Because TRP channels are modulated by second messenger molecules and the phosphorylation state of the receptor, it is essential to keep the interior of the neuron intact without dialysis. Therefore, the perforated patch-clamp technique is used.[28] In this technique, access to the cell interior is achieved by adding nystatin or amphoteracin B (250 µg/mL) to the pipette solution, which forms pores that allow passage of only monovalent cations, thus preventing dialysis of the cell interior.

Current-Clamp Recording: To determine the change in the membrane potential, the current-clamp mode can be used in the patch-clamp technique. The current-clamp mode allows for the study of the resting membrane potential and depolarization caused by activation of various TRP channels. If the depolarization is sufficient to reach the threshold, action potentials are generated. In the current-clamp mode, injection of depolarizing currents elicits action potentials. Using the firing pattern in response to depolarizing pulse, it is possible to categorize the type of neuron from which recordings are made (DRG versus DH neurons). For current-clamp recording, the pipette solution contains (in mM): 130 K-gluconate/$KMeSO_3$, 10 NaCl, 1 $MgCl_2$, 5 or 10 EGTA, 2 K_2ATP, and 10 HEPES, titrated to pH 7.35 with KOH. The bath solution contains (in mM): 140 NaCl, 4 KCl, 2 $CaCl_2$, 1 $MgCl_2$, and 10 HEPES, titrated to pH 7.35 with NaOH.

13.4.1.3 Single-Channel Recordings

Gigaseal patch-clamp techniques[27,28] are used to record single-channel currents. For single-channel recording using the cell-attached mode, NaCl in the bath solution is replaced by KCl in order to neutralize the membrane potential and also to identify the K^+ channels that may interfere with the recording (at 0 mV, there should not be any TRP channel activity with Na-gluconate in the pipette because TRP channels are nonselective cation channels). For cell-attached patches, the bath solution contains (in mM): 140 K-gluconate, 2.5 KCl, 5 HEPES, and 1.5 EGTA, titrated to pH 7.35 with NaOH. The patch pipettes are made from glass capillaries (Drummond, Microcaps, Broomall, PA, USA), coated with sylgard (Dow Corning, Midland, MI, USA) to minimize the pipette capacitance, and are filled with a solution that contains (in mM): 140 Na-gluconate, 10 NaCl, 2 $MgCl_2$, 1.5 EGTA, and 5 HEPES, titrated to pH 7.35 with NaOH. Ca^{2+} free extracellular solutions are used to avoid desensitization and tachyphylaxis. While studying the effects of extracellular Ca^{2+}, EGTA is eliminated, and 2 mM $CaCl_2$ is added to the solution; the concentration of $MgCl_2$ is reduced to 1 mM. For inside-out patches, the bath solution contains (in mM): 140 Na-gluconate, 10 NaCl, 1 or 2 $MgCl_2$, 10 EGTA, 10 HEPES, 2 K_2ATP, and 0.25 GTP, titrated to pH 7.3 with NaOH, and the pipette solution contains the same solution as the cell-attached patch. For outside-out patches, the pipette solution contains (in mM): 140 Na-gluconate, 10 NaCl, 1 or 2 $MgCl_2$, 10 EGTA, 10 HEPES, 2 K_2ATP, and 0.25 GTP, titrated to pH 7.3 with NaOH/KOH, and the bath solution contains (in mM): 140 Na-gluconate, 10 NaCl, 2 $MgCl_2$, 1.5 EGTA, and 5 HEPES, titrated to pH 7.35 with NaOH. Using these solutions, it is possible to eliminate K^+ and Cl^- currents contaminating the recording. For recording currents from a large number of channels in isolated patch configurations, the macropatch technique[29] can be used. Data are collected with an open filter in the amplifier, digitized (VR-10B), and stored on videotapes. For analysis of amplitude and open probability (P_o), the data are filtered at 2.5 kHz and digitized at 5 kHz; for kinetic analysis, the data are filtered at 10/20 kHz and digitized at 50/100 kHz using pCLAMP (Molecular Devices) based hardware and software. Single-channel current amplitude and P_o are estimated from all-point current-amplitude histograms (pCLAMP) and fitted to Gaussian densities (Microcal Origin). For current-voltage relationships, 10–50 single-channel openings are grouped, and the amplitude is determined by fitting a Gaussian curve. P_o is determined using unedited segments of data, which is typically 1–5 min long. For multiple-channel patches, mean P_o is measured as NP_o divided by N (N is the number of channels in the patch). Chord conductance is measured at +60 or –60 mV, and the slope conductance is determined by plotting a current-voltage curve and fitting the inward and outward currents with linear functions.

Patches that apparently have a single TRP channel (assessed by the lack of overlapping events at +60 mV, when P_o is >0.8) are used for dwell-time analysis. However, this criterion is not valid when lower concentrations of agonists are used. The number of channels in the patch can be determined by exposing the patch to higher agonist concentrations at the end of the experiment. Single-channel currents are idealized (10-kHz bandwidth) using a modified Viterbi algorithm (QUB software, www.qub.buffalo.edu). Dwell-time distributions are fitted with mixtures of exponential densities using a method of maximum likelihood. Additional exponential components are incorporated only if the maximum log likelihood increases more than 2 log likelihood units.[30–35] A dead time (τ_d) of 50 μs is imposed retrospectively; events shorter than 50 μs are ignored.

13.4.1.4   Mechanosensitivity

Determining the mechanosensitivity of ion channels has proven to be a daunting task. Mechanosensitive ion channels have been studied extensively, but the mechanosensitivity of TRP channels is still unclear. In order to conclusively determine the type of mechanical stimulus a channel responds to, it is necessary to use several types of mechanical stimuli with multiple controls.

Mechanosensitive channels can be activated by changes in osmolarity, so hypotonic solutions are often used to study mechanosensitivity. The standard extracellular solution for electrophysiological measurements contains (in mM): 150 NaCl, 6 CsCl, 1 $MgCl_2$, 5 $CaCl_2$, 10 glucose, and 10 HEPES, titrated to pH 7.4 with NaOH. The osmolarity of this solution, as measured with a vapor pressure Vapro 5520 osmometer (Wescor, Inc, Logan, Utah, USA), is around 320 milliosmolar. For measuring currents due to a change in osmolarity, an isotonic solution is used that contains (in mM): 105 NaCl, 6 CsCl, 5 $CaCl_2$, 1 $MgCl_2$, 10 HEPES, 90 D-mannitol, and 10 glucose titrated to pH 7.4 with NaOH (~320 milliosmolar). Hypoosmolar solution is obtained by omitting mannitol from the solution (240 milliosmolar).

Mechanical stress induced by changing osmolarity represents a diffused mechanical stimulus. However, responses to mechanical stimuli can be due to punctate indentation of the membrane. To study this properly, a piezo-controlled probe (EXFO Life Sciences Group)[36,37] that advances in 0.5–2 µm/ms steps can be used. The stimulus is applied for a duration of 200–500 ms. The probe is advanced until a change in membrane conductance is observed.

In order to determine the direct mechanical sensitivity, the pressure-clamp technique developed by McBride and Hamill[38,39] can be used. The positive and negative pressure is applied by opening a piezo-driven valve. The extent of valve opening regulates the final pressure applied. The pressure is measured using a pressure transducer that has been calibrated. The pressure is changed in a ramp-like fashion from 0 to 100 mmHg over a period of 20 s. The ramp protocol changes the pressure over time and enables recording of the single-channel activity in response to increasing magnitude of pressure. Patch pipettes are pulled on a Narishige vertical puller to have a tip diameter of ~2 µm. Single mechanosensitive channels are analyzed as described above.

13.4.2   Function of TRP Channels in the Central Terminals of Sensory Neurons

13.4.2.1   Synaptic Current Recordings from DH Neurons in DRG-DH Co-Cultures

Spontaneous and Miniature Synaptic Currents: Cover slips with the co-cultured neurons are mounted on the stage of an Olympus IMT-2 microscope for patch-clamp recordings. The DH neurons are distinguished from the DRG neurons based on their morphology and electrophysiological properties. DH neurons are fusiform, pyramidal, or multipolar in shape, in contrast to the rounded pseudounipolar characteristics of the DRG neurons. Moreover, when a DRG neuron is voltage clamped, no miniature excitatory postsynaptic currents (mEPSCs) are observed, and capsaicin application may result in an inward whole-cell current. However, mEPSCs in controls and upon TRP channel agonist application can be recorded when a DH neuron is voltage clamped. The bath solution contains (in mM): 150 Na-gluconate/NaCl, 5 KCl, 2 $MgCl_2$, 0.1 $CaCl_2$, 10 glucose, and 10 HEPES, titrated to pH 7.4 with NaOH. The recording pipettes are made from borosilicate glass, which has a resistance of 3–10 MΩ when filled with a solution that contains (in mM): 140 Cs-Gluconate/CsMeSO$_3$, 5 CsCl, 10 HEPES, 5 $MgCl_2$, 10 EGTA, 5 Mg-ATP, and 1 Li-GTP, titrated to pH 7.4 with CsOH. Experiments are performed at the desired temperature by adjusting the temperature of the flow solution. In order to record EPSCs, DH neurons are voltage clamped (EPC10) at –60 mV (close to E_{Cl}). In order to record inhibitory postsynaptic currents (IPSCs), the currents are recorded at 0 mV. The capacitance and the series resistance are compensated, and the input resistance of the cell is monitored. The data are filtered at 2.5 kHz and digitized at 5 kHz. Adequate voltage clamp is confirmed by recording the mEPSCs at different voltages and determining their reversal potentials. A new cover slip should be used for every experiment to prevent rundown of mEPSCs and the activation of second messenger pathways.

Neurons are perfused with standard bathing solution and lidocaine (10 mM), strychnine (1 μM), and bicuculline (10 μM) to study glutamatergic transmission, and CNQX (20 μM) and APV (50 μM) to study GABAergic and glycinergic transmission. A high concentration of lidocaine is used to block both TTX-sensitive and TTX-resistant Na^+ channels. TRP channel agonists/antagonists are bath-applied for 30 s, and changes in mEPSCs are recorded continuously thereafter. Effects of phosphorylation can be studied using kinase activators such as phorbol 12,13-dibutyrate (1 μM).

Evoked Synaptic Currents: DRG and DH neurons are voltage-clamped simultaneously using a dual-electrode voltage-clamp technique. Monosynaptic EPSCs are recorded in DH neurons in the voltage-clamp mode by generating an action potential in DRG neurons in the current-clamp mode. The bath solution contains (in mM): 150 NaCl, 5 KCl, 2 $MgCl_2$, 0.5 $CaCl_2$, 10 glucose, and 10 HEPES, titrated to pH 7.2 with NaOH. The pipette solution for the DRG neuron contains (in mM): 140 KCl, 10 HEPES, 5 $MgCl_2$, 10 EGTA, 5 Mg-ATP, and 1 Li-GTP, titrated to pH 7.4 with KOH. The pipette solution for DH neurons contains $CsSO_4$ instead of KCl. Monosynaptic connections are confirmed by the constant latency of the EPSCs. It is necessary to monitor the membrane capacitance (C_m), membrane resistance (R_m), and access resistance (R_a) continuously to avoid erroneous interpretation of the results.

13.4.2.2 Synaptic Current Recording from Spinal Cord Slices

Spontaneous and Miniature Synaptic Currents: Spinal cord or CSTN slices are placed on the stage of an upright near-infrared differential interference contrast microscope, Olympus BX-50wi, for patch-clamp recording.[40,41] In these preparations, only the TG and DRG neuronal terminals are intact without the cell bodies. Pipettes with tips of 2–5 MΩ are pulled from thick-walled borosilicate glass (1B150F-4, World Precision Instruments) with a horizontal pipette puller P-97 (Sutter Instruments). Slices are visualized with an upright microscope and Gibralter stage using a 40× objective and infrared filter (Olympus). The image is sent to a TV monitor. For voltage-clamp studies, the pipette solution contains (in mM): 140

CsMeSO$_3$, 10 EGTA, 10 HEPES, 5 CsCl, 5 MgCl$_2$, 5 MgATP, and 1 LiGTP, titrated to pH 7.4 with CsOH. For current-clamp studies, a pipette solution contains (in mM): 130 K-gluconate, 10 NaCl, 1 MgCl$_2$, 0.2 EGTA, 1 K$_2$ATP, and 10 HEPES, titrated to pH 7.4 with NaOH. Whole-cell currents are recorded using a patch-clamp amplifier (EPC 10/Axopatch 200B) and acquisition software (Pulse 8.6/pCLAMP). The data are collected with the filter set at 2.5 kHz (−3 dB frequency with an eight-pole low-pass filter, LP10), digitized at 5 kHz (VR-10B), and stored on a hard disk. The recording chamber (PH1, Warner Instrument) is perfused at 4 mL/min with an extracellular solution, artificial cerebrospinal fluid (aCSF) that contains (in mM): 126 NaCl, 2.5 KCl, 1.4 NaH$_2$PO$_4$, 1.2 MgCl$_2$, 2.4 CaCl$_2$, 25 NaHCO$_3$, and 10 glucose, titrated to pH 7.4 (290 mOSM), and is aerated with a 95% O$_2$–5% CO$_2$ gas mixture. Temperature is held at 28°C using a glass water jacket and circulating water bath (VWR 1130, VWR, Batavia, IL, USA). The tip potential is cancelled before forming a gigaseal. Junction potential can be determined and corrected according to Barry and Lynch.[22] The activities of GABA, glycine, AMPA, NMDA, and voltage-gated Na$^+$ channels are blocked when necessary with 30 μM bicuculline, 5 μM strychnine, 100 μM CNQX, 20 μM APV, and 500 nM TTX (all dissolved in the extracellular solution), respectively. The lidocaine derivative (QX-314) is included to prevent action potentials in the recording neuron.

Evoked Synaptic Currents: To obtain evoked EPSCs or IPSCs, the neurons are voltage-clamped at −60 or 0 mV. A concentric bipolar stimulating electrode is placed on the spinal trigeminal or Lissauer tract. For horizontal slices of CSTN, the electrode is placed caudal to the recording site for orthograde transmission. A Grass Stimulator (S88) with stimulus isolation unit PSIU 6 (Grass Technologies, West Warwick, RI, USA) triggered by a Master 8 (A.M.P.I., Jerusalem, Israel) is used to stimulate a concentric bipolar electrode (Rhodes Medical Instruments, Tujunga, CA) placed on the sensory fiber tract. Stimulus duration is 100 μs, and half maximal stimulus intensity is used (less than 800 μA, usually 200–800 μA for C-fibers). Spontaneous and evoked EPSCs are low-pass-filtered at 2.5 kHz and digitized at 5 kHz. The digitized signal is stored on a hard drive on a PC-compatible computer. Fast and slow capacitance compensation is performed in Pulse 8.6/pCLAMP. Input resistance and series resistance are measured every 2–5 min in the voltage-clamp mode with three small (ΔV 10 mV, 150 ms) hyperpolarizing voltage steps. Cells showing greater than 20% change in series resistance should not be included for further analysis. Offline data analysis is done with pCLAMP software. sEPSCs are analyzed using the Mini Analysis Program (Synaptosoft, Decatur, GA, USA). The threshold for event detection (usually 10 pA) is at least 3 times baseline noise levels.

Blind Patch-Clamp Technique: Using the blind patch-clamp technique, a randomized unbiased sample of neurons in a given preparation can be accomplished without sophisticated equipment. Electrophysiological recordings can be made in spinal cord slices using a standard patch-clamp amplifier, dissecting microscope, and recording chamber. In addition, some preparations require thick slices, and direct visualization is not possible.[42]

Sharp Electrode Technique: In situations where slices from adult animals are used, it is difficult to form gigaseals routinely. Therefore, the sharp electrode technique (single electrode voltage-clamp or current-clamp)[43] is used, in which the neurons are impaled using a manipulator (PCS-5100, EXFO Life Sciences Group), and the potential inside the cell membrane can be recorded with minimal effect on the ionic constitution of the intracellular fluid. The electrodes are like those for patch clamp (pulled from glass capillaries), but the tip size is much smaller (10 to 100 s MΩ). Signals from impaled neurons are amplified with an Axoclamp 2A amplifier (Molecular Devices). Neurons are recorded under current-clamp or voltage-clamp conditions. While using voltage-clamp conditions, it is necessary to make sure that adequate voltage-clamp is achieved because of high-resistance electrodes. The data collection and analysis are similar to those described for patch-clamp techniques.

13.5   EX VIVO PREPARATIONS TO STUDY TRP CHANNEL FUNCTION IN SENSORY NEURONS

13.5.1   SKIN-NERVE PREPARATION

The details of skin-nerve preparation have been described previously.[44–46] Rats/mice are anesthetized with isoflurane and killed by cervical dislocation. The saphenous nerve and its innervated skin are located in the medial-dorsal side of the hind paw and exposed by removing the hair. The nerve and skin are carefully excised. The skin is placed corium side up in the in vitro perfusion chamber and is superfused with a modified Krebs-Hensleit solution containing (in mM): 110.9 NaCl, 4.8 KCl, 2.5 $CaCl_2$, 1.2 $MgSO_4$, 1.2 KH_2SO_4, 24.4 $NaHCO_3$, and 20 glucose, which is saturated with 95% O_2–5% CO_2. The temperature of the bath is maintained at 34°C. The saphenous nerve is drawn through a small hole into the recording chamber, which is filled with a layer of paraffin oil. The nerve is placed on a mirror, and individual nerve fibers are separated to record single-unit activity.

Fiber types are categorized by conduction velocity. Units conducting slower than 1.2 m/s are classified as unmyelinated C-fibers; those conducting between 1.2 and 10 m/s are classified as thinly myelinated Aδ-fibers; and units conducting faster than 10 m/s are classified as myelinated Aβ-fibers. Mechanically sensitive Aβ-fibers are further categorized as slowly adapting (SA) if they responded throughout a sustained force of 10-s duration or rapidly adapting (RA) if they responded only at the onset or offset of force. Mechanically sensitive Aδ-fibers are classified as A-mechanoreceptor (AM) fibers if they exhibit slowly adapting responses to sustained force, or as down-hair (D-hair) receptors if they are rapidly adapting.[45,46]

Receptive fields are identified by probing with a blunt glass rod in the corium side of skin. Conduction velocity of the fiber is determined by monopolar electrical stimulation (variable intensity, 0.2 Hz, and 2–3 ms duration) into the receptive field. The distance between the receptive field and the recording electrode (conduction distance) is divided by the latency of the action potential. The mechanical threshold of units is tested with a set of calibrated von Frey hairs made from nylon filaments (Stoelting, Wood Dale, IL, USA). To test the heat sensitivity of the localized receptors, radiant heat stimulation from a halogen lamp (150 W) is applied to the epidermal surface. A thermocouple is placed at the corium side in the receptive field for feedback control. The temperature is increased linearly from 32° to 47°C in 15 s. For applying chemical solutions, a metal ring is used to isolate the receptive field.

Action potentials are amplified, filtered, and displayed on an oscilloscope and continuously recorded on videotape using an analog–digital converter (VR-10B). The data are analyzed offline using pCLAMP software. The magnitude of the responses of a C-fiber nociceptor is determined by counting the total impulses (action potentials) evoked during the 5 min after onset of superfusion. For counting the total impulses, spontaneous discharges during the 60-s control period are multiplied by 5 and then subtracted from the 5-min count after drug addition.[46]

13.5.2 SKIN-NERVE-DRG-SPINAL CORD PREPARATION

Sensory neurons project to specific regions of the spinal DH and are correlated with specific modality sensations. The ontogenic changes and the changes that occur in different disease conditions can be studied using an ex vivo somatosensory preparation that includes the skin, nerve, DRG, and the spinal cord.[47] Adult rat/mice are anesthetized with a mixture of ketamine and xylazine (85 and 5 mg/kg, respectively, I.P.). Following the conformation of anesthesia, animals are perfused transcardially with chilled (14°C) and oxygenated (95% O_2–5% CO_2) modified aCSF, which consists of (in mM): 253.9 sucrose, 1.9 KCl, 1.2 KH_2PO_4, 1.3 $MgSO_4$, 2.4 $CaCl_2$, 26.0 $NaHCO_3$, and 10.0 D-glucose. The right hind limb and the spinal cord are excised and placed in a constantly circulating bath. The area of the skin innervated by the saphenous nerve and the corresponding DRGs and spinal cord are dissected. The skin is spread flat on a platform and allowed to dry by adjusting the fluid level of the recording chamber. The cell bodies of L2 and L3 DRG neurons are impaled with sharp quartz microelectrodes (Vector Laboratories). The peripheral response properties of the neurons are assessed using controlled thermal, mechanical, and chemical stimuli as described in the previous section.

13.6 IN VIVO BEHAVIORAL STUDIES TO EVALUATE THE FUNCTION OF DRG/TG NEURONS

13.6.1 MEASUREMENT OF THERMAL SENSITIVITY

13.6.1.1 Paw Withdrawal Latency

Thermal nociceptive responses are determined using a plantar test instrument (Ugo Basile, Camerio, Italy), as described previously.[48] Rats/mice are habituated to the apparatus (1 hour per day for 5 days). A mobile radiant heat source is located under the table and focused onto the desired paw. Paw withdrawal latencies (PWLs) are recorded three times for each hind paw, and the average is taken as the baseline value. A timer is automatically activated with the light source, and response latency is defined as the time required for the paw to show an abrupt withdrawal. The apparatus can be calibrated to give a PWL of approximately 6–20 s. In order to prevent tissue damage, a cutoff at 20 s must be set. Rats are accustomed to the test conditions 1 hour per day for 5 days. At least 4 days should elapse between two consecutive tests.

13.6.1.2   Tail-Flick Test for Thermal Hyperalgesia

Tail flick latencies to radiant heat are assessed using methods described.[49,50] Tail flick latencies are determined every 5 min four times before and after systemic or intrathecal drug administration.

13.6.1.3   Carrageenan-Induced Thermal Hyperalgesia

After obtaining baseline values of PWL to radiant heat, the animals receive an intraplantar injection of carrageenan (1%, 100 µL) into the left hind paw.[51] After the induction of inflammation (assessed by thickness and volume), PWL to thermal stimuli is determined after 24 to 48 hours.

13.6.1.4   Complete Freund's Adjuvant (CFA) Induced
Inflammatory Thermal Hyperalgesia

Chronic peripheral inflammation is induced by injecting 100 µL CFA (1 mg myco-bacterium per 0.85 mL paraffin oil and 0.15 mL of mannide monooleate) (Sigma) in the left hind paw of animals. The inflammation develops slowly over a period of a few days and lasts for a few weeks. Hyperalgesia to noxious thermal stimulation is measured by determining the PWL 3 days after CFA administration as described above. Data are plotted to contrast thermal hyperalgesia between inflamed and control paws of each animal in the two groups.

13.6.2   Measurement of Chemical Sensitivity by Nocifensive Behavior

Chemical-evoked nocifensive behavior in rats is defined as lifting (guarding), licking, and shaking of the injected paw.[52] The number of times the rat exhibits guarding, licking, and shaking is counted, and the total duration of this behavior is measured over 5 min immediately after intraplantar administration of agonists.

13.6.2.1   Formalin-Induced Pain Behavior

Each rat is placed in an observation cage and is allowed to adapt to its environment for 10 minutes. Formalin (50 µL, 5%) is then injected intracutaneously into the hind paw, and the animal is immediately transferred to the observation cage. The number of flinches (rapid withdrawal of the injected hind paw) produced by the animal is counted for 60 min. Animals exhibit two phases of pain behavior (phase 1, first 10 min; phase 2, 10–60 min). Data are plotted to represent the number of flinches produced in each 5-min period following formalin injection.

13.6.3   Measurement of Mechanosensitivity

13.6.3.1   von Frey Filament Test

Weekly calibrated nylon monofilaments are used (Stoelting). Filaments 2 to 9 (0.015–1.3 g) are firmly applied to the plantar surface of the mid hind paw (alternating the side of the body being tested) until they bowed for 5 s.[53] Vigorous vertical and/or horizontal movement of the hind paw away from the fiber is considered as a withdrawal response. Fibers are applied in random order; no more than two measurements are made per behavioral state per fiber per rat/mouse.

Mechanical sensitivities are also assessed using a dynamic plantar aesthesiometer instrument using von Frey probe (Ugo Basile, Camerio, Italy).[54] A 0.5-mm-diameter von Frey probe is applied to the plantar surface of the rat hind paw with pressure increasing by 0.05 Newtons/s. The pressure at which a paw withdrawal occurs is recorded, and this is taken as the paw withdrawal threshold (PWT). For each hind paw, the procedure is repeated 3 times, and the average pressure to produce withdrawal is calculated. Successive stimuli are applied to alternating paws at 5-min intervals.

13.6.3.2   Randall–Selitto Paw Pressure Test

The nociceptive flexion reflex is quantified with a Randall–Selitto paw pressure device[55] (Analgesymeter, Stoelting) that applies a linearly increasing mechanical force to the dorsum of the rat's hind paw.[56] The nociceptive threshold is defined as the force in grams at which the rat withdraws its paw. Rats are familiarized in the testing procedure at 5-min intervals for a period of 1 hour/day for 3 days in the week preceding the experiments. Baseline PWT is defined as the mean of six readings.

13.6.4   CHRONIC CONSTRICTION INJURY (CCI) MODEL, BENNETT–XIE MODEL

Male SD rats (250–300 g) are anesthetized with ketamine and xylazine (85 and 5 mg/kg, respectively, I.P.). The skin of the middle thigh is cut, and the muscle is separated to expose the left sciatic nerve. Four ligatures are loosely tied around the nerve by using 4-0 braided silk thread with a 1.5-mm interval. The surgical incision in the skin is closed. Amikacin (10 mg/kg, s.c.) is injected every day for one week to prevent infection. Thermal, mechanical, and chemical sensitivities are tested to determine the changes that occur following nerve injury.

13.6.5   INTRATHECAL ADMINISTRATION OF DRUGS BY CATHETER IMPLANTATION

13.6.5.1   Intrathecal Administration by Catheter

In order to target the TRP channels expressed at the central terminals of the sensory neurons, chronic intrathecal catheters are implanted.[57] Male SD rats (225–250 g) are anesthetized with ketamine and xylazine (85 and 5 mg/kg, respectively, I.P.). When they no longer respond to the paw pinch test, the neck area is shaved, and the skin is swabbed with betadine followed by 70% alcohol. A small incision is made in the skin, and the muscles are separated to expose the atlanto-occipital membrane. A small incision is made in the membrane to allow a polyethylene-10 catheter with 0.9% sterile saline to be inserted into the subarachnoid space. The catheter is gently threaded through the space as far as the lumbar enlargement (approximately 7.5 cm). The catheter is then sutured in place with the muscles, and the incision is closed. About 5 cm of catheter is exposed externally for injections. The external port is sealed with Parafilm to prevent flow of cerebrospinal fluid. Rats are allowed to recover for 7 days after surgery. To prevent infection, 10 mg/kg of kanamycin is injected subcutaneously every day for 5 days during recovery. Agents to be tested are administered in 20-μL volumes by slow infusion followed by 20-μL saline using a Hamilton syringe.[58]

13.6.5.2   Intrathecal Administration of Drugs by Lumbar Puncture

All drugs or their appropriate vehicles are injected intrathecally in a volume of 3 µL by lumbar puncture using a Hamilton syringe and 30-gauge needle.[59] The injection is made at the space between L5 and L6 where the spinal cord ends and the cauda equina begins.[60] A 27-gauge needle is used to puncture the muscle, spinal process, and the dura until a characteristic brief motor response of tail or hind limbs is observed indicating the penetration of the dura.

13.6.6   Behavioral Assays Involving TG Neurons

13.6.6.1   Catheter Implantation

In order to study the changes in peripheral terminals of TG neurons, the nerve innervations of the dura mater can be manipulated.[61] Male SD rats (225–250 g) are anesthetized with ketamine and xylazine (85 and 5 mg/kg, respectively, I.P.). When they no longer respond to the paw pinch test, the hair is removed, and the skin is swabbed with betadine followed by 70% alcohol. The bregma is exposed with a 2-cm incision and retracting the skin. An electric drill is used to bore bilateral (3–4 mm to midsaggital suture) troughs (2-mm diameter and 8–10 mm long). The catheters are constructed using PE-10 tubing (Instech Laboratories, Inc., Plymouth Meeting, PA, USA) and inserted through the troughs about 4–5 mm; they are placed above the occipital lobe. The catheter is then sutured in place with the skin while closing the incision. About 5 cm of the catheter is exposed externally for injections. The external port is sealed with Parafilm to prevent flow of cerebrospinal fluid. Rats are allowed to recover for 7 days after surgery. To prevent infection, 10 mg/kg of kanamycin is injected subcutaneously every day for 5 days during recovery.

In order to sensitize the peripheral terminals of TG neurons, pro-inflammatory agents to be tested, namely, bradykinin (1 µM), histamine (0.1 mM), prostaglandin E2 (2 µM), and serotonin (0.1 mM) (inflammatory soup), are administrated in 5–6 µL volumes by slow infusion followed by 5–6 µL saline.[61]

13.6.6.2   Facial Allodynia

Following the injection of inflammatory soup into the skull, facial allodynia can be assessed using von Frey filaments. All testing is done in a blinded fashion. Areas around the eye, the cheek, the shoulder, and the forearm are tested. There are two methods that can be used: (1) The gram force filament in the middle of the logarithmic range is tested, and then the force is increased or decreased by using different filaments until the animal responds reliably.[62] (2) The gram force filament to which the majority of control rats respond versus the gram force filament to which the majority of allodynic rats respond is determined.[62] Using the method described by Ren,[62] absolute thresholds can be recorded. With the method described by Tawfik et al.,[63] absolute thresholds cannot be determined. Following 5 days of general habituation to handling (~5 min/day), the baseline behavior is assessed. A logarithmic series of 10 calibrated von Frey hairs (Stoelting) are applied as described by Ren.[62] The rat is stimulated 5 times with each filament for 2 s on either side of the face. The filament size is increased if the rat responds 5 times to the filament. These data are then analyzed as previously described.[64]

13.7 STATISTICAL METHODS FOR DATA ANALYSIS

Data are expressed as mean ± S.E.M. When comparing the means of only two groups, Student's t test is used. For all tests, a p value lower than 0.05 ($p < 0.05$) is considered significant. Statistical procedures are implemented using SPSS 14.0 (SPSS, Chicago, IL, USA).

For in situ hybridization and immunohistochemical data, analysis of variance (ANOVA) is used to determine if differences in background-adjusted mean grain density or staining can be attributed to a change. Tests subsequent to the ANOVA are carried out using Bonferroni corrections to control the overall type I error rate.

For analysis of miniature and spontaneous synaptic events, Kolmogorov–Smirnov (KS) test is used to compare the cumulative probability curves for inter-event intervals and amplitudes.

For conduction velocity, temperature thresholds, and the number of spikes induced by different manipulations, statistical comparisons between two groups are made using an unpaired two-tailed t-test. Fisher's exact probability test is used to compare the percentage change of C-fiber responsiveness to various TRP channel agonists. The magnitudes of responses are compared using a nonparametric Mann–Whitney U-test.

For experiments that involve manipulation of one of the legs, data are normalized for each animal as the maximum possible effect (MPE). This value is calculated as follows: MPE = (PDR – IBR)/(CBR – IBR), where PDR is the postdrug response of the ipsilateral paw, IBR is the ipsilateral paw baseline response, and CBR is the contralateral paw baseline response. The nociceptive thresholds are calculated as mean values obtained from both hind paws. The data are subjected to a one-way ANOVA followed, when significant, by a post hoc Dunnett's t-test. Data for the touch-evoked agitation test are compared using the Mann–Whitney U-test because the data are not continuously distributed. von Frey Filament data are analyzed by performing linear regression applied to percentage withdrawal responses obtained across the entire fiber range and by solving the linear equation for a 50% response. Significance is established by nonoverlapping 95% confidence intervals.

13.8 CONCLUDING REMARKS

In recent years, there have been several discoveries regarding the molecular structures that respond to specific stimuli and their transduction mechanisms in sensory neurons. Several members of the TRP channel family have become important players in sensory transduction. A rich array of techniques is available to study the expression and function of TRP channels in sensory neurons. It is always desirable to use multiple techniques to infer definitive conclusions. While studying the expression of TRP channels, the specificity of antibodies must be confirmed. With lack of change in the message of a particular TRP channel, one must consider the possibility of RNA stabilization and posttranslational modifications before arriving at conclusions. Whole-cell and single-channel recordings can be used to characterize the conductance and kinetic properties and to identify the mechanism of rectification. However, caution should be exercised to prevent activation of channels other than the TRP channel of interest. Agonist and antagonist specificities are important to infer accurate conclusions. TRP channel-mediated responses exhibit Ca^{2+}-dependent desensitization and tachyphylaxis. Therefore, responses obtained with repeated agonist application must be analyzed carefully. While studying evoked synaptic responses, a decrease in the amplitude of evoked responses to application of TRP channel agonists may be due to presynaptic depolarization block.[58] While using in vivo or ex vivo approaches, changes in other channels (e.g., voltage-gated Na^+ channels) may induce an exaggerated response to the application of TRP channel agonists because activation of TRP channels causes the release of pro-inflammatory agents that can sensitize other channels. Therefore, sensitization of TRP channels as compared to sensitization of other channels must be taken into consideration. In most TRP channels, both N- and C-termini are intracellular, and generally, agonist binding sites are located at the N-terminus. Therefore, when using membrane impermeable agonists or second messenger molecules, the membrane permeability of the applied compounds must be known. Finally, appropriate statistical methods must be adopted to analyze the data for significance to arrive at conclusions.

ACKNOWLEDGMENTS

I thank Lauren Hughes and Deshou Cao for their help with editing the manuscript. This work was supported by grants from the National Institutes of Health (NS042296 and DK065742 and DA028017) and EAM award from SIUSOM.

REFERENCES

1. Clapham, D. E. 2003. TRP channels as cellular sensors. *Nature* 426: 517–524.
2. Minke, B., and B. Cook. 2002. TRP channel proteins and signal transduction. *Physiological Reviews* 82: 429–472.
3. Nilius, B., G. Owsianik, T. Voets, and J. A. Peters. 2007. Transient receptor potential cation channels in disease. *Physiological Reviews* 87: 165–217.
4. Cortright, D. N., and A. Szallasi. 2009. TRP channels and pain. *Current Pharmaceutical Design* 15: 1736–1749.
5. Lewin, G. R., and R. Moshourab. 2004. Mechanosensation and pain. *Journal of Neurobiology* 61: 30–44.
6. Lewin, G. R., Y. Lu, and T. J. Park. 2004. A plethora of painful molecules. *Current Opinion in Neurobiology* 15: 129.
7. Patapoutian, A., S. Tate, and C. J. Woolf. 2009. Transient receptor potential channels targeting pain at the source. *Nature Reviews Drug Discovery* 81: 55–68.
8. Wood, J. N. 2007. Recent advances in understanding molecular mechanisms of primary afferent activation. *Gut* 53 (Suppl. 2): 9–12.
9. Gottlieb, P., J. Folgering, R. Maroto et al. 2008. Revisiting TRPC1 and TRPC6 mechanosensitivity. *Pflügers Archiv* 455: 1097–1103.
10. Caterina, M. J., and D. Julius. 2001. The vanilloid receptor: a molecular gateway to the pain pathway. *Annual Review of Neuroscience* 24: 487–517.

11. Julius, D., and A. I. Basbaum. 2001. Molecular mechanisms of nociception. *Nature* 413: 203–210.

12. Premkumar, L. S., and P. Sikand. 2008. TRPV1: a target for next generation analgesics. *Current Neuropharmacology* 6: 151–163.

13. Grudt, T. J., and J. T. Williams. 1994. mu-Opioid agonists inhibit spinal trigeminal substantia gelatinosa neurons in guinea pig and rat. *Journal of Neuroscience* 14: 1646–1654.

14. Monyer, H., and P. Jonas. 1995. Polymerase chain reaction analysis of ion channel expression in single neurons in brain slices. Chapter 16 in *Single-Channel Recording*, 2nd ed., eds. B. Sakmann and E. Neher, 357–373. New York: Plenum Press.

15. Ramos-Vara, J. A. 2005. Technical aspects of immunohistochemistry. *Veterinary Pathology* 42: 405–426.

16. Hofler, H. 1990. Principles of in situ hybridization. In *In Situ Hybridization: Principle and Practice*, eds. J. M. Polak and J. O'D. McGee, 15–30. Oxford: Oxford University Press.

17. Kus, L., S. B. Mazzone, G. Paxinos, and D. P. Geraghty. 1998. Autoradiographic localisation of substance P (NK1) receptors in human primary visual cortex. *Brain Research* 794: 309–312.

18. Szallasi, A., and P. M. Blumberg. 1990. Resiniferatoxin and its analogs provide novel insights into the pharmacology of the vanilloid (capsaicin) receptor. *Life Sciences* 47: 1399–1408.

19. Szallasi, A., and P. M. Blumberg. 1990. Specific binding of resiniferatoxin, an ultrapotent capsaicin analog, by dorsal root ganglion membranes. *Brain Research* 524: 106–111.

20. Tognetto, M., S. Amadesi, S. Harrison et al. 2001. Anandamide excites central terminals of dorsal root ganglion neurons via vanilloid receptor-1 activation. *Journal of Neuroscience* 21: 1104–1109.

21. Zygmunt, P. M., J. Peterson, D. A. Anderson, H. Chuang, M. Sorgard, and V. Di Marzo. 1999. Vanilloid receptors on sensory nerves mediate the vasodilator action of anandamide. *Nature* 400: 452–457.

22. Barry, P. H., and J. W. Lynch. 1991. Liquid junction potentials and small cell effects in patch clamp analysis. *Journal of Membrane Biology* 121: 101–117.

23. Wollmuth, L. P., and B. Sakmann. 1998. Different mechanisms of Ca^{2+} transport in NMDA and Ca^{2+}-permeable AMPA glutamate receptor channels. *Journal of General Physiology* 112: 623–636.

24. Schneggenburger, R., Z. Zhou, A. Konnerth, and E. Neher. 1998. Fractional contribution of calcium to the cation current through glutamate receptor channels. *Neuron* 11: 133–143.

25. Frings, S., D. H. Hackos, T. Dzeja, T. Ohyama, V. Hagen, U. B. Kaupp, and J. I. Korenbrot. 2000. Determination of fractional calcium ion current in cyclic nucleotide-gated channels. *Methods Enzymology* 315: 797–817.

26. Egan, T. M., and B. S. Khakh. 2004. Contribution of calcium ions to P2X channel responses. *Journal of Neuroscience* 24: 3413–3420.

27. Hamill, O. P., A. Marty, E. Neher, B. Sakmann, and F. J. Sigworth. 1981. Improved patch-clamp techniques for high-resolution current recording from cells and cell-free membrane patches. *Pflügers Archiv* 391: 85–100.

28. Sakmann, B., and E. Neher, eds. 1995. *Single-Channel Recording*, 2nd ed. New York: Plenum Press.

29. Hilgemann, D. W. 1995. The giant membrane patch. Chapter 13 in *Single-Channel Recording*, eds. B. Sakmann and E. Neher, 307–326. 2nd ed. New York: Plenum Press.

30. Chung, S. H., J. B. Moore, L. G. Xia, L. S. Premkumar, and P. W. Gage. 1990. Characterization of single channel currents using digital signal processing techniques based on Hidden Markov Models. *Philosophical Transactions of the Royal Society of London. Series B, Biological Sciences* 329: 265–285.

31. Qin, F., A. Auerbach, and F. Sachs. 1997. Maximum likelihood estimation of aggregated Markov processes. *Proceedings Biological Sciences* 264: 375–383.
32. Premkumar, L. S., F. Qin, and A. Auerbach. 1990. Subconductance states of a mutant NMDA receptor channel kinetics, calcium, and voltage dependence. *Journal of General Physiology* 109: 181–189.
33. Premkumar, L. S., and G. P. Ahern. 2000. Induction of vanilloid receptor channel activity by protein kinase C. *Nature* 408: 985–990.
34. Premkumar, L. S., S. Agarwal, and D. Steffen. 2002. Single-channel properties of native and cloned rat vanilloid receptors. *Journal of Physiology* 545: 107–117.
35. Raisinghani, M., R. Pabbidi, and L. S. Premkumar. 2005. Activation of TRPV1 by Resiniferatoxin: Implications in chronic pain conditions. *Journal of Physiology* 567: 771–786.
36. Drew, L. J., D. K. Rohrer, M. P. Price, K. E. Blaver, D. A. Cockayne, P. Cesare, and J. N. Wood. 2004. Acid-sensing ion channels ASIC2 and ASIC3 do not contribute to mechanically activated currents in mammalian sensory neurons. *Journal of Physiology* 556: 691–710.
37. Hu, J., and G. R. Lewin. 2006. Mechanosensitive currents in the neurites of cultured mouse sensory neurons. *Journal of Physiology* 577: 815–828.
38. McBride, D. W., Jr., and O. P. Hamill. 1999. Simplified fast pressure-clamp technique for studying mechanically gated channels. *Methods Enzymology* 294: 482–489.
39. McBride, D. W., Jr., and O. P. Hamill. 1995. A fast pressure-clamp technique for studying mechanogated channels. Chapter 14 in *Single-Channel Recording*, 2nd ed., eds. B. Sakmann and E. Neher, 329–339. New York: Plenum Press.
40. Edwards, F. A., A. Konnerth, B. Sakmann, and T. Takahashi. 1989. A thin slice preparation for patch clamp recordings from neurons of the mammalian central nervous system. *Pflügers Archiv* 414: 600–612.
41. Eilers, J., R. Schneggenburger, and A. Konnerth. 1995. Patch clamp and calcium imaging in brain slices. Chapter 9 in *Single-Channel Recording*, 2nd ed., eds. B. Sakmann and E. Neher, 213–227. New York: Plenum Press.
42. Casteñada-Castellanos, D. R., A. C. Flint, and A. R. Kriegstein. 2006. Blind patch clamp recordings in embryonic and adult mammalian brain slices. *Nature Protocols* 1: 532–542.
43. Sontheimer, H. 1995. Whole-cell patch clamp recordings. In *Patch-Clamp Applications and Protocols, Neuromethods Series*, vol. 26, eds. A. A. Boulton, G. B. Baker, and W. Wolfgang, 37–73. New York: Springer-Verlag.
44. Reeh, P. W. 1986. Sensory receptors in mammalian skin in an in vitro preparation. *Neuroscience Letters* 66: 141–146.
45. Kress, M., M. Koltzenburg, P. W. Reeh, and H. O. Handwerker. 1992. Responsiveness and functional attributes of electrically localized terminals of cutaneous C-fibers in vivo and in vitro. *Journal of Neurophysiology* 68: 581–595.
46. Stucky, C. L., T. DeChiara, R. M. Lindsay, G. D. Yancopoulos, and M. Koltzenburg. 1998. Neurotrophin 4 is required for the survival of a subclass of hair follicle receptors. *Journal of Neuroscience* 18: 7040–7046.
47. Koerber, H. R., and C. J. Woodbury. 2002. Comprehensive phenotyping of sensory neurons using an ex vivo somatosensory system. *Physiology and Behavior* 77: 589–594.
48. Hargreaves, K., R. Dubner, F. Brown, C. Flores, and J. Joris. 1988. A new and sensitive method for measuring thermal nociception in cutaneous hyperalgesia. *Pain* 32: 77–88.
49. D'Amour, G. E., and D. L. Smith. 1941. A method for determining loss of pain sensation. *Journal of Pharmacology and Experimental Therapeutics* 72: 74–79.
50. Watkins, L. R., D. Martin, P. Ulrich, K. J. Tracey, and S. F. Maier. 1997. Evidence for involvement of spinal cord glia in subcutaneous formalin induced hyperalgesia in the rat. *Pain* 71: 225–235.

51. Winter, C. A., E. A. Risley, and G. W. Nuss. 1962. Carrageenin-induced edema in hind paw of the rat as an assay for anti-inflammatory drugs. *Proceedings of the Society for Experimental Biology and Medicine* 111: 544–547.
52. Gilchrist, H. D., B. L. Allard, and D. A. Simone. 1996. Enhanced withdrawal responses to heat and mechanical stimuli following intraplantar injection of capsaicin in rats. *Pain* 67: 179–188.
53. von Frey, M. 1922. Zur physiologie der juckemfindung. *Archives Neerland Physiologies* 7: 142–145.
54. Chaplan, S. R., F. W. Bach, J. W. Pogrel, J. M. Chung, and T. L. Yaksh. 1994. Quantitative assessment of tactile allodynia in the rat paw. *Journal of Neuroscience Methods* 53: 55–63.
55. Randall, L. O., and J. J. Selitto. 1957. A method for measurement of analgesic activity on inflamed tissue. *Archives internationals de pharmacodynamie et de thérapie* 111: 409–419.
56. Taiwo, Y. O., T. J. Coderre, and J. D. Levine. 1989. The contribution of training to sensitivity in the nociceptive paw-withdrawal test. *Brain Research* 487: 148–151.
57. Yaksh, T. L., and T. A. Rudy. 1976. Analgesia mediated by a direct spinal action of narcotics. *Science* 192: 1357–1358.
58. Jeffry J. A., S. Q. Yu, P. Sikand, A. Parihar, M. S. Evans, and L. S. Premkumar. 2009. Selective targeting of TRPV1 expressing central terminals of spinal cord for long lasting analgesia. *PLoS One* 4: e7021.
59. Hylden, J. L., and G. L. Wilcox. 1980. Intrathecal morphine in mice: a new technique. *European Journal of Pharmacology* 67: 313–316.
60. Sidman, R. L., J. B. Angevine, and E. T. Pierce. 1971. *Atlas of the Mouse Brain and Spinal Cord*. Cambridge, MA, USA: Harvard University Press.
61. Burstein, R., H. Yamamura, A. Malick, and A. M. Strassman. 1998. Chemical stimulation of the intracranial dura induces enhanced responses to facial stimulation in brain stem trigeminal neurons. *Journal of Neurophysiology* 79: 964–982.
62. Ren, K. 1999. An improved method for assessing mechanical allodynia in the rat. *Physiology and Behavior* 67: 711–716.
63. Tawfik, V. L., N. Nutile-McMenemy, M. L. Lacroix-Fralish, and J. A. Deleo. 2007. Efficacy of propentofylline, a glial modulating agent, on existing mechanical allodynia following peripheral nerve injury. *Brain, Behavior, and Immunity* 21: 238–246.
64. Milligan, E. D., K. A. O'Connor, K. T. Nguyen et al. 2001. Intrathecal HIV-1 envelope glycoprotein gp120 induces enhanced pain states mediated by spinal cord proinflammatory cytokines. *Journal of Neuroscience* 21: 2808–2819.

14 Time-Resolved Activation of Thermal TRP Channels by Fast Temperature Jumps

Feng Qin

CONTENTS

14.1 INTRODUCTION

Humans recognize thermal stimuli with distinct sensations of being noxious cold, cold, warm, and noxious hot. Recordings from peripheral nerve fibers have demonstrated the existence of thermally active neurons known as thermal receptors. When skin temperature is raised above 30°C, thermal receptors specialized for detection of warmth start to fire action potentials,[1] and when the temperature is further increased to above about 45°C, the so-called nociceptors become active causing perception of pain. The activity of warm receptors reaches maximum discharge at or below 45°C and then decreases abruptly as the skin is warmed further.[1] Thus, warm receptors cannot generate signals for differentiating noxious from innocuous hot stimuli. Cold receptors, on the other hand, show peak responses at about 25°–27°C,[1,2] while cold nociceptors have an activation threshold below about 20°C. The response of cold receptors covers a relatively wide range of temperatures, which overlap with the response of warm receptors at near 37°C.[1] Cold receptors become inactive upon warming. They are poor indicators of absolute temperature but are extremely sensitive to localized temperature changes, and their responses are slowly adapting.[3]

Despite extensive demonstration of thermally sensitive nerve fibers in vivo, the molecular entities for thermal transduction have remained enigmatic until the recent discovery of TRP channels (see, e.g., Ref. 4). Several TRP channels from different subfamilies show temperature-dependent activation. These include TRPV1-4 from the vanilloid subfamily, TRPM8 from the melastatin subfamily, and TRPA1 from the TRPA subfamily. TRPV1 was first shown to be a long-sought ion channel in the nociceptive sensory neurons that responded to capsaicin, the hot pungent ingredient of chili peppers.[5] In addition to its vanilloid sensitivity, the cloned TRPV1 receptors, when heterologously expressed, are activated at temperatures above 42°C, consistent with the known temperature threshold of nociception.[6] TRPV2-4 were identified as homologues of TRPV1.[7–11] TRPV2 has an activation temperature threshold of above 50°C and is expressed in medium- to large-diameter myelinated neurons of the dorsal root ganglia (DRG). Both its temperature threshold and expression patterns support it as a candidate for the high-threshold heat response of Aδ fibers in vivo. TRPV3 and TRPV4 are activated by innocuous heat with reported temperature threshold ranges of 30°–40°C and ~25°–34°C, respectively. TRPV3 is also sensitive to compounds that induce warm feeling such as oregano, savory, and thyme, consistent with its role for warmth detection.[12] Common to both TRPV3 and TRPV4 is their very low expression levels in the sensory neurons; instead, they are prominently found in skin keratinocytes.[8–11] TRPV4 is also strongly expressed in the kidney, where it has been implicated in regulating the body fluid level as an osmosensor owing to its mechanical sensitivity to hypotonicity.[13,14]

Parallel to heat sensation, cold sensation can be chemically mimicked by plant-derived cooling compounds such as menthol from mint oil. TRPM8 is a molecular target of menthol in sensory neurons[15,16] and is activated by innocuous cool temperatures below 25°C, although its responsiveness continues into the noxious cold range. TRPA1 has been suggested to mediate noxious cold transduction,[17] but its cold sensitivity remains a highly debated issue (see, e.g., Ref. 18). Nevertheless, its functions in nociception are supported by its co-expression with TRPV1 in nociceptors and its chemosensitivity to compounds such as mustard oil.[19] Studies of the knockout of TRPV1, TRPV, and TRPM8 genes in mice have corroborated their contributions to thermal sensation (TRPV3 and TRPM8) or heat-related hyperalgesia (TRPV1),[20–25] while reports on the TRPA1 knockouts have invigorated the controversy regarding its involvement in cold sensation.[26,27]

The discoveries of the thermally active ion channels have now made it possible to study thermal sensation at the molecular level, just like the studies of other senses quite a long time ago. Importantly, these channels also raise biophysical questions about the mechanisms of thermal gating. Historically, ion channels have been mostly studied for gating by other variables such as voltage and chemical agonists.

In contrast, the temperature is fundamentally different from these stimuli because thermal energy interacts with proteins globally. It is understood that voltage may be sensed by localized charges in a low dielectric environment (e.g., membranes), while agonists can be detected through stereochemically specific binding sites. However, it is not obvious whether such a "key-and-lock" mechanism also works for detection of thermal energy. The activation of thermal TRP channels by temperature has been observed in excised membrane patches (e.g., Ref. 28), indicating that their thermal sensitivity is membrane delimited. The temperature coefficients of the responses are high (for review, see Ref. 29), and single-channel analysis shows that the temperature has a localized effect primarily on the long closures between opening bursts, in spite of complex gating kinetics comprising multiple closed and open states.[28]

Both heat activation of TRPV1 and cold activation of TRPM8 exhibit large but compensatory enthalpy and entropy changes.[28,30] As thermally gated TRP channels are also sensitive to voltage, Voets et al.[31] suggested that temperature opens the channel through voltage sensors by shifting the voltage-dependent gating curve toward more physiological membrane potentials. Alternatively, a change of membrane potential can also alter thermal or agonist activation. Brauchi et al.[30] and Matta and Ahern[32] showed that such mutual dependence between stimuli can be better explained by a Monod–Wyman–Changeux (MWC)-type allosteric model. For cold receptors TRPM8, Latorre et al.[33] also reported a temperature-independent Cole–Moore shift of voltage activation, suggesting that voltage sensors move separately from thermal sensors. Common to these models is that the energies from different stimuli contribute in an additive manner to stabilize the open conformations of the channel. Lately, Yao et al.[34] showed instead that the thermal sensitivity interacts with voltage or agonist sensitivity in a nonadditive manner based on direct measurements of temperature responses of TRPV1. Their data support that the activation of thermal sensors is coupled to agonist binding or charge movement of voltage sensors.

Structurally, thermal TRP channels have a membrane topology similar to that of voltage-gated channels, consisting of six transmembrane segments (S1–S6) and a reentrant pore loop between S5 and S6. However, despite their apparent activation by membrane depolarization, they lack a highly charged S4 segment, which is common to voltage-gated channels and acts as a voltage sensor. They also have relatively large N- and C-termini, which harbor a variety of regulatory sites and protein–protein interaction domains, such as the ankyrin-like repeats found in TRPV and TRPA channels. The sequence homology of TRP channels is generally limited across subfamilies. The structural basis of the temperature sensitivity has been revealed in several mutagenesis studies. First, exchanges of the C-terminal domains between TRPV1 and TRPM8 reversed their hot and cold sensitivity.[35] Second, mutations in the inner pore of TRPV1 also influenced its heat responses,[36] while single residue mutations in the outer pore region impacted heat activation of TRPV3.[37] Thus, it appears that the thermal sensitivity of these channels may spread over multiple regions of the proteins. In addition, both chemical and thermal activators induce similar structural changes in the S6 gate region,[38] suggesting that a common gate is shared by different stimuli.

Studies of thermal TRP channels require controlled temperature perturbation. This chapter presents a review of thermal control methods for patch-clamp experiments and their applications for measuring the temperature dependence of thermal TRP channels. The technique of laser diode irradiation is emphasized. This approach offers a time resolution far superior to conventional thermoelectric heating while allowing for rapid modulation of laser output to clamp temperature. The instrumentation design of a submillisecond temperature clamp apparatus is described. Applicability of this system is demonstrated with the heat-gated TRPV1 channel by measurements of its activation rates and energy landscape of temperature gating. The results show that the time-resolved fast temperature jump experiments can provide new insights into the temperature gating mechanisms of thermal TRP channels.

14.2   THERMODYNAMIC PRINCIPLES OF TEMPERATURE GATING

Regardless of the structural basis of thermal gating, it is imperative to understand its thermodynamic basis. For a channel existing between two states, its open probability is dictated by a Boltzmann relationship

$$P_o = \frac{1}{1 + \exp\left[\dfrac{\Delta G}{RT}\right]} \tag{14.1}$$

where ΔG represents the free energy change of the system from closed to open. At rest, the free energy has a profile with $\Delta G > 0$ so that the closed state is favored. Application of a stimulus alters the profile to favor the open state instead. For voltage or agonist gating, the external electrical or chemical energy is added to reduce the free energy difference ΔG. For temperature gating, the perturbation results merely from thermal energy. The gating by temperature therefore relies on the entropy of the system to offset the free energy barriers of activation.

In terms of enthalpy and entropy, the Boltzmann relationship becomes

$$P_o = \frac{1}{1 + \exp\left[\dfrac{\Delta H}{RT} - \dfrac{\Delta S}{R}\right]} \tag{14.2}$$

where ΔH and ΔS are, respectively, the enthalpy and entropy changes between the closed and open states, and R and T have their standard definitions. A few observations can be inferred from this relationship. First, the temperature dependence of opening is strictly determined by the enthalpy change (ΔH). In an analogy to voltage gating, the inverse of temperature ($1/T$) is equivalent to voltage, while the enthalpy change plays a similar role to the gating charges. The enthalpy change thus determines the slope of the Boltzmann curve of temperature responses. Second, the entropy change (ΔS) has no effect on the temperature dependence of opening, but it affects the midpoint of the Boltzmann curve of opening ($T_{1/2} = \Delta H/\Delta S$). Third, the sign of the enthalpy change determines the polarity of temperature sensitivity as to whether the channel is gated by heat or cold.

In theory, Equations 14.1 and 14.2 imply that temperature gating may occur to any ion channel, although the activation temperature range may be too high or too low, beyond the melting point of the channel protein. In this regard, the thermal TRP channels are unique in that they have evolved with an accessible range of temperatures for activation (overlapping with physiological temperatures) and temperature dependence that is adequately high so that appreciable activity occurs when the ambient temperature becomes only slightly above (or below) the threshold.

The temperature coefficient, or Q_{10}, is commonly used to characterize the temperature dependence of the kinetics of a process. It measures the change of a rate when temperature is increased by 10°C and is related to the activation enthalpy (ΔH) by

$$Q_{10} = \exp\left[-\frac{\Delta H}{R}\left(\frac{1}{T+10} - \frac{1}{T}\right)\right]$$
$$\approx \exp\left[-\frac{\Delta H}{RT^2}\right]$$

(14.3)

where the approximation holds since $T \gg 10$. Conversely, for a given Q_{10} value, the activation enthalpy is determined by

$$\Delta H = \frac{RT(T+10)}{10}\ln Q_{10}.$$

(14.4)

Figure 14.1 plots the relationship of ΔH versus Q_{10} on both linear and logarithmic scales. Enzymatic reactions typically have Q_{10} values between 2 and 3, which corresponds to an enthalpy of <20 kcal/mol. In this range, the enthalpy is sharply dependent on Q_{10}. Thus, the term Q_{10} provides a sensitive measure of the temperature

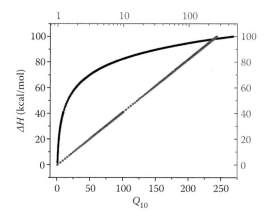

FIGURE 14.1 Q_{10} versus activation enthalpy (ΔH) plotted on both linear (black; left-bottom axes) and logarithmic (dark gray; right-top axes) scales. Typical ion channel gating has Q_{10} of 2–3.

dependence of the reaction. For large Q_{10} values (e.g., >20), however, the rate of enthalpy increase becomes slow. Thermal TRP channels can have temperature dependence as high as $\Delta H \sim 100$ kcal/mol,[34] and in this range, the enthalpy becomes nearly insensitive to changes in Q_{10}. In this regard, the term Q_{10} is not a sensitive descriptor for the temperature dependence of thermal TRP channels; the enthalpy itself seems to be more appropriate. By definition, Q_{10} is only pertinent to the rate of a process. In the studies of TRP channel gating, the temperature dependence is often evaluated from (pseudo) steady-state responses (the kinetics are hard to measure). The temperature coefficients calculated in this way reflect the equilibrium enthalpy change between the closed and open states (Equation 14.2) rather than the activation enthalpy between the closed and transition states (Equations 14.2 and 14.3).

14.3 MEASUREMENT OF TEMPERATURE DEPENDENCE

Because gating by temperature is governed by the thermodynamic properties of the channel, it is important to have precise estimates of the energetics for understanding the underlying mechanisms of gating. Below, we briefly summarize the experimental assessments of the temperature dependence of thermal TRP channels as they have been reported in the literature. The estimates of the temperature dependence sometimes exhibit significant variations with different types of measurements, which presumably could have resulted from the complexity of the experiments as well as the susceptibility of these channels to modulation by uncontrolled variables.

14.3.1 PSEUDO-EQUILIBRIUM ANALYSIS

By far, temperature ramps have appeared to be the most common protocol for measuring temperature dependence. In a typical experiment, whole-cell currents are continuously recorded while the temperature of the bath media is slowly changed. The ambient temperature of the bath solution may be changed through direct heating/cooling of the chamber or by superfusion with solutions that pass through an inline solution heater/cooler. Between the two, the inline solution heating/cooling appears to be more commonly used. Using a small-diameter tube coated with a platinum heating element, Dittert et al.[39] demonstrated a remarkable heating rate (100°C/s). Reid et al.[40] achieved a bipolar temperature change rate up to 4°C/s by restricting superfusion areas. More recently, Dittert et al.[41] further significantly improved the speed up to −40° to 60°C/s through a combination of resistive heating and thermoelectric heating and cooling at the very end of the perfusion outlet. With commercially available inline solution heaters/coolers (e.g., SC-20/CL-100, Warner Instruments), a rate of ~1°C/s appears to be more common. The rise of temperature is generally nonlinear, typically with a time course being relatively linear during the onset and becoming slower (sublinear) when approaching the set point.

For analysis, data (currents) are often plotted versus temperature ($1/T$) on a semilog scale. The plot may be considered as a variant of the Arrhenius plot for rate constants, but unlike the Arrhenius plot, which is linear against $1/T$, the current plot is generally nonlinear. It is typically biphasic, beginning with a slow increase at low temperature followed by a rapid increase above a certain threshold. The slow component arises from leak currents, while the fast component results from channel opening. The temperature coefficient of the channel activity is determined from the slope of the linear fit of the second component. According to the Boltzmann equation, when the open probability (P_o) is low, the current–temperature relationship is approximately

$$I \approx I_{max} \exp\left(\frac{\Delta H}{RT} - \frac{\Delta S}{R}\right), \tag{14.5}$$

and on the log scale, it becomes

$$\ln I \approx \frac{\Delta H}{RT} + \left(\ln I_{max} - \frac{\Delta S}{R}\right). \tag{14.6}$$

The slope of the fit thus gives rise to $\Delta H/R$. The asymptotic slope of the Boltzmann relationship consequently determines the enthalpy change between closed and open states. Once the enthalpy change is known, Q_{10} can be evaluated from Equation 14.3.

Although the approximation in Equation 14.5 is applicable only at low P_o, the linearity of the Boltzmann relationship can extend to P_o as high as ~0.2, as illustrated in Figure 14.2. Nevertheless, in practice, the choice of such a linear region may be less certain. The inevitable leak current will cause the linear asymptote to be actually curvilinear at low P_o. The lack of explicit knowledge of P_o also leaves the upper bound of the region largely empiric to determine. A small error in the choice of the region can lead to a large deviation in the fitting results because the activity of the channel is sharply dependent on temperature. Also of concern is that the analysis assumes that the gating of the channel can be considered at equilibrium as temperature is changed. The assumption is appropriate at high temperatures where the gating

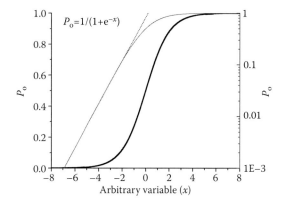

FIGURE 14.2 Boltzmann function plotted on both linear (black, left axes) and logarithmic (dark gray, right axes) scales. On the log scale, the function is asymptotically linear at small values of P_o (~<0.2).

kinetics are relatively fast, but around the threshold temperature, the activation of the channel is slow relative to the rate of temperature change. Because the fitting is made within this range, the assumption that the channel is at equilibrium may fail if the rate of temperature change is not carefully controlled.

14.3.2 SINGLE-CHANNEL ANALYSIS

Without fast temperature jumps, single-channel analysis provides an alternative to resolve the kinetics of temperature gating. Because data are recorded at equilibrium, the control of temperature is relatively easy. With two miniature thermistors placed at the front and the back of a pipette tip, the temperature of the patch can be precisely monitored and clamped.[28] Although recordings are made at equilibrium, their combinations across temperatures can lead to determination of the kinetic effects of temperature on gating. Single-channel analysis of TRPV1 shows that the apparent thermal sensitivity of whole-cell responses results predominantly from gating.[28] The unitary conductance of the channel has temperature dependence similar to that of aqueous diffusion of electrolytes. The gating by temperature involves multiple states, at least three closed and three open states, similar to the gating by agonists (e.g., capsaicin and low pH).[42,43] The channel in its open conformation is relatively independent of temperature. Instead, temperature mainly drives the long closures between opening bursts. As temperature is elevated, these closures become shortened so that the overall open probability increases. Statistical analysis of these long closures gives a temperature coefficient of $Q_{10} > 9$ (at +60 mV). By analogy to heat-induced protein unfolding, these events involve an enthalpy change similar to the denaturation of a globular protein of ~100 amino acids, suggesting that the gating by temperature may be accompanied with large structural rearrangements.

Single-channel experiments are technically challenging. For reliable analysis, good data quality is essential. For example, the patches have to be ensured to contain single channels. They have to be also extremely stable so that they can sustain a relatively long exposure to high temperatures (e.g., >50°C for TRPV1). Furthermore, because the response of thermal TRP channels such as TRPV1 often exhibits considerable variability across patches,[34] it is important to have recordings for multiple experimental conditions (e.g., different temperatures) from the same patches so as to minimize the patch-related variations. In practice, the patches that can meet all these criterions are scarce, making the success rate of the experiments quite low. In addition, while single-channel data reveal gating events at high resolutions, their analysis can be difficult owing to the complex gating behavior of the channels.

14.3.3 ANALYSIS OF THERMAL SENSITIVITY OF VOLTAGE RESPONSES

Because thermal TRP channels are also activated by strong membrane depolarization, the thermal sensitivity of voltage responses provides another means for assessing the temperature dependence of the channels.[30–32] In this approach, the voltage-driven responses of the channel are measured at several ambient temperatures, which are typically below the threshold of thermal activation. The change of temperature affects the voltage responses by shifting the midpoint of channel opening versus voltage. The slope of the curve, which is interpreted as the effective gating charges, is independent of temperature. Because the voltage-dependent opening of the channels is observed even at room temperature, the determination of its thermal sensitivity requires temperature to be perturbed only at a moderate range (e.g., <40°C for TRPV1). The approach thus has the advantage of alleviating high-temperature exposures, which are experimentally challenging. On the other hand, it is also of concern that the measurements obtained at the moderate temperatures may contain a limited amount of information on temperature gating because the temperature sensors are only minimally activated at temperatures below the activation threshold.

The voltage-driven measurements from these experiments combine contributions from both voltage- and temperature-dependent gating. As a result, the interpretation of the data requires explicit modeling to separate them. At a minimum, it is necessary to know how the two activation pathways are interrelated. In one model, Voets et al.[31] assume that all stimuli are *directly* coupled to the "gate" of the channel in an energetically *additive* manner. This is equivalent to a two-state model in which the opening or closing rate is driven simultaneously by all stimuli. A simple Boltzmann equation can describe the interplay between temperature and voltage gating,

$$P_o = \frac{1}{1 + \exp\left(\dfrac{\Delta H - T\Delta S - qFV}{RT}\right)} \tag{14.7}$$

where qFV accounts for the electrical energy of voltage sensors (q is the gating charge, V is the membrane potential, and F is Faraday's constant), and other terms have the same definitions as in Equation 14.1.

An alternative model is the allosteric MWC scheme as mentioned above. In this model, the channel is assumed to possess distinct structural domains as sensors for different stimuli, and these domains are *independent* of each other but all *allosterically* coupled to a common "gate." A stimulus by itself does not evoke conformational changes for opening; instead, it serves to only increase the equilibrium constants of the intrinsic opening. Assuming a single sensor for each stimulus, the model gives rise to an open probability that depends on voltage and temperature by

$$P_o = \frac{1}{1 + \dfrac{1}{L} \cdot \dfrac{1 + K_V}{1 + cK_V} \cdot \dfrac{1 + K_T}{1 + dK_T}} \tag{14.8}$$

where L, K_V, and K_T are, respectively, the equilibrium constants of intrinsic opening and activation of voltage and temperature sensors, and c and d are the corresponding allosteric coupling factors. The temperature dependence of P_o arises from that of K_T.

Voets et al.[31] resolved the voltage and temperature dependence in Equation 14.7 by fitting voltage activation time courses. Brauchi et al.[30] and Matta and Ahern[32] determined the MWC model from the steady-state measurements (i.e., the conductance–voltage curves at multiple temperatures). Despite different mechanistic assumptions, the two models give rise to very similar estimates for both temperature dependence and voltage dependence. They both predict that changing the voltage shifts the midpoint or threshold of thermal activation and, conversely, changing the temperature shifts the midpoint of the conductance-voltage curves. The slopes of these gating curves are generally not affected.

14.4 FAST TEMPERATURE JUMPS

One of the important unknowns for temperature gating has been the rate of activation. This information is essential to understanding the gating mechanisms of the channels. Single-channel data show that temperature-independent events have lifetimes on the order of milliseconds.[28] This implies that the activation by temperature can be as fast as in a few milliseconds. Thus, to resolve the time course of thermal activation would require "instant" temperature changes at a submillisecond resolution. As reviewed below, several efforts have been undertaken toward development of fast temperature jumps.

14.4.1 T-Jumps by Solution Superfusion

One strategy to improve the speed of temperature perturbation is to exploit fast solution exchangers. Solutions at control and test temperatures pass through separate perfusate delivery tubes. A fast switching mechanism is then used to select which perfusion tube is directed at the sample. Hayes et al.[44] demonstrated the technique for thermal activation of TRPV1 using a step motor-driven exchanger (VC-6, Warner Instruments). The exchanger has a time resolution of ~30 ms for a maximal ~700-mm step. Temperature control is achieved by preheating test solutions through an inline heater (TC-324B, Warner Instruments). With this approach, they obtained a temperature rise from room temperature to ~50°C in ~1 s (equivalent to a rate of 25°C/s). The approach is potentially applicable for bipolar temperature perturbations, but its time resolution remains inadequate relative to the activation rate of the channels.

Another attempt to speed up temperature changes by solution superfusion is to minimize the volume of the solution to be heated. Microfluidic technology allows for fabrication of submillimeter compartments for precise control and manipulation of fluids. Pennell et al.[45] have developed such a microfluidic chip with a channel of 250 μm wide and 25 μm deep. A platinum heater and a thin-film resistive sensor are placed along the channel on the outlet end for inline heating and temperature monitoring. The inlet end of the channel is connected to an open solution reservoir. With a flow rate of ~10 μL/min, the chip is capable of increasing the solution temperature at the outlet end from bath temperature (20°C) to 80°C at an optimum heating rate of 0.5°C/ms. The device therefore still has an inadequate time resolution.

14.4.2 T-Jumps by Laser Irradiation

Laser irradiation has long been used for thermal perturbation of biological processes including gating of ion channels (e.g., Ref. 46) and in vivo pain sensation (e.g., Ref. 47).

For resolving the rate of biomolecular reactions, ultrafast temperature jumps on the order of nanoseconds have been demonstrated using high-energy pulsed lasers, such as an Nd-YAG laser with a pulse energy of a few hundreds of milliwatts and a pulse duration of ~10 ns.[48] Ultrafast laser T-jumps generated in this way have proven useful for studying protein folding events in the microsecond regime, where they have provided critical data on the time scales of such elementary processes as formation of secondary protein structures.[49,50]

However, the high-energy pulsed lasers are expensive and offer a time resolution unnecessarily high for studying a process such as ion channel gating that occurs in milliseconds. In addition, the nanosecond temperature jumps produced by a pulsed YAG laser are not steady on a time scale of milliseconds or longer, although the decay of temperature may be negligible on a microsecond time scale (where the ultrafast folding of small proteins or secondary structures takes place). The flash lamp pumped lasers have an intrinsically low repetition rate, typically a few tens of hertz. As a result, feedback modulation of these lasers for producing constant temperature steps is impossible. Applicability of this approach to electrophysiological experiments involving live cells has also not been demonstrated.

For moderately fast temperature jumps in a micro- to minisecond regime, the requirement on laser power becomes considerably less demanding. Semiconductor laser diodes may therefore provide a more cost-effective substitution to the high-energy pulsed YAG lasers. Solid-state infrared laser diodes have steadily grown in power in recent years and become increasingly popular for their industrial applications. In the field of nociception, they have been exploited as a heat irradiation tool to stimulate peripheral nerves in vivo and cultured nociceptive neurons in vitro. For example, Baumann and Maternson[51] reported the use of a laser diode with 500 mW at 970 nm for heating a blackened tip of an optical fiber placed near the preparation. They were able to record heat-evoked action potentials in cultured trigeminal ganglion neurons, which provide the first direct evidence for a thermal transduction mechanism in such sensory neurons. Miura and Kawatani[52] and Jimbo et al.[53] also explored the effects of laser irradiation on cultured nodose ganglion and DRG neurons, although the power of their lasers was too low (16–150 mW) to elicit significant thermal effects. To obtain a high output power, Greffrath et al.[54] combined six 980-nm laser diode outputs into a single fiber output, so that they could obtain a maximum irradiation power of 15 W. With this device, they demonstrated heat-evoked currents in isolated DRG neurons. The laser irradiation results in an approximately sublinear temperature ramp, which reaches a temperature jump of ~30°C in 400 ms (~75°C/s). The whole-cell currents evoked by such heat stimuli exhibited a half activation time of $t_{1/2} \approx 28$ ms.

14.5   A SUBMILLISECOND TEMPERATURE JUMP SYSTEM

The previous attempts with either solution superfusion or laser diode heating have not been able to obtain a time resolution necessary to time resolve the activation of thermal TRP channels. The failure raises questions about the fundamental feasibility of these approaches. In the following, a theoretical evaluation of the laser diode irradiation approach is first presented, followed by a description of a laser diode heating system we have recently developed. The system can be considered as a temperature clamp apparatus—it achieves for the first time a submillisecond time resolution and also the capability of holding temperature steady after the initial rise.

14.5.1 FEASIBILITY OF LASER DIODE HEATING

The heating performance with laser irradiation may be assessed by numerical simulations.[55] The analysis also provides insights on the choice of laser diodes in terms of wavelength and necessary optical power. Briefly, the temperature rise of the solution resulting from illumination by a laser beam can be modeled by the standard heat conduction equation, i.e.,

$$c_p \frac{\partial T}{\partial t} = \kappa \nabla T + \varepsilon \times u(x, y, z) \tag{14.9}$$

where c_p is the specific heat capacity of water, κ is the thermal conductivity, ε is the optical absorption coefficient, and u is the spatial distribution of laser power. For a first-order approximation, one may assume a laser source producing a collimated emission beam, so that the spatial distribution of power has a simple cylindrical geometry, i.e.,

$$u(x, y, z) = \begin{cases} P/\pi R^2, & \sqrt{x^2 + y^2} < R \\ 0, & \text{otherwise} \end{cases} \tag{14.10}$$

where R is the beam radius, and P is the total output power. Other assumptions that may be invoked to further simplify the equation include a negligible loss of power of the laser beam due to water absorption and a constant specific heat capacity of water independent of temperature. Under these conditions, the heat conduction equation may be readily solved by a partial differential equation solver.

High-power laser diodes that are applicable for heating typically have wavelengths around either 980 or 1460 nm. Today, a single emitter diode of 980 nm can output an optical power of >10 W. The longer wavelength diode generally has much less power (e.g., <3 W). The 980-nm wavelength overlaps with a minor absorption peak of water, while the 1460-nm wavelength is close to a major peak in the near-infrared region. The simulation suggests that with a diode of 3 W at 1460 nm, a temperature jump from room temperature to 60°C (saturating temperature for thermal TRP channels) could be achieved in <150 μs. The rise of temperature was almost linear with a rate of ~280°C/ms. On the other hand, with a 10-W 980-nm laser diode, the same temperature jump requires a considerably longer irradiation time (~4 ms), whereas the rise of temperature becomes sublinear. Thus, a diode of 1460 nm, albeit with a lower power, is much more efficient. Its superior performance occurs because of a higher absorption coefficient of water at longer wavelengths (0.46/cm at 980 nm and 30/cm at 1460 nm). The simulation supports the conclusion that a submillisecond temperature jump should be possible using a single emitter diode with a wavelength at ~1500 nm or an array of twenty 980-nm diodes.

The irradiation area is also an important factor for consideration of heating performance. A larger area is experimentally preferred but limits the brightness of the laser beam. The use of optical fibers for illumination typically limits the size to be around 100 μm in diameter. This size is large enough to cover most mammalian cells, while giving a light intensity strong enough for a submillisecond temperature rise. With the 1460-nm diode, the temperature within the laser beam (±50 μm) is nearly uniform, and outside the beam, it falls sharply. Thus, the spatial profile of the temperature changes resulting from laser diode irradiation is useable for single-cell experiments.

14.5.2 LASER DIODE INSTRUMENTATION

The theoretical simulations suggest that it is possible to produce submillisecond temperature jumps using a single emitter laser diode with an appropriate wavelength. One implementation of such a system is diagramed in Figure 14.3, which employs a 1060-nm diode with a maximum 4-W output and is capable of a temperature jump from room temperature to ~60°C in 0.75 ms (Figure 14.4).[55] The laser beam emitted from the diode is first collimated and then launched into a multimode fiber, which has a diameter of 100 μm and a numeric aperture NA = 0.22. Simple single-element lenses, such as an aspheric lens, are adequate for both collimation and fiber

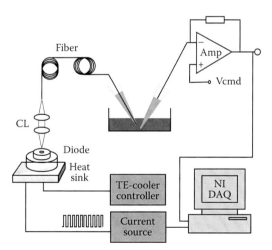

FIGURE 14.3 Schematic drawing of a submillisecond temperature jump system using a single emitter laser diode. The laser beam was launched into a multimode fiber with a core diameter of 100 μm, and the other end of the fiber was placed close to the samples with a micromanipulator. The diode was powered by either a pulsed or CW current source controlled by a computer. (Adapted from Yao, J. et al., *Biophysical Journal* 96, 3611–3619, 2009. With permission.)

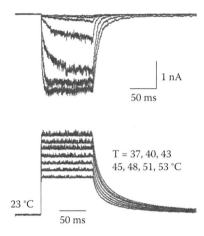

1 nA

50 ms

T = 37, 40, 43
45, 48, 51, 53 °C

23 °C

50 ms

FIGURE 14.4 Activation of TRPV1 by fast temperature jumps. Top: Macroscopic current responses of the channel from an outside-out patch ($V_h = -100$ mV). Bottom: A family of temperature jumps. (Adapted from Yao, J. et al., *Biophysical Journal* 96, 3611–3619, 2009. With permission.)

coupling. A good launch efficiency (>75%) is readily achievable using a common XY translation lens mount (0.25 mm/revolution) and a flexcture Z-axis translator (1 µm/revolution).

The diode needs to be powered by a constant current source with a relatively high output current (>10 A) and a low compliance voltage (1–2 V). Such high-power laser diode drivers usually come in two configurations: continuous wave (CW) or pulsed. For ease of control and modulation, a quasi-pulsed controller is preferred, which allows for separate controls on current amplitude and output pulsing. A separate voltage power supply may be used for quick dial of the output current. The pulsing can be controlled through a computer. A computer-based instead of a microprocessor-based autonomous control for pulsing is preferred because of the need to store a relatively large amount of data as discussed below. To maintain a consistent optical output and also prevent overheating, the diode needs to operate at a constant temperature, typically room temperature by thermoelectrical cooling.

Feedback control on the diode output is implemented by modulating the pulse input of the power supply. A quasi-CW controller allows for a variable pulse width ranging from CW to a minimal duration as limited by the pulse rise times. Thus, both pulse frequency (duty cycle) and durations can be exploited for modulation using a common feedback control algorithm such as PID. For quick prototyping, a simple on-and-off modulation scheme proves adequate for maintaining a constant temperature. When the temperature is above its set point, the laser is pulsed off; otherwise, the laser is on. The control algorithms are readily implemented with any programmable digital processors, although a computer-based control through a National Instrument (NI) multifunctional acquisition card has been used in our implementation. NI provides both visual Labview and object-oriented library for low-level controls of its boards, thus making programming possible in high-level languages such as C/C++.

14.5.3 TEMPERATURE CALIBRATION

The actual temperature at the beam spot may be calibrated using an open pipette with the tip placed at the beam center. The pipette is filled with a normal saline solution. A small voltage is applied to the pipette to induce a measurable current. Laser irradiation of the pipette tip causes an increase of the current. The temperature rise is related to the current change by

$$\frac{1}{T} - \frac{1}{T_0} = \frac{\ln(I/I_0)}{\ln Q_{10}} \cdot \left(\frac{1}{T_0 + 10} - \frac{1}{T_0} \right)$$

(14.11)

where T_0 and T are the temperatures before and after laser irradiation, respectively, I_0 and I are the corresponding currents, and Q_{10} is the temperature coefficient of the electrolyte conductivity at T_0. The value of Q_{10} may be predetermined by placing an open electrode in a hot bath solution. As the solution is cooled, the current through the pipette and the ambient temperature are simultaneously recorded. The current is then plotted on a log scale against $1/T$ (Arrhenius plot). The plot is generally linear in the temperature range of interest. The slope of the plot then gives rise to the temperature dependence of the electrolyte conductance. With our normal saline buffer, we typically obtained a Q_{10} value of 1.2–1.4.

The laser beam spot on the cover slip can be located with the help of fluorescent cells. The red (mRFP) or yellow (YFP) fluorescent proteins are preferable choices because the green lasers, which are readily available at low cost, can be exploited for excitation of these fluorescent proteins. The green laser line also needs to be coupled to the illumination fiber. Once the beam area is identified, the center of the beam can be estimated. A video camera can greatly help the identification and also allow for more precise controls. The resultant center of the beam is marked on a computer screen for subsequent experiments.

Presently, it is technically difficult to simultaneously perform patch-clamp recording and real-time monitoring of temperature at the patches during actual experimentation. Miniature thermal sensors are either too bulky in size or too slow in response time. Thermally sensitive dyes are also difficult to use and suffer from problems such as bleaching. A more practical alternative we have found is to use a two-step procedure in which the desired temperature jumps are first generated with an open electrode and then applied to patched cells or excised patches by playing back the diode control protocols that were stored during the first step. The temperature jumps obtained in the first step appear to be quite reproducible in the second step. In practice, it is found that the reproducibility does not have a very high sensitivity to the patch positions, presumably owing to a relatively uniform temperature distribution within the beam. Proper alignments are readily achievable by positioning the patches at the beam center as previously marked on the computer screen during the first step. The vertical positions of the patch can be determined from the focal distance of the open pipette tip.

14.5.4 RAPID COOLING

To generate a square temperature pulse, it is necessary to rapidly cool the sample following a temperature jump. Passive cooling by heat dissipation is slow, taking hundreds of milliseconds to return to the ambient temperature. To accelerate cooling, it is necessary to superfuse cells with solutions at the control temperature. One possible solution is to use a fast solution exchanger in conjunction with the laser. The two can be synchronized so that during heating, the solution stream is outside the laser beam, whereas when the laser is switched off, the exchanger is simultaneously activated to move the solution stream to the beam center. The solution exchanger can be attached to a third manipulator mounted on the back of the microscope stage to produce a solution flow perpendicular to the fiber and the patch pipette. An exchange time <1 ms can be obtained with a double-barrel capillary installed on a piezo actuator.[34]

14.6   APPLICATIONS OF FAST TEMPERATURE JUMPS

The application of fast temperature jumps to study thermal TRP channels is still at a fledgling stage. One immediate application is to resolve the time course of temperature gating (Figure 14.4). It turns out that although a large amount of energy (~100 kcal/mol) is involved, the activation by temperature is surprisingly fast. With TRPV1, it occurs on the order of a few milliseconds. It is significantly faster than activation by chemical agonists. Capsaicin (100 µM) involves a half-activation time over a hundred milliseconds. Low pH is faster but still takes tens of milliseconds. Thus, temperature is the fastest stimulus to activate the channel. In contrast to its fast kinetics, the temperature response shows a more limited steady-state dynamic range, which is only approximately half of the capsaicin response at the same temperature.[34,56] Temperature alone does not seem to fully open the channel.

With fast temperature jumps, one can expose patches to elevated temperatures only for a short period. This is important because the reduced thermal stress makes experimentation at high temperatures more feasible, for example, for measuring temperature response at hyperpolarizing membrane potentials, which occurs at a temperature range of 40°–55°C. The experiments at these temperatures provide information on the gating of the channel mainly by temperature (i.e., in absence of charge movement).

At hyperpolarizing potentials, TRPV1 shows relatively simple activation kinetics. In response to a temperature jump, the activation of current follows a single exponential time course. The temperature dependence of both steady-state response and time course can be described by a simple two-state model where the opening rate is driven by temperature while the closing has nominal but negative temperature dependence (i.e., sensitive to cold instead of heat). By analysis of thermal sensitivity of voltage gating, Voet et al.[31] also reached the conclusion that the opening of the channel is mainly temperature dependent, although the temperature dependence of the closing rate from these experiments was found to be the opposite.

The energetics of thermal gating obtained from temperature jump measurements are considerably larger than those inferred from voltage-driven responses at different temperatures. Between closed and open, the enthalpy of the channel increases by ~100 kcal/mol. This energy is equivalent to an electrical energy of moving ~71 unit charges across 60 mV (i.e., about 5 times the energy of voltage-gated channels) or melting of 10 phospholipids.[57] The gating by temperature thus evolves from a low enthalpy and entropy closed state to a high enthalpy and strongly entropic (disordered) open state. The opening of the channel involves an activation enthalpy of ~85 kcal/mol, which accounts for most of the large equilibrium enthalpy change between closed and open. Thus, the open state of the channel is energetically similar to the transition state. The channel becomes open only at the very end of the large energetic change. Despite this large amount of energy involved, the activation by temperature (thermal energy) is made fast because the entropy change mostly cancels the enthalpy change leaving the free energy difference quite moderate.

14.7 SUMMARY

Thermal TRP channels are activated on a time course of milliseconds and over a broad temperature range from a few Celsius degrees to >50°C. Such fast time responsiveness in conjunction with the wide temperature dynamics has brought challenges to instrumentation of rapid temperature controls for time-resolved measurements of these channels in live cells. Thermoelectric or resistive approaches are slow in time, thereby limiting their uses mostly to steady-state temperature-dependent measurements. They also inevitably incur prolonged thermal stresses on patches and channels, which can give rise to unintended thermal effects such as leak currents and/or rundown of channels, thereby complicating analysis of data. Optical approaches such as laser diode irradiation are more efficient for rapid temperature perturbation. With a proper choice of wavelength, a single-emitter laser diode is capable of producing a temperature jump with a rise time in submilliseconds. Implementation of constant temperature jumps is also possible by exploiting the capability of laser diodes for rapid modulation. A practical, useable system by laser irradiation can be readily implemented using off-shelf optics and electronic drivers. With the availability of such systems, we are now in a position for time-resolved measurements of temperature responses.

REFERENCES

1. Spray, D. C. 1986. Cutaneous temperature receptors. *Annual Review of Physiology* 48: 625–638.
2. Dubner, R., R. Sumino, and W. I. Wood. 1975. A peripheral "cold" fiber population responsive to innocuous and noxious thermal stimuli applied to monkey's face. *Journal of Neurophysiology* 38: 1373–1389.
3. Hensel, H. 1974. Thermoreceptors. *Annual Review of Physiology* 36: 233–249.
4. Clapham, D. E. 2003. TRP channels as cellular sensors. *Nature* 426: 517–524.
5. Caterina, M. J., M. A. Schumacher, M. Tominaga, T. A. Rosen, J. D. Levine, and D. Julius. 1997. The capsaicin receptor: a heat-activated ion channel in the pain pathway. *Nature* 389: 816–824.
6. Cesare, P., and P. McNaughton. 1996. A novel heat-activated current in nociceptive neurons and its sensitization by bradykinin. *Proceedings of the National Academy of Sciences of the United States of America* 93: 15435–15439.

7. Caterina, M. J., T. A. Rosen, M. Tominaga, A. J. Brake, and D. Julius. 1999. A capsaicin-receptor homologue with a high threshold for noxious heat. *Nature* 398: 436–441.
8. Smith, G. D., J. Gunthorpe, R. E. Kelsell et al. 2002. TRPV3 is a temperature-sensitive vanilloid receptor-like protein. *Nature* 418: 186–190.
9. Xu, H. X., I. S. Ramsey, S. A. Kotecha et al. 2002. TRPV3 is a calcium-permeable temperature-sensitive cation channel. *Nature* 418: 181–186.
10. Peier, A. M., A. J. Reeve, D. A. Andersson et al. 2002. A heat-sensitive TRP channel expressed in keratinocytes. *Science* 296: 2046–2049.
11. Guler, A. D., H. Lee, T. Iida, I. Shimizu, M. Tominaga, and M. Caterina. 2002. Heat-evoked activation of the ion channel, TRPV4. *Journal of Neuroscience* 22: 6408–6414.
12. Xu, H., M. Delling, J. C. Jun, and D. E. Clapham. 2006. Oregano, thyme and clove-derived flavors and skin sensitizers activate specific TRP channels. *Nature Neuroscience* 9: 628–635.
13. Strotmann, R., C. Harteneck, K. Nunnenmacher, G. Schultz, and T. D. Plant. 2000. OTRPC4, a nonselective cation channel that confers sensitivity to extracellular osmolarity. *Nature Cell Biology* 2: 695–702.
14. Liedtke, W., Y. Choe, M. A. Marti-Renom et al. 2000. Vanilloid receptor-related osmotically activated channel (VR-OAC), a candidate vertebrate osmoreceptor. *Cell* 103: 525–535.
15. McKemy, D. D., W. M. Neuhausser, and D. Julius. 2002. Identification of a cold receptor reveals a general role for TRP channels in thermosensation. *Nature* 416: 52–58.
16. Peier, A. M., A. Moqrich, A. C. Hergarden et al. 2002. A TRP channel that senses cold stimuli and menthol. *Cell* 108: 705–715.
17. Story, G. M., A. M. Peier, A. J. Reeve et al. 2003. ANKTM1, a TRP-like channel expressed in nociceptive neurons, is activated by cold temperatures. *Cell* 112: 819–829.
18. Latorre, R. 2009. Perspectives on TRP channel structure and the TRPA1 puzzle. *Journal of General Physiology* 133: 227–229.
19. Jordt, S. E., D. M. Bautista, H. H. Chuang et al. 2004. Mustard oils and cannabinoids excite sensory nerve fibres through the TRP channel ANKTM1. *Nature* 427: 260–265.
20. Caterina, M. J., A. Leffler, A. B. Malmberg et al. 2000. Impaired nociception and pain sensation in mice lacking the capsaicin receptor. *Science* 288: 306–313.
21. Davis, J. B., J. Gray, M. J. Gunthorpe et al. 2000. Vanilloid receptor-1 is essential for inflammatory thermal hyperalgesia. *Nature* 405: 183–187.
22. Moqrich, A., S. W. Hwang, T. J. Earley et al. 2005. Impaired thermosensation in mice lacking TRPV3, a heat and camphor sensor in the skin. *Science* 307: 1468–1472.
23. Bautista, D. M., J. Siemens, J. M. Glazer et al. 2007. The menthol receptor TRPM8 is the principal detector of environmental cold. *Nature* 448: 204–208.
24. Colburn, R. W., M. L. Lubin, D. J. Stone, Jr. et al. 2007. Attenuated cold sensitivity in TRPM8 null mice. *Neuron* 54: 379–386.
25. Dhaka, A., A. N. Murray, J. Mathur, T. J. Earley, M. J. Petrus, and A. Patapoutian. 2007. TRPM8 is required for cold sensation in mice. *Neuron* 54: 371–378.
26. Bautista, D. M., S. E. Jordt, T. Nikai et al. 2006. TRPA1 mediates the inflammatory actions of environmental irritants and proalgesic agents. *Cell* 124: 1269–1282.
27. Kwan, K. Y., A. J. Allchorne, M. A. Vollrath et al. 2006. TRPA1 contributes to cold, mechanical, and chemical nociception but is not essential for hair-cell transduction. *Neuron* 50: 277–289.
28. Liu, B., K. Hui, and F. Qin. 2003. Thermodynamics of heat activation of single capsaicin ion channels VR1. *Biophysical Journal* 85: 2988–3006.
29. Benham, C. D., M. J. Gunthorpe, and J. B. Davis. 2003. TRPV channels as temperature sensors. *Cell Calcium* 33: 479–487.

30. Brauchi, S., P. Orio, and R. Latorre. 2004. Clues to understanding cold sensation: thermo-dynamics and electrophysiological analysis of the cold receptor TRPM8. *Proceedings of the National Academy of Science of the United States of America* 101: 15494–15499.

31. Voets, T., G. Droogmans, U. Wissenbach, A. Janssens, V. Flockerzi, and B. Nilius. 2004. The principle of temperature-dependent gating in cold- and heat-sensitive TRP channels. *Nature* 430: 748–754.

32. Matta, J. A., and G. P. Ahern. 2007. Voltage is a partial activator of rat thermosensitive TRP channels. *Journal of Physiology* 585: 469–482.

33. Latorre, R., S. Brauchi, G. Orta, C. Zaelzer, and G. Vargas. 2007. ThermoTRP channels as modular proteins with allosteric gating. *Cell Calcium* 42: 427–438.

34. Yao, J., B. L. Liu, and F. Qin. 2010. Kinetic and energetic analyses of thermally activated TRPV1 channels. *Biophysical Journal* 99: 1743–1753.

35. Brauchi, S., G. Orta, M. Salazar, E. Rosenmann, and R. Latorre. 2006. A hot-sensing cold receptor: C-terminal domain determines thermosensation in transient receptor potential channels. *Journal of Neuroscience* 26: 4835–4840.

36. Susankova, K., R. Ettrich, L. Vyklicky, J. Teisinger, and V. Vlachova. 2007. Contribution of the putative inner-pore region to the gating of the transient receptor potential vanilloid subtype 1 channel (TRPV1). *Journal of Neuroscience* 27: 7578–7585.

37. Grandl, J., H. Hu, M. Bandell, B. Bursulaya, M. Schmidt, M. Petrus, and A. Patapoutian. 2008. Pore region of TRPV3 ion channel is specifically required for heat activation. *Nature Neuroscience* 11: 1007–1013.

38. Salazar, H., A. Jara-Oseguera, E. Hernandez-Garcia et al. 2009. Structural determinants of gating in the TRPV1 channel. *Nature Structural and Molecular Biology* 16: 704–710.

39. Dittert, I., V. Vlachova, H. Knotkova, Z. Vitaskova, L. Vyklicky, M. Kress, and P. W. Reeh. 1998. A technique for fast application of heated solutions of different composition to cultured neurones. *Journal of Neuroscience Methods* 82: 195–201.

40. Reid, G., B. Amuzescu, E. Zech, and M. L. Flonta. 2001. A system for applying rapid warming or cooling stimuli to cells during patch clamp recording or ion imaging. *Journal of Neuroscience Methods* 111: 1–8.

41. Dittert, I., J. Benedikt, L. Vyklicky, K. Zimmermann, P. W. Reeh, and V. Vlachova. 2006. Improved superfusion technique for rapid cooling or heating of cultured cells under patch-clamp conditions. *Journal of Neuroscience Methods* 151: 178–185.

42. Hui, K. Y., B. Y. Liu, and F. Qin. 2003. Capsaicin activation of the pain receptor, VR1: Multiple open states from both partial and full binding. *Biophysical Journal* 84: 2957–2968.

43. Ryu, S. J., B. Y. Liu, and F. Qin. 2003. Low pH potentiates both capsaicin binding and channel gating of VR1 receptors. *Journal of General Physiology* 122: 45–61.

44. Hayes, P., H. J. Meadows, M. J. Gunthorpe et al. 2000. Cloning and functional expression of a human orthologue of rat vanilloid receptor-1. *Pain* 88: 205–215.

45. Pennell, T., T. Suchyna, J. Wang, J. Heo, J. D. Felske, F. Sachs, and S. Z. Hua. 2008. Microfluidic chip to produce temperature jumps for electrophysiology. *Analytical Chemistry* 80: 2447–2451.

46. Moore, L. E., J. P. Holt, Jr., and B. D. Lindley. 1972. Laser temperature-jump technique for relaxation studies of the ionic conductances in myelinated nerve fibers. *Biophysical Journal* 12: 157–174.

47. Bromm, B., H. Neitzel, A. Tecklenburg, and R. D. Treede. 1983. Evoked cerebral potential correlates of C-fibre activity in man. *Neuroscience Letters* 43: 109–114.

48. Kubelka, J. 2009. Time-resolved methods in biophysics. 9. Laser temperature-jump methods for investigating biomolecular dynamics. *Photochemistry and Photobiology Science* 8: 499–512.

49. Williams, S., T. P. Causgrove, R. Gilmanshin, K. S. Fang, R. H. Callender, W. H. Woodruff, and R. B. Dyer. 1996. Fast events in protein folding: helix melting and formation in a small peptide. *Biochemistry* 35: 691–697.

50. Munoz, V., P. A. Thompson, J. Hofrichter, and W. A. Eaton. 1997. Folding dynamics and mechanism of beta-hairpin formation. *Nature* 390: 196–199.

51. Baumann, T. K., and M. E. Martenson. 1994. Thermosensitivity of cultured trigeminal neurons. *Society of Neuroscience Abstracts* 20: 1379.

52. Miura, A., and M. Kawatani. 1996. Effects of diode laser irradiation on sensory ganglion cells from the rat. *Pain Research* 11: 175–183.

53. Jimbo, K., K. Noda, K. Suzuki, and K. Yoda. 1998. Suppressive effects of low-power laser irradiation on bradykinin evoked action potentials in cultured murine dorsal root ganglion cells. *Neuroscience Letters* 240: 93–96.

54. Greffrath, W., M. I. Nemenov, S. Schwarz, U. Baumgartner, H. Vogel, L. Arendt-Nielsen, and R. D. Treede. 2002. Inward currents in primary nociceptive neurons of the rat and pain sensations in humans elicited by infrared diode laser pulses. *Pain* 99: 145–155.

55. Yao, J., B. Liu, and F. Qin. 2009. Rapid temperature jump by infrared diode laser irradiation for patch-clamp studies. *Biophysical Journal* 96: 3611–3619.

56. Tominaga, M., M. J. Caterina, A. B. Malmberg et al. 1998. The cloned capsaicin receptor integrates multiple pain-producing stimuli. *Neuron* 21: 531–543.

57. Heimburg, T. 2007. *Thermal Biophysics of Membranes*. Weinheim, Germany: Wiley-VCH.

15 Fluorescence Microscopy in ThermoTRP Channel Research

Fan Yang and Jie Zheng

CONTENTS

15.1 A BRIEF REVIEW OF FLUORESCENCE MICROSCOPY

Fluorescence microscopy utilizes the well-known physical phenomenon: fluorescence. When a fluorophore molecule absorbs the energy carried by a photon of the excitation light, an electron of the particular chemical structure (termed "chromophore") jumps from the ground state to one of the excited states. After dwelling

319

in the excited state for a brief period of time, the excited electron falls back to the ground state and gives out the captured energy in the form of a photon. The fluorescence lifetime, determined by the dwell time in the excited state, is usually within the nanosecond (ns) range. Processes that shorten the electron's dwell in the excited state, e.g., contact of the chromophore with a quenching molecule, affect the fluorescence lifetime accordingly. Energy contained in an excited electron can also be dissipated in forms of molecular rotation, vibration, or heat, leading to loss of energy in nonfluorescence forms. Therefore, the emitted photon carries less energy than the absorbed photon. The resulting increase in the wavelength of the emission light compared to that of the excitation light is known as Stokes shift.[1] With an understanding of these physical properties of fluorescence, experiments can be better designed and performed.

Successful fluorescence experiments are built upon a number of components. Selection of a fluorophore is the first key component in designing fluorescence experiments. A broad spectrum of fluorophores is commercially available. Important factors to be considered in choosing fluorophores include spectral properties, sensitivity to local environment, and the way of fluorophore labeling. Each of these factors will be discussed in detail below.

First, choosing a fluorophore with the right excitation and emission spectra is essential. The spectra of commonly used fluorophores in temperature-sensitive transient receptor potential (thermoTRP) channel research are within the visible wavelength range (400 to 750 nm).[2] A basic consideration is that the spectral properties of a fluorophore should match the excitation and emission filters of the microscope. In addition, ample spectra separation is often advantageous in fluorescence experiments where multiple types of fluorophores are employed. For example, in the colocalization study of TRPV1 and TRPV2 channels, fluorescein isothiocyanate and rhodamine were used to label the two channel types.[3] The emission peaks for these two fluorophores, at 519 and 590 nm, respectively, are well separated. At the rhodamine emission peak, the contribution of fluorescein is less than 10%. Because of this, reliable separation of emissions can be achieved conveniently with band pass filters. Certain fluorescence experiments have special requirements for the spectral properties of fluorophores. Fluorescence resonance energy transfer (FRET) measurements, for example, require a substantial overlap of the donor emission spectrum and the acceptor excitation spectrum.

Fluorescence emission is sensitive to the local environment. Factors such as temperature, pH, and presence of quenchers or ligands such as calcium can substantially change their fluorescence properties.[1] This feature allows the design of a wide range of experiment strategies. For example, Ca^{2+} indicator dyes such as Fura-2 are designed based on Ca^{2+} affinity binding to fluorescent chelators. Ca^{2+} binding near the chromophore shifts the Fura-2 excitation peak wavelength from 340 nm in the Ca^{2+}-free state to 360 nm, which can be sensitively detected from the change of the emission intensity ratio when excited at these two wavelengths, respectively. Fluorophores can be quenched by certain ions upon direct contact. Conformational changes affecting the accessibility of the quenching ions to the fluorophore will change the efficiency of quenching. On the basis of this principle, quenching of Alexa488 by iodide was used to detect local conformational changes in the fluorophore labeling site on cyclic nucleotide-gated (CNG) channels.[4] Fluorescence emission is temperature sensitive. For example, an increase in temperature leads to a small but detectable decrease in the emission intensity of rhodamine and fluorescein. When these fluorophores are employed in thermoTRP channel research where temperature is changed, proper control experiments are needed to correct for this temperature-induced change in fluorescence intensity.[5]

Fluorescent proteins and chemical fluorophores not only differ in size and structure but also require different strategies to label target proteins.[6] Green fluorescent protein (GFP) and its derivatives can be genetically fused with target proteins,[6–8] yielding stoichiometrical labeling. In this way, there is no unlabeled target and no free fluorophore, which substantially increases the signal-to-noise ratio. GFPs are approximately barrel-shaped with an axial length of 4.2 nm and a diameter of 2.4 nm.[9] Attaching such a large molecule to a target protein may alter its functions. For this reason, GFPs are often introduced at the peripheral region of the target protein, for example, the N- and C-termini. In comparison, chemical fluorophores such as tetramethylrhodamine-5-maleimide (TMRM) are advantageous in their relatively smaller sizes. Their attachment introduces less structural perturbation to the target protein. A common approach to attach chemical fluorophores is to covalently modify native or engineered cysteine residues in the target protein with sulfhydryl-reactive fluorophores.[10–13] A caveat of this method is that fluorophores may react with cysteine residues in endogenous proteins. If nonspecific labeling occurs, the signal-to-noise ratio is decreased. So the *pros* and *cons* of choosing each type of fluorophores should be balanced in the context of the biological questions to be addressed by the fluorescence measurements.

Detecting a fluorescence signal is another key component of fluorescence experiments. Emission light from fluorophores is usually much lower in intensity than the excitation light.[1] A fluorescence microscopy system employs a series of measures to make sure that the emission light is separated from the excitation light and recorded faithfully. Epi-fluorescence has been the default configuration for modern fluorescence microscopes, in which excitation and emission light pass though the same objective but in opposite directions.[1] With this configuration, most excitation light travels away from the detector. Nonetheless, there is still a significant portion of excitation light that is either reflected or scattered back into the emission light collecting pathway. Additional measures are required.

One of these measures is the application of filter sets. For an epi-fluorescence microscope, the light generated by a mercury lamp is of multiwavelength. To excite a particular fluorophore, the light of a certain wavelength must be selected. This is achieved by using a band pass (BP) excitation filter. Band pass filters allow light within a wavelength range to pass, while long pass (LP) filters allow light with wavelengths longer than a threshold value to pass. The transmission efficiency of desired light lies in the range from 70 to over 90%,[1] while 0.01% of unwanted light is transmitted.[14] As mentioned before, owing to Stokes shift, emission light is usually longer in wavelength than excitation light.[1] Employing this property of fluorescence, a dichroic mirror is used to separate excitation and emission light. The transmission efficiency of dichroic mirrors for desired light is above 90%, while up to 10% of unwanted light can pass.[1,14] When light passed by an excitation filter arrives at a dichroic mirror tilted at a 45° angle, it is reflected off the mirror, reaching and exciting fluorophores in a sample. When emission light hits the dichroic mirror, it is able to pass through because of its longer wavelength. The emission filter further ensures that only the component of the desired wavelength in emission light is recorded by a detecting devise such as a digital camera. The combination of a dichroic mirror and an emission filter thus allows the selection of emission light over excitation light by a ratio of at least 1:100,000, which makes recording of the much weaker emission light possible. The filter sets need to be carefully built to ensure this designed selection process. Excitation filter, dichroic mirror, and emission filter are usually fitted into a filter cube. A modern fluorescence microscope often contains multiple filter cubes for different fluorophores.

Fluorescence microscopy has many advantages pertaining to thermoTRP channel research. First, fluorophore labeling makes highly selective recordings possible. GFP and derivatives can be genetically linked to target channels by linking the cDNA sequence for GFP to a cDNA encoding the channel subunit. Fluorescent antibodies can be used to tag a particular type of thermoTRP channels. Maleimide-containing fluorophores like fluorescein-5-maleimide (FM) react chemically with cysteine residues with high specificity, so these fluorophores can be covalently linked to desired sites in a channel protein. The high selectivity associated with fluorescence recordings allows the detection of fluorescence signals from a particular population of channel proteins or a specific part of the channel. Moreover, imaging more than one type of a channel simultaneously is possible, as demonstrated by the study of TRPV1 and TRPV2 channel colocalization with fluorescence imaging of antibodies labeling these two channels with different colors.[3,15]

In addition, fluorescence microscopy techniques such as Ca^{2+} imaging greatly facilitate high-throughput screening. Ca^{2+} imaging allows simultaneous monitoring of channel activity in a large number of cells. The search for the capsaicin receptor, for example, was greatly facilitated by Ca^{2+} imaging with eight-well chambers.[16] Another advantage of fluorescence microscopy over traditional optical microscopy is the enhancement of contrast. The theoretical spatial resolution limit of optical microscopy of a few hundred nanometers, resulting from light diffraction, is, in general, hard to achieve owing to the lack of contrast in biological samples.[1] Specific labeling with fluorophores greatly enhances the contrast and achievable spatial resolution. Certain fluorescence microscopy techniques such as FRET and super-resolution imaging further improve the resolution. FRET, for example, reports the distance between fluorophores in the range of <10 nm,[17–20] which is comparable to the size of many channel molecules.[21] Therefore, it is ideal for monitoring channel assembly and trafficking protein–protein interactions, as well as protein structural dynamics. One such example is the detection of heteromeric thermoTRP channels by FRET.[22,23]

While fluorescence techniques have been widely used in TRP channel research, in this chapter, their applications will be reviewed using example studies of thermoTRP channels. Special attention will be given to technical considerations. Advantages and significance of these techniques, as well as their drawbacks and caveats, will also be discussed.

15.2 DISCOVERY OF THERMOTRP CHANNELS

15.2.1 HISTORICAL PERSPECTIVES OF THERMOTRP CHANNEL DISCOVERY

Historically, it was the search for the capsaicin receptor that led to the discovery of thermoTRP channels. Long before the molecular identities of temperature sensors were discovered, it was known that sensory neurons responsible for pain sensation, or the "nociceptors," can be activated by noxious chemicals like capsaicin and by thermal stimuli.[16,24] In the search for the capsaicin receptor, it was observed that when the nociceptors were activated by capsaicin, intracellular Ca^{2+} concentration ($[Ca^{2+}]_i$) increased.[24–27] In membrane patches of sensory neurons, nonselective and Ca^{2+} permeable single-channel currents were recorded in response to capsaicin.[28] Therefore, the Julius group employed Ca^{2+} imaging as a tool to screen for the capsaicin receptors.[16] Because DRG neurons have long been known as nociceptors where capsaicin receptors locate,[24] they first divided a cDNA library prepared from rat DRG neurons into 144 pools. Each pool that contains 16,000 clones was used to transfect human embryonic kidney (HEK) 293 cells. Transfected cells were loaded with the Ca^{2+} sensitive fluorophore Fura-2 acetoxymethyl ester (Fura-2 AM). The cells were challenged with capsaicin. Cells expressing the receptor showed a change in the fluorescence signal owing to receptor activation and the following Ca^{2+} influx. The pool that gave positive signals was further divided to subgroups. The same experimental strategy was repeated until a single clone that transferred capsaicin sensitivity to HEK293 cells was identified in 1997, which was initially named vanilloid receptor 1 and later changed to TRPV1.[16]

With electrophysiological recording, it was shown that TRPV1 can be activated by noxious heat above 40°C.[16] Gene knockout studies later confirmed TRPV1's role as a molecular temperature sensor.[29,30] Identification of TRPV1 opened up the research field of temperature sensing to molecular biologists. Within 6 years, five additional thermoTRP channels were discovered by either similar Ca^{2+} imaging-based strategy (TRPV4,[31–33] TRPM8,[34,35] and TRPA1[36]) or homology cloning (TRPV2[37] and TRPV3[38–40]).

15.2.2 CALCIUM IMAGING

Cloning of TRPV1 demonstrated the power of high-throughput fluorescence techniques like Ca^{2+} imaging. Principally, this technique measures changes in $[Ca^{2+}]_i$. At the resting condition, free $[Ca^{2+}]_i$ is very low, ranging from 30 to 150 nanomoles (nM).[41] In comparison, $[Ca^{2+}]_o$ is about 10,000 times higher, which is often further elevated under experimental conditions to enhance Ca^{2+} signals.[41] This steep concentration gradient would drive a fast influx of Ca^{2+} upon the increase in membrane permeability. ThermoTRPs are Ca^{2+}-permeable ion channels. Their extended activation leads to an increase in $[Ca^{2+}]_i$, which is reported by Ca^{2+}-binding fluorescent dyes. These dyes possess two important properties: (1) they bind Ca^{2+} with a desirable affinity, and (2) Ca^{2+} binding changes their fluorescence properties. The affinity of Ca^{2+} binding is measured by the dissociation constant (K_d), which is the $[Ca^{2+}]$ at which half of the dye molecules are bound. On the basis of dissociation constant values, Ca^{2+} indicators can be categorized as high affinity and low affinity dyes (see Ref. 1). For example, Fura-2 has a high Ca^{2+} affinity with K_d values of 135 nM at 20°C and 224 nM at 37°C,[42] which are comparable to the resting $[Ca^{2+}]_i$. Mag-Fura-2 has a low Ca^{2+} affinity with a K_d value of 53 μM,[43,44] which is more appropriate for measuring $[Ca^{2+}]$ in the endoplasmic reticulum (ER), where Ca^{2+} levels are high.[45,46] Therefore, depending on local $[Ca^{2+}]$, choosing indicators with a proper K_d is critical for the success of Ca^{2+} imaging experiments. Specifically for thermoTRP studies, one should keep in mind that K_d (as well as fluorescence) is a temperature-dependent parameter.

Ca^{2+} indicator dyes are also categorized as either single wavelength or ratiometric. For single wavelength indicators, such as quin2, binding of Ca^{2+} changes their fluorescence intensity but not excitation or emission spectra.[42] Monitoring the amplitude of the fluorescence emission peak allows the study of intracellular Ca^{2+} dynamics. However, besides $[Ca^{2+}]$, fluorescence intensity is also affected by other factors, such as dye concentration, excitation light intensity, and emission light collecting efficiency. If these parameters are not well controlled, which is not uncommon for in vivo experiments, the $[Ca^{2+}]_i$ measurement could be largely compromised.

For ratiometric dyes, Ca^{2+} binding quantitatively changes their excitation and/or emission spectra while maintaining a stable high quantum yield. They work in a dye concentration-independent manner (see below), which is advantageous in cellular applications. Fura-2 AM is the most widely used ratiometric Ca^{2+} indicator in thermoTRP channel research. While the emission peak wavelength remains at 505 nm, its excitation peak shifts from 340 nm in the Ca^{2+}-bound state to 360 nm in the calcium-free state.[42] Because of the lack or low influence by dye concentration and light properties of the instrument, ratiometric dyes are generally preferred in practice.

While Ca^{2+} indicators in slat- or dextran-conjugated forms may be introduced into the cell by invasive techniques such as microinjection or diffusion from patch-clamp pipettes,[43] the more common approach is to use acetoxymethyl (AM) esters.[43] The AM group is hydrophobic, making the AM-fused dyes membrane permeable. Once inside, the endogenous esterases cleave the AM off, leaving the dye trapped in the cytosol. This also concentrates the dye, e.g., 1 to 5 μM of Fura-2 AM in bath solution can lead to a concentration above 100 μM in cytosol.[43] Therefore, with AM-esterification, dye loading is not only convenient but also more efficient, both in terms of the number of cells loaded and the dye concentration achieved inside the cell.

Calcium imaging has been done both in vitro and in vivo. In vitro calcium imaging can be performed with a conventional fluorescence microscope.[16] For example, cells expressing TRPV1 may be cultured in polyorinithine-coated glass cover slips or eight-well chambers and loaded with 1–10 μM Fura-2 AM at 37°C for 30 min in a physiological buffer. After 30 min, free Fura-2 AM molecules are washed off twice with the same buffer. Using an inverted fluorescence microscope equipped with a CCD camera, proper 340- and 380-nm excitation filters, and 510-nm emission filters, as well as a filter wheel and a dichroic mirror for wavelengths >500 nm, images at alternating 340- and 380-nm excitation wavelengths will be acquired at intervals of 0.5 to 5s, depending on the speed of the camera and the experimental needs. In some systems, a monochromator, instead of excitation filters in a filter wheel, may be used to provide the alternating excitation lights. Ratio images of 340/380 calculated by computer software are acquired first for basal conditions and then during drug (capsaicin for TRPV1) stimulation. An increase in $[Ca^{2+}]_i$ will lead to a higher 340/380 intensity ratio.

In addition to wavelength considerations, the temporal and spatial resolution of a Ca^{2+} imaging experiment should be considered. Depending on whether steady-state or dynamic Ca^{2+} behavior is to be examined, the required time resolution will differ. Similarly, whether the Ca^{2+} behavior is global or local will determine the spatial requirement. In most studies with TRPV1, the capsaicin-induced $[Ca^{2+}]_i$ increase is long-lasting and occurs throughout the cells. This steady-state measurement has low requirements of temporal and spatial resolution for an imaging system. However, $[Ca^{2+}]_i$ changes are often transient and/or local, such as Ca^{2+} sparks in vascular smooth muscle cells, which occur briefly (1 to 50 ms) in a very restricted scale (1 to 5 μm).[41] To detect such a brief Ca^{2+} signal, the recording device should have a matching high speed and a high sensitivity to record the expected low fluorescence signal.

The spatial resolution of a fluorescence microscope system is determined by the objective and the pixel size of the detector. To achieve high spatial resolution, one should use a more powerful, high numerical aperture, color- and focus-corrected aspheric objective. When recording the amplified image, a large CCD chip with a high number of small pixels is preferred. Furthermore, in vivo Ca^{2+} imaging is possible with confocal and two-photon microscopy.[47,48] Two-photon microscopy provides confined excitation volume and deeper penetration into the tissue.[49–51] Using this technique, Svoboda et al. observed a large Ca^{2+} transient in dendritic spines elicited by current injection.[52]

With Ca^{2+} imaging, not only the dynamic change in concentration can be monitored in real time but also the absolute concentration can be measured after proper calibration. For ratiometric dyes like Fura-2, absolute Ca^{2+} concentration is calculated with the following equation:[42,53]

$$\left[Ca^{2+}\right] = \frac{R - R_{min}}{R_{max} - R} \times S_f \times K_d \tag{15.1}$$

in which R is the 340/380 intensity ratio, and R_{min} and R_{max} are intensity ratios measured from conditions in which $[Ca^{2+}]$ is zero or saturating, respectively. The K_d for Fura-2-Ca^{2+} complex is 224 nM at 37°C and 135 nM at 20°C.[42] S_f is a scaling factor describing the ratio of fluorescence intensities excited by 380-nm light at zero and saturating $[Ca^{2+}]$. Readers are referred to Ref. 53 for details on in vivo and in vitro calibration for experiments using Fura-2.

There are some limitations and caveats with Ca^{2+} imaging. First, the linearity of the recording device in an imaging system, such as a CCD camera, should be examined. A high linearity is required for Ca^{2+} imaging, which means that the digital output values of a CCD should be linearly dependent on the amount of incident light. For high-performance CCDs, deviations from linearity are less than a few tenths of 1% over 5 orders of incident light magnitude.[54] Without high linearity, such fluorescence intensity measurement as 340/380 ratio of Fura-2 will not faithfully report the $[Ca^{2+}]_i$.

Second, Ca^{2+} imaging reports the channel function indirectly. A series of events, such as Ca^{2+} diffusion and binding with fluorescent dyes, has to happen before a detectable fluorescence signal is collected. Any process that interacts with any of these events could potentially alter the fluorescence signal and affect its interpretation. For example, the total amount of Ca^{2+} entering a cell depends on the integral of the channel open probability, single-channel conductance, and the length of recording time. Moreover, there may be other pathways for Ca^{2+} to enter the cytosol, including the release from the endoplasmic reticulum. These processes may be sensitive to the activity of the channel under study and thereby complicate the interpretation of the Ca^{2+} imaging results.

Third, restricted by its K_d for Ca^{2+} and stoichiometry for binding, a fluorescent dye has a specific Ca^{2+} sensitivity range. When strong channel activation generates a large Ca^{2+} influx, it may elevate the local $[Ca^{2+}]$ to high levels that saturate the dye's capacity, resulting in underestimation of the channel function.

Furthermore, the dye molecules may disturb cellular physiology. In essence, all Ca^{2+} dyes are Ca^{2+} buffers. Altering intracellular Ca^{2+} buffering with loaded fluorescent dyes can cause unwanted effects, which for thermoTRP channels could mean perturbation of interactions with modulatory molecules such as calmodulin in the short term, as well as effects on Ca^{2+}-dependent trafficking and expression in the long term.

Last, similar to other Ca^{2+} indicators, once in the cytosol, Fura-2 can be trapped in membrane-bound vacuoles, a process known as compartmentalization. Using dextran-conjugated dyes can effectively reduce compartmentalization. The $[Ca^{2+}]$ in different cellular compartments varies considerably, reaching about 100 to 1000 μM in the endoplasmic reticulum.[46] Therefore, the compartmentalized Fura-2 will likely give false high readings of $[Ca^{2+}]_i$.

15.3   SUBUNIT ASSEMBLY OF THERMOTRP CHANNELS

Subunit composition is a major determinant of channel properties.[21] In the TRP superfamily, heteromeric channels have been widely observed, at least in TRPC[55–62] and TRPM[63,64] subfamilies. For example, TRPC1 and TRPC5 subunits can form heteromeric channels with distinct biophysical properties, such as altered voltage-dependent activation and single-channel conductance compared to homomers.[65] The thermoTRP channels belong to three different TRP subfamilies (TRPV, TRPM, and TRPA). It is generally agreed that they are tetrameric channels based on biochemical and functional studies.[66–68] ThermoTRPs exhibit great functional diversity. Therefore, whether thermoTRP subunits can form heteromultimers is of particular interest and importance.

Heteromultimerization of thermoTRPVs has been examined by a number of fluorescence techniques. When TRPV3 was cloned, Ca^{2+} imaging experiments showed that its co-expression with TRPV1 enhanced the cellular response to capsaicin or proton.[39] With confocal microscopy, extensive colocalization of TRPV2 and TRPV1 in rat cerebral cortex and DRG neurons was observed by fluorescent antibody staining. Association between these subunits was further confirmed by co-immunoprecipitation.[3,15] FRET has also been extensively applied in the study of heteromultimer channels.[22,23] These fluorescence techniques will be discussed below.

15.3.1 FLUORESCENT ANTIBODY LABELING

Fluorescent antibody labeling is a widely used technique in ion channel research. Principally, fluorophore-conjugated antibodies are developed to target specific channel protein. After antibody-antigen binding, the spatial distribution of labeled channels is illustrated by the fluorescence signal.[69] Experimentally, there are three major categories of fluorescent antibody labeling techniques: direct fluorescent antibody (DFA), indirect fluorescent antibody (IFA), and its derivative, streptavidin-biotin fluorescent antibody labeling.

For DFA labeling, fluorophore is directly conjugated on the primary antibody that recognizes the target channel protein. The advantage of DFA is that the experimental procedure is straightforward, as only one type of antibody is required. However, a major issue of DFA labeling is that it requires the preparation of specific fluorophore-linked antibodies for each channel type, which could be difficult and inefficient to do. In addition, because antibody-antigen binding is not 100% specific, fluorophore-linked antibody may recognize and label nonspecific proteins, decreasing the signal-to-noise ratio of fluorescence labeling.

For IFA labeling, the fluorophore is conjugated to a secondary antibody, which recognizes the antigen-specific primary antibody. Experimentally, a nonfluorescent primary antibody for the target channel is always used in conjunction with a fluorescent secondary antibody. For example, to study colocalization of TRPV2 and TRPV1, rat DRG and brain samples were fixed in 4% paraformaldehyde overnight and stored in 12% sucrose for 1 day. These samples were then sectioned to 20-μm-thick slides, which were blocked with a 10% normal goat serum and 0.1% Triton-X100 overnight at 4°C. The rabbit polyclonal TRPV2 antibody and the guinea pig polyclonal TRPV1 antibody were applied sequentially as primary antibodies at 1:100 dilution in a 10% normal goat serum and 0.1% Triton-X100 for 12 hours at 4°C. Next, the rhodamine-conjugated anti-rabbit antibody (1:600 diluted, 1.5 mg/mL) was applied to detect the TRPV2-primary antibody complex, while the fluorescein isothiocyanate (FITC)-conjugated anti-guinea pig antibody (1:100 diluted, 13 μg/mL) was used to label the TRPV1 containing complex. The secondary antibody labeling step lasted for 2 hours at room temperature.[3]

There are several advantages of IFA. First, only a limited number of standard fluorophore-linked secondary antibodies are needed, which will work with numerous nonfluorescent primary antibodies. The cost is significantly reduced, while detecting sensitivity is increased, because more than one fluorophore-conjugated secondary antibody can bind with one primary antibody.[69–71] In addition, IFA can be easily modified to further increase detection sensitivity. For example, biotin molecules can be conjugated to the secondary antibody instead of fluorophores. Then fluorophore-linked avidin or streptavidin molecules are added to bind with the biotinylated secondary antibody. The binding affinity of avidin for biotin is very high ($K_d \approx 10^{-15}$ M)[72] so that avidin is able to bind with biotinylated proteins even if the concentration of the latter is low. Because multiple biotin molecules can be conjugated to one protein molecule, a biotinylated protein can be bound by several fluorophore-labeled avidins, which amplifies the fluorescence signal and further enhances detection sensitivity. In some applications, streptavidin is preferred over avidin because, unlike the positively charged avidin at physiological pH, it is electrically neutral so that nonspecific binding with negatively charged proteins is substantially reduced.[73] Because of these advantages, IFA labeling has been widely used.

For colocalization studies with IFA labeling, it is experimentally preferred but not required to choose two fluorophore-conjugated antibodies with similar excitation wavelength. However, they should have well-separated emission spectra. In the TRPV2–TRPV1 colocalization study mentioned above, emission peaks of rhodamine and FITC are at 590 and 520 nm, respectively. Emissions from these fluorophores can be selected either by band pass filters or by spectroscopic measurements. The fluorescence signals can be viewed in sequence or, with approaches to divide the light into two paths (for example, dual-view microscopy), simultaneously.

Colocalization experiments can be carried out with genetically fused fluorescence indicators. Hellwig et al. inserted cyan fluorescence protein (CFP) or yellow fluorescence protein (YFP) cDNA into vectors containing the thermoTRPV channel cDNA, so that when the tagged channels were expressed in HEK293 cells, CFP or YFP was attached at the C-terminus of the channel protein. By co-expressing two channel vectors, they found that only TRPV1/V2 and TRPV2/V3 pairs showed significant CFP and YFP fluorescence colocalization.[23] Expression of genetically tagged channel subunits avoids nonspecific antibody labeling. However, all channel subunits are fluorescently tagged, no matter whether they are properly folded and trafficked to the plasma membrane.

A general concern about fluorescence signal colocalization is the limited spatial resolution, which is determined by a number of factors. First, the spatial resolution is limited optically by light diffraction. The Rayleigh criterion states that the spatial resolution (R) of a microscope is determined by the emission light wavelength (λ) and the numeric aperture (NA) of the objective lens:

$$R = \frac{1.22 \times \lambda}{2 \times NA} \tag{15.2}$$

Equation 15.2 indicates that the spatial resolution of the colocalization experiment is at best a few hundred nanometers. However, the dimension of an ion channel protein is generally in the order of 10 nm,[21] which is well below the spatial resolution of the colocalization technique.

Moreover, the spatial resolution is limited by the size of the antibodies. Antibodies are gamma globulin proteins, which are approximately 10 nm in length and 6 × 7 nm in cross-section area.[74,75] With such a large protein, the fluorophore labeled on an antibody could be 10 nm away from the target ion channel. In IFA labeling, fluorophores are farther away from the target because of the stacking of two antibodies. Therefore, the low spatial resolution of fluorescence antibody staining limits the interpretation of colocalization results.

15.3.2 Confocal Microscopy

In conjunction with other fluorescence imaging techniques, confocal microscopy is designed to improve the signal-to-noise ratio and achieve three-dimensional imaging.[76] "Confocal" refers to the two pinholes used in the experimental setup. The first pinhole is placed in front of the light source, while the second is in front of the detecting device such as a digital camera or a photomultiplier tube. By adjusting the relative position of these pinholes, only fluorescence emission from the focal point can pass the second pinhole and be recorded. Any fluorescence from outside of the focal point will be blocked. Because of the reduction in the out-of-focus fluorescence signal, the signal-to-noise ratio and spatial resolution are substantially enhanced over conventional wide-field fluorescence microscopy.[77] Nonetheless, the spatial resolution of confocal microscopy is still limited by light diffraction.

To take advantage of the enhanced imaging quality, fluorescent antibody labeling and colocalization experiments are usually performed using a confocal microscope. For example, confocal imaging revealed the presence of TRPV1 in the trachea of guinea pigs,[78] human bladder,[79] and rat skeletal muscle;[80] TRPV2 in human bladder;[81] and TRPV4 in rat cortical astrocytes[82] and female mouse reproductive organs.[83] The colocalization of TRPV1 and TRPV2,[3] as well as other thermoTRPV channels,[23] was also observed with confocal microscopy.

Besides an improved signal-to-noise ratio, it is possible to achieve three-dimensional (3D) imaging with confocal microscopy. With a confocal microscope, the image of a sample is constructed by moving the focal point pixel by pixel and scanning the entire x–y plane. As the axial position of the focal point is adjustable, multiple images of the x–y plane at different z depth can be recorded. From this stack of x–y plane images, a 3D image of the sample can be reconstructed.[84] A 3D image of a cell or tissue provides valuable information on the spatial distribution of target channels.

15.3.3 FRET

Although a colocalization experiment is relatively easy to set up, its spatial resolution is limited by light diffraction as well as the size of antibodies (or fluorescent proteins). Therefore, this technique alone cannot prove or disprove the existence of heteromeric subunit assembly.

Another fluorescence technique with the required spatial resolution, fluorescence resonance energy transfer (FRET), has been employed to investigate heteromeric channel assembly. FRET is a physical process where the energy of an excited "donor" fluorophore is transferred to a nearby "acceptor" fluorophore via nonradiative dipole–dipole coupling.[85] The energy transfer rate is highly sensitive to the distance between donor and acceptor fluorophores, dropping to virtually zero when the fluorophores are 10 nm or more apart. The dimension of an ion channel is in the same order, which makes FRET an ideal tool to measure intra- or intersubunit distances.[17–20,85,86]

When FRET occurs, the donor fluorescence intensity decreases while the acceptor fluorescence intensity increases. The energy transfer efficiency between a single pair of donor–acceptors is determined by

$$E = \cfrac{1}{1 + \left(\cfrac{R}{R_0}\right)^6} \tag{15.3}$$

where R is the distance between a donor and an acceptor, and R_0 is the characteristic distance for a specific pair of fluorophores at which the FRET efficiency is 50%. R_0 is determined by quantum yield of the donor (Φ_D), relative orientation (κ^2), refractive index of the medium (n), and spectra overlap of the fluorophores ($J(\lambda)$), as shown in

$$R_0^6 = 8.8 \times 10^{-28} \cdot \Phi_D \cdot \kappa^2 \cdot n^{-4} \cdot J(\lambda) \tag{15.4}$$

Useful FRET pairs normally have an R_0 value around 3–6 nm (see Table 15.1).

When fluorophores are selected for FRET experiment, there are several considerations:

1. The emission spectrum of the donor overlaps with the excitation spectrum of the acceptor. If there is no such spectral overlap, $J(\lambda)$ would be zero, and no FRET would occur regardless of the distance between the fluorophores. For example, the emission spectrum of CFP overlaps extensively with the excitation spectrum of YFP around 500 nm, so they are able to form an efficient donor–acceptor pair.
2. The match between the R_0 value of a FRET pair and the distance to be measured is an important consideration for the design of a FRET experiment. This is because FRET efficiency is most sensitive to distance change when $R = R_0$. For the commonly used FRET pair CFP and YFP, the R_0 value is 4.9 nm.[87] This characteristic distance is ideal for many ion channel studies because within the scale of channel protein (<10 nm), the CFP–YFP pair will show a significant FRET signal and be sensitive to distance changes.
3. The labeling of target proteins and the fluorophore sizes meet the requirement of a particular FRET experiment. Chemical FRET pairs, such as fluorescein-5-maleimide (FM) and TMRM, can react covalently with cysteine residues on channel proteins. This allows them to report local distance changes at specific channel positions. Green fluorescent protein and its derivatives have greatly facilitated application of the FRET technique in ion channel research.[7] CFP and YFP are the most commonly used FRET pair. By inserting their cDNA into a target channel vector and expressing in cells, these fluorescent proteins can be genetically linked to either the N- or C-terminus or even within the channel protein. Given their large size, fluorescence proteins are more suitable for addressing questions concerning domain movements, subunit assembly, protein–protein interaction, and channel trafficking and distribution.

Heteromeric assembly of thermoTRPV channels has been investigated by FRET with the CFP–YFP pair. In 2005, Hellwig and colleagues reported that subunits from different thermoTRPV species in general do not co-assemble in HEK293 cells.[23]

TABLE 15.1
R_0 Values of Commonly Used FRET Pairs

Donor	Acceptor	R_0 (Å)	Reference
Fluorescein	TMRM	55	1
EDANS	Dabcyl	33	1
Fluorescein	QSY 7 and QSY 9 dyes	61	1
Alexa Fluor 350	Alexa Fluor 488	50	1
Alexa Fluor 488	Alexa Fluor 546	64	1
Alexa Fluor 555	Alexa Fluor 594	47	1
Alexa Fluor 594	Alexa Fluor 647	85	1
Blue fluorescence protein	Cyan fluorescence protein	38	2
Cyan fluorescence protein	Yellow fluorescence protein	49	2
Green fluorescence protein	Yellow fluorescence protein	56	2

Note: Ref. 1, Invitrogen, *Molecular Probes: The Handbook*; Ref. 2, G. H. Patterson et al., *Analytical Biochemistry*, 284(2), 438–440, 2000.

However, on the basis of the same expression system, Cheng and colleagues reported in 2007 that the four thermoTRPV subunit types are able to form heteromeric channels.[22] The difference illustrated that, while the FRET measurement is conceptually straightforward, a number of pitfalls in FRET practice need to be carefully considered. The experiment setups and methods to calculate FRET efficiency will be discussed below.

FRET efficiency can be calculated experimentally in a number of ways.[18,19] Here we limit the discussion to fluorescence intensity-based methods, as they are commonly used in channel studies including the two thermoTRP studies mentioned above. For more detailed discussion of FRET methods, readers are referred to reviews.[19,20,88–90] FRET will decrease the donor fluorescence intensity while increasing the acceptor fluorescence intensity, so three general experimental approaches can be devised: (1) donor dequenching, (2) acceptor-enhanced emission, and (3) donor–acceptor emission peak ratio.[20]

15.3.3.1 Donor Dequenching

The donor dequenching method measures the increase in donor emission intensity after the acceptor is destroyed by photobleaching. Removal of the acceptor terminates energy transfer. As a result, donor emission is expected to increase. For example, in the study of the heteromeric assembly of TRPV subunits, Hellwig et al. co-expressed C-terminal CYF- and YFP-tagged TRP channel subunits in HEK293 cells and performed donor-dequenching FRET experiments on an epifluorescence microscope with proper filters for CFP/YFP excitation and emission. The fluorescence intensity of CFP and YFP before YFP photobleaching was recorded with a 15-cycle protocol, which consisted of a 40-ms/cycle exposure for CFP excitation at 410 nm and 8 ms/cycle for YFP excitation at 515 nm. Photobleaching of YFP was performed in the following 60-cycle procedure, in which an additional 2.1-s/cycle exposure for YFP was included. YFP was bleached at a rate of about 12% per cycle.[23] During this process, the CFP intensity was seen to gradually increase. With the recorded CFP intensity before (F_D) and after (F_D') YFP photobleaching, FRET efficiency is calculated as

$$E = 1 - \frac{F_D}{F_D'}$$
(15.5)

Besides being conceptually simple, donor dequenching might be less prone to the detection of pseudo positive FRET signals because an increase in donor emission is less likely to be achieved without a decrease in FRET.[91]

In donor-dequenching experiments, FRET is measured from the donor fluorophore. It is strongly preferable to measure from cells with a high acceptor-to-donor intensity ratio. Under this condition, energy transfer from donors is maximized by acceptors. Upon photobleaching, the emission recovery of donors will be more significant. Hellwig et al. used a high YFP-to-CFP cDNA ratio in cell transfection. During FRET imaging, cells with an acceptor-to-donor intensity ratio larger than 0.8 were chosen to measure FRET efficiency.[23]

There are a few caveats in, but not limited to, donor-dequenching FRET measurement. First, because photobleaching is irreversible, measurement can only be taken once from each sample. Second, if acceptor photobleaching is not complete, measured FRET efficiency will be an underestimate of the true value.[89] Upon excitation, fluorescent proteins may move into a reversible nonfluorescent state instead of the nonreversible bleached state. Third, care needs to be taken to ensure that the donor is not inadvertently bleached during acceptor photobleaching.

Importantly, choosing a proper region on a cell to measure FRET warrants serious consideration. Although thermoTRP channels are membrane proteins, it is well known that they are widely distributed inside the cell. Translocation of intracellular TRP channels to the cell surface is thought to be an important mechanism underlying cellular responses to inflammation and nociceptive stimuli.[92,93] More generally, overexpression of exogenous proteins often leaves a large amount of proteins trapped inside the cell. Factors that affect the intracellular pool of channel proteins are not well understood. In certain intracellular compartments, assembly of subunits may not be completed. Conversely, a fraction of the intracellular channel proteins might be in the recycling pathway and partially degraded. The different environment may also affect the fluorescence from these compartments. It is also likely that proteins trapped inside the cell are misfolded. Therefore, in comparison to the intracellular channel proteins, those in the plasma membrane are thought to better represent properly assembled, functional channel complexes. Indeed, a direct comparison of FRET collected from the plasma membrane and the cytosol of the same cells showed substantial differences.[22]

15.3.3.2 Acceptor-Enhanced Emission

This method measures the increase in acceptor emission intensity when FRET occurs. Energy from the excited donor is transferred to the acceptor, causing the acceptor emission intensity to increase. The acceptor-enhanced emission method was applied in the study by Cheng et al., where heteromeric thermoTRPV channels were observed.[22] Similar to the donor-dequenching FRET experiment discussed above, C-terminal CFP- and YFP-tagged TRPV subunits were expressed in HEK293 cells and imaged for FRET. In this study, a specific form of acceptor-enhanced emission measurement was performed, known as "spectra FRET."

In an acceptor-enhanced emission FRET experiment, two major problems should be considered.[20] The first is cross-talk of the fluorophores' excitation. As shown in Figure 15.1, cross-talk refers to the phenomenon that acceptor fluorophores are partially excited by the donor excitation light. Cross-talk occurs because the excitation spectrum of the acceptor is usually quite broad. The excitation light for the donor fluorophore will inadvertently excite the acceptor. When the CFP–YFP FRET pair is used, 15 to 25% of the YFP emission could be due to cross-talk.[94] The second problem is bleed-through of donor emission (Figure 15.1), which means the donor emission light is detected in the acceptor emission wavelength, which is caused by the broad emission spectrum of the donor fluorophore. For CFP, its emission at 530 nm (YFP emission peak) can be as large as 50% of its own emission peak at 480 nm.[20] In addition to the two issues listed above, if the concentrations of fluorophores are high, the probability of finding a pair of a donor and an acceptor within 10 nm by chance is increased. This random adjacent positioning of a donor and an acceptor can generate nonspecific FRET, which is a contamination to the real FRET signal. At very high fluorophore densities, re-absorption of fluorescence emission may appear to be FRET. This, however, is unlikely to happen in live cells.

To overcome these problems, several experimental strategies are designed to measure acceptor-enhanced emission, such as "three-cube FRET"[95] and "netFRET."[88]

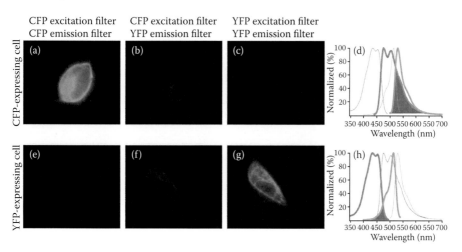

FIGURE 15.1 **(See color insert.)** Two types of major errors in FRET measurements: cross-talk and bleed-through. (a–c), a cell expressing TRPV1-CFP; (e–g), a cell expressing TRPV1-YFP. Filter combinations are labeled on top. (d) Emission spectra of CFP and YFP are in thick lines, while their excitation spectra are in thin lines. Shaded area indicates bleed-through of CFP emission. (h) Excitation spectra of CFP and YFP are in thick lines, while their emission spectra are in thin lines. Shaded area indicates cross-talk of excitation.

In the 2007 study, "spectra FRET" was employed.[22] The experiment setup of spectra FRET was discussed in details previously.[90] An epifluorescence microscope, a shutter-controlled excitation light source, and two filter cubes are required. For the CFP–YFP pair, one filter cube contains an excitation filter for CFP at 436 nm and a dichroic mirror; another filter cube has an excitation filter for YFP at 500 nm and a corresponding dichroic mirror. There is no emission filter in either cube, which allows the emission spectra of CFP and YFP to be recorded in a wide wavelength range by a spectrograph attached to the output port of the microscope. A CCD camera is attached to the output port of the spectrograph to record spectral images.

Experimentally, spectra FRET requires three standard spectral images and two for each experimental cell. One standard spectral image from the donor-only cell is taken with the donor excitation light. The emission spectrum of the donor is constructed from this image. Two standard spectral images from the acceptor-only cell are taken with the donor and acceptor excitation light, respectively. The ratio of fluorescence intensities measured from these two images is termed "RatioA0," which is used to correct for cross-talk. Two spectral images are taken from cells co-expressing donor- and acceptor-labeled subunits, with the donor and acceptor excitation light, respectively, as shown in Figure 15.2a to f. To calculate FRET efficiency, the donor emission spectrum from image 1 mentioned above is normalized to the donor emission peak in the spectrum with donor excitation (Figure 15.2g and h) and subtracted to correct for donor bleed-through. Next, the ratio between the subtracted spectrum and the total acceptor emission spectrum with acceptor excitation is calculated and termed "RatioA." Finally, the apparent FRET efficiency is computed as

$$E = \frac{\varepsilon_A}{\varepsilon_D} \times \left(\frac{\text{RatioA}}{\text{RatioA0}} - 1 \right) \quad (15.6)$$

Contrary to donor dequenching, in the acceptor enhanced emission approach, it is preferable to take measurements from cells with a high donor-to-acceptor intensity ratio. Under this condition, each acceptor is coupled with one or more donors. The extra donors are no problem because they are removed from the calculation when the donor emission component is subtracted. Specifically for thermoTRP research, there are two concerns about the relationship between the measured apparent FRET efficiency and the true efficiency. First, because TRP channels are tetramers, more than one pair of donor and acceptor fluorophores will be presented in an assembled channel complex. Interpretation of distances between fluorophores within such a channel complex using the apparent FRET efficiency would be different from the single donor–acceptor pair case, e.g., the CLC-0 chloride channel.[96] Second, as the expression levels of CFP- and YFP-tagged subunits can vary substantially from cell to cell,[20,22,96] the apparent FRET efficiency fluctuates according to the degree of donor–acceptor pairing. While adjusting the cDNA concentration ratio does help to alter the overall expression level, it does not ensure equal expression in each cell. To address these concerns, models describing the dependence of FRET and channel subunit stoichiometry have been developed.[22,96,97]

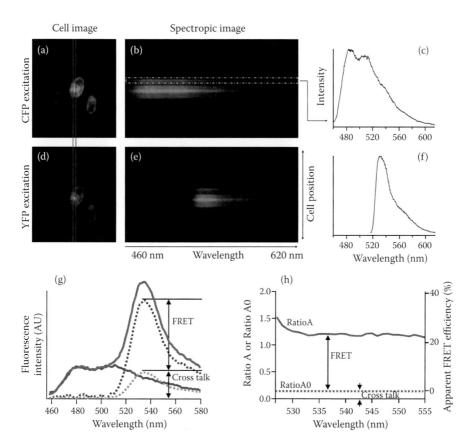

FIGURE 15.2 **(See color insert.)** Spectra FRET (adopted from Refs. 51 and 73). A single HEK293 cell co-expressing TRPV3-CFP and TRPV3-YFP was observed with (a) CFP or (d) YFP excitation. Spectroscopic images from the region under the slit (indicated by the rectangle) were recorded with each excitation (b, e). In the spectroscopic images, the *Y* axis represents the position of the cell, and the *X* axis represents the wavelength. The fluorescence intensity values measured from the upper membrane region (indicated by a box) are shown in c and f. Notice the bright strips in b and e between the membrane signals, which represent fluorescence from intracellular sources. (g) Removal of the bleed-through and cross-talk contaminations. A CFP emission spectrum (blue curve) is normalized to the CFP peak of an emission spectrum from a cell expressing CFP–YFP tandem dimers (red curve). The extracted YFP spectrum (green curve, which represents the difference between the red curve and the blue curve) contains both the FRET emission and the cross-talk emission (yellow curve). (h) The difference between RatioA (solid line) and RatioA0 (dotted line) indicates the existence of the FRET signal. (From Takanishi, C.L. et al., *Brain Research* 1091(1), 132–139, 2006. With permission.)

As discussed above, choosing the proper region of the cell to measure FRET is critical. In spectra FRET, the spectroscope's narrow input slit is placed over the cell. When a cell image is projected along the wavelength axis as shown in Figure 15.2a and b, the fluorescence from the plasma membrane of the cell appears as two bright bands at the edges. It provides a convenient way to separate (within the limit of optical resolution as discussed above) intracellular fluorescent signals and the plasma membrane fluorescent signal. In the 2007 study,[22] positive FRET signals were detected from the plasma membrane, which is in support of the existence of heteromeric thermoTRPV channels. This conclusion is further confirmed by single-channel recordings, which revealed that channels having different subunit compositions exhibited different gating and conductance properties. In this study, it was also observed that the FRET efficiency value calculated from cytosolic fluorescence was significantly lower than that estimated from fluorescence originating from the plasma membrane of the same cells.[22] This observation may help explain the discrepancy between the 2005 and 2007 studies.[22,23]

15.3.3.3   Donor–Acceptor Emission Ratio

Donor–acceptor emission ratio measurement allows rapid monitoring of FRET changes without calculating the efficiency value. When FRET occurs, the donor emission decreases while the acceptor emission increases. Donor–acceptor emission ratio measurement incorporates these two changes into one number. As a result, its sensitivity to the change in FRET efficiency is enhanced.[91]

Compared with donor dequenching and acceptor emission enhance methods, emission ratio measurement is easier to implement experimentally. For example, in the study of TRPV1 outer pore rearrangement during channel activation, subunits containing cysteines in the outer pore region were expressed in HEK293 cells. A mixture of FM and TMRM was used to covalently label the cysteines. On an epifluorescence microscope, the 488-nm laser light was used to excite FM and initiate FRET. Emission spectra containing both donor emission (F_D) and acceptor emission (F_A) were recorded.[98] After correcting for the background signal, FRET_index is calculated as

$$\mathrm{FRET}_{\mathrm{index}} = \frac{F_A}{F_D} \qquad (15.7)$$

It was observed that when the channel was activated by an increase in temperature, the FRET index value increased, suggesting that the channel outer pore regions moved closer to each other.[98] With proper equipment, donor–acceptor emission ratio measurement can be done in real time, permitting studies of dynamic FRET changes.

While the donor–acceptor emission ratio is very sensitive to FRET efficiency changes,[20] from this ratio, it is very difficult to calculate FRET efficiency. Unlike the two previously discussed methods, in emission ratio measurements, both donor and acceptor intensities contribute to FRET_index. Matching amounts of the donor and the acceptor may be a good starting point to set up optimal conditions for FRET detection. Applications of emission ratio measurement in thermoTRP research will be further discussed in the following sections.

15.4   TRAFFICKING OF THERMOTRP CHANNELS

The physiological functions of assembled thermoTRPs are highly regulated by protein trafficking events such as transportation, membrane insertion, and internalization.[99] For example the cell surface expression level of a channel type is regulated by modulating the channel insertion or retrieval rate.[100] Indeed, trafficking of thermoTRP channels to the cell surface is found to be a cellular response to environmental stimuli.[92,93] In conjunction with biochemistry and cell biology techniques, fluorescence microscopy has been widely applied in thermoTRP channel trafficking research. Using total internal reflection fluorescence (TIRF) microscopy, Stein et al. observed an increase in the number of TRPV1 channels on the plasma membrane after nerve growth factor (NGF) stimulation.[101] By fluorescent antibody staining, Stokes et al. suggested that surface expression of TRPV2 is enhanced when co-expressed with recombinase gene activator protein.[102]

To regulate trafficking processes, the interaction between channel proteins and a network of regulators, such as cytoskeleton proteins, protein kinases, and scaffolding proteins, is of vital importance. Fluorescence techniques have been applied in this field of research as well. For example, combining FRET and the fluorescence recovery after photobleaching (FRAP) technique, Becker et al. found that TRPV4 interacts with actin in cell volume regulation.[103] In this section, TIRF and FRAP will be discussed.

15.4.1   TIRF MICROSCOPY

TIRF microscopy measures the fluorescence signal from a very thin layer of sample cells adjacent to the cover glass.[104–106] When light travels from one medium with a high refractive index (n_1) to another with a low refractive index (n_2) at an angle (θ_1), refraction occurs. The angle of refracted light (θ_2) is determined by Snell's law:

$$n_1 \times \sin \theta_1 = n_2 \times \sin \theta_2 \tag{15.8}$$

Because $n_1 > n_2$, it is expected that $\theta_2 > \theta_1$. Therefore, when the angle of incident light is larger than a critical value the incident light will be totally reflected at the interface of two media. However, because of the near field effect, a tiny fraction of incident light energy will penetrate into the other medium as an "evanescent wave."[107] The intensity of the evanescent wave decays exponentially along the axis perpendicular to the interface, as shown in

$$I_z = I_0 \times \exp^{\frac{-z}{d_p}} \tag{15.9}$$

$$d_p = \frac{\mu}{4 \cdot \pi \cdot \sqrt{n_1^2 \cdot \sin \theta_1^2 - n_2^2}} \tag{15.10}$$

In these equations, μ is the wavelength of the incident light, I_z is the evanescent wave intensity at distance Z from the interface, I_0 is the intensity at the interface, and d_p is the penetration depth, a parameter of the TIRF system determined by Equation 15.10. For a typical TIRF system, the penetration depth is less than 100 nm, which is much thinner than the optical section generated by confocal microscopy (500–800 nm).[105] Therefore, TIRF microscopy can substantially increase the signal-to-noise ratio because the shallow penetration depth of the evanescent light ensures that only fluorophores in the plasma membrane and the nearby region are excited without touching the bulk of fluorophores in cytosol.

Experimentally, a TIRF setup is built upon conventional fluorescence microscopy. Two types of TIRF configurations are available: prism type and objective type.[105] In the prism configuration, a prism is placed above sample cells. When excitation light travels through the prism, the evanescent wave is generated at the surface of the prism facing sample cells. However, this configuration suffers two major limitations. First, because the prism is placed above the sample, it is difficult to access and manipulate cells simultaneously during imaging. In addition, with this configuration, the top of the sample, which is near the prism, is illuminated by the evanescent wave. Most modern fluorescence microscopes are designed as inverted microscopes. Emission fluorescence from the top of samples needs to go through the bulk of the cell before being captured by the objective lens.

Objective-type configuration requires only one objective lens to generate the evanescent wave and collect fluorescence emission. With this configuration, the space above sample cells is open, so other manipulation of cells such as patch clamping can be performed simultaneously with TIRF imaging. To reach the critical angle of the incident light, an objective lens with NA equal to or larger than 1.4 is required.[108] For example, in the TRPV1 and NGF study mentioned above, an objective-type TIRF microscope with a 60× oil immersion lens (NA = 1.45) was used.[101] Moreover, the excitation light is fed through the objective in a designed manner. For laser excitation, it should be focused at the periphery of the lens' rear aperture. For other excitation light such as the arc-discharge lamp, an opaque disk should be placed in the middle of the lens' rear aperture to ensure that the excitation light goes through its periphery.[109] Objective-type TIRF is generally preferred over prism-type TIRF.

The TIRF technique is advantageous in the study of membrane protein trafficking. In the objective-type configuration, the plasma membrane and intracellular space next to the cover glass will be illuminated. The plasma membrane is very thin (<10 nm),[110] so when fluorescent molecules are transported from the cytosol toward the membrane, the excitation evanescent field intensity gradually increases. As a result, fluorescence emission is expected to increase. Gordon and colleagues observed that after NGF incubation, the fluorescence intensity from the cell membrane and/or its nearby region was substantially enhanced, indicating that NGF-stimulated TRPV1-YFP was trafficking toward the plasma membrane.[101]

Practically, several caveats should be considered in designing and performing TIRF experiments. First, because the intensity of the evanescent wave decays exponentially along the vertical axis, a slight change in the distance could lead to a substantial change in excitation evanescent field strength. The associated variation in the emission intensity complicates quantitative interpretation of imaging data. Therefore, the relative positioning of sample cells on the cover glass, such as the degree of cell attachment and the flatness of the cell–glass interface, should always be considered in TIRF experiments. In addition, although the typical refractive index of cytosol is 1.38,[109] this parameter may be variable inside a cell. Furthermore, scattering of the evanescent wave may occur at the boundaries of different refractive indexes. This scattering will cause a deeper penetration and decrease in resolution.[108,111,112]

15.4.2 FRAP

Photobleaching is a photochemical process by which a fluorophore is irreversibly destroyed. On the basis of this process the FRAP technique measures the mobility of fluorophore-labeled biomolecules. After a small region of a cell is photobleached by high-power excitation light, the intact fluorophores nearby can diffuse into the bleached region, leading to a time-dependent recovery of the fluorescence intensity.[113] Information on the mobility of fluorophore-labeled proteins is extracted from the rate and the degree of fluorescence recovery.

To quantify FRAP measurement, two parameters are calculated: mobile fraction (M_f) and diffusion constant (D). M_f measures the extent to which fluorophores can move. It is calculated as

$$M_f = \frac{F_r - F_0}{F_i - F_0} \tag{15.11}$$

in which F_r is the fluorescence intensity after full recovery, F_0 is the fluorescence immediately after photobleaching, and F_i is the fluorescence before photobleaching.[103] A larger M_f indicates more fluorophores are mobile.

The parameter D measures the rate of fluorophore diffusion. To calculate this parameter, an exponential equation is used to fit the time course of fluorescence recovery. With time constant (τ_D) from fitting, the following equation is applied to a small spot bleaching scenario:

$$D = \frac{\omega^2 \cdot \gamma}{4 \cdot \tau_D} \tag{15.12}$$

in which ω is the radius of light beam, and γ is a correlation factor.[114–116] Diffusion constant has a unit of $\mu m^2/s$.

Experimentally, FRAP measurement can be carried out using a conventional fluorescence microscope. To photobleach fluorophores, a high-power laser (usually a few hundred milliwatts)[115] is used as a light source. GFP and its derivatives are ideal fluorophores for FRAP; owing to their barrel-shape structure, less reactive intermediates are generated during photobleaching, so it is less damaging to the cell.[113,117,118]

Using the FRAP technique, Jendrach et al. studied the interaction between TRPV4 and actin. GFP was attached to TRPV4 as the fluorescence reporter. With a Leica SP 5 microscope and proper filters, GFP was excited at 488 nm. Fluorescence emission was recorded at 500 to 540 nm. The lamellipodia region of TRPV4-GFP expressing cells was photobleached by a high-power laser. To record fluorescence recovery, the bleached region was imaged every 3 s for 3–5 min at 37°C. It was observed that when latrunculin A (an actin-destabilizing reagent) was applied to the cell at a final concentration of 0.5 µM for 20 min, the M_f value was significantly increased.[103] This observation indicates that more TRPV4-GFP molecules are mobile after latrunculin A treatment, suggesting that TRPV4-GFP interacts with actin. This interaction was further confirmed by donor-dequenching FRET measurement using CFP-labeled TRPV4 and YFP-labeled actin.[103]

15.5 STRUCTURE–FUNCTION RELATIONSHIP IN THERMOTRP CHANNELS

It is well recognized that the biophysical properties and physiological functions of a protein are ultimately determined by its three-dimensional structure.[2,21,119] There are techniques to reveal the high-resolution structure of a protein, such as X-ray crystallography and nuclear magnetic resonance (NMR) spectroscopy. However, these techniques are limited in that they can only capture a "snap shot" of the channel structure at a particular state.[120,121] Proteins carry out their functions by undergoing conformational rearrangements. Fluorescence techniques[122,123] have been widely applied to understand dynamic changes in conformation of functional proteins under physiological conditions.

For thermoTRP channels, their structural information is still quite limited.[124–126] While X-ray crystallography has revealed the structure of many potassium channels and chloride channels at atomic resolution,[127–130] only parts of a few thermoTRP channels have been studied with crystallography, namely, the N-terminal ankyrin-like domains of TRPV1 and TRPV2.[126,131–133] Low-resolution 3D structures of full-length TRPV1 and TRPV4 have been investigated by cryo-electron microscopy.[134,135] Advance in structural research is expected to greatly facilitate the study of thermoTRP protein dynamics and structure–function relationships.

15.5.1 Patch-Clamp Fluorometry

Both voltage-clamp fluorometry[10] and patch-clamp fluorometry[4] are methods designed to simultaneously record fluorescence and current signals from functional channels.[13] The basic idea is to attach fluorophores to the moving parts of an ion channel protein as molecular sensors. Movement of the labeled channel structure is recorded from the change in the fluorescence signal due to changes in local environment, proximity to other fluorophores, or accessibility by quenching molecules. When combined with simultaneous current recording, which reports the functional state of the channel, rich information about dynamic conformational rearrangement can be gained. These approaches have been demonstrated as powerful tools in studying ion channel structure–function relationships.[13,122,123,136]

A patch-clamp fluorometry system is basically a combination of a patch-clamp setup and a fluorescence microscope, the latter being either a regular epifluorescence microscope[13,136] or a TIRF microscope.[137] The electrophysiological recording, either by patch clamping[13,136] or two-electrode voltage clamping,[138,139] is synchronized with fluorescence recording. Synchronization can be achieved by linking data acquisition software. Fluorescence recording devices should be chosen in accordance with the requirement of specific experiments. For example, when the spatial resolution of the sample is required, a CCD camera should be considered. When high-frequency (above kHz) fluorescence recording is required to monitor rapid changes in protein structure, a photomultiplier tube (PMT) or avalanche photodiodes should be used.[4,12,136,137]

Using voltage-clamp fluorometry, the Isacoff group and the Bezanilla group investigated the movements of the voltage sensor domain in potassium channels,[10,11,137,140–145] which greatly advanced our understanding of the voltage gating mechanism.[122,123] In research on cyclic nucleotide-gated (CNG) channels, the Zagotta group recorded a fluorescence signal from an inside-out patch.[4,136,146–149] These approaches have been applied to the study of other channel proteins as well.[12,150–152]

In a thermoTRP channel study, Yang et al. observed movements of TRPV1 pore turret during temperature-driven activation.[5] TRPV1 has two native cysteine residues in the pore turret sequence. They can be labeled by extracellularly applied sulfhydryl-reactive fluorophores such as TMRM and FM. Experimentally, after a whole-cell patch was formed, a mixture of FM and TMRM at a concentration ratio of 1:1 was perfused onto the cell, and the change in the current amplitude due to attachment of fluorophores to the cysteine residues followed. Free fluorophores were washed off from the bath solution after the current amplitude reached a stable level, indicating the completion of fluorescence labeling.

In this experiment, the imaging system was built upon a Nikon TE2000-U microscope equipped with a 40× oil-immersion objective (NA 1.30). The excitation light was generated by an Ar laser. The duration of light exposure was controlled by a computer-driven mechanical shutter. Spectral measurement was achieved with a spectrograph in conjunction with a CCD camera. Both the shutter and the camera were controlled by the MetaMorph software, which was synchronized with PatchMaster through a TTL trigger. In this way, the current and the fluorescence image were recorded simultaneously. From the fluorescence image the TMRM/FM intensity ratio was calculated as a FRET efficiency index, as previously described. Yang et al. observed an increase in this ratio induced by heat activation only, but not by voltage- or ligand-gated activation. This finding suggests that the pore turret of TRPV1 is an important component of the temperature-sensing apparatus, which moves closer to each other during temperature-driven activation.

Practically, there are two major concerns about techniques using simultaneous fluorescence and current recording.[13] First, the size of the fluorophore should be considered. If the fluorophore is large, its attachment to the channel protein may incur substantial perturbation to the channel function. In addition, the fluorophore size and the linker length between chromophore and the labeling site on the channel may complicate the interpretation of FRET results. In this regard, chemical fluorophores such as FM and TMRM are better than fluorescent proteins.

Second the fluorescence signal-to-noise ratio should be optimized. When chemical labeling is applied, besides target proteins, fluorophores may react with cysteines in native membrane proteins, stick in the lipid bilayer, or stay in the recording chamber. For example, in a patch fluorometry experiment on CNG channels, it was estimated that less than half of the total fluorescence was from target labeling sites.[4] If possible, a nonfluorescent reagent should be used to mask background cysteines prior to fluorescent labeling. From this perspective, genetically linked fluorescent proteins would be superior to chemical fluorophores. Doing experiments in a dark room and turning off the computer monitor can effectively reduce background light contamination, which helps reduce the noise level. In addition, using an objective lens with a large NA value will increase light-collecting efficiency and, in turn, enhances the recorded fluorescence signal intensity.

15.6 EMERGING OPTICAL TECHNIQUES

Optic techniques evolve constantly and rapidly. With emerging tools such as single-molecule imaging, super-resolution imaging, and two-photon microscopy, the power of fluorescence microscopy has been greatly enhanced. For temperature-driven activation of thermoTRPs, infrared laser has been employed as a fast and reliable heat source (see Chapter 14). Although these new techniques are still in their infancy, their potentials are promising.

15.6.1 SINGLE-MOLECULE IMAGING

Single-molecule imaging allows for detection of the fluorescence signal from individual protein molecules, which generates rich information on behaviors that could otherwise be buried in ensemble-averaged fluorescence from a population of molecules. The power of single-molecule recording is demonstrated by the contribution of single-channel recording and analysis of ion channel functions.[153] With single-molecule imaging, the structural rearrangements associated with these random transitions can be revealed.[154] In conjunction with TIRF microscopy, single-molecule imaging also sheds light on the study of channel protein assembly and trafficking.[106]

15.6.2 SUPER-RESOLUTION IMAGING

The spatial resolution of conventional fluorescence microscopy (a few hundred nanometers) is limited by light diffraction. Super-resolution imaging breaks this limitation, bringing spatial resolution up to less than one hundred nanometers.[155–157] Spatial resolution beyond the diffraction limit has been achieved in multiple ways. For example, stimulated emission depletion (STED) microscopy uses two beams of laser, one to excite fluorophores and the other to suppress the excitation of fluorophores peripheral to the focal point.[158,159] The spatial resolution is determined by the power of the suppression laser. In practice, ~30-nm resolution has been achieved.[160] With such a high resolution, it is possible to perform multicolor colocalization, 3D structure imaging, etc.

15.6.3 TWO-PHOTON EXCITATION MICROSCOPY

Two-photon excitation microscopy allows imaging fluorophores in intact tissues.[49,50] In this technique, two photons with lower energy (longer wavelength) cooperate to excite a fluorophore as if their energy is combined. The long wavelength allows the excitation light to penetrate tissues deeper; in addition, excitation occurs only in a tiny volume that is in focus.[161] This technique has already been applied to the study of ligand activation of TRPV1. For this purpose, caged vanilloid ligands are developed and loaded into nociceptive neurons. Two-photon excitation was used to photo-release the caged ligands and activate TRPV1 channels in situ.[162]

15.6.4 LASER HEATING

Lasers have become an increasingly useful tool in biological research. The effect of lasers on biological samples can be photochemical, photothermal, photomechanical, and photoelectrical, depending on the wavelength of the laser and the properties of the sample. Among these effects, heating by laser, because of its high maneuverability and power density, is emerging as a promising experimental tool. Rapid heating by infrared laser is possible because water molecules have high absorption efficiency. With a 980-nm laser, Liang et al. reported that the electrical activities of hippocampal neurons can be modulated as a result of local temperature rise.[163] Similar studies have appeared in the literature.[164–167]

With infrared laser heating, temperature can be elevated to a high level in milliseconds. This advantage in heating speed allows the study of channel activation kinetics. Using a diode laser at 1460 nm, Yao et al. were able to elevate the temperature of a membrane patch from room temperature to over 50°C within one millisecond[168] (also see Chapter 14). They observed that the time constant of heat activation of TRPV1 can be as short as about 5 ms. The rate of gating is noticeably faster than the heating rate of most commercial heating apparatus. It is expected that more experiments with laser heating will be performed to characterize the kinetic properties of thermoTRP activation.

15.6.5 OTHERS

Fluorescence microscopy is improving as better fluorophores are developed. Derivatives of GFP with improved properties for diverse measurements continue to contribute to better experimental designs.[6,8] To reduce the size of the fluorophore, amino acids with artificial fluorescent side chains are genetically incorporated to target proteins and serve as fluorescence reporters.[169–171] In addition, fluorescence correlation spectroscopy (FCS) is applied to study diffusion of proteins and membrane properties.[172,173] Fluorescence lifetime imaging microscopy (FLIM) combined with FRET has been used to detect TRPV4-microfilament interactions.[174]

15.7 CONCLUSION

Fluorescence microscopy has been a very powerful experimental approach in thermoTRP channel research. Ca²⁺ imaging contributed to cloning of thermoTRP channels and their confirmation as molecular thermosensors. Subunit assembly was investigated by fluorescent antibody labeling and FRET, methods that enabled heteromeric channels to be observed. TIRF microscopy detects fluorescence signals from cell membranes and nearby regions and has greatly advanced our knowledge of channel trafficking. Mobility of channel proteins in cells was measured by FRAP, demonstrating the interaction between thermoTRP channels and cytoskeleton proteins. Simultaneous recording of current and fluorescence led to correlation of functional states of a channel to its structural rearrangements. With emerging optical techniques such as single-molecule imaging and laser heating, our understanding of thermoTRP channels should be further advanced.

REFERENCES

1. Lakowicz, J. R. 2006. *Principles of Fluorescence Spectroscopy*. New York: Springer.
2. Boron, W. F., and E. L. Boulpaep. 2005. *Medical Physiology*. Philadelphia, Pennsylvania: Elsevier Saunders.
3. Liapi, A., and J. N. Wood. 2005. Extensive co-localization and heteromultimer formation of the vanilloid receptor-like protein TRPV2 and the capsaicin receptor TRPV1 in the adult rat cerebral cortex. *European Journal of Neuroscience* 22(4): 825–834.
4. Zheng, J., and W. N. Zagotta. 2000. Gating rearrangements in cyclic nucleotide-gated channels revealed by patch-clamp fluorometry. *Neuron* 28(2): 369–374.
5. Yang, F., Y. Cui, K. Wang, and J. Zheng. 2010. Thermosensitive TRP channel pore turret is part of the temperature activation pathway. *Proceedings of the National Academy of Science of the United States of America* 107(15): 7083–7088.
6. Lippincott-Schwartz, J., and G. H. Patterson. 2003. Development and use of fluorescent protein markers in living cells. *Science* 300(5616): 87–91.
7. Tsien, R. Y. 1998. The green fluorescent protein. *Annual Review of Biochemistry* 67: 509–544.
8. Wang, Y., J. Y. Shyy, and S. Chien. 2008. Fluorescence proteins, live-cell imaging, and mechanobiology: Seeing is believing. *Annual Review of Biomedical Engineering* 10: 1–38.
9. Ormo, M., A. B. Cubitt, K. Kallio et al. 1996. Crystal structure of the Aequorea victoria green fluorescent protein. *Science* 273(5280): 1392–1395.
10. Mannuzzu, L. M., M. M. Moronne, and E. Y. Isacoff. 1996. Direct physical measure of conformational rearrangement underlying potassium channel gating. *Science* 271(5246): 213–216.
11. Cha, A., and F. Bezanilla. 1997. Characterizing voltage-dependent conformational changes in the shaker K$^+$ channel with fluorescence. *Neuron* 19(5): 1127–1140.
12. Chanda, B., and F. Bezanilla. 2002. Tracking voltage-dependent conformational changes in skeletal muscle sodium channel during activation. *Journal of General Physiology* 120(5): 629–645.
13. Zheng, J. 2006. Patch fluorometry: Shedding new light on ion channels. *Physiology* 21: 6–12.
14. Reichman, J. 2007. *Handbook of Optical Filters for Fluorescence Microscopy*. Brattleboro, Vermont: Chroma Technology.
15. Rutter, A. R., Q. P. Ma, M. Leveridge, and T. P. Bonnert. 2005. Heteromerization and colocalization of TrpV1 and TrpV2 in mammalian cell lines and rat dorsal root ganglia. *Neuroreport* 16(16): 1735–1739.
16. Caterina, M. J., M. A. Schumacher, M. Tominaga et al. 1997. The capsaicin receptor: a heat-activated ion channel in the pain pathway. *Nature* 389(6653): 816–824.
17. Stryer, L. 1978. Fluorescence energy transfer as a spectroscopic ruler. *Annual Review of Biochemistry* 47: 819–846.
18. Clegg, R. M. 1992. Fluorescence resonance energy transfer and nucleic acids. *Methods Enzymology* 211: 353–388.

19. Selvin, P. R. 1995. Fluorescence resonance energy transfer. *Methods Enzymology* 246: 300–334.
20. Takanishi, C. L., E. A. Bykova, W. Cheng, and J. Zheng. 2006. GFP-based FRET analysis in live cells. *Brain Research* 1091(1): 132–139.
21. Hille, B. 2001. *Ion Channels of Excitable Membranes*. Sunderland, Massachusetts: Sinauer Associates.
22. Cheng, W., F. Yang, C. L. Takanishi, and J. Zheng. 2007. Thermosensitive TRPV channel subunits coassemble into heteromeric channels with intermediate conductance and gating properties. *Journal of General Physiology* 129(3): 191–207.
23. Hellwig, N., N. Albrecht, C. Harteneck, G. Schultz, and M. Schaefer. 2005. Homo- and heteromeric assembly of TRPV channel subunits. *Journal of Cell Science* 118(Pt 5): 917–928.
24. Caterina, M. J., and D. Julius. 2001. The vanilloid receptor: A molecular gateway to the pain pathway. *Annual Review of Neuroscience* 24: 487–517.
25. Williams, J. T., and W. Zieglgansberger. 1982. The acute effects of capsaicin on rat primary afferents and spinal neurons. *Brain Research* 253(1–2): 125–131.
26. Baccaglini, P. I., and P. G. Hogan. 1983. Some rat sensory neurons in culture express characteristics of differentiated pain sensory cells. *Proceedings of the National Academy of Science of the United States of America* 80(2): 594–598.
27. Heyman, I., and H. P. Rang. 1985. Depolarizing responses to capsaicin in a subpopulation of rat dorsal root ganglion cells. *Neuroscience Letters* 56(1): 69–75.
28. Oh, U., S. W. Hwang, and D. Kim. 1996. Capsaicin activates a nonselective cation channel in cultured neonatal rat dorsal root ganglion neurons. *Journal of Neuroscience* 16(5): 1659–1667.
29. Caterina, M. J., A. Leffler, A. B. Malmberg et al. 2000. Impaired nociception and pain sensation in mice lacking the capsaicin receptor. *Science* 288(5464): 306–313.
30. Davis, J. B., J. Gray, M. J. Gunthorpe et al. 2000. Vanilloid receptor-1 is essential for inflammatory thermal hyperalgesia. *Nature* 405(6783): 183–187.
31. Guler, A. D., H. Lee, T. Iida et al. 2002. Heat-evoked activation of the ion channel, TRPV4. *Journal of Neuroscience* 22(15): 6408–6414.
32. Watanabe, H., J. Vriens, S. H. Suh et al. 2002. Heat-evoked activation of TRPV4 channels in a HEK293 cell expression system and in native mouse aorta endothelial cells. *Journal of Biological Chemistry* 277(49): 47044–47051.
33. Chung, M. K., H. Lee, and M. J. Caterina. 2003. Warm temperatures activate TRPV4 in mouse 308 keratinocytes. *Journal of Biological Chemistry* 278(34): 32037–32046.
34. McKemy, D. D., W. M. Neuhausser, and D. Julius. 2002. Identification of a cold receptor reveals a general role for TRP channels in thermosensation. *Nature* 416(6876): 52–58.
35. Peier, A. M., A. Moqrich, A. C. Hergarden et al. 2002. A TRP channel that senses cold stimuli and menthol. *Cell* 108(5): 705–715.
36. Story, G. M., A. M. Peier, A. J. Reeve et al. 2003. ANKTM1, a TRP-like channel expressed in nociceptive neurons, is activated by cold temperatures. *Cell* 112(6): 819–829.
37. Caterina, M. J., T. A. Rosen, M. Tominaga, A. J. Brake, and D. Julius. 1999. A capsaicin-receptor homologue with a high threshold for noxious heat. *Nature* 398(6726): 436–441.
38. Peier, A. M., A. J. Reeve, D. A. Andersson et al. 2002. A heat-sensitive TRP channel expressed in keratinocytes. *Science* 296(5575): 2046–2049.
39. Smith, G. D., M. J. Gunthorpe, R. E. Kelsell et al. 2002. TRPV3 is a temperature-sensitive vanilloid receptor-like protein. *Nature* 418(6894): 186–190.
40. Xu, H., I. S. Ramsey, S. A. Kotecha et al. 2002. TRPV3 is a calcium-permeable temperature-sensitive cation channel. *Nature* 418(6894): 181–186.
41. Knot, H. J., I. Laher, E. A. Sobie et al. 2005. Twenty years of calcium imaging: Cell physiology to dye for. *Molecular Interventions* 5(2): 112–127.

42. Grynkiewicz, G., M. Poenie, and R.Y. Tsien. 1985. A new generation of Ca^{2+} indicators with greatly improved fluorescence properties. *Journal of Biological Chemistry* 260(6): 3440–3450.

43. Paredes, R. M., J. C. Etzler, L. T. Watts, W. Zheng, and J. D. Lechleiter. 2008. Chemical calcium indicators. *Methods* 46(3): 143–151.

44. Hofer, A. M., W. R. Schlue, S. Curci, and T. E. Machen. 1995. Spatial distribution and quantification of free luminal [Ca] within the InsP3-sensitive internal store of individual BHK-21 cells: Ion dependence of InsP3-induced Ca release and reloading. *FASEB Journal* 9(9): 788–798.

45. Mogami, H., A. V. Tepikin, and O. H. Petersen. 1998. Termination of cytosolic Ca^{2+} signals: Ca^{2+} reuptake into intracellular stores is regulated by the free Ca^{2+} concentration in the store lumen. *EMBO Journal* 17(2): 435–442.

46. Park, M. K., A. V. Tepikin, and O. H. Petersen. 2002. What can we learn about cell signalling by combining optical imaging and patch clamp techniques? *Pflügers Archiv* 444(3): 305–316.

47. Duffy, S., and B. A. MacVicar. 1995. Adrenergic calcium signaling in astrocyte networks within the hippocampal slice. *Journal of Neuroscience* 15(8): 5535–5550.

48. Helmchen, F., and J. Waters. 2002. Ca^{2+} imaging in the mammalian brain in vivo. *European Journal of Pharmacology* 447(2–3): 119–129.

49. So, P. T., C. Y. Dong, B. R. Masters, and K. M. Berland. 2000. Two-photon excitation fluorescence microscopy. *Annual Review of Biomedical Engineering* 2: 399–429.

50. Rubart, M. 2004. Two-photon microscopy of cells and tissue. *Circulation Research* 95(12): 1154–1166.

51. Diaspro, A., G. Chirico, and M. Collini. 2005. Two-photon fluorescence excitation and related techniques in biological microscopy. *Quarterly Review of Biophysics* 38(2): 97–166.

52. Svoboda, K., F. Helmchen, W. Denk, and D. W. Tank. 1999. Spread of dendritic excitation in layer 2/3 pyramidal neurons in rat barrel cortex in vivo. *Nature Neuroscience* 2(1): 65–73.

53. Barreto-Chang, O. L., and R. E. Dolmetsch. 2009. Calcium imaging of cortical neurons using Fura-2 AM. *Journal of Visualized Experiments* 23: 1067.

54. Photometrics. 2009. Linearity. Available at http://www.photomet.com/learningzone/linearity.php (accessed December 9, 2010).

55. Amiri, H., G. Schultz, and M. Schaefer. 2003. FRET-based analysis of TRPC subunit stoichiometry. *Cell Calcium* 33(5–6): 463–470.

56. Strubing, C., G. Krapivinsky, L. Krapivinsky, and D. E. Clapham. 2003. Formation of novel TRPC channels by complex subunit interactions in embryonic brain. *Journal of Biological Chemistry* 278(40): 39014–39019.

57. Hofmann, T., M. Schaefer, G. Schultz, and T. Gudermann. 2002. Subunit composition of mammalian transient receptor potential channels in living cells. *Proceedings of the National Academy of Science of the United States of America* 99(11): 7461–7466.

58. Goel, M., W. G. Sinkins, and W. P. Schilling. 2002. Selective association of TRPC channel subunits in rat brain synaptosomes. *Journal of Biological Chemistry* 277(50): 48303–48310.

59. Lintschinger, B., M. Balzer-Geldsetzer, T. Baskaran et al. 2000. Coassembly of Trp1 and Trp3 proteins generates diacylglycerol- and Ca^{2+}-sensitive cation channels. *Journal of Biological Chemistry* 275(36): 27799–27805.

60. Xu, X. Z., H. S. Li, W. B. Guggino, and C. Montell. 1997. Coassembly of TRP and TRPL produces a distinct store-operated conductance. *Cell* 89(7): 1155–1164.

61. Poteser, M., A. Graziani, C. Rosker et al. 2006. TRPC3 and TRPC4 associate to form a redox-sensitive cation channel. Evidence for expression of native TRPC3-TRPC4 heteromeric channels in endothelial cells. *Journal of Biological Chemistry* 281(19): 13588–13595.

62. Liu, X., B. C. Bandyopadhyay, B. B. Singh, K. Groschner, and I. S. Ambudkar. 2005. Molecular analysis of a store-operated and 2-acetyl-sn-glycerol-sensitive non-selective cation channel. Heteromeric assembly of TRPC1-TRPC3. *Journal of Biological Chemistry* 280(22): 21600–21606.

63. Chubanov, V., S. Waldegger, M. Mederos y Schnitzler et al. 2004. Disruption of TRPM6/TRPM7 complex formation by a mutation in the TRPM6 gene causes hypomagnesemia with secondary hypocalcemia. *Proceedings of the National Academy of Science of the United States of America* 101(9): 2894–2899.

64. Li, M., J. Jiang, and L. Yue. 2006. Functional characterization of homo- and heteromeric channel kinases TRPM6 and TRPM7. *Journal of General Physiology* 127(5): 525–537.

65. Strubing, C., G. Krapivinsky, L. Krapivinsky, and D. E. Clapham. 2001. TRPC1 and TRPC5 form a novel cation channel in mammalian brain. *Neuron* 29(3): 645–655.

66. Jahnel, R., M. Dreger, C. Gillen et al. 2001. Biochemical characterization of the vanilloid receptor 1 expressed in a dorsal root ganglia derived cell line. *European Journal of Biochemistry* 268(21): 5489–5496.

67. Kedei, N., T. Szabo, J. D. Lile et al. 2001. Analysis of the native quaternary structure of vanilloid receptor 1. *Journal of Biological Chemistry* 276(30): 28613–28619.

68. Kuzhikandathil, E. V., H. Wang, T. Szabo et al. 2001. Functional analysis of capsaicin receptor (vanilloid receptor subtype 1) multimerization and agonist responsiveness using a dominant negative mutation. *Journal of Neuroscience* 21(22): 8697–8706.

69. Moyes, R. B. 2009. Fluorescent staining of bacteria: Viability and antibody labeling. *Current Protocols of Microbiology* Appendix 3: Appendix 3K.

70. Xu, H. S., N. C. Roberts, L. B. Adams et al. 1984. An indirect fluorescent-antibody staining procedure for detection of vibrio-cholerae serovar 01 cells in aquatic environmental-samples. *Journal of Microbiological Methods* 2(4): 221–231.

71. Tong, C. Y., G. M. Samuda, W. K. Chang, and C. Y. Yeung. 1989. Direct and indirect fluorescent-antibody staining techniques using commercial monoclonal antibodies for detection of respiratory syncytial virus. *European Journal of Clinical Microbiology and Infectious Diseases* 8(8): 728–730.

72. Green, N. M. 1963. Avidin. 1. The use of (14-C)biotin for kinetic studies and for assay. *Biochemistry Journal* 89: 585–591.

73. Hollinshead, M., J. Sanderson, and D. J. Vaux. 1997. Anti-biotin antibodies offer superior organelle-specific labeling of mitochondria over avidin or streptavidin. *Journal of Histochemistry and Cytochemistry* 45(8): 1053–1057.

74. Harris, L. J., S. B. Larson, K. W. Hasel et al. 1992. The three-dimensional structure of an intact monoclonal antibody for canine lymphoma. *Nature* 360(6402): 369–372.

75. Harris, L. J., E. Skaletsky, and A. McPherson. 1995. Crystallization of intact monoclonal antibodies. *Proteins* 23(2): 285–289.

76. Minsky, M. 1988. Memoir on inventing the confocal scanning microscope. *Scanning* 10(4): 128–138.

77. Erie, J. C., J. W. McLaren, and S. V. Patel. 2009. Confocal microscopy in ophthalmology. *American Journal of Ophthalmology* 148(5): 639–646.

78. Watanabe, N., S. Horie, D. Spina et al. 2008. Immunohistochemical localization of transient receptor potential vanilloid subtype 1 in the trachea of ovalbumin-sensitized Guinea pigs. *International Archives of Allergy and Immunology* 146(Suppl 1): 28–32.

79. Ost, D., T. Roskams, F. Van Der Aa, and D. De Ridder. 2002. Topography of the vanilloid receptor in the human bladder: More than just the nerve fibers. *Journal of Urology* 168(1): 293–297.

80. Xin, H., H. Tanaka, M. Yamaguchi et al. 2005. Vanilloid receptor expressed in the sarcoplasmic reticulum of rat skeletal muscle. *Biochemical and Biophysical Research Communications* 332(3): 756–762.

81. Caprodossi, S., R. Lucciarini, C. Amantini et al. 2008. Transient receptor potential vanilloid type 2 (TRPV2) expression in normal urothelium and in urothelial carcinoma of human bladder: Correlation with the pathologic stage. *European Urology* 54(3): 612–620.

82. Benfenati, V., M. Amiry-Moghaddam, M. Caprini et al. 2007. Expression and functional characterization of transient receptor potential vanilloid-related channel 4 (TRPV4) in rat cortical astrocytes. *Neuroscience* 148(4): 876–892.

83. Teilmann, S. C., A. G. Byskov, P. A. Pedersen et al. 2005. Localization of transient receptor potential ion channels in primary and motile cilia of the female murine reproductive organs. *Molecular Reproduction and Development* 71(4): 444–452.

84. Shotton, D., and N. White. 1989. Confocal scanning microscopy: Three-dimensional biological imaging. *Trends in Biochemical Sciences* 14(11): 435–439.

85. Förster, T. 1948. Zwischenmolekulare Energiewanderung und Fluoreszenz. *Annalen der Physik* 437(1–2): 55–75.

86. Lilley, D. M., and T. J. Wilson. 2000. Fluorescence resonance energy transfer as a structural tool for nucleic acids. *Current Opinion in Chemical Biology* 4(5): 507–517.

87. Patterson, G. H., D. W. Piston, and B. G. Barisas. 2000. Forster distances between green fluorescent protein pairs. *Analytical Biochemistry* 284(2): 438–440.

88. Gordon, G. W., G. Berry, X. H. Liang, B. Levine, and B. Herman. 1998. Quantitative fluorescence resonance energy transfer measurements using fluorescence microscopy. *Biophysical Journal* 74(5): 2702–2713.

89. Berney, C., and G. Danuser. 2003. FRET or no FRET: A quantitative comparison. *Biophysical Journal* 84(6): 3992–4010.

90. Zheng, J. 2006. Spectroscopy-based quantitative fluorescence resonance energy transfer analysis. *Methods in Molecular Biology* 337: 65–77.

91. Miyawaki, A., and R. Y. Tsien. 2000. Monitoring protein conformations and interactions by fluorescence resonance energy transfer between mutants of green fluorescent protein. *Methods Enzymology* 327: 472–500.

92. Vetter, I., W. Cheng, M. Peiris et al. 2008. Rapid, opioid-sensitive mechanisms involved in transient receptor potential vanilloid 1 sensitization. *Journal of Biological Chemistry* 283(28): 19540–19550.

93. Schmidt, M., A. E. Dubin, M. J. Petrus, T. J. Earley, and A. Patapoutian. 2009. Nociceptive signals induce trafficking of TRPA1 to the plasma membrane. *Neuron* 64(4): 498–509.

94. Zheng, J., M. C. Trudeau, and W. N. Zagotta. 2002. Rod cyclic nucleotide-gated channels have a stoichiometry of three CNGA1 subunits and one CNGB1 subunit. *Neuron* 36(5): 891–896.

95. Erickson, M. G., B. A. Alseikhan, B. Z. Peterson, and D. T. Yue. 2001. Preassociation of calmodulin with voltage-gated Ca^{2+} channels revealed by FRET in single living cells. *Neuron* 31(6): 973–985.

96. Bykova, E. A., X. D. Zhang, T. Y. Chen, and J. Zheng. 2006. Large movement in the C terminus of CLC-0 chloride channel during slow gating. *Nature Structural and Molecular Biology* 13(12): 1115–1119.

97. Moss, F. J., P. I. Imoukhuede, K. Scott et al. 2009. GABA transporter function, oligomerization state, and anchoring: correlates with subcellularly resolved FRET. *Journal of General Physiology* 134(6): 489–521.

98. Yang, F., Y. Y. Cui, K. W. Wang, and J. Zheng. 2010. Thermosensitive TRP channel pore turret is part of the temperature activation pathway. *Proceedings of the National Academy of Sciences of United States of America* 107(15): 7083–7088.

99. Ambudkar, I. S. 2007. Trafficking of TRP channels: determinants of channel function. *Handbook of Experimental Pharmacology* 179: 541–557.

100. Royle, S. J., and R. D. Murrell-Lagnado. 2003. Constitutive cycling: A general mechanism to regulate cell surface proteins. *Bioessays* 25(1): 39–46.

101. Stein, A. T., C. A. Ufret-Vincenty, L. Hua, L. F. Santana, and S. E. Gordon. 2006. Phosphoinositide 3-kinase binds to TRPV1 and mediates NGF-stimulated TRPV1 trafficking to the plasma membrane. *Journal of General Physiology* 128(5): 509–522.

102. Stokes, A. J., C. Wakano, K. A. Del Carmen, M. Koblan-Huberson, and H. Turner. 2005. Formation of a physiological complex between TRPV2 and RGA protein promotes cell surface expression of TRPV2. *Journal of Cell Biochemistry* 94(4): 669–683.

103. Becker, D., J. Bereiter-Hahn, and M. Jendrach. 2009. Functional interaction of the cation channel transient receptor potential vanilloid 4 (TRPV4) and actin in volume regulation. *European Journal of Cell Biology* 88(3): 141–152.

104. Thompson, N. L., and B. C. Lagerholm. 1997. Total internal reflection fluorescence: Applications in cellular biophysics. *Current Opinion in Biotechnology* 8(1): 58–64.

105. Toomre, D., and D. J. Manstein. 2001. Lighting up the cell surface with evanescent wave microscopy. *Trends in Cell Biology* 11(7): 298–303.

106. Schneckenburger, H. 2005. Total internal reflection fluorescence microscopy: Technical innovations and novel applications. *Current Opinion in Biotechnology* 16(1): 13–18.

107. Axelrod, D., E. H. Hellen, and R. M. Fulbright. 1992. Total internal reflection fluorescence. In *Topics in Fluorescence Spectroscopy: Biochemical Applications*, Chapter 7, p. 291. New York: Plenum Press.

108. Oheim, M. 2001. Imaging transmitter release. II. A practical guide to evanescent-wave imaging. *Lasers in Medical Science* 16(3): 159–170.

109. Ross, S. T., S. Schwartz, T. J. Fellers, and M. W. Davidson. 2009. Total Internal Reflection Fluorescence (TIRF) microscopy. Available at http://www.microscopyu.com/articles/fluorescence/tirf/tirfintro.html (accessed December 9, 2010).

110. Andersen, O. S., and R. E. Koeppe, 2nd. 2007. Bilayer thickness and membrane protein function: an energetic perspective. *Annual Review of Biophysical and Biomolecular Structures* 36: 107–130.

111. Oheim, M., and W. Stuhmer. 2000. Tracking chromaffin granules on their way through the actin cortex. *European Biophysical Journal* 29(2): 67–89.

112. Rohrbach, A., and W. Singer. 1998. Scattering of a scalar field at dielectric surfaces by Born series expansion. *Journal of the Optical Society of America A: Optics, Image Science, and Vision* 15(10): 2651–2659.

113. Lippincott-Schwartz, J., E. Snapp, and A. Kenworthy. 2001. Studying protein dynamics in living cells. *Nature Reviews. Molecular Cell Biology* 2(6): 444–456.

114. Axelrod, D., D. E. Koppel, J. Schlessinger, E. Elson, and W. W. Webb. 1976. Mobility measurement by analysis of fluorescence photobleaching recovery kinetics. *Biophysical Journal* 16(9): 1055–1069.

115. Ellenberg, J., E. D. Siggia, J. E. Moreira et al. 1997. Nuclear membrane dynamics and reassembly in living cells: Targeting of an inner nuclear membrane protein in interphase and mitosis. *Journal of Cell Biology* 138(6): 1193–1206.

116. Sciaky, N., J. Presley, C. Smith et al. 1997. Golgi tubule traffic and the effects of brefeldin A visualized in living cells. *Journal of Cell Biology* 139(5): 1137–1155.

117. Yang, F., L. G. Moss, and G. N. Phillips, Jr. 1996. The molecular structure of green fluorescent protein. *Nature Biotechnology* 14(10): 1246–1251.

118. Prendergast, F. G. 1999. Biophysics of the green fluorescent protein. *Methods in Cell Biology* 58: 1–18.

119. Sachs, J. N., and D. M. Engelman. 2006. Introduction to the membrane protein reviews: The interplay of structure, dynamics, and environment in membrane protein function. *Annual Review of Biochemistry* 75: 707–712.

120. Wuthrich, K. 1990. Protein structure determination in solution by NMR spectroscopy. *Journal of Biological Chemistry* 265(36): 22059–22062.

121. Joachimiak, A. 2009. High-throughput crystallography for structural genomics. *Current Opinion in Structural Biology* 19(5): 573–584.

122. Bezanilla, F. 2000. The voltage sensor in voltage-dependent ion channels. *Physiology Review* 80(2): 555–592.

123. Tombola, F., M. M. Pathak, and E. Y. Isacoff. 2006. How does voltage open an ion channel? *Annual Review of Cell and Developmental Biology* 22: 23–52.

124. Latorre, R. 2009. Perspectives on TRP channel structure and the TRPA1 puzzle. *Journal of General Physiology* 133(3): 227–229.

125. Moiseenkova-Bell, V. Y., and T. G. Wensel. 2009. Hot on the trail of TRP channel structure. *Journal of General Physiology* 133(3): 239–244.

126. Gaudet, R. 2009. Divide and conquer: High resolution structural information on TRP channel fragments. *Journal of General Physiology* 133(3): 231–237.

127. Doyle, D. A., J. Morais Cabral, R. A. Pfuetzner et al. 1998. The structure of the potassium channel: Molecular basis of K$^+$ conduction and selectivity. *Science* 280(5360): 69–77.

128. Jiang, Y., A. Lee, J. Chen et al. 2003. X-ray structure of a voltage-dependent K$^+$ channel. *Nature* 423(6935): 33–41.

129. Dutzler, R., E. B. Campbell, M. Cadene, B. T. Chait, and R. MacKinnon. 2002. X-ray structure of a ClC chloride channel at 3.0 A reveals the molecular basis of anion selectivity. *Nature* 415(6869): 287–294.

130. Long, S. B., E. B. Campbell, and R. Mackinnon. 2005. Crystal structure of a mammalian voltage-dependent Shaker family K$^+$ channel. *Science* 309(5736): 897–903.

131. Jin, X., J. Touhey, and R. Gaudet. 2006. Structure of the N-terminal ankyrin repeat domain of the TRPV2 ion channel. *Journal of Biological Chemistry* 281(35): 25006–25010.

132. McCleverty, C. J., E. Koesema, A. Patapoutian, S. A. Lesley, and A. Kreusch. 2006. Crystal structure of the human TRPV2 channel ankyrin repeat domain. *Protein Science* 15(9): 2201–2206.

133. Lishko, P. V., E. Procko, X. Jin, C. B. Phelps, and R. Gaudet. 2007. The ankyrin repeats of TRPV1 bind multiple ligands and modulate channel sensitivity. *Neuron* 54(6): 905–918.

134. Moiseenkova-Bell, V. Y., L. A. Stanciu, I. I. Serysheva, B. J. Tobe, and T. G. Wensel. 2008. Structure of TRPV1 channel revealed by electron cryomicroscopy. *Proceedings of the National Academy of Science of the United States of America* 105(21): 7451–7455.

135. Shigematsu, H., T. Sokabe, R. Danev, M. Tominaga, and K. Nagayama. 2009. A 3.5-nm structure of rat TRPV4 cation channel revealed by zernike phase-contrast cryo-EM. *Journal of Biological Chemistry* 285: 11210–11218.

136. Zheng, J., and W. N. Zagotta. 2003. Patch-clamp fluorometry recording of conformational rearrangements of ion channels. *Science STKE* 2003(176): PL7.

137. Blunck, R., D. M. Starace, A. M. Correa, and F. Bezanilla. 2004. Detecting rearrangements of shaker and NaChBac in real-time with fluorescence spectroscopy in patch-clamped mammalian cells. *Biophysics Journal* 86(6): 3966–3980.

138. Claydon, T. W., M. Vaid, S. Rezazadeh, S. J. Kehl, and D. Fedida. 2007. 4-aminopyridine prevents the conformational changes associated with p/c-type inactivation in shaker channels. *Journal of Pharmacology and Experimental Therapeutics* 320(1): 162–172.

139. Vaid, M., T. W. Claydon, S. Rezazadeh, and D. Fedida. 2008. Voltage clamp fluorimetry reveals a novel outer pore instability in a mammalian voltage-gated potassium channel. *Journal of General Physiology* 132(2): 209–222.

140. Cha, A., and F. Bezanilla. 1998. Structural implications of fluorescence quenching in the Shaker K$^+$ channel. *Journal of General Physiology* 112(4): 391–408.

141. Cha, A., P. C. Ruben, A. L. George, Jr., E. Fujimoto, and F. Bezanilla. 1999. Voltage sensors in domains III and IV, but not I and II, are immobilized by Na$^+$ channel fast inactivation. *Neuron* 22(1): 73–87.

142. Cha, A., G. E. Snyder, P. R. Selvin, and F. Bezanilla. 1999. Atomic scale movement of the voltage-sensing region in a potassium channel measured via spectroscopy. *Nature* 402(6763): 809–813.

143. Glauner, K. S., L. M. Mannuzzu, C. S. Gandhi, and E. Y. Isacoff. 1999. Spectroscopic mapping of voltage sensor movement in the Shaker potassium channel. *Nature* 402(6763): 813–817.

144. Mannuzzu, L. M., and E. Y. Isacoff. 2000. Independence and cooperativity in rearrangements of a potassium channel voltage sensor revealed by single subunit fluorescence. *Journal of General Physiology* 115(3): 257–268.

145. Chanda, B., O. K. Asamoah, R. Blunck, B. Roux, and F. Bezanilla. 2005. Gating charge displacement in voltage-gated ion channels involves limited transmembrane movement. *Nature* 436(7052): 852–856.

146. Zheng, J., M. D. Varnum, and W. N. Zagotta. 2003. Disruption of an intersubunit interaction underlies Ca^{2+}-calmodulin modulation of cyclic nucleotide-gated channels. *Journal of Neuroscience* 23(22): 8167–8175.

147. Taraska, J. W., and W. N. Zagotta. 2007. Structural dynamics in the gating ring of cyclic nucleotide-gated ion channels. *Nature Structural and Molecular Biology* 14(9): 854–860.

148. Islas, L. D., and W. N. Zagotta. 2006. Short-range molecular rearrangements in ion channels detected by tryptophan quenching of bimane fluorescence. *Journal of General Physiology* 128(3): 337–346.

149. Trudeau, M. C., and W. N. Zagotta. 2004. Dynamics of Ca^{2+}-calmodulin-dependent inhibition of rod cyclic nucleotide-gated channels measured by patch-clamp fluorometry. *Journal of General Physiology* 124(3): 211–223.

150. Li, M., R. A. Farley, and H. A. Lester. 2000. An intermediate state of the gamma-aminobutyric acid transporter GAT1 revealed by simultaneous voltage clamp and fluorescence. *Journal of General Physiology* 115(4): 491–508.

151. Harms, G. S., G. Orr, M. Montal et al. 2003. Probing conformational changes of gramicidin ion channels by single-molecule patch-clamp fluorescence microscopy. *Biophysics Journal* 85(3): 1826–1838.

152. Dahan, D. S., M. I. Dibas, E. J. Petersson et al. 2004. A fluorophore attached to nicotinic acetylcholine receptor beta M2 detects productive binding of agonist to the alpha delta site. *Proceedings of the National Academy of Science of the United States of America* 101(27): 10195–10200.

153. Hamill, O. P., A. Marty, E. Neher, B. Sakmann, and F. J. Sigworth. 1981. Improved patch-clamp techniques for high-resolution current recording from cells and cell-free membrane patches. *Pflügers Archiv* 391(2): 85–100.

154. Borisenko, V., T. Lougheed, J. Hesse et al. 2003. Simultaneous optical and electrical recording of single gramicidin channels. *Biophysics Journal* 84(1): 612–622.

155. Huang, B. 2009. Super-resolution optical microscopy: Multiple choices. *Current Opinion in Chemical Biology* 14(1): 10–14.

156. Huang, B., M. Bates, and X. Zhuang. 2009. Super-resolution fluorescence microscopy. *Annual Review of Biochemistry* 78: 993–1016.

157. Lippincott-Schwartz, J., and S. Manley. 2009. Putting super-resolution fluorescence microscopy to work. *Nature Methods* 6(1): 21–23.

158. Hell, S. W., and J. Wichmann. 1994. Breaking the diffraction resolution limit by stimulated emission: Stimulated-emission-depletion fluorescence microscopy. *Optics Letters* 19(11): 780–782.

159. Klar, T. A., and S. W. Hell. 1999. Subdiffraction resolution in far-field fluorescence microscopy. *Optics Letters* 24(14): 954–956.

160. Westphal, V., and S. W. Hell. 2005. Nanoscale resolution in the focal plane of an optical microscope. *Physical Review Letters* 94(14): 143903.

161. Svoboda, K., and R. Yasuda. 2006. Principles of two-photon excitation microscopy and its applications to neuroscience. *Neuron* 50(6): 823–839.

162. Zhao, J., T. D. Gover, S. Muralidharan et al. 2006. Caged vanilloid ligands for activation of TRPV1 receptors by 1- and 2-photon excitation. *Biochemistry* 45(15): 4915–4926.

163. Liang, S., F. Yang, C. Zhou et al. 2009. Temperature-dependent activation of neurons by continuous near-infrared laser. *Cell Biochemistry and Biophysics* 53(1): 33–42.

164. Vogel, A., and V. Venugopalan. 2003. Mechanisms of pulsed laser ablation of biological tissues. *Chemistry Review* 103(2): 577–644.

165. Hirase, H., V. Nikolenko, J. H. Goldberg, and R. Yuste. 2002. Multiphoton stimulation of neurons. *Journal of Neurobiology* 51(3): 237–247.

166. Wells, J., C. Kao, P. Konrad et al. 2007. Biophysical mechanisms of transient optical stimulation of peripheral nerve. *Biophysics Journal* 93(7): 2567–2580.

167. Wells, J., C. Kao, K. Mariappan et al. 2005. Optical stimulation of neural tissue in vivo. *Optics Letters* 30(5): 504–506.

168. Yao, J., B. Liu, and F. Qin. 2009. Rapid temperature jump by infrared diode laser irradiation for patch-clamp studies. *Biophysics Journal* 96(9): 3611–3619.

169. Saks, M. E., J. R. Sampson, M. W. Nowak et al. 1996. An engineered Tetrahymena tRNAGln for in vivo incorporation of unnatural amino acids into proteins by nonsense suppression. *Journal of Biological Chemistry* 271(38): 23169–23175.

170. Nowak, M. W., J. P. Gallivan, S. K. Silverman et al. 1998. In vivo incorporation of unnatural amino acids into ion channels in Xenopus oocyte expression system. *Methods in Enzymology* 293: 504–529.

171. Cohen, B. E., T. B. McAnaney, E. S. Park et al. 2002. Probing protein electrostatics with a synthetic fluorescent amino acid. *Science* 296(5573): 1700–1703.

172. Bates, I. R., P. W. Wiseman, and J. W. Hanrahan. 2006. Investigating membrane protein dynamics in living cells. *Biochemical and Cell Biology* 84(6): 825–831.

173. Chiantia, S., J. Ries, and P. Schwille. 2009. Fluorescence correlation spectroscopy in membrane structure elucidation. *Biochimica et Biophysica Acta* 1788(1): 225–233.

174. Ramadass, R., D. Becker, M. Jendrach, and J. Bereiter-Hahn. 2007. Spectrally and spatially resolved fluorescence lifetime imaging in living cells: TRPV4-microfilament interactions. *Archives of Biochemistry and Biophysics* 463(1): 27–36.

16 Regulation of TRP Channels by Osmomechanical Stress

Min Jin, Jonathan Berrout, and Roger G. O'Neil

CONTENTS

16.1 INTRODUCTION

16.1.1 OSMOMECHANICAL SENSING AND TRANSDUCTION

Most prokaryotic and eukaryotic cells have the ability to sense and respond to alterations in various mechanical stimuli. This is most apparent in higher organisms (eukaryotic cells), which have developed sensory systems with specialized sensory cells for detecting a wide range of stimuli including mechanical distortions of the cell membrane. However, the basic mechanisms of sensing (mechanosensation) and transducing (mechanotransduction) mechanical stimuli are apparent throughout evolutionary biology with bacteria already displaying mechanosensitive channels that are activated in response to osmotic stimuli (hyperosmolarity).[1–3] Mechanosensitive channel opening in these prokaryotes is induced upon the generation of tension within the cell membrane. In eukaryotes, much more elaborate systems of mechanosensation have been developed in association with the evolution of an extensive scaffolding network, the actin cytoskeleton, with sites of tethering to the plasma membrane and to external sites through integrin-extracellular matrix attachments.[4–6] As a result, changes in cell volume, shape, or tension within the cell membranes can rapidly lead to induction of ion channels or biochemical cascades as part of the transduction process.

Most mammalian cells can sense and respond to mechanical stimuli. This is particularly true in systems with continuously changing mechanical stresses, such as those found in the cardiovascular and renal systems. Indeed, in addition to the cardiomyocytes and vascular smooth muscle cells, the vascular endothelial cells and renal tubular epithelial cells are well known to be sensitive to alterations in fluid flow, hydrostatic pressures, contractile forces, and/or cell shape changes.[7–10] Dysfunctional control of the sensing and transduction machinery can lead to numerous pathophysiological states ranging from hypertension, to atherosclerosis, and to altered fluid and electrolyte balance, to name a few. The underlying alteration in osmomechanical stresses is often associated with an early change in calcium signaling and downstream calcium-dependent processes. The source of the mechanosensitive calcium signals and the downstream effector pathways has only recently begun to be elucidated. Increasing evidence points to a key role of calcium-permeable TRP channels in this process.[7,10]

This chapter will focus on the methods and techniques of applying and studying the effect of osmotic and mechanical stresses on a functional group of TRP channels dubbed the "osmomechanical TRP channels." This is a group of TRP channels that are known to be activated when exposed to osmotic and mechanical stresses.[10–12] This chapter is not meant to be an exhaustive review on osmomechanical TRP channels, as many excellent reviews are available that cover various functional aspects of these channels.[7,10,12–14] The methods outlined below will focus on application of mechanical stresses to whole cells in the cardiovascular and renal systems that will include osmotic stresses, shear stresses, or fluid flow stresses and cell membrane stretch, stresses these cells typically experience on a minute-to-minute basis.

16.1.2 OSMOMECHANICAL-SENSITIVE TRP CHANNELS

Transient receptor potential (TRP) proteins are a family of nonselective cation permeable channels, most of which are permeable to calcium ions.[12,15,16] As first described in *Drosophila*, TRP channels have been shown to be expressed in a multitude of animal cells and tissues. In mammals, the TRP channels expressed belong to six subfamilies: classical (TRPC1-7), vanilloid (TRPV1-6), melastatin (TRPM1-8), polycystin (TRPP1-4), mucolipin (TRPML1-3), and ankyrin (TRPA1). A subunit of TRP channels is composed of cytosolic carboxyl and amino-termini with multiple protein-to-protein interaction sites, six transmembrane spanning segments, and an ion conductive "pore loop" between the fifth and sixth transmembrane segments. Functional channels are believed to be formed by a homo/heterotetramer configuration of TRP proteins. Furthermore, many TRP channels seem to be activated by a broad range of stimuli including osmomechanical stress. The group of TRP channels that are sensitive to osmomechanical stress includes proteins from multiple subfamilies as outlined below (see Table 16.1). However, there is considerable diversity in channel activation. For example, a member of the TRPC subfamily, TRPC1, was demonstrated to be activated when tension was applied to the lipid bilayer.[17] Through a series of experiments where TRPC1 was expressed in liposomes and CHO-K1 cells, activation evaluated using the patch-clamp technique showed that the channel is likely gated by bilayer-dependent tension. Alternatively, a member of the TRPV subfamily, TRPV4, has been shown to respond to mechanostimuli delivered by shear/hypotonic stress.[17,18] However, while direct activation of TRPV4 by mechanical tension in a bilayer has not been forthcoming, accumulating evidence points to an indirect control of the channel via arachidonic acid and epoxyeicosatrienoic acids, the latter being metabolites of the P450 (CYP) epoxygenase pathway.[19]

TABLE 16.1
TRP Channels Regulated by Osmomechanical Stresses

Species	TRP Channel	Family Subtype	Activation by Osmo- and Mechanical Stresses
Caenorhabditis elegans			
	OSM-9	TRPV	Osmolality (hyper), touch
	OCR-2	TRPV	Osmolality (hyper), touch
	NOMPC	TRPN	Stretch
Drosophila			
	NAN	TRPV	Osmolality (hypo), stretch
	IAV	TRPV	Osmolality (hypo), stretch
Mammals			
	TRPV1	TRPV	Osmolality (hyper), stretch
	TRPV2	TRPV	Osmolality (hypo), stretch, shear stress
	TRPV4	TRPV	Osmolality (hypo), shear stress
	TRPC1	TRPC	Osmolality (hypo), stretch
	TRPC6	TRPC	Stretch
	TRPP2	TRPP	Osmolality (hypo), stretch
	TRPP3	TRPP	Osmolality (hypo)
	TRPA1	TRPA	Stretch
	TRPM3	TRPM	Osmolality (hypo)
	TRPM4	TRPM	Stretch
	TRPM7	TRPM	Osmolality (hypo), stretch, shear stress

16.1.3   Mechanisms of Channel Activation by Osmomechanical Stresses

Mechano- or osmomechano-sensitive channels are believed to have multiple mechanisms of activation.[6-8] The first model of mechano-sensitivity supposes channels to be activated in a lipid bilayer-dependent manner. In this mechanism, tension applied deforms/bends the membrane, causing the membrane lipids to "pull away" from the channel protein. As a result, tension is increased on the hydrophobic sites of the channel (increasing hydrophobic forces), which then leads to a conformational change of the channel and possible activation. A second model supposes channels to be linked to accessory molecules, such as extracellular matrix/cytoskeletal proteins, and that mechanical forces acting on these molecules may produce tension to activate the channel in a lipid bilayer-independent manner.

A third model of channel activation is an indirect model where the channel is not directly activated by mechanical forces producing tension on the channels but rather is activated by processes involving secondary signaling components (lipases, kinases, G-proteins) that are initiated as a result of mechanical stimulus. This is a widespread scenario that underlies the activation of many "mechano-sensitive" processes in cells, for example, osmotic activation of TRPV4 by epoxyeicosatrienoic acid metabolites of the P450 (CYP) epoxygenase pathway.[19]

Accumulating evidence indicates that osmotic activation of some TRP channels may occur via one of two mechanisms: (1) Swelling-induced activation of PLA_2/arachidonic acid and/or the PLC/diacylglycerol/PKC pathway, whose downstream metabolites then activate TRP channels; (2) PDZ scaffolding proteins such as ezrin/radixin/moesin act as the mediators between some TRP channels and F-actin cytoskeleton.[20-22] Thus, disruption or de-polymerization of actin cytoskeleton can activate some TRP channels via interaction with these PDZ scaffolding proteins.

16.2   OSMOMECHANICAL REGULATION OF TRP CHANNELS

There are a number of methods available to study the effect of mechanical stresses on TRP channel properties. Not all methods can be utilized for all cells or for studying the effect of specific osmomechanical stresses, as the application of certain stresses requires defined methods and study configurations that may not always allow a particular technique to be employed. In this section, we first outline the most typical experimental approaches for studying the effect of osmomechanical stresses on TRP channels, followed by the methods for studying the application of three types of specific osmomechanical stresses on cells grown on cover slips: osmotic stress, shear stress (fluid flow), and cell membrane stretch.

16.2.1   Assessing the Activity of Calcium-Permeable TRP Channels

16.2.1.1   Patch-Clamp Techniques

The patch-clamp techniques provide the most direct measure of channel activity, but the methods are often the most difficult to apply. These methods allow one to directly measure the ionic current carried by the channel of interest. Typically, the magnitude of channel currents (ionic fluxes) is measured under voltage clamp conditions using the standard patch-clamp technique.[23,24] The patch electrode is lowered onto the cell surface, and a high resistance seal (giga-ohm seal) between the pipette tip and the cell membrane is established, usually requiring a brief application of negative pipette pressure (suction). Once a giga-ohm seal is achieved, small unitary currents from single channels in the patched membrane (single-channel currents) or integrated currents from the whole cell (whole-cell currents) can often be measured. Success can be elusive because not all cells can be successfully patched with formation of giga-ohm seals. Success is best achieved when the cells in culture have "clean" cell surfaces with minimal expression of extracellular glycoproteins/glycocalyx.

There are several configurations of the patch-clamp technique that allow one to measure single-channel currents or whole-cell currents, the latter being the integrated sum of all channel currents activated in a cell. Single-channel currents can be measured in cell-attached patches of an intact cell or in excised patches where a small area of the cell membrane that is attached to the patch pipette tip is literally pulled away (excised) from the rest of the cell by rapid pipette withdrawal (with significant mechanical perturbation). Excised patches have the advantage of using a "simplified" membrane preparation with currents flowing across only a small membrane surface that is isolated from the rest of the cell and many of its signaling components. Both inside-out patches (membrane cytoplasmic side facing the bathing media) and outside-out patches (membrane extracellular side facing the bathing media) allow access to both the cytoplasmic face of the membrane and the extracellular face of the membrane, respectively, giving the investigator access to either face of the channel. However, many of the cellular components and signaling cascades may be missing in the excised patch configuration, components that may be essential for the regulation and control of the channel. These need to be evaluated before proceeding. Hence, the basic properties of the channel are typically first evaluated with an on-cell patch configuration, often in the whole-cell mode, before progressing to the excised patch configuration to establish the basic properties of the channel. The details of these various techniques have been described in various reviews[23,24] and will not be discussed in depth in this chapter.

Use of the patch-clamp technique to study the effect of osmomechanical stimuli on channel activation, including for TRP channels, has both advantages and disadvantages. Most TRP channels have been successfully evaluated with the various configurations of the patch-clamp technique, with the whole-cell patch configuration being the most widely used method. The use of the patch-clamp technique to study osmomechanical TRP channel activation is limited to the types of stimuli applied to the cells as outlined below.

Membrane Stretch. As heretofore noted, various patch-clamp configurations have been employed to study the effect of stretching the cell membrane upon channel activation. The most direct approach is to apply a hydrostatic pressure pulse, either negative (suction; membrane flexing into the pipette) or positive (pressure; membrane flexing away from the pipette), to the back of the patch pipette to "stretch" the membrane attached to the tip of the patch pipette.[23,25] The first stretch-activated channels were identified using this approach. Typical pressures applied are from −30 to +30 mm Hg for mammalian channels. This approach must be utilized with extreme caution and testing, as the pressure pulses often will lead to loss of the seal or generation of "seal leaks" that may appear like mechanosensitive currents. An alternative approach to apply membrane stretch is to induce a mechanical stretch of the whole cell. This can be accomplished by growing cells on elasticized membranes, then inducing "cell stretch" by stretching of the elasticized membrane (see below). This approach is not readily applicable to patch-clamp techniques because the abrupt movement of the cell during stretching will normally lead to loss of the pipette seal with the cell membrane. In this case, calcium imaging may be a more appropriate technique for calcium-permeable TRP channels as discussed below.

Osmotic Stress (Osmotic-Induced Cell Volume Changes). Alterations in the osmolarity of the media surrounding the cell will lead to changes in cell volume as water (and solute) leaves or enters the cell to maintain osmotic balance. As cells shrink or swell, this leads to cell membrane tension at sites of cytoskeletal network attachment and/or sites of extracellular matrix attachment that can result in channel activation. This is typically evaluated by first establishing a giga-ohm seal in a whole-cell configuration and then exposing the cell to an osmotic challenge (typically exposing the cell to a hypoosmotic media, 200–270 mOsm/kg, to induce cell swelling). Again, care must be used to assess for generation of seal leak currents that may be mistaken as tension-induced channels. Further consideration must also be given to the effect of osmolarity itself, independent of cell volume changes, and this should be tested separately.

Shear Stress (Fluid Flow). Application of a defined shear stress using the patch-clamp technique is not generally attempted. Indeed, a shear stress stimulus normally employs a defined stress of known magnitude in an enclosed compartment of known geometry and fluid flow properties. For example, cells grown on cover slips can be inserted into a parallel plate chamber, which includes an upper cover slip or cover over the cells with a known distance between the cells (lower cover slip) and the upper cover slip (typically 100–300 μm) and a known chamber width and length so that shear stress associated with a defined flow through the chamber can be precisely defined (see below). Such configurations are not readily adaptable to patch-clamp analysis because the cells are not easily accessible with the patch-clamp electrode owing to the presence of an upper cover. In some cases, an open-top chamber is employed and fluid flow applied to generate a shear stress at the cell membrane surface, but the magnitude of the shear stress is not defined.

16.2.1.2 Calcium Imaging

Calcium imaging allows for the monitoring of calcium ion dynamics within living cells in real time. The methods do not provide a direct measure of TRP channel activity but rather provide an integrated calcium signal that can reflect all sources of calcium entry into the cytoplasm. This includes both calcium entry from the extracellular space and any release of calcium from internal stores. Hence, care must be taken to determine what component of the calcium signal reflects calcium influx through the TRP channel (or any other calcium-permeable channel) of interest. This can be accomplished by performing the experiment in the presence of extracellular calcium and then, again, in the absence of extracellular calcium to abolish all calcium influx. The difference in intracellular calcium levels between the two conditions is typically used as an index of calcium influx, but not a quantitative index because the absolute rates of calcium influx cannot be determined with this approach.

Typically, intracellular calcium levels are determined with the use of fluorescent molecular probes, which upon binding/interacting with calcium ions, undergo a chemical/structural change leading to an alteration of their fluorescent signal. The growing number of available calcium-sensitive molecular probes can be divided into two groups: genetically encoded calcium indicators[26] and small molecule calcium indicators.[27] Genetically encoded calcium indicators are produced by the translation of a nucleic acid sequence, typically incorporated into the cell by a gene transfer technique. An advantage of using genetically encoded calcium indicators is that the cellular localization and the expression level of the indicator can be controlled by the incorporation of a signal sequence and a promoter sequence in the cDNA construct. On the other hand, small molecule calcium indicators do not require translation by the cell but rather are introduced into cells by diffusion or disruption of the lipid bilayer of the membrane. Use of these indicators generally provides greater sensitivity, a greater ability to measure rapid calcium dynamics, and no need for transfection. An example of a commonly utilized small molecule calcium indicator is Fura-2. Fura-2 is a useful ratiometric dye because of its ability to undergo an absorption shift from 380 nm in its Ca^{2+}-free state to a shorter wavelength of 340 nm in its Ca^{2+}-bound state. Additionally, it offers the advantages of having a high affinity for Ca^{2+} ($K_d \approx 225$ nM), little affinity for other ions, no detectable binding to membrane, and a wide Ca^{2+} sensitivity ranging from approximately 20 nM to 10 μM. Furthermore, the attachment of an acetoxymethyl (AM) ester group to the carboxyl tail allows Fura-2/AM to freely diffuse through the plasma membrane, making its loading into cells a simple procedure. Once inside the cell, esterases cleave off the AM tail from Fura-2, which leaves the indicator in its charged form and therefore unable to cross the cellular membranes, i.e., it becomes trapped in the cytosol.

The calcium imaging techniques can generally be applied to study TRP channel activity in a variety of experimental setups. Typically, cells are grown on cover slips or elasticized membranes (silastic membranes that can be stretched) and then attached to the bottom of a flow chamber that can be mounted on an inverted microscope that allows ready imaging of the cells. Cells can then be exposed to the various osmomechanical stresses while simultaneously monitoring intracellular calcium: osmotic stresses, shear stresses (or fluid flow), and cell membrane stretch. The methods are outlined in more detail below for each of the conditions.

16.2.1.3 Channel Abundance and Localization

As complementary approaches to patch-clamp and calcium imaging, Western blotting and immunocytochemistry can be used to study TRP channel regulation under different osmomechanical stimuli. These methods provide measures of channel abundance and localization within the cell, respectively. Whole-cell lysates and plasma membrane proteins are harvested, and standard Western blotting (protocol, AbCam) is performed to characterize and quantify the TRP channel protein of interest. Although the biotin-streptavidin method (protocol, Thermo Scientific) has been used extensively to pull down biotinylated plasma membrane proteins, cytosolic proteins, especially cytoskeleton proteins, readily come down as a contaminant with the plasma membrane proteins. Therefore, one has to be careful with the purity of plasma membrane proteins, depending on the goal of the investigation. The sucrose gradient differential centrifugation method[28] is by far the cleanest way to extract plasma membrane proteins; however, the low yield is always a concern. With the internal loading control, for instance, pan cadherin in the plasma membrane preparation, one can perform densitometry (Image J) to quantify the abundance of a TRP channel in the plasma membrane sample.

Quantification of a functional TRP on the plasma membrane is a must to understand TRP channel trafficking and recycling under osmomechanical stimuli. Immunostaining techniques are widely used to study the subcellular distribution of TRP channels in various cell and tissue types. Usually, a primary antibody against a TRP channel and another primary antibody against a possible partner of the TRP channel are carefully selected. Then incubation of the primary antibodies and the matching secondary antibodies with fluorescence tags is performed according to standard immunostaining protocols (Millipore). Confocal microscopy is then utilized to visualize the TRP channel distribution and their relations to other target proteins of interest.

16.2.2 Osmotic Stress/Cell Volume

There have been numerous studies showing that certain TRP channels can be activated by hypoosmotic stress, leading to cell swelling as heretofore noted. Some of the examples are TRPV2,[29,30] TRPV4,[11,31–33] TRPM3,[34] TRPM7,[35] TRPC1,[36] and TRPP2[37] (see Table 16.1). However, there are only limited studies assessing the effect of hyperosmotic stress on TRP channels, with few channels identified as sensitive to shrinkage. A notable exception is an apparent N-terminal truncated version of TRPV1. Ciura and Bourque have shown that TRPV1 truncated protein is activated by hyperosmotic stress, and TRPV1−/− mice show a reduced drinking response under hypertonic challenge.[38]

Subsequent to hypoosmotic-induced cell swelling, most cells undergo a regulatory volume decrease (RVD) where the cell volume is returned toward the basal isotonic volume. Although there is little evidence that TRP channels directly activate RVD, it is proposed that TRP channels regulate RVD indirectly by activating the Ca^{2+}-activated K^+ channel via calcium influx.[17,35,39] Indeed, it has been shown recently that TRPV4 may be involved in volume regulation. For example, CHO cells gain RVD capability when overexpressed with TRPV4.[40] Interactions between TRPV4 and other channels like AQP5 and CFTR may also affect RVD.[39,41]

The application of hypoosmotic stress to cells has been widely used to study cell volume-dependent phenomena for many years. In a typical experiment, hypotonic stress is usually achieved by applying an extracellular bathing solution with an osmolarity typically ranging from 200 to 270 mOsm (with reduced NaCl concentration) to the cells. Figure 16.1a shows the diagram of the expected cell volume changes upon exposure of the cells to a hypotonic media (HYPO). The volume flow of water into the cells can be expressed as

$$J_v = \sigma K_f (\Delta[\text{Osm}]),$$

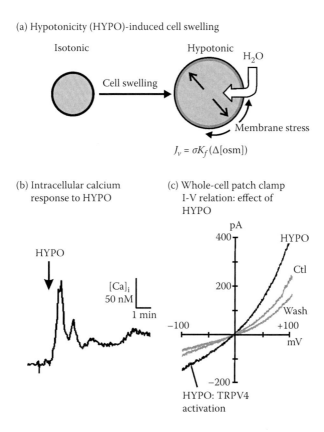

(a) Hypotonicity (HYPO)-induced cell swelling

Isotonic   Hypotonic

$J_v = \sigma K_f (\Delta[\text{osm}])$

(b) Intracellular calcium response to HYPO

(c) Whole-cell patch clamp I-V relation: effect of HYPO

HYPO: TRPV4 activation

FIGURE 16.1 Cellular responses upon exposure of cortical collecting duct M-1 cells to hypoosmotic stress. (a) Diagram of the expected cell volume changes upon exposure to a hypotonic media (HYPO). The volume flow of water into the cells is shown as $J_v = \sigma K_f$ (Δ[Osm]) (see text for details). (b) An example of the influence of hypotonic stress on intracellular calcium levels based on ratiometric calcium imaging. HYPO induced rapid activation of calcium influx. (c) A representative current-voltage plot (I-V plot) from a whole-cell patch clamp study demonstrating activation of TRPV4 upon cell swelling (HYPO) in M-1 cells.

where J_v is the volume flow, σ is the reflection coefficient of the membrane (typically 1 for cell membranes), K_f is the membrane filtration coefficient, and Δ[Osm] is the applied osmotic gradient across the cell membrane. Figure 16.1b gives an example of the influence of hypotonic stress on intracellular calcium levels based on ratiometric calcium imaging, and Figure 16.1c gives a representative current-voltage plot (I-V plot) from a whole-cell patch-clamp study demonstrating activation of TRPV4 upon cell swelling (HYPO). Similarly, hypertonic stress is achieved by adding mannitol or NaCl to the extracellular media to make it hypertonic, typically in the range of 310–500 mOsm. The patch-clamp technique is utilized to directly measure TRP channel activity or that of associated channels, i.e., the Ca^{2+}-activated K^+ channel, when the bath solution osmolality is changed. Standard patch-clamp techniques for single-channel and whole-cell current recording are described above. Calcium imaging can be used in combination with the patch clamp to monitor the changes in intracellular calcium concentration following activation of the TRP channel during cell volume changes.

Actual measurements of the cell volume are complicated. Scanning laser confocal microscopy (SLCM) has been utilized to estimate the cell volume. Basically, a Z-stack of thin optical slices is obtained using SLCM and a three-dimensional (3D) image of a cell constructed using imaging/morphometric software.[42–44] However, laser scanning can readily induce photodynamic damage to the biological sample and must therefore be used with care, thus greatly limiting its application for cell volume measurements. Newly developed techniques, particularly scanning probe microscopy (SPM), may be more promising for cell volume measurements. Scanning ion-conductance microscopy (SICM) is one type of SPM with great promise. SICM utilizes a glass micropipette filled with electrolyte solution. The micropipette tip is positioned over, but very close to, the cell surface for scanning. The gap between the tip of the microelectrode probe and the sample strongly affects the ion current through the pipette, which decreases as the gap diminishes. Changes in the ion current are used in a feedback configuration to control the positioning system to keep the distance between the sample surface and microelectrode probe constant. Thus, the tip of the pipette traces the surface of the cell during scanning from which an actual cell volume can be calculated.[45–48]

A more widely used technique for measuring the cell volume uses a cytosolic fluorescent probe, e.g., calcein-AM,[49,50] to indirectly monitor the cell volume. Calcein-AM is calcein-acetoxymethyl ester, a cell-permeable calcein derivative that is cleaved and trapped in the cytosol once loaded into the cells, similar to that observed for Fura-2/AM loading. Hamann and coworkers first described an approach to measure the cell volume by calcein-AM (at μM concentration) fluorescence self-quenching due to collisional quenching and dimerization of calcein molecules within the cell cytoplasm.[49] Subsequently, Solenov and coworkers demonstrated that rather than calcein self-quenching, intracellular proteins would effectively quench calcein fluorescence as the protein concentration increased upon hyperosmotic challenge. For example, when cells shrink in response to hyperosmotic media, the intracellular protein concentration increases, which enhances calcein quenching; when cells swell, calcein quenching is reduced due to decreased intracellular protein concentration.[50] Calcein can be excited at 488 nm, and its emission between 503 and 530 nm recorded and converted to a digital signal using a photomultiplier detector and a 14-bit analog-to-digital converter (or a digital camera). Both the Hamann group and the Solenov group have shown that the change in calcein fluorescence is approximately linear to the relative changes in the cell volume.[49,50] Mitchell et al. have used an indirect approach to distinguish between swelling and shrinkage by monitoring the cell area as an index of the cell volume in calcein-loaded cells.[51–54] A brief experimental procedure is as the following.

Cells are loaded with 2 μM calcein-AM and 0.2% Pluronic for 30 to 40 min and subsequently superfused with hypertonic or hypotonic solution for 5 to 30 min while acquiring fluorescent data. The cell area is determined by the number of pixels detected above the preset threshold within the region of interest. Because calcein goes through bleaching when excited at 488 nm, one should always take into consideration the best frequency for data acquisition and balance this against the rate of calcein photobleaching.

16.2.3 SHEAR STRESS/FLUID FLOW

Many cells of the body are constantly exposed to fluid mechanical forces arising from the flow of fluid over the cell surface, giving rise to shear stress. This is particularly true for the endothelial cells lining blood vessels and the epithelial cells lining renal tubules where cells are continuously exposed both to circumferential stretch due to the effects of hydrostatic pressures and to shear stress, a frictional force.[8,9,55] In recent years, it has also become apparent that fluid flow over the cell surface can also induce bending or mechanical tension in immotile cilia (primary cilia) of endothelial and renal epithelial cells.[56] The bending of the cilia can lead to mechanical activation of numerous signaling processes, including TRP channels. This section of the chapter will focus on the frictional shear stress that any cell would experience with alternations in fluid flow over the cell surface.

Shear stress is the force per unit area that is generated on the cell surface due to fluid flow over the surface. The force is parallel to the luminal cell surface. Endothelial and epithelial cells are known to respond to such forces with changes in morphology, cell signaling, and gene expression.[8,9,55] Studying the effect of these forces on TRP channels in vitro requires application of these forces in a controlled environment where channel properties can be measured. Typical shear stresses on endothelial cells in the vasculature range from near 2–20 dyn/cm^2 in capillaries and small vessels but can increase to more than 30–100 dyn/cm^2 near arterial branches or regions of sharp wall curvature.[57,58] In the renal tubule, where tubular fluid has a much lower viscosity owing to protein-free tubular fluid, shear stresses in the distal nephron and collecting ducts, the most flow-sensitive regions, can also approach high values (due to high flow rates) with typical values ranging from near 0.3 to 25 dyn/cm^2, although higher values beyond this range are likely in various pathophysiological states.[18]

The effects of shear stress can be studied in vitro using a parallel plate chamber. The most straightforward method to study the effects of shear stress on TRP channel activity is to grow model culture cells in vitro on cover slips that can be mounted to the bottom of a parallel plate chamber. Parallel plate chambers are constructed with defined dimensions that permit calculation of flow rates, velocities, and shear stresses (Figure 16.2). The cover slip forms the bottom portion of the chamber that allows cells to be visualized microscopically and assessed using fluorescence techniques. In the case of calcium-permeable TRP channels, assessment of channel activity can generally be accomplished by fluorescence analysis of changes in intracellular calcium levels, as outlined above, while applying shear stresses of defined magnitudes, simply by altering the rate of fluid flow, changing the fluid viscosity, or altering the temperature. Flow must be laminar, not turbulent, for estimating shear stress. A laminar flow chamber can be readily constructed as a parallel plate chamber of constant width, length, and height, with the cover slip (with the cells attached) forming the bottom wall. Commercial sources are available for purchase of parallel plate chambers (e.g., C&L Instruments, Inc., Hershey, PA, USA; Flexcell International Corp.,

(a) Parallel plate chamber

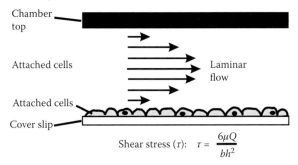

Shear stress (τ): $\tau = \dfrac{6\mu Q}{bh^2}$

(b) Intracellular Ca response to shear stress

FIGURE 16.2 Application of shear stress to TRPV4-transfected human embryonic kidney 293 cells grown on cover slips. (a) Diagram of the stress chamber for application of shear stress to cells on the cover slip while simultaneously monitoring intracellular calcium levels using fluorescence imaging. Increasing flow (laminar flow) through the chamber generates shear stress on the cells (see text for details). (b) A representative example of the effect of application of shear stress (15 dyn/cm²) on intracellular calcium levels showing shear stress-induced activation of calcium influx.

Hillsborough, NC, USA). Typical dimensions for such a parallel plate chamber are: 250 microns H × 1 cm W × 2–5 cm L (or similar). Fluid enters at one end and is collected at the opposite end. Typical flow rates through the chamber can be varied from 0 to 30 mL/min or more using a peristaltic pump to generate shear stresses of from 0 to upward of 30 dynes/cm². For a rectangular chamber, the shear stress at the middle of the chamber (laminar flow) can be calculated as

$$\tau = 6\mu Q/bh^2$$

where τ is the shear stress (dyn/cm²), μ is the fluid viscosity (using the fluid viscosity of water: 0.01002 Poise at 20°C and 0.006915 Poise at 37°C; note, 1 Poise = 1 dynes/cm²), Q is the flow rate (mL/s), b is the chamber width (cm), and h is the chamber height (cm).[57,58] With laminar flow, the shear stress is linearly dependent upon the fluid flow rate. The major determinates of the shear stress in this setting are the fluid flow rate, fluid viscosity, and temperature. The effect of temperature is primarily related to the dependence of fluid viscosity on temperature as noted above for the viscosity of water at 20° and 37°C.

An example of the effect of increasing shear stress on calcium signaling in TRPV4 transfected HEK cells is shown in Figure 16.2.[18,33] Intracellular calcium levels were obtained using standard ratiometric calcium imaging techniques (see methods above). As shown, at low shear stress, calcium levels are near basal values of 100 nM or less. Upon rapid application of the shear stress to 15 dyn/cm², intracellular calcium levels transiently rise to peak values of 200–300 nM, then relax to a low pseudo-plateau value over several minutes. While not shown here, removing extracellular calcium or knocking down TRPV4 by siRNA techniques largely abolishes the rise in intracellular calcium, demonstrating the dependence of TRPV4 on shear stress/fluid flow.[18,33] Similar shear stress effects have been demonstrated in the cortical collecting duct cell line, M-1, where TRPV4 is endogenously expressed.[18]

While the effect of a step change in the shear stress has long been studied in vascular and renal cells, it is well known today that vascular cells, in particular, may also be sensitive to pulsatile shear stress, in addition to steady-state shear stress. Endothelial cells of the vasculature typically experience pulsatile flow (and pulsatile shear stress) as blood flow varies during the cardiac cycle. Similar methods as outlined above can be used to study pulsatile and other variations of the shear stress on cells by regulating the perfusion flow rate in a pulsatile manner through the parallel plate chamber. Indeed, complicated shear stress profiles can be generated using computer-controlled perfusion systems and well-designed algorithms for controlling flow (see Flexcell International).

16.2.4 MEMBRANE STRETCH/CELL STRETCH

16.2.4.1 Stretch-Induced Stresses and Cell Injury

Stretch-induced regulation of vascular and renal cells can be induced by mechanical stress that occurs rapidly, such as in traumatic injuries (brain, spinal cord, etc.), or slowly, such as pulsatile stretch that accompanies blood pressure and flow changes over the course of the normal cardiac cycle or during development of hypertension. Cellular and tissue stretch occurs rapidly, typically over milliseconds, in traumatic injury events and slowly, typically minutes to hours to days, during pulsatile or blood pressure alterations. Because the time periods for the stretch-induced effect can differ greatly, different instrumentation is normally used to permit assessment of different time periods and magnitudes of stretch-induced processes.

During trauma-induced stretching events, cells typically experience rapid alterations in mechanical stress. Such trauma-induced stress is thought to be equivalent to a linear tensile strain applied on a cell, causing a change in its length, i.e., stretching. This type of mechanical stress has been recognized to be a component of many traumatic injuries stemming primarily from generation of high angular acceleration.[59–61] Many mechanical strain instruments and methods have been described to generate such stress-induced stretch for studying cells and tissues in culture. The strain systems outlined below utilize deformation of a flexible membrane to generate stress on cultured cells. However, it should be noted that these strain units do not predict the strain generated on the cells but rather rely on measurements of strain generated on the flexible membrane the cells are grown on to predict injury to cells. This section will aim to highlight key features of a few in vitro strain systems with the goal of presenting the reader with a broad view of stretch models designed to be mounted on a microscope and facilitate couplings with standard fluorescent imaging techniques.

16.2.4.2 Membrane Deformation Models

The tension generated on the cell by stretching has been described to be distributed either equally in two dimensions (pulling the cell in all directions) or unequally in one dimension (pulling the cell in one direction more than others). The former is known as biaxial stretch injury, while the latter is regarded as uniaxial stretch injury. Both forms of stretch injury have been frequently utilized in studies of traumatic injury.[59,61]

Uniaxial Stretch. Lusardi et al. describe a strain system by which rapid uniaxial stretch injury is applied to cell cultures[62] as a method for application of stress that could be used as a model of traumatic brain injury. In the strain unit, cell cultures are grown on a flexible elastic membrane that is attached to a stainless-steel well, so that the membrane occupies the center of the steel chamber just above the well. A closed chamber is formed around the cells by attaching a cover plate over the steel well. The cover plate is designed with openings for pressure input and pressure measurement. The delivery of a pressure pulse applied into the closed upper portion of the chamber causes the elastic substrate to deform rapidly downward, resulting in stretching of the cells. Furthermore, a steel support plate may be placed below the elastic membrane to restrict the amount of membrane deformation that can occur. The use of underlying support plates, with holes in their center of different geometric shape/dimension, allows for generation of varying degrees of stretch. To induce uniaxial stretching, a support plate with a narrow rectangular shape (length 8 times greater than width) is placed over the central opening, thereby favoring longitudinal stretch. This model allows for a >50% strain of the elastic membrane; strain beyond this point can produce rupture of the elastic membrane.[62]

Biaxial Stretch. Biaxial strain units are available, or can be built, to allow application of either rapid or slow/static stretching of cells. Ellis and coworkers developed an in vitro model with a Cell Injury Controller unit designed to deliver rapid biaxial stretch to induce trauma on cell cultures.[63] The unit is commercially available from Virginia Commonwealth University (Virginia Commonwealth University Medical Center, Department of Radiology, Richmond, VA, USA). We are using this unit in our own laboratory to mimic traumatic brain injury (TBI) of brain microvessel endothelial cells. In this model, stretch injury is induced by downward deformation of a flexible silastic membrane on which cell cultures have been grown (see Figure 16.3). The chamber unit calls for the use of six-well tissue culture flex plates, with silastic membranes, from Flexcell International. A removable plug fits on top of an individual well creating an airtight seal. Downward deformation of the membrane, into a spherical cap, is produced by a burst of gas from above. As the membrane is deformed, a measurement of pressure is taken inside the well that corresponds to the amount of biaxial stretch delivered. Any noncorrosive gas such as nitrogen or air can be utilized to create the pressure burst. The model allows for up to a 72% elastic membrane strain ratio and stretch duration rates of 1–99 ms, allowing for examination of various rapid injury conditions. The parameters of the stretch-induced (mild,

(a) Stretch unit for applying cell stretch

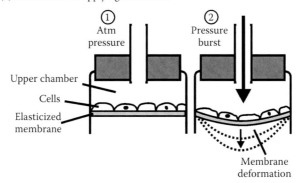

(b) Intracellular calcium response to rapid stretch

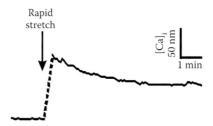

FIGURE 16.3 Membrane/cell stretch on brain microvessel endothelial cell line, bEnd3 cells. (a) Stretch injury unit where a cell monolayer is grown on top of an elastic membrane (a1). A burst of pressure causes the elastic membrane to deform (stretch) and generates stress on cells (a2). (b) Representative trace of intracellular calcium levels in endothelial bEnd3 cells following exposure to stretch (35%) stress for 50 ms. Dashed line represents gap in imaging acquisition. Rapid stretch leads to a rapid increase in intracellular calcium levels.

moderate, and severe) injury were assessed in this model. Elastic membrane deformation of 5.7, 6.3, and 7.5 mm generates cell injuries typical of mild, moderate, and severe traumatic injury corresponding to membrane stretch of 20, 35, and 55%, respectively.

The rapid strain units outlined above all induce injury via deformation of the elasticized membrane, and therefore, all lack the ability to maintain cells at a continuous focus plane during induction of stretch. This presents a fundamental complication in the ability to gain complete injury data with microscopic tools, such as those used for standard fluorescence imaging. Other stretch models, like that described by Hung and coworkers, produce stretch tension on cells while maintaining cells at a continuous focus plane.[64] The model utilizes a polysulfone well designed with two grooves at the bottom (inner and outer) and an open area at the center (cell growth area) for cell visualization. At the bottom of the well, an elastomeric circular membrane, on which a monolayer of cells has been grown, is attached to the outer grove by an O-ring. Cells situated in the growth area of the well then will receive radial stretching as a ring located below the membrane is raised vertically, causing indentation of the membrane into the inner groove of the well. The indentation is designed so that while some membrane is raised vertically, the membrane of the cell growth area remains in the same plane. Ring indentation of 0.1–3.5 mm was calculated to achieve strain on the elastic membrane surface in the range of 0.04–8.0% strain.

It should be mentioned that a growing appreciation exists for the rate of strain induced in stretch models, as several studies have documented that the magnitude of injury is dependent on the strain rate.[65–67] Thus far, stretch-induced stress models have proven a useful tool in the study of various physiological and pathological states, such as the effects of strain on bone or stretch on cells in traumatic brain injury. Yet in many instances, it is assumed that stretch alone can mimic the physiological stress state, when in reality it is the combination of multiple stress factors. To this end, it seems likely that further evolution of these mechanical models will aim at the combining of osmomechanical stimuli in pursuit of developing a more accurate representation of the physiological stress states facing cells.

Chamber study units are also available for a much slower application of cell stretch to better mimic the pulsatile or static stretch normally present in the mammalian vasculature. Flexcell International has developed elasticized membrane systems in six-well plate formats (Flex I plates) to promote the assessment of pulsatile or static stretch on cells grown on the elasticized membranes, using pressure-pulse methods to apply stretch to the membrane as outlined above for the rapid application models. Such systems allow simple pulsatile stretch of cells but can also be programmed to provide more complex waveforms that include combined static and pulsatile components. The chamber can be attached to a microscope stage to allow application of fluorescent techniques to the cells, such as for measurement of intracellular calcium levels, while undergoing cycles of stretching. Flexcell International has developed both culture plates (Flex I culture plates) and computer-controlled strain systems (e.g., Flexcell 4000 Strain Unit) to perform such analysis. Investigators interested in their systems should visit their Web site for more details (http://www.flexcellint .com/; Flexcell International Corp, Hillsborough, NC, USA).

16.3 APPLICATION OF OSMOMECHANICAL STRESSES: PRACTICAL CONSIDERATIONS

16.3.1 PATCH-CLAMP VERSUS CALCIUM IMAGING METHODS

This chapter has described the basic methods for studying the effects of osmomechanical stresses on the properties of TRP channels. The methods employed to study the channel properties have limitations owing to the fact that application of mechanical stresses often leads to movement of cells in the in vitro study chambers or requires the use of mechanical units that restrict access to the cells. This is particularly applicable to the two methods of studying TRP channel properties described here: patch-clamp versus ratiometric calcium imaging. The patch-clamp technique can be very limited for this type of study because the application of mechanical stresses will typically lead to movement of the cells and loss of the electrical seal with the patch pipette. Further, some strain units, such as for application of shear stress, have an upper cover that restricts access to the cells with a patch pipette.

In contrast to patch-clamp techniques, calcium imaging can usually be used to study calcium influx in cells that are attached to glass cover slips of the study chambers during the application of stress. While some methods for applying strain, such as with the trauma injury models with cells on elasticized membranes, can lead to limitations in maintaining cells of interest in focus, this can be largely overcome with specialized chambers as outlined above. Alternatively, if continuous imaging is not needed during stretching, one can immediately refocus the microscope over a few seconds after application of injury and continue imaging of the cells, although a small degree of lateral movement may be introduced. We have also tested the use of automatic infrared focus systems on today's newer microscopes, but the elasticized membrane is not compatible with the infrared focusing beam. Hence, the investigator will typically either have to refocus after the stretch stimulus or invest in more sophisticated strain units that reduce the focusing problem (never fully eliminating the focus issue).

16.3.2 Cell Adhesion

A common problem when applying osmomechanical stimuli is maintenance of cell adhesion during and after the stimulus. This is particularly true for elasticized membranes that are physically displaced during the application of strain but can also be an issue with the fluid flow rate or upon cell swelling with cells grown on cover slips. Acid washing the cover slips will improve adhesion (1 N HCl for 4–8 hours followed by extensive washing in very clean distilled water). A more widely used approach for both cover slips and elasticized membranes is to precoat the membranes or cover slips with an adhesive-type material. This can be accomplished by treating the membranes or cover slips with polyamino acids such as poly-L-lysine. The protocol is as follows: First, briefly acid wash the membranes or cover slips, then apply a small droplet (adjust based on the area desired for cells to adhere) of poly-L-lysine (1 mg/mL; use 150 kDa size or larger) to the middle of the cover slip, let stand for at least 30 min, and wash with clean water. This should be followed by rinsing the cover slip with 100% ethanol and then resting it on the edge in an open tissue culture dish in a laminar flow hood until dry. Cells can then be plated in the usual manner. Finally, various extracellular matrix substances, including collagens, laminins, fribronectin, gelatin, and others, can also be pre-plated onto the cover slips and elasticized membranes to both improve adhesion and provide the cells with a more "natural" adhesion substrate. Most manufacturers of the extracellular matrix molecules also provide defined protocols for use of these substrates as adhesion substrates in cell culture (see Sigma-Aldrich).

16.4 CONCLUSIONS

The current chapter outlines typical methods and instrumentation that can be used to study the effects of osmomechanical stress on mechanically sensitive TRP channels. Methods appropriate for studying TRP channel properties under these conditions are presented including both patch-clamp analysis of channel currents and ratiometric calcium imaging analysis of calcium influx under various states of mechanical stress. Further, the application of various mechanical stresses is outlined with an emphasis on the osmotic and mechanical stresses that are typically experienced by cells of the cardiovascular and renal systems.

ACKNOWLEDGMENTS

This work was supported by the National Institutes of Health awards to R. G. O'Neil, grant numbers DK70950 and R21 DE018522.

REFERENCES

1. Shepherd, V. A., M. J. Beilby, and T. Shimmen. 2002. Mechanosensory ion channels in charophyte cells: the response to touch and salinity stress. *European Biophysical Journal* 31(5): 341–355.
2. Edwards, M. D., I. R. Booth, and S. Miller. 2004. Gating the bacterial mechanosensitive channels: MscS a new paradigm? *Current Opinions in Microbiology* 7(2): 163–167.
3. Koprowski, P., and A. Kubalski. 2001. Bacterial ion channels and their eukaryotic homologues. *Bioessays* 23(12): 1148–1158.
4. Sukharev, S., and D. P. Corey. 2004. Mechanosensitive channels: multiplicity of families and gating paradigms. *Science STKE* 2004(219): re4.
5. Bianchi, L. 2007. Mechanotransduction: touch and feel at the molecular level as modeled in Caenorhabditis elegans. *Molecular Neurobiology* 36(3): 254–271.
6. Sachs, F., W. E. Brownell, and A. G. Petrov. 2009. Membrane electromechanics in biology, with a focus on hearing. *MRS Bulletin* 34(9): 665.
7. Inoue, R., Z. Jian, and Y. Kawarabayashi. 2009. Mechanosensitive TRP channels in cardiovascular pathophysiology. *Pharmacology Therapeutics* 123(3): 371–385.
8. Ando, J., and K. Yamamoto. 2009. Vascular mechanobiology: endothelial cell responses to fluid shear stress. *Circulation Journal* 73(11): 1983–1992.
9. Davies, P. F. 2009. Hemodynamic shear stress and the endothelium in cardiovascular pathophysiology. *Nature Clinical Practice. Cardiovascular Medicine* 6(1): 16–26.
10. O'Neil, R. G., and S. Heller. 2005. The mechanosensitive nature of TRPV channels. *Pflügers Archiv* 451(1): 193–203.
11. Liedtke, W. 2007. TRPV channels' role in osmotransduction and mechanotransduction. *Handbook in Experimental Pharmacology* 179: 473–487.
12. Christensen, A. P., and D. P. Corey. 2007. TRP channels in mechanosensation: direct or indirect activation? *Nature Reviews. Neuroscience* 8(7): 510–521.
13. Giamarchi, A., F. Padilla, M. Crest et al. 2006. TRPP2: Ca^{2+}-permeable cation channel and more. *Cellular and Molecular Biology* 52(8): 105–114.
14. Woudenberg-Vrenken, T. E., R. J. Bindels, and J. G. Hoenderop. 2009. The role of transient receptor potential channels in kidney disease. *Nature Reviews. Nephrology* 5(8): 441–419.
15. Montell, C. 2005. TRP channels in Drosophila photoreceptor cells. *Journal of Physiology* 567(Part 1): 45–51.
16. Nilius, B., G. Owsianik, T. Voets et al. 2007. Transient receptor potential cation channels in disease. *Physiological Reviews* 87(1): 165–217.
17. Maroto, R., A. Raso, T. G. Wood et al. 2005. TRPC1 forms the stretch-activated cation channel in vertebrate cells. *Nature Cell Biology* 7(2): 179–185.
18. Wu, L., X. Gao, R. C. Brown et al. 2007. Dual role of the TRPV4 channel as a sensor of flow and osmolality in renal epithelial cells. *American Journal of Physiology Renal Physiology* 293(5): F1699–F1713.

19. Vriens, J., H. Watanabe, A. Janssens et al. 2004. Cell swelling, heat, and chemical agonists use distinct pathways for the activation of the cation channel TRPV4. *Proceedings of the National Academy of Sciences of the United States of America* 101(1): 396–401.

20. Lockwich, T., B. B. Singh, X. Liu et al. 2001. Stabilization of cortical actin induces internalization of transient receptor potential 3 (Trp3)-associated caveolar Ca^{2+} signaling complex and loss of Ca^{2+} influx without disruption of Trp3-inositol trisphosphate receptor association. *Journal of Biological Chemistry* 276(45): 42401–42408.

21. Bretscher, A., K. Edwards, and R. G. Fehon. 2002. ERM proteins and merlin: integrators at the cell cortex. *Nature Reviews. Molecular and Cellular Biology* 3(8): 586–599.

22. Wu, K. L., S. Khan, S. Lakhe-Reddy et al. 2004. The NHE1 Na^+/H^+ exchanger recruits ezrin/radixin/moesin proteins to regulate Akt-dependent cell survival. *Journal of Biological Chemistry* 279(25): 26280–26286.

23. Brown, A. L., B. E. Johnson, and M. B. Goodman. 2008. Patch clamp recording of ion channels expressed in Xenopus oocytes. *Journal of Visualization Experiments* 20.

24. Jurkat-Rott, K., and F. Lehmann-Horn. 2004. The patch clamp technique in ion channel research. *Current Pharmaceutical Biotechnology* 5(4): 387–395.

25. Hamill, O. P. 2006. Twenty odd years of stretch-sensitive channels. *Pflügers Archiv* 453(3): 333–351.

26. McCombs, J. E., and A. E. Palmer. 2008. Measuring calcium dynamics in living cells with genetically encodable calcium indicators. *Methods* 46(3): 152–159.

27. Paredes, R. M., J. C. Etzler, L. T. Watts et al. 2008. Chemical calcium indicators. *Methods* 46(3): 143–151.

28. Navarre, C., H. Degand, K. L. Bennett, J. S. Crawford, E. Mørtz, and M. Boutry. 2002. Subproteomics: identification of plasma membrane proteins from the yeast *Saccharomyces cerevisiae*. *Proteomics* 2(12): 1706–1714.

29. Beech, D. J., K. Muraki, and R. Flemming. 2004. Non-selective cationic channels of smooth muscle and the mammalian homologues of Drosophila TRP. *Journal of Physiology* 559(Part 3): 685–706.

30. Muraki, K., Y. Iwata, Y. Katanosaka et al. 2003. TRPV2 is a component of osmotically sensitive cation channels in murine aortic myocytes. *Circulation Research* 93(9): 829–838.

31. Liedtke, W., and J. M. Friedman. 2003. Abnormal osmotic regulation in trpv4$^{-/-}$ mice. *Proceedings of the National Academy of Sciences of the United States of America* 100(23): 13698–13703.

32. Nilius, B., J. Prenen, U. Wissenbach et al. 2001. Differential activation of the volume-sensitive cation channel TRP12 (OTRPC4) and volume-regulated anion currents in HEK-293 cells. *Pflügers Archiv* 443(2): 227–233.

33. Gao, X., L. Wu, and R. G. O'Neil. 2003. Temperature-modulated diversity of TRPV4 channel gating: activation by physical stresses and phorbol ester derivatives through protein kinase C-dependent and -independent pathways. *Journal of Biological Chemistry* 278(29): 27129–27137.

34. Grimm, C., R. Kraft, S. Sauerbruch et al. 2003. Molecular and functional characterization of the melastatin-related cation channel TRPM3. *Journal of Biological Chemistry* 278(24): 21493–21501.

35. Numata, T., T. Shimizu, and Y. Okada. 2007. TRPM7 is a stretch- and swelling-activated cation channel involved in volume regulation in human epithelial cells. *American Journal of Physiology Cell Physiology* 292(1): C460–C467.

36. Chen, J., and G. J. Barritt. 2003. Evidence that TRPC1 (transient receptor potential canonical 1) forms a Ca^{2+}-permeable channel linked to the regulation of cell volume in liver cells obtained using small interfering RNA targeted against TRPC1. *Biochemistry Journal* 373(Part 2): 327–336.

37. Montalbetti, N., Q. Li, S. Gonzalez-Perrett et al. 2005. Effect of hydro-osmotic pressure on polycystin-2 channel function in the human syncytiotrophoblast. *Pflügers Archiv* 451(1): 294–303.
38. Ciura, S., and C. W. Bourque. 2006. Transient receptor potential vanilloid 1 is required for intrinsic osmoreception in organum vasculosum lamina terminalis neurons and for normal thirst responses to systemic hyperosmolality. *Journal of Neuroscience* 26(35): 9069–9075.
39. Arniges, M., E. Vazquez, J. M. Fernandez-Fernandez et al. 2004. Swelling-activated Ca^{2+} entry via TRPV4 channel is defective in cystic fibrosis airway epithelia. *Journal of Biological Chemistry* 279(52): 54062–54068.
40. Becker, D., C. Blase, J. Bereiter-Hahn et al. 2005. TRPV4 exhibits a functional role in cell-volume regulation. *Journal of Cell Science* 118(Part 11): 2435–2440.
41. Liu, X., B. C. Bandyopadhyay, T. Nakamoto et al. 2006. A role for AQP5 in activation of TRPV4 by hypotonicity: concerted involvement of AQP5 and TRPV4 in regulation of cell volume recovery. *Journal of Biological Chemistry* 281(22): 15485–15495.
42. Guilak, F. 1994. Volume and surface area measurement of viable chondrocytes in situ using geometric modelling of serial confocal sections. *Journal of Microscopy* 173(Part 3): 245–256.
43. Zhu, Q., P. Tekola, J. P. Baak et al. 1994. Measurement by confocal laser scanning microscopy of the volume of epidermal nuclei in thick skin sections. *Analytical and Quantitative Cytology and Histology* 16(2): 145–152.
44. Errington, R. J., M. D. Fricker, J. L. Wood et al. 1997. Four-dimensional imaging of living chondrocytes in cartilage using confocal microscopy: a pragmatic approach. *American Journal of Physiology* 272(3 Pt 1): C1040–C1051.
45. Gorelik, J., Y. Zhang, A. I. Shevchuk et al. 2004. The use of scanning ion conductance microscopy to image A6 cells. *Molecular and Cellular Endocrinology* 217(1–2): 101–108.
46. Korchev, Y. E., C. L. Bashford, M. Milovanovic et al. 1997. Scanning ion conductance microscopy of living cells. *Biophyics Journal* 73(2): 653–658.
47. Korchev, Y. E., J. Gorelik, M. J. Lab et al. 2000. Cell volume measurement using scanning ion conductance microscopy. *Biophysics Journal* 78(1): 451–457.
48. Korchev, Y. E., M. Milovanovic, C. L. Bashford et al. 1997. Specialized scanning ion-conductance microscope for imaging of living cells. *Journal of Microscopy* 188(Part 1): 17–23.
49. Hamann, S. 2002. Measurement of cell volume changes by fluorescence self-quenching. *Journal of Fluorescence* 12(2): 139–145.
50. Solenov, E., H. Watanabe, G. T. Manley et al. 2004. Sevenfold-reduced osmotic water permeability in primary astrocyte cultures from AQP-4-deficient mice, measured by a fluorescence quenching method. *American Journal of Physiology Cell Physiology* 286(2): C426–C432.
51. Mitchell, C. H., J. C. Fleischhauer, W. D. Stamer et al. 2002. Human trabecular meshwork cell volume regulation. *American Journal of Physiology Cell Physiology* 283(1): C315–C326.
52. Do, C. W., C. W. Kong, and C. H. To. 2004. cAMP inhibits transepithelial chloride secretion across bovine ciliary body/epithelium. *Investigative Ophthalmology and Visual Science* 45(10): 3638–3643.
53. Do, C. W., K. Peterson-Yantorno, and M. M. Civan. 2006. Swelling-activated Cl- channels support Cl- secretion by bovine ciliary epithelium. *Investigative Ophthalmology and Visual Science* 47(6): 2576–2582.
54. Do, C. W., K. Peterson-Yantorno, C. H. Mitchell et al. 2004. cAMP-activated maxi-Cl(-) channels in native bovine pigmented ciliary epithelial cells. *American Journal of Physiology Cell Physiology* 287(4): C1003–C1011.

55. Chiu, J. J., S. Usami, and S. Chien. 2009. Vascular endothelial responses to altered shear stress: pathologic implications for atherosclerosis. *Annals of Medicine* 41(1): 19–28.

56. Nauli, S. M., F. J. Alenghat, Y. Luo et al. 2003. Polycystins 1 and 2 mediate mechanosensation in the primary cilium of kidney cells. *Nature Genetics* 33(2): 129–137.

57. Dewey, C., Jr. 1979. Fluid mechanics of arterial flow. *Advances in Experimental Medicine and Biology* 115: 55–103.

58. Levesque, M. J., and R. M. Nerem. 1985. The elongation and orientation of cultured endothelial cells in response to shear stress. *Journal of Biomechanical Engineering* 107(4): 341–347.

59. Kumaria, A., and C. M. Tolias. 2008. In vitro models of neurotrauma. *British Journal of Neurosurgery* 22(2): 200–206.

60. Zhang, L., K. H. Yang, and A. I. King. 2001. Biomechanics of neurotrauma. *Neurological Research* 23(2–3): 144–156.

61. Geddes-Klein, D. M., K. B. Schiffman, and D. F. Meaney. 2006. Mechanisms and consequences of neuronal stretch injury in vitro differ with the model of trauma. *Journal of Neurotrauma* 23(2): 193–204.

62. Lusardi, T. A., J. Rangan, D. Sun et al. 2004. A device to study the initiation and propagation of calcium transients in cultured neurons after mechanical stretch. *Annals of Biomedical Engineering* 32(11): 1546–1558.

63. Ellis, E. F., J. S. McKinney, K. A. Willoughby et al. 1995. A new model for rapid stretch-induced injury of cells in culture: characterization of the model using astrocytes. *Journal of Neurotrauma* 12(3): 325–339.

64. Hung, C. T., and J. L. Williams. 1994. A method for inducing equi-biaxial and uniform strains in elastomeric membranes used as cell substrates. *Journal of Biomechanics* 27(2): 227–232.

65. Barbee, K. A. 2005. Mechanical cell injury. *Annals of the New York Academy of Sciences* 1066: 67–84.

66. LaPlaca, M. C., and L. E. Thibault. 1997. An in vitro traumatic injury model to examine the response of neurons to a hydrodynamically-induced deformation. *Annals of Biomedical Engineering* 25(4): 665–677.

67. Cargill, R. S., 2nd, and L. E. Thibault. 1996. Acute alterations in $[Ca^{2+}]_i$ in NG108-15 cells subjected to high strain rate deformation and chemical hypoxia: an in vitro model for neural trauma. *Journal of Neurotrauma* 13(7): 395–407.

17 Study of TRP Channels in Cancer Cells

V'yacheslav Lehen'kyi and Natalia Prevarskaya

CONTENTS

17.1 INTRODUCTION

Many proteins in cancer cells exhibit increased or decreased expression compared to their levels in normal cells. Some of these proteins, i.e., those encoded by oncogenes and tumor suppressor genes, play key roles in tumorigenesis and in the development of metastases, while others, most likely including those involved in intracellular Ca^{2+} homeostasis (reviewed in Refs. 1 and 2), are associated with cancer progression but are not causative in further development of the tumor and/or malignant cells (reviewed in Ref. 3). Most cancers are heterogeneous with respect to rates of growth and degrees of aggression. This most likely reflects the fact that, for a given cancer, there can be different combinations of oncogenes and tumor suppressor genes that are mutated, different sequences in which these mutations occur, and variations in the time over which the mutations accumulate.[3]

Some of the most important signaling pathways altered in tumorigenesis enhance cell proliferation and inhibit apoptosis. Ca^{2+} homeostasis controls these cellular processes, including proliferation, apoptosis, gene transcription, and angiogenesis.[4] Ca^{2+}

signaling is thus required for cell proliferation in all eukaryote cells, while some transformed cells and tumor cell lines show reduced dependence on Ca^{2+} to maintain proliferation.[5,6] Furthermore, the regulation of cell cycles, apoptosis, or proliferation depends on the amplitude and temporal-spatial aspects of the Ca^{2+} signal,[7,8] thus highlighting the importance of Ca^{2+} signaling components such as Ca^{2+} channels. Indeed, dysfunctions in Ca^{2+} channels are involved in tumorigenesis because increased expression of plasma membrane Ca^{2+} channels amplifies Ca^{2+} influx with consequent promotion of Ca^{2+}-dependent proliferative pathways.[4,7,9]

Transient receptor potential (TRP) channels contribute to changes in intracellular Ca^{2+} concentrations, either by acting as Ca^{2+} entry pathways in the plasma membrane or via changes in membrane polarization, modulating the driving force for Ca^{2+} entry mediated by alternative pathways,[10] as well as the activity of voltage-gated Ca^{2+} channels. In addition, TRP channels are expressed on the membranes of internal Ca^{2+} stores[11–13] where they may act as triggers for enhanced proliferation, aberrant differentiation, and impaired ability to die, leading to the uncontrolled expansion and invasion characteristic of cancer.

All these normal as well as abnormal misregulations are within the scope of this chapter, which is aimed at the study of TRP channels at all known levels, such as mRNA transcription, splicing variants, mRNA translation into protein, further protein processing in Golgi, regulation of channel trafficking to the plasma membrane, and the final modulation of TRP channel activity at the plasma membrane.

17.1.1   TRP CHANNELS AND CANCER

The extent to which TRP channels are associated with cancer has been increasingly clarified in recent years. The approximately 30 TRPs identified to date are classified in six different families: TRPC (canonical), TRPV (vanilloid), TRPM (melastatin), TRPML (mucolipin), TRPP (polycystin), and TRPA (ankyrin transmembrane protein).[14] The expression levels of members of the TRPC, TRPM, and TRPV families are correlated with the emergence and/or progression of certain epithelial cancers.[15–18] It has not yet been established whether these expression changes are drivers, required to sustain the transformed phenotype. Usually, the progression of cells from a normal, differentiated state to a tumorigenic, metastatic state involves the accumulation of mutations in multiple key signaling proteins, encoded by oncogenes and tumor suppressor genes, together with the evolution and clonal selection of more aggressive cell phenotypes.

Several recent works have shown that changes in the expression of TRP channels contribute to malignancy. The first evidence that TRP channel expression is correlated with different types of cancers came from the analysis of the expression of TRPM1. The expression of the TRPM1 gene is inversely correlated with the aggressiveness of melanoma malignant cells, which suggests that TRPM1 may behave as a tumor suppressor gene.[18,19] A second member of the TRPM subfamily, TRPM5, was shown to be responsible for the Beckwith-Wiedemann syndrome, a disease characterized by a childhood predisposition to tumors.[20] TRPC channels are often coupled to a membrane receptor with which they work in synergy. Thus, it was demonstrated in prostate cancer cells that a Ca^{2+} signal can promote either cell proliferation or apoptosis,[12,15] depending on the type of TRPC channel involved: Ca^{2+} entry via TRPC6 channels stimulates cell proliferation, whereas TRPC1 and TRPC4 are mostly involved in apoptosis induction.

In addition, TRP channels could also play a key role in cancer progression. Indeed, this seems to be the case for TRPM8 and TRPV6 channels in prostate cancer. TRPM8 has originally been cloned from cancerous prostate tissue[16] and was thereafter identified as an ion channel responding to cold stimuli.[21] TRPM8 is expressed in normal prostate; however, its expression is increased in ADCaP. TRPV6 is strongly expressed in advanced prostate cancer, with little or no expression in healthy and benign prostate tissues.[17,22]

Thus, to date, most changes involving TRP proteins do not involve mutations in the TRP gene but rather increased or decreased expression levels of the wild-type TRP protein, depending on the stage of the cancer. Table 17.1 summarizes these changes in cancer and metastatic cells.

17.1.2 TRP Channels as Potential Pharmaceutical Targets

Two aspects of the properties of TRP proteins and the association of increased or decreased expression of a given TRP protein with cancer and the progression of cancer have been used to try to develop strategies to kill cancer cells. One approach

---

TABLE 17.1
Expression Profile of TRPs in Cancer

			Expression			
Channel	Cancer Type	Isoforms	Healthy/ Benign	Tumor	Invasive	Reference
TRPV1	Bladder	Yes				19
TRPV1	Bladder	Yes	Yes	↓	Loss	
	Glioma		Yes	↑		
TRPV2	Bladder	Full	Yes	↑	↑	20
		Short	Yes	↓	Loss	
	Prostate	Full	Yes	↑	↑	27
TRPV6	Breast	ND	Yes	↑	↑	28
	Ovarian	ND	Yes	↑	↑	17, 21
	Thyroid	ND	Yes	↑		29
	Prostate	ND	Yes	↑	↑	25, 30
	Colon	ND		↑		31
TRPM1	Melanoma	Yes	Yes	↓	Loss	18, 23
TRPM8	Melanoma	Yes	Yes	↑	Loss	11, 16, 24
	Lung			↑		16
	Colon			↑		16
TRPC6	Breast	Yes	Yes	↑	↑	26

uses Ca^{2+} and Na^+ entry through TRP channels expressed in cancer cells, which leads to a sustained high $[Ca^{2+}]_{cyt}$ and cytoplasmic Na^+ concentration ($[Na^+]_{cyt}$), a condition that kills cells by apoptosis and necrosis. This strategy requires the selective expression and activation of a given TRP channel in the targeted cancer cells. New strategies for killing cancer cells by activation of the apoptotic pathway are valuable because for many cancer cells, including androgen-insensitive prostate cancer cells, the normal pathways of apoptosis are inhibited, and the cells are resistant to apoptosis.[23–25] The other aspect makes use of the high expression of some TRP channels in cancer cells to provide a target for delivering a toxic payload (e.g., a radioactive nuclide or toxic chemical) to the cancer cells. Recognition of the TRP protein could be achieved through a tight-binding agonist or an anti-TRP antibody (reviewed in Ref. 26).

Studies with some other cancers also suggest that TRPV1 may be a useful target for killing cancer cells through a sustained increase in $[Ca^{2+}]_{cyt}$ and $[Na^+]_{cyt}$. Resiniferatoxin, an analogue of capsaicin and an agonist of TRPV1, was shown to cause inhibition of mitochondrial function and induce apoptosis in pancreatic cancer cells, presumably via endogenous TRPV1 channels in the plasma membrane.[27] It was suggested that vanilloids might be used to treat pancreatic cancer.[27]

Lignesti and colleagues tested the ability of various plant cannabinoids, which bind to the cannabinoid receptors (CB) and TRPV1 channels, to inhibit tumor cell growth.[32] Using a panel of tumor cell lines as well as a xenograft mouse model of breast cancer (MDA-MB-231 cells), these investigators found that cannabinoids, of which cannabidiol was the most potent, inhibited cell and tumor growth. They suggested that cannabinoids may act through CB2 receptors and the TRPV1 channels. Endogenous cannabinoids play an important role in the neuronal control of the digestive tract.[33] It has been suggested that the pharmacological administration of cannabinoids, which in part act through TRPV1 channels, could be used to treat colon cancer.[33]

As discussed above, TRPM8 is expressed in prostate cancer cells, and its expression decreases as the cancer progresses to a more metastatic state. Hence, TRPM8 is considered potentially useful both for the diagnosis of prostate cancer and as a target for cancer therapy. The treatment of prostate cancer would be greatly enhanced by better prediction of the course of the disease, including the likelihood of the development of androgen insensitivity and metastases, and by new strategies to kill androgen-insensitive prostate cancer cells, which, as mentioned above, are resistant to apoptosis.[34,35]

Recently, a menthol analogue and TRPM8 agonist, WS-12, has been synthesized and characterized. WS-12 has an affinity for the TRPM8 menthol binding site, which is about 2000 times higher than that for menthol itself.[36] Incorporation of a fluorine atom into WS-12 resulted in an analogue (WS-12F) that activated TRPM8 by 75% of the activation induced by WS-12 and retained a high affinity for TRPM8. It has been suggested that WS-12 and WS-12F offer potential possibilities in the detection of micro metastases and in killing prostate cancer cells. Thus, the incorporation of ^{18}F into WS-12 may permit radio-imaging of micrometastases, and/or the delivery of a radionuclide, which could kill target cells, to specific locations of cancer cells in both the prostate and in metastases.[36]

17.2 ROLE OF TRP CHANNEL ISOFORM EXPRESSION AND FUNCTION IN THE STUDY OF THE CANCER INITIATION AND PROGRESSION

TRP proteins form tetrameric channel complexes, and at least the closely related members of one subfamily are capable of building heteromeric channels.[37] The diversity of native TRP-related channels might be considerably increased by combining different TRP channel subunits to build a common ion-conducting pore. There is growing evidence that transcriptional regulation and alternative mRNA processing also contribute to the diversity of TRP channels. Some TRPs are expressed in two or more short splice variants, which may also exhibit different expression profiles in cancer as compared to the full-length forms. Alternative splicing enables the same gene to generate multiple mature mRNA types for translation, resulting in multiple-channel proteins. This leads to functional diversity, which, in turn, may have consequences for cellular function. Alternative splicing generates protein isoforms with different biological properties, such as a change in functionality, protein/ protein interaction, or subcellular localization.[38] Many of the splice variants are not functional and may not even be efficiently translated, so they may be considered negligible populations of incomplete or aberrantly spliced transcripts. Nevertheless, alternative splicing, as a regulatory process, contributes to biological complexity, not only by proteome expansion but also through its ability to control the expression of functional proteins. This may be achieved by producing nonfunctional isoforms of the gene by altering the domains necessary for TRP channel opening, membrane localization, or association.

This is the case for TRPV2, which expresses two transcripts in normal human urothelial cells and bladder tissue specimens: full-length TRPV2 and a short-splice variant, s-TRPV2. Analysis of *TRPV2* gene and protein expressions in distinct superficial and invasive grades and stages indicates that the mRNA of TRPV2 increases gradually at increasing grades and stages, while that of s-TRPV2 gradually decreases.[20] The authors suggested that the differences observed in the short/full TRPV2 ratio during tumorigenesis implied that s-TRPV2 was lost as an early event in bladder carcinogenesis, whereas the enhanced expression of full-length TRPV2 in high-stage muscle-invasive urothelial cancer is a secondary event. In a similar study, a different s-TRPV2, lacking the pore-forming region and the fifth and sixth transmembrane domains, was characterized in human macrophages.[39] As for TRPV1, these naturally occurring alternative splice variants may act as dominant-negative mutants[40] by forming a heterodimer with TRPV2 and inhibiting its trafficking and translocation to the plasma membrane.

In the case of TRPV1, the short isoform (TRPV1β) is produced by alternative splicing of the *TRPV1* gene, with 10 amino acids missing near the end of the cytoplasmic N-terminus.[40] TRPV1β does not form a functional channel when it is heterologously expressed alone, but exerts a dominant-negative effect on TRPV1 when they are co-expressed. Stability is affected when TRPV1β is assembled with the full-length channel, making less TRPV1 protein available at the plasma membrane. The residual amount of TRPV1β on the plasma membrane is not activated by factors known to stimulate TRPV1, but there are two other possibilities.[40] Either the residual proteins are not properly assembled into tetrameric channels or channels that contain TRPV1β subunits cannot be opened. It should be noted, however, that TRPV1 Western blot analysis of the urothelium revealed two bands of equal intensity at 100 and 95 kDa, which decreased as the cancer progressed.[19] Further investigation is required to determine whether these are the two splice TRPV1 isoforms and to analyze their expression regulation as cancer progresses. A similar mechanism is present in normal and benign melanocytes, which express the full-length TRPM1 mRNA, along with some shorter products.[18] Heterologous co-expression of the full-length and short TRPM1 isoforms results in retention of the full-length channel in the endoplasmic reticulum (ER).[13] However, it is currently unclear whether TRPM1 expression in metastasizing lines inhibits their growth. Metastatic melanomas lack the full-length transcript, but express several short fragments of TRPM1,[18] probably owing to proteolysis of the full-length protein.[41]

TRPM8 also encodes some splice variants, comprising an altered N-terminus cloned from lung epithelia[42] and cancerous prostate.[43] The lung epithelia splice variant localizes preferentially to the ER, and its activation controls cell responses to cold air-induced inflammation.[42] It has not yet been clarified whether this newly identified variant is implicated in cancer, whereas it may constitute a regulatory mechanism for the full-length TRPM8 in tissues where they both localize, such as liver, colon, and testis.[42] Little information is available concerning the cancerous prostate TRPM8 isoform. It has a truncated N-terminus[43] and may serve as a dominant negative regulator of full-length TRPM8, as suggested for TRPM1 truncated variants.[13] Furthermore, a recent study by Bidaux and coworkers identified two TRPM8 isoforms with different androgen sensitivity and distinct localization on the plasma and ER membranes.[11] Figure 17.1 shows schematically the *TRPM8* gene, TRPM8 mRNA, and the position of Q-PCR primers used to distinguish between the two isoforms. The relative ratio between the two PCR products is used to calculate the predominance of one isoform over the other. This strategy may be also used to selectively knock down either of the isoforms. The differential regulation of TRPM8 activity may be due to complex regulation of the two isoforms by androgen receptors: An alternative *TRPM8* gene promoter may make the ER TRPM8 isoform less sensitive to androgens. However, this ER localization may also result from a variation in the primary sequence leading to the appearance of an ER retention signal or the involvement of other associated proteins affecting its trafficking. It should be noted that there is a controversy in the literature concerning the localization of ER TRPM8. Two studies proposed a TRPM8-independent ER Ca^{2+} release mechanism in LNCaP [44] and PC3[45] cells when using high doses of menthol (3 mM)[44] versus the ER TRPM8 activation with 100–250 μM menthol.[12,46]

Furthermore, immunocytochemistry experiments in LNCaP revealed contradictory results concerning the presence of TRPM8 on the ER.[12,47] Two scenarios may explain this incongruity: firstly, as TRPM8 is under androgenic control, culture conditions of LNCaP cells, especially with respect of the serum used, may be critical for channel expression and localization, and, secondly, the putative ER TRPM8 isoform is not necessarily detected by the different antibodies used in these studies. In any case, the presence of ER TRPM8 was demonstrated in freshly isolated primary epithelial prostate cancer cells.[11] Consequently, to clarify whether TRPM8 localizes into

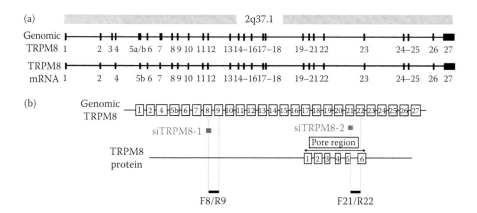

FIGURE 17.1 The TRPM8 gene encodes for classical TRPM8 channel and a putative truncated TRPM8 splice variant. (a) The human TRPM8 gene localizes on chromosome 2 in position 37.1. (b) Genomic map of TRPM8 (not to scale; numbered boxes denote exons) with its associated protein structure (boxes 1–6 represent putative transmembrane domains). The two pairs of PCR primers and siTRPM8-1 and -2 are aligned with their matching exons.

the ER, it is necessary to clone this putative ER-specific TRPM8 isoform and identify its distinguishing features, as compared to the two previously cloned variants.

Therefore, the abundant short or long mRNA forms in some cancers arise from a regulatory mechanism that produces either spliced or partially degraded nonproductive RNAs. These spliced transcripts form multimers and regulate targeting to the plasma or ER membrane, and consequently protein activity. Changes in TRP localization may have a causal or promoting role in cancer. The increases in constitutively active channels, such as TRPV6, in the plasma membrane of prostate cancer cells[17,30] may augment Ca^{2+} in the cytosol, thus promoting Ca^{2+}-dependent proliferative pathways. The same may hold true for TRPM1 in melanocytes[13,18,23] because Ca^{2+} imaging experiments on transfected HEK293 cells revealed an increase in intracellular Ca^{2+} concentrations in comparison to the nontransfected cells.[13] However, in the absence of electrophysiological data, it would be premature to conclude that TRPM1 is a constitutively active channel. Alternatively, altered expression of the channels localizing on the internal stores such as the membranes of the ER may be an adaptive response or may offer a survival advantage, such as resistance to apoptosis. In that respect, the decrease in urothelial TRPV1[19] and prostatic TRPM8[12,46] in intracellular stores in aggressive tumors probably reduces the Ca^{2+} release content and confers resistance to apoptosis.

17.3 TRP PROTEIN TRANSLATION FROM ALTERNATIVE START CODONS: AN UNKNOWN MECHANISM OF CANCER CELL-ENHANCED RESISTANCE AND SURVIVAL

There is not so much known so far about TRP protein translation from alternative start codons (ATGs) situated downstream of the first methionine codon of the predicted full-length sequence. These ATG codons are preceded by the Kozak consensus sequence (gcc)gccRccAUGG, where R is a purine (adenine or guanine), three bases upstream of the start codon (AUG), which is followed by another "G."[48] The Kozak consensus sequence occurs on eukaryotic mRNAs and plays a major role in the initiation of the translation process.[49] In vivo, this site is often not matched exactly on different mRNAs, and the amount of protein synthesized from a given mRNA is dependent on the strength of the Kozak sequence. There are examples in vivo of each type of Kozak consensus, and they probably evolved as yet another mechanism of gene regulation. Lmx1b is an example of a gene with a weak Kozak consensus sequence.[50] For initiation of translation from such a site, other features are required in the mRNA sequence in order for the ribosome to recognize the initiation codon. It seems likely that the above mechanisms may be engaged by the cancer cell to enhance its survival and resistance to apoptosis. It may also be an alternative mechanism to increase the cancer cell plasticity in the accelerated process of evolution and adaptation to the environment.

The cloning and expression pattern of bCCE 1Δ514, a 5′ truncated splice variant of the bovine TRPC4 (formally bCCE1, as supported by the evidence that it functioned as a capacitative calcium entry channel), provided the first example of native TRP protein expression using a downstream ATG codon,[51] although it was shown to represent an alternatively spliced product of the TRPC4 gene that gives rise to an ~1.9-kb transcript rather than protein translation from an alternative ATG. It is interesting that in contrast to the six transmembrane segments predicted to be present in the full-length bTRPC4, bCCE 1Δ514 contains only three hydrophobic segments corresponding to transmembrane segments 5 and 6 and the putative pore-forming region in between. This membrane topology is reminiscent to that of the inward rectifier-type K$^+$ channel (K$_{ir}$) gene family, which also encodes proteins of less than 500 amino acids with two putative transmembrane spanning domains M1 and M2 and a hydrophobic segment in between contributing to ion conducting pore formation.[52] Accordingly, the ion conducting properties of the bTRPC4 protein might still be preserved in bCCE 1Δ514, although with different regulation mechanisms arising from the lack of the greater part of the N-terminus of bTRPC4.

By utilizing recently developed full-length cDNA technologies, large-scale cDNA sequencing is carried out by several cDNA projects. Now full-length cDNA resources cover the major part of the protein-coding human genes. Comprehensive analyses of the collected full-length cDNA data reveal not only the complete sequences of thousands of novel gene transcripts but also novel alternatively spliced isoforms of hitherto identified genes. However, it is not as easy as expected to deduce their encoded amino acid sequences based solely on the full-length cDNA sequences. It is neither always the case that the longest open reading frame corresponds to the real protein coding region nor that the first ATG should be the translation initiation codon. Also, proteome-wide mass spectrometry analysis has shown that there is an unexpectedly large population of small proteins, encoded by so-called upstream open reading frames, within the cell (for review see Ref. 53). Figure 17.2 shows the methodological approach to study the expression of the TRPV6 channel from the cloned cDNA. The TRPV6 molecule contains several ATG codons preceded by the functional consensus Kozak sequences theoretically capable of producing different size proteins. As

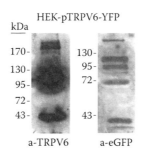

FIGURE 17.2 TRPV6-YFP protein expression from different ATGs. HEK293 cells were transfected with the pTRPV6-YFP plasmid, the total protein lysates were subject to SDS-PAGE, blotted, and revealed with either anti-TRPV6 or anti-GFP antibodies.

can be seen from the blots, at least five alternatively translated protein molecules may be produced. Site-directed mutagenesis should be used thereafter to prove that putative protein molecules result from proper alternative ATG codons. This method may also be used to verify the specificity of the antibody. Some other examples follow hereafter.

Sequence analysis of various human papillomavirus types associated with particular clinical outcomes has revealed that L1 protein sequences of the major cervical cancer-associated viruses generally possess the ability to encode a longer translation product whilst the non-cancer-causing viruses do not.[54] Equally intriguing, the upstream initiation codon is always separated by 78 nucleotides from the initiation codon that produces L1 protein, which efficiently assembles into viral particles. The authors conclude that the longer L1 protein could play a role in the development of cervical carcinoma and that human papillomaviruses with the potential to cause cervical cancer may be identified by the presence of an in-frame ATG situated 78 nucleotides upstream.

A novel form of the E2F-3 protein, termed E2F-3B, has been identified.[55] In contrast to the full-length E2F-3, which is expressed only at the G1/S boundary, E2F-3B is detected throughout the cell cycle with peak levels in G0 where it is associated with Rb. Transfection and in vitro translation experiments demonstrated that a protein identical to E2F-3B in size and isoelectric point is produced from the E2F-3 mRNA via the use of an alternative translational start site. Owing to this alternative initiation codon, the E2F-3B is missing 101 N-terminal amino acids relative to the full-length E2F-3. This region includes a moderately conserved sequence of unknown function that is present only in the growth-promoting E2F family members, including E2F-1, E2F-2, and full-length E2F-3. These observations make E2F-3B the first example of an E2F gene giving rise to two different protein species and also suggest that E2F-3 and E2F-3B may have opposing roles in the cell cycle control.[55]

Therefore, the alternative initiation of translation may represent an important evolutionary mechanism for a cancer cell to survive, to escape apoptosis, and to invade the body. TRP channels represent a potential "cancer target" because they are involved in all of them and therefore are subject to additional variability and regulation.

17.4 TRP PROTEIN POSTTRANSLATIONAL MODIFICATIONS: DIRECT OR INDIRECT MODULATION OF THE CHANNEL ACTIVITY

Protein posttranslational modifications are the chemical modifications of a protein after its translation. These modifications are largely used in the cell to regulate not only the protein activity but also its trafficking and stability at the plasma membrane in the case of the channel. Several types of posttranslational modifications exist, including the addition of functional groups such as acetylation, glycosylation, and phosphorylation; the addition of other proteins and peptides such as SUMOylation and ubiquitination; modifications involving changing the chemical nature of amino acids such as deamidation and deimination, as well as modifications involving structural changes such as disulfide bridges, proteolytic cleavage, etc. Modification and/ or misregulation of posttranslational modifications may play a significant role in carcinogenesis and therefore should be considered in the case of TRP channels and cancer.

Nothing is known so far as to whether particular posttranslational modifications are explicitly used by the cancer cell for its needs. However, some data exist as to the posttranslational changes the channels are subjected to in cancer. In the case of Morris hepatoma H5123 cells, the cell-to-cell channel protein, connexin-43 (Cx43), is little expressed, and these cells lack gap junctional communication.[56] The authors found that the inhibition of glycosylation by tunicamycin induced open channels in these cells. Although tunicamycin caused the formation of open channels, channels were not found aggregated into gap junctional plaques, as they are when they have been induced by elevation of intracellular cAMP. The results suggest that although Cx43 itself is not glycosylated, other glycosylated proteins influence Cx43 posttranslational modification and the formation of Cx43 cell-to-cell channels.[56] Below, we will consider the two most frequent posttranslational modifications, namely, glycosylation and phosphorylation.

17.4.1 GLYCOSYLATION TO STABILIZE CHANNEL EXPRESSION AT THE PLASMA MEMBRANE

N-linked glycosylation is considered to be important for channel trafficking to the plasma membrane. For instance, in human two-pore domain K$^+$ channel TRESK subunits, one or two N-glycosylation consensus sites were identified.[57] Using site-directed mutagenesis and Western immunoblotting, a single residue of both orthologues was found to be glycosylated upon heterologous expression. Two-electrode voltage-clamp recordings from *Xenopus* oocytes revealed that current amplitudes of N-glycosylation mutants were reduced by 80% as compared to the wild-type TRESK, so that their lower current amplitudes substantially result from inadequate, very low surface expression of the channel.[57]

Glycosylation is the covalent addition of sugar, or saccharide, moieties to a macromolecule via enzymatic action; glycation, in contrast, is its nonenzymatic counterpart. Mammalian cells can call upon a repertoire of nine distinct monosaccharides to enzymatically modify proteins and lipids.[58] N-linked oligosaccharides serve many functions: they promote folding of glycoproteins; help target proteins reach the correct cellular compartments; contribute to protein–protein interactions and other ligand recognition processes; stabilize proteins against denaturation and proteases; increase protein solubility; facilitate proper orientation in membranes; confer structural rigidity; influence protein turnover; and modify the charge and isoelectric point of proteins.[58] Among membrane-associated glycoproteins in epithelial cells, a role for glycosylation in intracellular trafficking has received the most attention. The plasma membrane is comprised of two functional compartments: the apical membrane compartment, facing the lumen, and the basolateral membrane compartment, generally apposed to a capillary. Targeting of cell surface proteins (e.g., ion channels) to the basolateral membrane is usually mediated via specific sorting motifs encoded within the amino acid sequence itself; generally, these motifs are in cytosolic regions of the protein.

Posttranslational modification of parts of the cardiac L-type Ca^{2+} channel by N-glycosylation is an important determinant for the binding of the dihydropyridine type of antagonists to Ca^{2+} channel $\alpha 1$ subunit, which itself is not glycosylated.[59] The results suggest a participation of N-glycosylation in subunit assembling of the functional channel and/or its turnover. However, a possible effect of tunicamycin on the expression of the Ca^{2+} channel as an alternative mechanism cannot be excluded.

Cystic fibrosis transmembrane conductance regulator (CFTR) is a cAMP-regulated Cl^- channel. Malfunction of CFTR causes cystic fibrosis (CF).[60] CFTR belongs to the ATP-binding cassette transporter superfamily, which includes P-glycoprotein (Pgp), the molecule that is responsible for multidrug resistance in cancer cells. It has been suggested that the membrane targeting and insertion of CFTR and Pgp may take the same pathway, i.e., the signal recognition particle (SRP) dependent pathway, but the membrane folding mechanism of these two proteins in microsomal membranes is probably different.[60]

The best studied example of TRP channel N-linked glycosylation, and the most striking example of its potential physiological impact, began with a gene called *klotho*. Klotho is highly expressed in the distal convoluted tubule of the kidney and in the hormone-secreting cells of the parathyroid gland;[61] both are sites of active regulation of systemic Ca^{2+} balance. Like Klotho, the Ca^{2+} transporting channel TRPV5 is expressed in the distal convoluted tubule of the kidney where it is involved in Ca^{2+} resorption from the glomerular filtrate.[62] Chang et al. found that Klotho cleaves an N-linked oligosaccharide from TRPV5, thereby trapping the channel in the apical plasma membrane.[63] This deglycosylation of TRPV5 is accomplished by an N-terminal ectodomain of the membrane-associated Klotho enzyme that is itself cleaved and liberated into the urinary space, and into the plasma and cerebrospinal fluid.[63] This model of regulation by compartmentalization is well suited to a channel such as TRPV5, which exhibits constitutive activity in heterologous expression systems. Whereas the channel may remain perpetually active, it is only physiologically active (that is, reabsorbing urinary Ca^{2+}) when it is confined to the tubular apical membrane. Thus, the above regulation is crucial for the proper function of TRPV5. The very similar channel TRPV6, which possesses 75% of amino acid identity to TRPV5, has already been shown to be implicated in some cancers.[21,25]

The alignment of TRPV5 with all members of the TRPC family was performed to uncover other analogous glycosylation motifs within the first extracellular loop. Of note, for this alignment, only the membrane-spanning regions were subjected to CLUSTAL analysis (http://www.ebi.ac.uk/clustalw/#).[64] Most members of the TRPC family, in contrast, have eight hydrophobic regions. Vannier et al., investigating human TRPC3, surmised that the first hydrophobic region was not a transmembrane segment,[65] and this is supported by the work of Dietrich et al.[66] Dohke et al., analyzing the closely related TRPC1 channel, concluded that the third hydrophobic region may not be a true transmembrane segment.[67] Accurate assignment of the first transmembrane segment is essential for investigating potential glycosylation motifs in the first extracellular loop. For our protein sequence alignment, we excluded the first hydrophobic domain from the TRPC family members and included only domains two through eight; this resulted in alignment of all of the known "e1" glycosylation motifs (i.e., TRPV5, TRPC3, and TRPC6) and uncovered a similar site in TRPC7. The previously unrecognized TRPC7 potential glycosylation site is evolutionarily conserved, and although there are no direct experimental data, migration of the heterologously expressed channel is consistent with glycosylation (e.g., see Ref. 68). In the case of TRPC6, the extensive glycosylation has also been shown.[26,69] The authors used peptide N-glycosidase F (PNGase) to treat the protein sample to deglycosylate the TRPC6 channel and to show that its real size corresponds to the predicted one of 106 kDa.

Glycosylation has been reported to influence trafficking and/or function of a variety of voltage-gated ion channels. This protein superfamily includes the focus of this review, the TRP channels, as well as the two-pore segment (TPC), voltage-gated sodium (Na_V), voltage-gated calcium (Ca_V), hyperpolarization-activated cyclic nucleotide gated (HCN), cyclic nucleotide-gated (CNG), voltage-gated potassium (K_V), two-pore potassium (K_{2p}), calcium-activated potassium (K_{Ca}), and inwardly rectifying potassium (K_{ir}) channel families.[70] Of these, the architecture of the K_V, HCN, CNG, and K_{Ca} channels most closely resembles that of the TRP channels, with six membrane-spanning domains and a pore-forming loop between helices five and six. Interestingly, the most abundant evidence for a functional role of glycosylation comes from this TRP-like subgroup. Two members—HCN2 and the Human Ether-a-go-go Related Gene (HERG) potassium channel—share N-linked glycosylation sites, adjacent to the pore-forming loop, that influence membrane trafficking and are potentially analogous to those of TRPV1 and TRPV4. In HERG channels (also known as $K_{V11.1}$), mutation of this glycosylation site disrupts targeting.[71,72] A human mutation in this channel gene confers a heritable variant of the "long QT syndrome," which is associated with potentially lethal cardiac dysrhythmias.[73] In HCN2, a channel mutated for the putative glycosylation site similarly fails to traffic to the plasma membrane.[74] TRPV4 appears to be unique among this group in that membrane trafficking is down-regulated rather than facilitated by glycosylation.[75]

At least a subset of TRP channel proteins undergo regulatory N-linked glycosylation. The function and/or subcellular localization of TRP channels are influenced by glycosylation, and other members of the TRP family share the motif. It is likely that N-linked glycosylation, and the dynamic regulation of this process, will play major roles in the function and targeting of a wide range of TRP and closely related ion channels.

17.4.2   PHOSPHORYLATION OF TRP CHANNELS TO
DIRECTLY MODULATE CHANNEL ACTIVITY

Protein phosphorylation and dephosphorylation are common, reversible, posttranslational modifications that can regulate the structure and function of ion channels.
A particular phosphorylation/dephosphorylation state can modify channel activity and thus alter the electrophysiological properties of excitable and nonexcitable
cells. A few well-known examples include protein kinase G (PKG) regulation of
large conductance Ca^{2+}-dependent K^+ channels (BK_{Ca}) and the regulation of NMDA
receptors by tyrosine phosphorylation. The BK_{Ca} channel is composed of four
α-subunits that form the pore and a regulatory β-subunit; the channel is regulated
by voltage in a Ca^{2+}-dependent manner. PKG phosphorylates the α-subunit at Ser-
1072 near the C-terminus, shifting the Ca^{2+} sensitivity of the channel and producing
hyperpolarization.[76]

   In the past few years, rapid development in the field of TRP channel research has
demonstrated important roles for protein phosphorylation in the regulation of TRP channels, particularly for members of the TRPC, TRPV, TRPM, and TRPP subfamilies.

   The TRPC subfamily contains seven members, which can be further divided into
four subgroups: TRPC1, TRPC2, TRPC4,5, and TRPC3,6,7.[77] TRPC1 needs to form
heteromultimeric complexes with TRPC4 or TRPC5 for its proper trafficking to the
plasma membrane in order to form functional channels.[78] Zhang and Saffen provided
the first evidence that TRPC6 activity is negatively regulated by PKC. They found
that TRPC6, when overexpressed in CHO cells, was inhibited by PKC-activating
phorbol 12-myristate 13-acetate (PMA).[26] Recently, Trebak et al. identified Ser-712
in the TRPC3 amino acid sequences to be a specific PKC phosphorylation site.[79] A
point mutation at this site abolished the PKC phosphorylation on TRPC3 proteins and
also markedly reduced the inhibitory effect of PKC activation on TRPC3-mediated
Ca^{2+} influx. Another study by Zhu et al. found that PKC phosphorylates Thr-972 of
mouse TRPC5, causing channel desensitization.[80] TRPC6 and 7 are also desensi-
tized by PKC.[81] Channel desensitization is expected to cause an overall reduction in
Ca^{2+} influx. However, several lines of evidence suggest that different mechanisms
may govern the channel desensitization observed by Zhu et al.[80]

   PKG is another kinase capable of inhibiting TRPC3 activity.[82] Disruption of two
consensus PKG phosphorylation sites, Thr-11 and Ser-263, markedly reduces the
inhibitory effect of cGMP. These data indicate that PKG phosphorylates TRPC3
at Thr-11 and Ser-263 and, as a consequence, inactivates TRPC3.[82] The inhibitory
action of PKC and PKG on TRPC may represent important negative feedback mechanisms in the control of cytosolic Ca^{2+} levels, thereby influencing Ca^{2+}-dependent
processes in a variety of different cell types. In these negative feedback pathways,
the activation of TRPC results in Ca^{2+} entry; a rise in cytosolic Ca^{2+}, together with
elevated diacylglycerol (DAG) levels, stimulates PKC activity, which feeds back to
inactivate the TRPC channels.[83]

   It has been known for a long time that tyrosine kinases are involved in the activation of capacitative Ca^{2+} influx, of which TRPC channels are among the major
molecular candidates.[84] Vazquez et al. found that inhibition of Src kinases by
genistein and erbstatin abolished the receptor- and OAG (a DAG analog)-induced

activation of TRPC3.[85] In addition, OAG failed to activate TRPC3 in cells that were either Src-deficient or expressed a dominant-negative mutant of Src, and furthermore, OAG activation of TRPC3 was restored after the cells were transfected with a Src-expressing construct. These results indicate an obligatory requirement for Src kinase in DAG-induced activation of TRPC3. Note that Src may not directly act on TRPC3. Instead, a concerted role for both DAG and Src seems to be necessary for TRPC3 activation, perhaps through a mechanism involving Src-dependent phosphorylation and/or recruitment of a yet unknown accessory/regulatory protein within the vicinity of TRPC3.[85]

CaM-kinase II can activate TRPC6. In patch-clamp studies, Shi et al. found that TRPC6, expressed in HEK293 cells, was activated by extracellular Ca^{2+}, which could be prevented by either the organic CaM-kinase II inhibitor KN-62 or a CaM-kinase II-specific inhibitory peptide.[81] These results suggest that CaM-kinase II-mediated phosphorylation is an obligatory step for TRPC6 channel activation.

The TRPV subfamily contains six members (TRPV1–6). TRPV1–4 channels are temperature-sensitive. TRPV5 and TRPV6 are only distinctly related to TRPV1–4 with a 30–40% sequence homology. TRPV5 and TRPV6 have high selectivity for Ca^{2+} over Na^+ ($P_{Ca}/P_{Na} = 100/1$), are mainly expressed in Ca^{2+}-transporting epithelia, and are assumed to play an important role in Ca^{2+} (re)absorption by the kidney and intestine.[86]

Multiple kinases are known to regulate TRPV1. PKC phosphorylates Ser-502 and Ser-800 in rat TRPV1 and, as a result, either potentiates or sensitizes the responses of this channel to capsaisin, heat, and anandamide.[87,88] TRPV2 and TRPV4 are two temperature-sensitive channels with activation thresholds of $\geq$53 and $\geq$25°–27°C, respectively.[89] TRPV2 is a substrate of PKA. In mast cells, PKA interacts with TRPV2 through a PKA-binding protein named ACBD3.[90] PKA phosphorylation enhances TRPV2-mediated Ca^{2+} influx in response to heat.[90] On the other hand, the activity of TRPV4 is stimulated by PKC,[91] although it is unclear whether this stimulation is due to direct PKC phosphorylation on the TRPV4 proteins.

TRPV5 is activated by serum and glucocorticoid-inducible kinase, SGK1. This stimulatory effect is due to enhanced TRPV5 abundance in the plasma membrane, requiring the presence of the scaffold protein, NHERF2.[92] On the other hand, the activity of TRPV6 can be regulated by calmodulin and PKC. Binding of Ca^{2+}-dependent calmodulin to TRPV6 inactivates the channel, which is countered by PKC-mediated phosphorylation of TRPV6.[93] Thus, by altering the inactivation behavior of TRPV6, PKC-mediated phosphorylation acts as a switch to regulate the amount of Ca^{2+} influx through TRPV6.[94] TRPV6 can also be activated by the Src tyrosine kinase, which is counterbalanced by the protein tyrosine phosphatase 1B.[94] Taken together, TRPV6 activity is closely controlled by both the calmodulin/PKC system and the tyrosine kinase/phosphatase system.[93,94]

The TRPM subfamily consists of eight members (TRPM1–8). TRPM4 is a voltage-dependent, Ca^{2+}-impermeable cation channel. Opening of this channel depolarizes the membrane. While the channel is activated by intracellular Ca^{2+}, the currents decay rapidly due to decreased sensitivity of the channel to Ca^{2+}. PMA, an activator of PKC, increases the activity of TRPM4[95] by enhancing the sensitivity of TRPM4 to Ca^{2+}.[96]

The physiological significance of the regulation of TRP channels by phosphorylation is fascinating. TRP channels play diverse functional roles, including thermal sensation, nociception, mechanosensing, growth cone guidance, inflammatory responses, membrane potential control, and Mg^{2+} homeostasis. Various kinases and phosphatases may regulate the activities of different TRP channel isoforms, providing enormous control on diverse cellular processes.

17.5 CONCLUSIONS

The progression of cells from a normal, differentiated state to a tumorigenic, metastatic state involves the accumulation of mutations in multiple key signaling proteins, encoded by oncogenes and tumor suppressor genes, together with the evolution and clonal selection of more aggressive cell phenotypes. These events are associated with changes in the expression of numerous other proteins. To date, most changes involving TRP proteins do not involve mutations in the TRP genes but rather increased or decreased expression levels of the wild-type TRP proteins, depending on the stage of the cancer. On the other hand, several common tuning pathways lead to this divergence in expression. In this respect, TRP channels may be regulated at different levels: (1) transcriptional, (2) translational, (3) trafficking to the plasma membrane, or (4) direct channel modulation on the plasma membrane. Modulation of TRP expression/activity on one of these levels affects intracellular Ca^{2+} signaling and, consequently, the processes involved in carcinogenesis, such as proliferation, apoptosis, and migration.

REFERENCES

1. Prevarskaya, N., R. Skryma, and Y. Shuba. 2004. Ca^{2+} homeostasis in apoptotic resistance of prostate cancer cells. *Biochemical and Biophysical Research Communications* 322: 1326–1335.
2. Rosado, J. A., P. C. Redondo, J. A. Pariente, and G. M. Salido. 2004. Calcium signalling and tumorigenesis. *Cancer Therapy* 2: 263–270.
3. Weinberg, R. A. 2006. Multi-step tumorigenesis in the biology of cancer. In *The Biology of Cancer*, 399–461. New York: Garland Science.
4. Roderick, H. L., and S. J. Cook. 2008. Ca^{2+} signalling checkpoints in cancer: remodelling Ca^{2+} for cancer cell proliferation and survival. *Nature Reviews. Cancer* 8: 361–375.
5. Cook, S. J., and P. J. Lockyer. 2006. Recent advances in Ca^{2+}-dependent Ras regulation and cell proliferation. *Cell Calcium* 39: 101–112.
6. Whitfield, J. F. 1992. Calcium signals and cancer. *Critical Reviews in Oncogenesis* 3: 55–90.
7. Berridge, M. J., M. D. Bootman, and H. L. Roderick. 2003. Calcium signalling: dynamics, homeostasis and remodelling. *Nature Reviews. Molecular Cell Biology* 4: 517–529.
8. Rizzuto, R., P. Pinton, D. Ferrari et al. 2003. Calcium and apoptosis: facts and hypotheses. *Oncogene* 22: 8619–8627.
9. Monteith, G. R., D. McAndrew, H. M. Faddy, and S. J. Roberts–Thomson. 2007. Calcium and cancer: targeting Ca^{2+} transport. *Nature Reviews. Cancer* 7: 519–530.
10. Nilius, B., G. Owsianik, T. Voets, and J. A. Peters. 2007. Transient receptor potential cation channels in disease. *Physiological Reviews* 87: 165–217.

11. Bidaux, G., M. Flourakis, S. Thebault et al. 2007. Prostate cell differentiation status determines transient receptor potential melastatin member 8 channel subcellular localization and function. *Journal of Clinical Investigation* 117: 1647–1657.

12. Thebault, S., L. Lemonnier, G. Bidaux et al. 2005. Novel role of cold/menthol-sensitive transient receptor potential melastatine family member 8 (TRPM8) in the activation of store-operated channels in LNCaP human prostate cancer epithelial cells. *Journal of Biological Chemistry* 280: 39423–39435.

13. Xu, X. Z., F. Moebius, D. L. Gill, and C. Montell. 2001. Regulation of melastatin, a TRP-related protein, through interaction with a cytoplasmic isoform. *Proceedings of the National Academy of Science of the United States of America* 98: 10692–10697.

14. Montell, C., L. Birnbaumer, V. Flockerzi et al. 2002. A unified nomenclature for the superfamily of TRP cation channels. *Molecular Cell* 9: 229–231.

15. Thebault, S., M. Flourakis, K. Vanoverberghe et al. 2006. Differential role of transient receptor potential channels in Ca^{2+} entry and proliferation of prostate cancer epithelial cells. *Cancer Research* 66: 2038–2047.

16. Tsavaler, L., M. H. Shapero, S. Morkowski, and R. Laus. 2001. TRP-p8, a novel prostate-specific gene, is up-regulated in prostate cancer and other malignancies and shares high homology with transient receptor potential calcium channel proteins. *Cancer Research* 61: 3760–3769.

17. Wissenbach, U., B. A. Niemeyer, T. Fixemer et al. 2001. Expression of CaT-like, a novel calcium-selective channel, correlates with the malignancy of prostate cancer. *Journal of Biological Chemistry* 276: 19461–19468.

18. Duncan, L. M., J. Deeds, J. Hunter et al. 1998. Down-regulation of the novel gene melastatin correlates with potential for melanoma metastasis. *Cancer Research* 58: 1515–1520.

19. Lazzeri, M., M. G. Vannucchi, M. E. Spinelli et al. 2005. Transient receptor potential vanilloid type 1 (TRPV1) expression changes from normal urothelium to transitional cell carcinoma of human bladder. *European Urology* 48: 691–698.

20. Caprodossi, S., R. Lucciarini, C. Amantini et al. 2007. Transient receptor potential vanilloid type 2 (TRPV2) expression in normal urothelium and in urothelial carcinoma of human bladder: correlation with the pathologic stage. *European Urology* 72: 440–448.

21. Fixemer, T., U. Wissenbach, V. Flockerzi, and H. Bonkhoff. 2003. Expression of the Ca^{2+}-selective cation channel TRPV6 in human prostate cancer: a novel prognostic marker for tumor progression. *Oncogene* 22: 7858–7861.

22. Prawitt, D., T. Enklaar, G. Klemm et al. 2000. Identification and characterization of MTR1, a novel gene with homology to melastatin (MLSN1) and the TRP gene family located in the BWS-WT2 critical region on chromosome 11p15.5 and showing allele-specific expression. *Human Molecular Genetics* 9: 203–216.

23. Fang, D., and V. Setaluri. 2000. Expression and up-regulation of alternatively spliced transcripts of melastatin, a melanoma metastasis-related gene, in human melanoma cells. *Biochemical and Biophysical Research Communications* 279: 53–61.

24. Fuessel, S., D. Sickert, and A. Meye. 2003. Multiple tumor marker analyses (PSA, hK2, PSCA, TRP-p8) in primary prostate cancers using quantitative RT-PCR. *International Journal of Oncology* 23: 221–228.

25. Lehen'kyi, V., M. Flourakis, R. Skrym, and N. Prevarskaya. 2007. TRPV6 channel controls prostate cancer cell proliferation via Ca^{2+}/NFAT-dependent pathways. *Oncogene* 26: 7380–7385.

26. Zhang, L., and D. Saffen. 2001. Muscarinic acetylcholine receptor regulation of TRP6 Ca^{2+} channel isoforms. Molecular structures and functional characterization. *Journal of Biological Chemistry* 276: 13331–13339.

27. Monet, M., D. Gkika, and V. Lehen'kyi. 2009. Lysophospholipids stimulate prostate cancer cell migration via TRPV2 channel activation. *Biochimica et Biophysica Acta* 1793: 528–539.

28. Bolanz, K. A., M. A. Hediger, and C. P. Landowski. 2008. The role of TRPV6 in breast carcinogenesis. *Molecular Cancer Therapeutics* 7: 271–279.

29. Zhuang, L., J. B. Peng, L. Tou et al. 2002. Calcium-selective ion channel, CaT1, is apically localized in gastrointestinal tract epithelia and is aberrantly expressed in human malignancies. *Lab Investigations* 82: 1755–1764.

30. Peng, J. B., L. Zhuang, U. V. Berger et al. 2001. CaT1 expression correlates with tumor grade in prostate cancer. *Biochemical and Biophysical Research Communications* 282: 729–734.

31. Peng, J. B., X. Z. Chen, U. V. Berger et al. 1999. Molecular cloning and characterization of a channel-like transporter mediating intestinal calcium absorption. *Journal of Biological Chemistry* 274: 22739–22746.

32. Ligresti, A., A. S. Moriello, K. Starowicz et al. 2006. Antitumour activity of plant cannabinoids with emphasis on the effect of cannabidiol on human breast carcinoma. *Journal of Pharmacology and Experimental Therapeutics* 318: 1375–1387.

33. Izzo, A. A., and A. A. Coutts. 2005. Cannabinoids and the digestive tract. *Handbook of Experimental Pharmacology* 168: 573–598.

34. Wertz, I. E., and V. M. Dixit. 2000. Characterization of calcium release-activated apoptosis of LNCaP prostate cancer cells. *Journal of Biological Chemistry* 275: 11470–11478.

35. Costa-Pereira, A. P., and T. G. Cotter. 1999. Molecular and cellular biology of prostate cancer—the role of apoptosis as a target for therapy. *Prostate Cancer Prostatic Diseases* 2: 126–139.

36. Beck, B., G. Bidaux, A. Bavencoffe et al. 2007. Prospects for prostate cancer imaging and therapy using high-affinity TRPM8 activators. *Cell Calcium* 41: 285–294.

37. Clapham, D. E., L. W. Runnels, and C. Strubing. 2001. The TRP ion channel family. *Nature Reviews. Neuroscience* 2: 387–396.

38. Stamm, S., I. Ben-Ari, Y. Rafalska et al. 2005. Function of alternative splicing. *Gene* 344: 1–20.

39. Nagasawa, M., Y. Nakagawa, S. Tanaka, and I. Kojima. 2007. Chemotactic peptide fMetLeuPhe induces translocation of the TRPV2 channel in macrophages. *Journal of Cell Physiology* 210: 692–702.

40. Wang, C., H. Z. Hu, C. K. Colton, J. D. Wood, and M. X. Zhu. 2004. An alternative splicing product of the murine TRPV1 gene dominant negatively modulates the activity of TRPV1 channels. *Journal of Biological Chemistry* 279: 37423–37430.

41. Zhiqi, S., M. H. Soltani, K. M. Bhat et al. 2004. Human melastatin 1 (TRPM1) is regulated by MITF and produces multiple polypeptide isoforms in melanocytes and melanoma. *Melanoma Research* 14: 509–516.

42. Sabnis, A. S., M. Shadid, G. S. Yost, and C. A. Reilly. 2008. Human lung epithelial cells express a functional cold-sensing TRPM8 variant. *American Journal of Respiratory Cell and Molecular Biology* 39: 466–474.

43. Lis, A., U. Wissenbach, and S. E. Philipp. 2005. Transcriptional regulation and processing increase the functional variability of TRPM channels. *Naunyn-Schmiedeberg's Archives of Pharmacology* 371: 315–324.

44. Mahieu, F., G. Owsianik, L. Verbert et al. 2007. TRPM8-independent menthol-induced Ca^{2+} release from endoplasmic reticulum and Golgi. *Journal of Biological Chemistry* 282: 3325–3336.

45. Kim, S. H., J. H. Nam, E. J. Park et al. 2008. Menthol regulates TRPM8-independent processes in PC-3 prostate cancer cells. *Biochimica et Biophysica Acta* 1792: 33–38.

46. Zhang, L., and G. J. Barritt. 2004. Evidence that TRPM8 is an androgen-dependent Ca²⁺ channel required for the survival of prostate cancer cells. *Cancer Research* 64: 8365–8373.

47. Harteneck, C., T. D. Plant, and G. Schultz. 2000. From worm to man: three subfamilies of TRP channels. *Trends in Neurosciences* 23: 159–166.

48. Kozak, M. 1987. An analysis of 5′-noncoding sequences from 699 vertebrate messenger RNAs. *Nucleic Acids Research* 15: 8125–8148.

49. De Angioletti, M., G. Lacerra, V. Sabato, and C. Carestia. 2004. Beta+45 G --> C: a novel silent beta-thalassaemia mutation, the first in the Kozak sequence. *British Journal of Haematology* 124, 224–231.

50. Dunston, J. A., J. D. Hamlington, J. Zaveri et al. 2004. The human LMX1B gene: transcription unit, promoter, and pathogenic mutations. *Genomics* 84: 565–576.

51. Freichel, M., U. Wissenbach, S. Philipp, and V. Flockerzi. 1998. Alternative splicing and tissue specific expression of the 5′ truncated bCCE 1 variant bCCE 1delta514. *FEBS Letters* 422: 354–358.

52. Doupnik, C. A., N. Davidson, and H. A. Lester. 1995. The inward rectifier potassium channel family. *Current Opinion in Neurobiology* 5: 268–277.

53. Suzuki, Y., and S. Sugano. 2006. Transcriptome analyses of human genes and applications for proteome analyses. *Current Protein and Peptide Science* 7: 147–163.

54. Webb, E., J. Cox, and S. Edwards. 2005. Cervical cancer-causing human papillomaviruses have an alternative initiation site for the L1 protein. *Virus Genes* 30: 31–35.

55. He, Y., M. K. Armanious, M. J. Thomas, and W. D. Cress. 2000. Identification of E2F-3B, an alternative form of E2F-3 lacking a conserved N-terminal region. *Oncogene* 19: 3422–3433.

56. Wang, Y., P. P. Mehta, and B. Rose. 1995. Inhibition of glycosylation induces formation of open connexin-43 cell-to-cell channels and phosphorylation and triton X-100 insolubility of connexin-43. *Journal of Biological Chemistry* 270: 26581–26585.

57. Egenberger, B., G. Polleichtner, E. Wischmeyer, and F. Döring. 2009. N-linked glycosylation determines cell surface expression of two-pore-domain K⁺ channel TRESK. *Biochemical and Biophysical Research Communications* 391: 1262–1267.

58. Ohtsubo, K., and J. D. Marth. 2006. Glycosylation in cellular mechanisms of health and disease. *Cell* 126: 855–867.

59. Henning, U., W. P. Wolf, and M. Holtzhauer. 1996. Influence of glycosylation inhibitors on dihydropyridine binding to cardiac cells. *Molecular Cell Biochemistry* 160–161: 41–46.

60. Chen, M., and J. T. Zhang. 1996. Membrane insertion, processing, and topology of cystic fibrosis transmembrane conductance regulator (CFTR) in microsomal membranes. *Molecular Membrane Biology* 13: 33–40.

61. Kuro-o, M., Y. Matsumura, H. Aizawa et al. 1997. Mutation of the mouse klotho gene leads to a syndrome resembling ageing. *Nature* 390: 45–51.

62. Hoenderop, J. G., A. W. van der Kemp, A. Harto et al. 1999. Molecular identification of the apical Ca²⁺ channel in 1, 25-dihydroxyvitamin D3-responsive epithelia. *Journal of Biological Chemistry* 274: 8375–8378.

63. Chang, Q., S. Hoefs, A. W. van der Kemp, C. N. Topala, R. J. Bindels, and J. G. Hoenderop. 2005. The beta-glucuronidase klotho hydrolyzes and activates the TRPV5 channel. *Science* 310: 490–493.

64. Thompson, J. D., D. G. Higgins, T. J. Gibson, and W. Clustal. 1994. Improving the sensitivity of progressive multiple sequence alignment through sequence weighting, position-specific gap penalties and weight matrix choice. *Nucleic Acids Research* 22: 4673–4680.

65. Vannier, B., M. X. Zhu, D. Brown, and L. Birnbaumer. 1998. The membrane topology of human transient receptor potential 3 as inferred from glycosylation-scanning mutagenesis and epitope immunocytochemistry. *Journal of Biological Chemistry* 273: 8675–8679.

66. Dietrich, A., M. Mederos y Schnitzler, J. Emmel, H. Kalwa, T. Hofmann, and T. Gudermann. 2003. N-linked protein glycosylation is a major determinant for basal TRPC3 and TRPC6 channel activity. *Journal of Biological Chemistry* 278: 47842–47852.

67. Dohke, Y., Y. S. Oh, I. S. Ambudkar, and R. J. Turner. 2004. Biogenesis and topology of the transient receptor potential Ca^{2+} channel TRPC1. *Journal of Biological Chemistry* 279: 12242–12248.

68. Maruyama, Y., Y. Nakanishi, E. J. Walsh, D. P. Wilson, D. G. Welsh, and W. C. Cole. 2006. Heteromultimeric TRPC6-TRPC7 channels contribute to arginine vasopressin-induced cation current of A7r5 vascular smooth muscle cells. *Circulation Research* 98, 1520–1527.

69. El Boustany, C., G. Bidaux, A. Enfissi, P. Delcourt, N. Prevarskaya, and T. Capiod. 2008. Capacitative calcium entry and transient receptor potential canonical 6 expression control human hepatoma cell proliferation. *Hepatology* 47: 2068–2077.

70. Yu, F. H., and W. A. Catterall. 2004. The VGL-chanome: a protein superfamily specialized for electrical signaling and ionic homeostasis. *Science STKE* 2004: re15.

71. Petrecca, K., R. Atanasiu, A. Akhavan, and A. Shrier. 1999. N-linked glycosylation sites determine HERG channel surface membrane expression. *Journal of Physiology* 515(Part 1): 41–48.

72. Gong, Q., C. L. Anderson, C. T. January, and Z. Zhou. 2002. Role of glycosylation in cell surface expression and stability of HERG potassium channels. *American Journal of Physiology. Heart and Circulatory Physiology* 283: H77–H84.

73. Satler, C. A., M. R. Vesely, P. Duggal, G. S. Ginsburg, and A. H. Beggs. 1998. Multiple different missense mutations in the pore region of HERG in patients with long QT syndrome. *Human Genetics* 102: 265–272.

74. Much, B., C. Wahl-Schott, X. Zong et al. 2003. Role of subunit heteromerization and N-linked glycosylation in the formation of functional hyperpolarization-activated cyclic nucleotide-gated channels. *Journal of Biological Chemistry* 278: 43781–43786.

75. Xu, H., Y. Fu, W. Tian, and D. M. Cohen. 2006. Glycosylation of the osmoresponsive transient receptor potential channel TRPV4 on Asn-651 influences membrane trafficking. *American Journal of Physiology. Renal Physiology* 290: F1103–F1109.

76. Fukao, M., H. S. Mason, F. C. Britton, J. L. Kenyon, B. Horowitz, and K. D. Keef. 1999. Cyclic GMP-dependent protein kinase activates cloned BKCa channels expressed in mammalian cells by direct phosphorylation at serine 1072. *Journal of Biological Chemistry* 274: 10927–10935.

77. Minke, B., and B. Cook. 2002. TRP channel proteins and signal transduction. *Physiological Reviews* 82: 429–472.

78. Strubing, C., G. Krapivinsky, L. Krapivinsky, and D. E. Clapham. 2001. TRPC1 and TRPC5 form a novel cation channel in mammalian brain. *Neuron* 29: 645–655.

79. Trebak, M., N. Hempel, B. J. Wedel, J. T. Smyth, G. S. J. Bird, J. W. Putney, Jr. 2005. Negative regulation of TRPC3 channels by protein kinase C-mediated phosphorylation of serine 712. *Molecular Pharmacology* 67: 558–563.

80. Zhu, M. H., M. Chae, H. J. Kim et al. 2005. Desensitization of canonical transient receptor potential channel 5 (TRPC5) by protein kinase C. *American Journal of Physiology* 289: C591–C600.

81. Shi, J., E. Mori, Y. Mori et al. 2004. Multiple regulation by calcium of murine homologues of transient receptor potential proteins TRPC6 and TRPC7 expressed in HEK293 cells. *Journal of Physiology* 561: 415–432.

82. Kwan, H.Y., Y. Huang, and X. Yao. 2004. Regulation of canonical transient receptor potential isoform 3 (TRPC3) channel by protein kinase G. Regulation of canonical transient receptor potential isoform 3 (TRPC3) channel by protein kinase G. *Proceedings of the National Academy of Sciences of the United States of America* 101: 2625–2630.

83. Venkatachalam, K., F. Zheng, and D. L. Gill. 2003. Regulation of canonical transient receptor potential channel function by diacylglycerol and protein kinase C. *Journal of Biological Chemistry* 278: 29031–29040.

84. Parekh, A. B., and J. W. Putney, Jr. 2005. Store depletion and calcium influx. *Physiological Reviews* 85: 757–810.

85. Vazquez, G., B. J. Wedel, B. T. Kawasaki, G. S. Bird, and J. W. Putney, Jr. 2004. Obligatory role of Src kinase in the signaling mechanism for TRPC3 cation channels. *Journal of Biological Chemistry* 279: 40521–40528.

86. Montell, C. 2005. The TRP superfamily of cation channels. The TRP superfamily of cation channels. *Science STKE* 272: re3.

87. Vellani, V., S. Mappleback, A. Moriondo, J. B. Davis, and P. A. McNaughton. 2001. Protein kinase C activation potentiates gating of the vanilloid receptor VR1 by capsaicin, protons, heat and anandamide. *Journal of Physiology* 534: 813–825.

88. Premkumar, L. S., and G. P. Ahern. 2000. Induction of vanilloid receptor channel activity by protein kinase C. *Nature* 408: 985–990.

89. Tominaga, M., and M. J. Caterina. 2004. Thermosensation and pain. *Journal of Neurobiology* 61: 3–12.

90. Stokes, A. J., L. M. Shimoda, M. Koblan-Huberson, C. N. Adra, and H. Turner. 2004. A TRPV2-PKA signalling module for transduction of physical stimuli in mast cells. *Journal of Experimental Medicine* 200: 137–147.

91. Xu, F., E. Satoh, and T. Iijima. 2003. Protein kinase C-mediated Ca^{2+} entry in HEK 293 cells transiently expressing human TRPV4. *British Journal of Pharmacology* 140: 413–421.

92. Embark, H. M., I. Setiawan, S. Poppendieck et al. 2004. Regulation of the epithelial Ca^{2+} channel TRPV5 by the NHE regulating factor NHERF2 and the serum and glucocorticoid inducible kinase isoforms SGK1 and SGK3 expressed in Xenopus oocytes. *Cellular Physiology and Biochemistry* 14: 203–212.

93. Niemeyer, B. A., C. Bergs, U. Wissenbach, V. Flockerzi, and C. Trost. 2001. Competitive regulation of CaT-like-mediated Ca^{2+} entry by protein kinase C and calmodulin. *Proceedings of the National Academy of Sciences of the United States of America* 98: 3600–3605.

94. Sternfeld, L., E. Krause, and A. Schmid. 2005. Tyrosine phosphatase PTP1B interacts with TRPV6 in vivo and plays a role in TRPV6-mediated calcium influx in HEK293 cells. *Cellular Signalling* 17: 951–960.

95. Guinamard, R., A. Chatelier, J. Lenfant, and P. Bios. 2004. Activation of the Ca^{2+}-activated nonselective cation channel by diacylglycerol analogues in rat cardiomyocytes. *Journal of Cardiovascular Electrophysiology* 15: 342–348.

96. Nilius, B., J. Prenen, J. Tang et al. 2005. Regulation of the Ca^{2+} sensitivity of the nonselective cation channel TRPM4. *Journal of Biological Chemistry* 280: 6423–6433.

18 In Vitro and in Vivo Assays for the Discovery of Analgesic Drugs Targeting TRP Channels

Jun Chen, Regina M. Reilly,
Philip R. Kym, and Shailen Joshi

CONTENTS

18.1 INTRODUCTION

Pain is the most common symptom of disease and the most frequent complaint presented to physicians. Current pain therapies, mostly nonsteroidal, anti-inflammatory drugs (NSAIDs) and opioids, often have limited efficacy and produce severe gastrointestinal and cardiovascular side effects. A major recent advancement in pain research is the identification of novel analgesic targets including several TRP channels. These channels (TRPV1-4, TRPM8, and TRPA1) are expressed in primary sensory neurons and involved in sensing temperature, mechanical, and chemical stimuli.[1] Among them, TRPV1 is the most extensively studied and has been the subject of extensive reviews.[1-3] As a polymodal receptor, TRPV1 responds to various noxious stimuli, including noxious heat, (>43°C), acidic pH, noxious chemicals (e.g., capsaicin) and endogenous lipid products.[2] Tissue injury and inflammation increase TRPV1 expression and function, causing heightened pain sensitivity. Reduction of TRPV1 function, as either a result of gene knockout or pharmacological blockade, affects heat sensation and produces analgesia in various animal models. It is estimated that >55 pharmaceutical companies have pursued TRPV1 modulators as therapeutics, with a combined investment of >$1 billion.[4] These efforts have identified numerous TRPV1 modulators, with several compounds currently in clinical trials.

TRPA1 is another validated target for pain. It colocalizes with TRPV1 and markers of peptidergic nociceptors.[5] The expression of TRPA1 is increased in inflammatory and neuropathic pain models, as well as in avulsion-injured human dorsal root ganglion (DRG) neurons.[6] TRPA1 is activated by noxious cold, intracellular Ca^{2+}, hypertonic solutions, and, most prominently, by numerous electrophilic compounds, including active ingredients of pungent natural products (e.g., allyl isothiocyanate or AITC), environmental irritants (acrolein), and endogenous molecules involved in pain, oxidative stress, and inflammation (e.g., 4-hydroxynonenal or 4-HNE, 15-d-PGJ2, H_2O_2).[7,8] TRPA1 knockout mice exhibit deficits in pain hypersensitivity to agonists and bradykinin; also, treatment with antisense oligodeoxynucleotides and small molecular antagonists attenuate pain in inflamed and neuropathic rats.[9-11] Collectively, these studies support the role of TRPA1 in sensory function and its utility as a pain target.

Although capsaicin, the active ingredient in chili pepper, has been used for thousands of years for treating ulcers, backache, cough, and other maladies, modern drug discovery targeting TRP channels started about a decade ago. Drug discovery is a lengthy and expensive process (>10 years and $1.5 billion on average for each new drug) and typically consists of a sequence of events including target identification, assay development, compound screening, optimization, preclinical characterization, and human clinical trials. Here we focus on several aspects of analgesic drug discovery, including expression systems to express TRP channels, in vitro assays to identify modulators, and in vivo assays to characterize pain efficacy and side effects in preclinical animal models. TRPA1 and TRPV1 are used as examples in the chapter. However, the methods can be adapted to other TRP channels and other therapeutic targets as well.

18.2   EXPRESSION SYSTEMS

In theory, in vitro assays can be based on endogenous expression in native cell lines and heterologous expression in bacteria, yeast, *Xenopus* oocytes, insect, and mammalian cells. In practice, expression in mammalian cells is the default platform, because it offers abundant gene-specific expression, an environment similar to native tissues, and opportunity for correct posttranslational modification. Furthermore, because most TRP studies are conducted using mammalian expression, this platform allows meaningful reference and comparison to literature data.[5] Expression in mammalian cells can be achieved either through stable cell lines or by transient expression. Baculovirus expression systems (e.g., BacMam) can also be used for in vitro assays, although extra steps and optimization are required.

18.2.1 STABLE CELL LINES

The generation of stable cell lines has become routine, and detailed protocols can be found in textbooks, manuals, and the Internet (see Chapter 1). In general, the gene of interest is cloned in a vector (e.g., pcDNA3) that contains a promoter to allow high-level expression in mammalian cells (e.g., CMV) and a resistance gene for selection (e.g., neomycin phosphotransferase). After being transfected into a mammalian cell line such as human embryonic kidney 293 (HEK293) or Chinese hamster ovary (CHO) cells, the target gene gets into the nucleus and integrates into the chromosomal DNA. Because chromosomal integration is a relatively rare event, cells have to be cultivated in a medium containing selective agents (e.g., Geneticin or G418) for rigorous selection. Only cells with integrated plasmid survive. In most scenarios, drug-resistant cell lines also carry the gene of interest. A procedure for establishing TRPA1 stable cell lines is described below.[12]

HEK293 cells are cultured in Dulbecco's modified Eagle's medium (Invitrogen, Carlsbad, CA, USA) supplemented with 10% fetal bovine serum and 1% penicillin–streptomycin in a humidified atmosphere of 5% CO_2 at 37 °C. The pcDNA3.1/V5-His/human TRPA1 plasmid is transfected using LipofectAMINE 2000 (Invitrogen). At 48 hours after transfection, cells are split to 1:4, 1:20 and cultured in the same medium as described above but supplemented with 1 mg/mL Geneticin. After 2 weeks, single colonies (1–2 mm in diameter) are picked by cloning cylinders, transferred to 24- or 48-well plates and cultured in medium supplemented with 300 µg/mL Geneticin. Clones are first screened for protein expression and AITC responses in a Ca^{2+} assay and then characterized in patch clamp recordings. Stable clones with robust and homogeneous (>90% positive) functional activities are chosen.

The process of generating, characterizing, and maintaining stable cell lines is labor-intensive and time-consuming (requiring at least 3 months). Constitutive expression of TRP channels often leads to cellular toxicity and degrading signals. Although inducible expression can be used,[5] it still requires generation and maintenance of stable cell lines. Moreover, induction of gene expression in a large scale can be tedious, expensive, and inconsistent.

18.2.2 TRANSIENTLY TRANSFECTED CELLS

Traditionally, transient transfection is conducted in a small scale (e.g., 10^7 cells) using adherent cells; therefore, its application is generally limited to characterization of the gene product, assay development, and pilot screens. Recently, large-scale transfection using suspension growth cell lines has become available, allowing transfection of >10^9 cells within a short time frame. The FreeStyle™ 293 Expression System

(Invitrogen) uses suspension HEK293 cells (HEK293-F) and suspension CHO cells. In our hands, this system has been proven to be extremely useful in advancing our TRP channel efforts. Described below is a routinely used protocol.[12]

Briefly, HEK293-F cells are grown in suspension in flasks (cell volume 30 mL to 1 L) or in a Wave Bioreactor (Wave Biotech, Somerset, NJ, USA) (6 L). To support high-density, suspension culture and transfection, cells are grown in FreeStyle™ 293 media, an optimized and serum-free formulation. Transfection is carried out using 293fectin™ (Invitrogen). To transfect 3×10^7 cells (30 mL volume), 30 μg of plasmid DNA and 40 μL 293fectin are used. For larger volumes, each reagent is scaled up proportionally. At 2 days posttransfection, cells (at a density of $2.5–3 \times 10^6$ cells/mL) are harvested by centrifugation ($1000 \times g$, 5 min) and resuspended to a density of 1.5×10^7 cells/mL in freezing medium (Freestyle media/10% fetal bovine serum/10% DMSO). Cells are transferred in 2-mL aliquots into cryovials, which are then placed in Nalgene Mr. Frosty slow-freeze devices (Sigma-Aldrich, St. Louis, MO, USA) at –80°C. As needed, vials are removed from –80°C and quickly thawed in a 37°C water bath. Cells are aseptically transferred into conical tubes containing FreeStyle media (10 mL/vial). After centrifugation at $1000 \times g$ for 3–5 min, the medium is removed by aspiration, and cells are re-suspended in FreeStyle medium again. Cells revived from frozen stock have a viability of ~90% as assessed by using VioCell (Beckman Coulter, Brea, CA, USA). Routinely, we generate $1.5–2.0 \times 10^{10}$ cells from a 6-L transfection, which is sufficient to prepare 1950–2600 384-well plates (2×10^4 cells/well).

18.2.3 TRANSIENT TRANSFECTION VERSUS STABLE CELL LINES

Compared to stable cell lines, large-scale transient transfection offers multiple advantages. It can generate samples in days instead of months required to establish and select stable cell lines. It circumvents instability and cellular toxicity associated with constitutive gene expression, especially during long-term passage of stable cell lines. It can generate a large quantity (e.g., $\geq 10^{10}$) of cells, therefore reducing labor and reagent costs dramatically. It also offers substantial flexibility because the cells can be stored at –80°C and thawed on demand for screening efforts. The relative low transfection rate (~60 versus ~90% positive in a high-quality stable cell line) could be a concern for some of the automated electrophysiology formats. However, this problem has been circumvented by the introduction of a population patch clamp, wherein multiple cells are simultaneously tested in a single well.[13]

18.3 IN VITRO ASSAYS FOR TRP CHANNELS

18.3.1 AN OVERVIEW

One of the major challenges in drug discovery is identification of lead compounds with optimal chemical and pharmacological properties. This is often achieved by high-throughput screening (HTS) and lead optimization (secondary screening). At the HTS stage, 100,000 to several millions of compounds from chemical libraries are screened against the target of interest. The major considerations are assay robustness, throughput, reagent supply, and cost. At the lead optimization or secondary screening stages, HTS hits and lead compounds (often in the thousands) are characterized, with data quality as the major consideration. Their activities are interrogated in greater detail (e.g., IC_{50} or EC_{50} determination), in different assay formats, against various orthologs (e.g., human, rat, and mouse) and against related family members. Therefore, a panel of in vitro assays is required.

Many assays previously used for other ion channel targets have been adapted to TRP channels.[14] Ligand binding assays identify compounds that recognize the same binding site as a known radio-labeled ligand. This method is amenable to HTS, but it does not distinguish agonism, antagonism, and allosteric effects. Therefore, cell-based functional assays, which monitor channel conductance to Ca^{2+}, Na^+, and K^+ ions, are most widely used. These assays differ in measurement parameters, information content, throughput, and cost and are suited for different stages of drug discovery (Table 18.1). The patch-clamp technique directly measures current flow through the channel and is considered as the gold standard. It is very useful for mechanistic studies and detailed characterization of advanced compounds, but its application in compound screen is limited by the requirement for high technical skills and notoriously low throughput. The recently introduced automated electrophysiology platforms (e.g., PatchXpress, Q-Patch, IonWorks) are amenable to higher throughput and are increasingly used for evaluation of some voltage-gated channels and ligand-gated channels.[15] Their application to TRP channels has been quite limited with only rare examples cited in the literature.[16] This is possibly due to several factors including requirement for fast perfusion, and current run-up and run-down commonly observed for TRP channels. Ion flux assays using radioactive isotopes (e.g., $^{45}Ca^{2+}$, $^{86}Rb^+$, and $[^{14}C^+]$-guanidine) have been successfully used to identify TRP ligands (e.g., capsaizepine).[17] Recently, a nonradioactive ion flux assay using Rb^+ and atomic absorption spectroscopy technology (ICR8000, Aurora Biomed, Vancouver, Canada) has been applied to TRP channels.[18] Compared to electrophysiology, the flux assay has increased throughput, although it offers no kinetic information and requires radioactivity or specialized equipment. In our opinion, the fluorescence-based Ca^{2+} assay and membrane potential assays are currently the predominant formats for TRP channel screening.

18.3.2 FLUORESCENCE CA^{2+} ASSAYS

Fluo-4 and its derivatives are the most commonly used intracellular Ca^{2+} indicators. When excited at a wavelength of 480 nm, these dyes give an emission at 525 nm, with intensities corresponding to free intracellular Ca^{2+} concentration ($[Ca^{2+}]_i$). In cells expressing TRP channels, addition of exogenous agonists activates TRP channels and significantly increases $[Ca^{2+}]_i$ (e.g., >10-fold). The rise in $[Ca^{2+}]_i$ causes a strong increase in fluorescence signals of Fluo-4, which can be detected by the CCD camera equipped in a FLIPR (Fluorescence Image Plate Reader, MDS Analytical Technologies, Sunnyvale, CA, USA) or FDSS instrument (Functional Drug Screening System, Hamamatsu Photonics, Hamamatsu, Shizuoka, Japan). As shown in Figure 18.1, capsaicin (Cap) elicits a robust, dose-dependent increase in fluorescence signals in TRPV1-expressing cells. When 30 nM capsaicin is used, the signal-to-background ratio is 21 and the Z factor is 0.79 ($n = 24$) (see Chapter 1 for

TABLE 18.1
In Vitro Assays for TRP Channels

Method	Measurement	Information Content	Throughput	Cost	Requirement	Example
Binding assay	Ligand competition	Low	High	Low	Radio-ligand	TRPV1[31]
Patch clamp	Currents	High	Very low	Very high	Skilled electrophysiologist	TRPA1[12]
Automated EP	Currents	High	Low-medium	High	PatchXpress, IonWorks, QPatch	TRPV1[16]
Ion flux-radio tracer	Ion flux	Medium	Medium	Low	$^{45}Ca^{2+}$, $^{86}Rb^+$, $[^{14}C]$-guanidine	TRPV1[17]
Ion flux-AAS	Rb^+ flux	Medium	Medium	Medium	ICR800	TRPA1[18]
Fluorescence Ca^{2+}	Intracellular Ca^{2+}	Medium	High	Low	FLIPR or FDSS	TRPA1[12]
Membrane potential	Membrane potential	Medium	High	Low	FLIPR or FDSS	TRPA1[32]
FRET-MP	Membrane potential	Medium	Medium-high	Low-medium	FLIPR-Tetra or FDSS	TRPA1

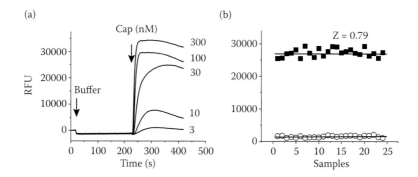

FIGURE 18.1 A Ca^{2+} assay for rTRPV1. (a) Representative traces of fluorescence changes evoked by different capsaicin (Cap) concentrations in FLIPR-based Ca^{2+} assay. Arrows indicate addition of buffer and Cap. Change in fluorescence is represented in relative fluorescence units (RFU). Large-scale transiently transfected cells expressing rTRPV1 are stored at $-80°C$ for ~10 months after transfection, thawed, and plated for the experiment. (b) Scatter plot of fluorescence signals obtained for addition of 30 nM CAP (■) or buffer (o). The average signals were 27277 ± 311 and 1245 ± 59 RFUs for 30 nM Cap and buffer, respectively ($n = 24$). Signal-to-background ratio is 21 and Z factor is 0.79.

definition of Z factor). Routinely, 20–30 microplates (96-, 384-, or 1535-well format) can be screened on a daily basis, amassing a large number of data points. Described below is a detailed protocol describing use of 96-well plates, which can be adapted easily to accommodate 384- or 1536-well formats.[19]

Frozen aliquots of transiently transfected cells are quickly thawed in a 37°C water bath, transferred to 50-mL conical polypropylene centrifuge tubes containing the 293 medium (10 mL/vial), and pelleted by low-speed centrifugation ($1000 \times g$ for 5 min). Supernatants are removed, and cells are resuspended in the 293 medium at a density of 10^6 cells/mL. Resuspended cells are seeded into black-walled clear-bottom Biocoat™ poly-D-lysine assay plates (10^5 cells/well for 96-well plate, or 2×10^4 cells/well for 384-well plate; BD Biosciences, Bedford, MA, USA) and incubated overnight at 37°C under a humidified 5% CO_2 atmosphere.

Ca^{2+} influx is measured using a FLIPR calcium assay kit (R8033; MDS Analytic Technology). The Ca^{2+} indicator dye is dissolved in Hanks' balanced salt solution supplemented with 20 mM HEPES buffer (HBSS/HEPES) according to manufacturer's instructions. Before initiating the assay, the medium is removed by aspiration, and cells are loaded with 100 μL Ca^{2+} dye for 2 to 3 hours at room temperature. For TRPV1, TRPM8, and TRPA1, capsaicin, menthol, or AITC is used to activate each channel, respectively. Solutions of the test compounds (4×) are prepared in HBSS/HEPES, and 50 μL is added to the cells at a delivery rate of 10 μL/s. Changes in fluorescence are measured over time in FLIPR. Two additions are made over the course of an experimental run. For agonist experiments, assay buffer is added at the 10 s time point, followed by addition of agonist at the 3 min 10-s time point. For antagonist experiments, the antagonist is added at the 10 s time point, followed by addition of the agonist 3 min later. Final assay volume for both the agonist and antagonist experiments is 200 μL. Total length of an experimental run is 6.5 min. For TRPV4, hypotonic solutions of 212 mOsm are used to activate the channel and evoke Ca^{2+} influx. For filter setting in FLIPR, an excitation wavelength of 485 nm and emission wavelength of 525 nm are used.

18.3.3 FLUORESCENCE MEMBRANE POTENTIAL ASSAYS

Opening of TRP channels elicit changes in membrane potentials, which can be detected by a combination of a membrane potential-sensitive dye and a fluorescence quencher. At hyperpolarized potentials, the negatively charged oxonol dye is localized at the outer layer of the lipid bilayer; upon excitation, the emission is absorbed by the quencher near the membrane. At depolarized potentials (i.e., opening of TRP channels), the oxonol dye moves to the inner layer of the lipid bilayer, and a strong fluorescent emission occurs. The fluorescence signals can be detected by the CCD camera in FLIPR or FDSS instruments. This assay has been applied to identify TRP modulators, with a procedure nearly identical to the Ca^{2+} assay.[19] In brief, we use a FLIPR membrane potential assay kit (R8034, MDS Analytic Technology), an excitation wavelength of 488 nm, and an emission wavelength of 550 nm. A more detailed protocol is covered in Chapter 1.

In addition to the fluorescence membrane potential assay, a fluorescence energy transfer-based membrane potential (FRET-MP) assay is also available. This assay is based on the fact that energy transfer between coumarin dye (donor) and oxonol dye (acceptor) increases at hyperpolarized potentials and deceases at depolarized potentials; therefore, the ratio of emission (460/580 nm) reflects membrane potential. This assay has been adapted to several TRP channels (J. Chen, R. M. Reilly, P. R. Kym, and S. Joshi, unpublished data). Although execution and data analysis for FRET-MP assays are more complex, its performance is generally superior to that of the fluorescence membrane potential assay in reducing well-to-well variation and incurrence of false-positive hits.

18.3.4 CONSIDERATIONS

Fluorescent Ca^{2+} and membrane potential assays are neither direct measurements nor linear correlates of channel activities. These assays can be prone to various artifacts, including those associated with compounds that are fluorescent, toxic, or those with the propensity to interact with dye or cellular pathways. Rigorous counterscreens should be performed to confirm that the compounds are specific to the intended target. For example, to confirm agonist activity, compounds should not have effects in mock-transfected cells, and signals should be sensitive to specific benchmark blockers. Lead compounds should be characterized, particularly for potency and selectivity in orthogonal assays. Compounds with adequate potency, selectivity, and pharmacokinetic profiles are subsequently evaluated in animal models.

18.4 PRECLINICAL PAIN AND SIDE EFFECT MODELS

The use of valid predictive preclinical animal models is critical for the identification and development of novel pharmacotherapies for treating human chronic pain syndromes. Here we describe some commonly used acute and pathological pain models in rats, which have proven to be valuable in understanding pain mechanisms and identifying novel drugs targeting TRP channels, particularly TRPV1 and TRPA1. Acute pain is assessed in normal animals, as measured by nocifensive withdrawal or escape behavior in response to painful stimuli including noxious temperatures, chemical reagents, and mechanical insult. Chronic pain models are used to study pathological pain transmission, validate pain targets, and characterize the potential analgesic profile of novel compounds. Historically, these models were developed to replicate inflammatory, osteoarthritic, and neuropathic pain in humans.[20,21] One of the major challenges in pain drug discovery is identification of agents that alleviate pathological pain without eliciting severe side effects. Therefore, several commonly used models for assessing motoric, CNS side effects, and core body temperature are also described.

18.4.1 Models of Pharmacological Antagonism (Capsaicin- or AITC-Induced Nocifensive Response)

TRPV1 and TRPA1 agonists (e.g., capsaicin and AITC/mustard oil, respectively) are known to produce pain behavior upon intrapaw injection by activating their respective receptors on primary afferents.[22,23] These acute pain models can therefore be used to assess whether a compound behaves as a TRPV1 or TRPA1 antagonist upon systemic or local administration in rats. For testing TRPV1 antagonists in the capsaicin-induced flinching model, rats are placed in individual observation cages. Following an acclimation period of 30 min, the test compound is administered. At the appropriate time following compound administration, 2.5 µg of capsaicin in a 10-µL solution of 10% ethanol/90% hydroxypropyl-β-cyclodextrin is injected subcutaneously into the dorsal aspect of the right hind paw. The observation cage is then suspended above mirrors in order to facilitate observation of the rat. Rats are observed for a continuous period of 5 min. The number of flinching behaviors of the injured paw is recorded during the 5 min observation period. Similarly, for testing TRPA1 antagonists, 5% mustard oil in 50-µL ethanol is injected subcutaneously into the dorsal aspect of the right hind paw. A TRPV1 or TRPA1 antagonist with adequate drug exposure should decrease the number of flinching behaviors induced by capsaicin or mustard oil, respectively. An example of a dose-related effect of a TRPV1 antagonist in attenuating capsaicin-induced flinching response is shown in Figure 18.2a.

18.4.2 Carrageenan Inflammatory Pain Model

Various models have been developed to induce a localized inflammatory reaction by injecting substances such as formalin, carrageenan, or complete Freund's adjuvant (CFA) into the paw or joints of rats. Following the initial injection, the inflammogens render the paw/joint very sensitive to thermal and mechanical stimuli; pain can be measured minutes to days later, either at the site or away from the site of injury.[24] Carrageenan produces a more subacute edema and inflammatory pain compared to CFA. In the carrageenan model, unilateral inflammation is induced by injecting 100 µL of a 0.5% solution of λ–carrageenan (Sigma, St. Louis, MO, USA) in 0.9%

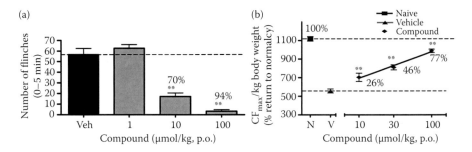

FIGURE 18.2 Antinociceptive effects of a TRPV1 antagonist. (a) In the capsaicin-induced nocifensive response model, a TRPV1 antagonist (compound) dose dependently attenuates the nocifensive flinching evoked by capsaicin. 2.5 μg of capsaicin is injected subcutaneously into the dorsal aspect of the right hind paw. The compound or vehicle (Veh) is dosed orally 30 min before testing; $n = 12$. (b) In the MIA-induced osteoarthritic model, pain is manifested as a decrease in hind limb grip force three weeks after MIA injection. The TRPV1 compound reverses the grip force in a dose-dependent manner, while vehicle has no effect. The compound is dosed 1 hour orally before behavior testing; $n = 12$.

sterile saline into the plantar surface of the right hind paw. Thermal hyperalgesia is assessed 2 hours later. The compound of interest is administered, and its effects are determined at the appropriate time depending on its pharmacokinetic attributes. Response to thermal stimulation is determined using a commercially available paw "hotbox" thermal stimulator (UARDG, Department of Anesthesiology, University of California, San Diego, CA, USA) modeled after that described previously.[24] Rats are placed in individually partitioned Plexiglas cubicles that are on a glass surface maintained at 30°C. A thermal stimulus, in the form of radiant heat emitted by a focused projection bulb, is applied to the plantar surface of each hind paw. The stimulus current is 4.5 amperes, and the maximum exposure time is limited to 20 s in order to avoid tissue damage from the stimulus. In each test session, both hind paws of each rat are tested in three sequential trials at approximately 5 min intervals. Paw withdrawal latencies are calculated as the mean of the two shortest latencies. A compound that alleviates thermal hyperalgesia in this model restores the paw withdrawal latency of the ipsilateral injured paw to that of the contralateral/uninjured paw.

18.4.3 Monoiodoacetate-Induced Osteoarthritic Pain Model

The development of preclinical osteoarthritic pain models in rodents has been particularly challenging owing to an inadequate understanding of the etiology of human osteoarthritic pain.[25] While it is clear that no single animal model can precisely replicate human osteoarthritis, a monosodium iodoacetate (MIA)-induced osteoarthritic model has been extensively used, based on the fact that robust pain behaviors in this model are amenable to pharmacological intervention.[26]

The procedures mentioned here for the induction and testing of osteoarthritic pain have been described previously.[27] Unilateral knee joint osteoarthritis is induced in rats by a single intra-articular (i.a.) injection of MIA (Sigma, St. Louis, MO, USA)

(3 mg in 0.05 mL sterile isotonic saline) into the joint cavity using a 26G needle under light (2–4%) isoflurane anesthesia. Following injection, the animals are allowed to recover from the effects of anesthesia (5–10 min) before being returned to their home cages. After approximately 21 days, testing is performed. Grip strength is assessed in osteoarthritic rats as a measure of activity-induced pain. Measurements of hind limb grip force are conducted by recording the maximum compressive force (CF_{max}) exerted on the hind limb strain gauge, in a commercially available grip force measurement system (Columbus Instruments, Columbus, OH, USA). Each rat is gently restrained and allowed to grasp the wire mesh frame attached to the strain gauge. The experimenter then moves the animal in a rostral-to-caudal direction until the grip is broken. Each rat is sequentially tested twice at an approximately 2–3 min interval to obtain a raw mean grip force (CF_{max} in gram force units). In order to account for the body weight differences among the rats, this raw mean grip force is normalized by the body weight of the rat (in kilograms) to generate a maximum hind limb compressive force for each animal [(CF_{max} in gram force)/kg body weight]. A group of age-matched naïve animals is included in each experiment. Data obtained from various dose groups for the test compound are compared with data from the naïve group. The vehicle control group is assigned a value of 0%, whereas the naïve group is assigned a value of 100%. The percent effects for each dose group are expressed as percent return to normal grip force as found in the naïve group, and calculated using the formula: (percent return to normalcy = [(Treatment CF_{max} − Vehicle CF_{max}) / (Naïve CF_{max} − vehicle CF_{max})] × 100%). An example of a dose-response curve with this protocol is shown in Figure 18.2b.

18.4.4 CHRONIC CONSTRICTION INJURY MODEL OF NEUROPATHIC PAIN

Neuropathic pain can result from inflammation around peripheral nerves and peripheral nerve compression. The chronic constriction injury (CCI, Bennett model) of the sciatic nerve model has been developed to mimic this clinical situation.[28] In this model, the right common sciatic nerve is isolated at the midthigh level and loosely ligated by four chromic gut ties separated by an interval of approximately 1 mm. All animals are allowed to recover for at least two weeks and no more than four weeks prior to testing of cold or mechanical allodynia. Cold allodynia is assessed following acetone application.[29] Rats are acclimated in a transparent plastic box for about 20 min before behavioral testing. Acetone (100 μL) is gently sprayed onto the plantar surface of the hind paw. For baseline recordings, acetone is sprayed three times on the right ipsilateral nerve-injured hind paw at 5 min intervals to identify allodynic animals. Animals included in the study are also tested on the contralateral paw. Following compound administration, acetone is sprayed 5 times at 5 min intervals. The paw flinching response after the acetone spray is considered as a positive response, and frequency of response to five acetone sprays is noted. The response frequency is calculated using the formula: ([number of positive responses following treatment/5] × 100%).

Mechanical allodynia testing is performed using calibrated von Frey monofilaments (Stoelting, Wood Dale, IL, USA). Paw withdrawal threshold is determined by increasing and decreasing stimulus intensity, and estimated using the Dixon's up-down method.[30] Rats are placed into inverted individual plastic containers (20 × 12.5 × 20 cm) on top of a suspended wire mesh with a 1-cm^2 grid to provide access to the ventral side of the hind paws and acclimated to the test chambers for 20 min. The von Frey filaments are then presented perpendicularly to the plantar surface of the selected hind paw, and held in this position for approximately 8 s with enough force to cause a slight bend in the filament. Positive responses include an abrupt withdrawal of the hind paw from the stimulus or flinching behavior immediately following removal of the stimulus. A 50% withdrawal threshold is determined using an up-down procedure.[30] The strength of the maximum filament used for von Frey testing is 15.0 g. Only rats with a baseline threshold score of less than 5 g are used in this study, and animals demonstrating motor deficit are excluded.

18.4.5   MODELS OF ACUTE THERMAL NOCICEPTION

It is important that a clinically used analgesic does not significantly impair the protective ability to sense acute noxious thermal or mechanical sensation. This is a particularly important issue for the modulators of TRPV1 and TRPA1, because these channels have been shown to play a role in sensing noxious heat and cold, respectively.[1] A tail immersion assay is used to assess effects on noxious heat sensation. In this assay, a circulating water bath is heated to 55°C. At an appropriate time post-dosing, the rats are handled for a few seconds to calm them down and then cupped with their back against the testers hand at a slight angle with the head facing away from the tester. With the rat in one hand and a 0.01 s stopwatch in the other hand, the tail is quickly immersed 6–8 cm in a water bath or to a distance leaving 2–3 cm of tail out of water. The timer is started simultaneously. When the rat flinches or attempts withdrawal, the timer is immediately stopped and the tail is quickly removed from the water bath. The response latency (in seconds) is recorded and repeated 3 times with 3–4 min between readings. A percent increase in the average response latency for tail withdrawal relative to a vehicle control is determined. Effects of compounds on noxious cold sensation can be assessed using cold plate apparatus (Columbus Instruments, Columbus, OH, USA) in rats or mice. The latency to first jump is measured after the animal is placed on the cold plate (0°C) with a cutoff time of 180 s.

18.4.6   LOCOMOTOR ACTIVITY AND ROTAROD CNS SIDE EFFECT ASSAYS

It is very important to evaluate compound effects on the locomotor activity or the motor system (rotarod assay) for at least two reasons. First, such side effects are undesirable from a clinical standpoint. Second, they could confound interpretation of observed efficacy in pain models. Locomotor activity is measured in an open field using photo beam activity monitors (AccuScan Instruments, Columbus, OH, USA). Rats are acclimated to the test room for 30 min. Following compound administration, the AccuScan activity boxes measure horizontal activity as photo beam breaks in the horizontal plane. Rotarod performance is measured using an accelerating Rotarod apparatus (Omnitech Electronics, Inc., Dartmouth, NS, Canada). Rats are placed on a 9 cm diameter rod with an increasing rotating speed from 0 to 20 rpm over a 6 s period. Each rat is given three training sessions. The Rotarod performance is determined by the total amount of time within 60 s that rats stayed on the rod without falling off (the maximum score is 60 s).

18.4.7 TELEMETRY FOR ASSESSMENT OF CORE BODY TEMPERATURE

TRPV1 is involved in regulating body temperature, and blockade of TRPV1 by some antagonists increases core body temperature.[4] Such effect can be assessed in rats implanted with telemetry catheters (TL11M2-C50PXT, Data Sciences International, St. Paul, MN, USA), which accurately measure core-body temperature. Anesthetized rats are placed on a heating pad and covered with a sterile surgical drape. A ventral midline abdominal incision is made, and sterile cotton tip applicators are used to gently move internal tissue and expose the abdominal aorta for implantation of the telemetry catheter. Blood flow is temporarily stopped to the lower extremities (5–7 min) with Diffenbach clamps to allow the insertion of the telemetry catheter into the abdominal aorta. A sterile cellulose patch is placed over the catheter/aorta and secured using a small amount of tissue adhesive (Vetbond, 3M, St. Paul, MN, USA). Once catheter placement is complete, the clamps are removed and blood flow is restored to the lower extremities. The transmitter is placed in the intraperitoneal cavity. The transmitter suture rib is sewn into the abdominal sutures to secure it in place. The skin is closed using sterile wound clips and the animal removed from Sevoflurane. Buprenex (0.01 mg/mL s.c.; Reckitt Benckiser Pharmaceuticals, Inc., Bristol, UK) is administered for post-operative analgesia. Animals are maintained on a heating pad until ambulatory and then individually housed with food and water *ad libitum*. Surgical staples are removed after 7–10 days postimplantation. Rats are allowed a two-week postsurgical recovery period before compound treatment.

Rats are randomly divided into study groups and administered either vehicle or test compound by oral dosing at 8:00 AM. Further entry into the telemetry room is minimal. Data collected during the time when staff is present are not included in the analysis. Figure 18.3 shows that the effect of a TRPV1 antagonist on core body

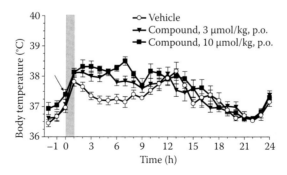

FIGURE 18.3 Effect of a TRPV1 antagonist on core body temperature in conscious-telemetry rats. After oral dosing at 3 and 10 μmol/kg, a TRPV1 antagonist increases core body temperature (approximately 1°C; lasting for ~7 hours) compared with vehicle. Arrow indicates dosing of the compound or vehicle. Gray bar indicates investigator in room; $n = 7$ for each group.

temperature. Clearly, at 3 and 10 µmol/kg dose, this compound increases body temperature (approximately 1°C). In addition to core body temperature, the telemetry device also measures cardiovascular functions such as mean arterial pressure and heart rate. All parameters are continuously recorded and analyzed using commercial software and a signal processing workstation (Dataquest Art v4.0, Data Sciences International).

18.5 CONCLUSION

The tremendous progress in TRP sensory biology has been greatly facilitated by the availability of potent and selective pharmacological tools. These agents are identified in in vitro assays and characterized in preclinical models to reveal the physiological function and therapeutic utility of the channel. As a recent example, evaluation of TRPV1 antagonists in telemetrized rodents uncovered the role of TRPV1 in body temperature regulation.[4] Currently, a variety of cell-based functional assays are available, each with its own utility in drug discovery. For example, the fluorescence Ca^{2+} assay and membrane potential assays are best suited for HTS, while manual patch clamp electrophysiology is best suited for detailed characterization of compounds. Emerging technologies, such as automated electrophysiology, will likely have an increasing impact. Preclinical behavior models have been, and will continue to be, essential for assessing analgesic efficacy and side effects, and their predictive values will be revealed when more compounds enter human clinical trials in the coming years. Through emerging technologies and concerted efforts across industry and academia, there is reason to believe novel TRP drugs are on the horizon.

REFERENCES

1. Wang, H., and C. J. Woolf. 2005. Pain TRPs. *Neuron* 46: 9.
2. Caterina, M. J., and D. Julius. 2001. The vanilloid receptor: a molecular gateway to the pain pathway. *Annual Review of Neuroscience* 24: 487.
3. Wong, G. Y., and N. R. Gavva. 2008. Therapeutic potential of vanilloid receptor TRPV1 agonists and antagonists as analgesics: Recent advances and setbacks. *Brain Research Reviews* 60: 267.
4. Gavva, N. R. 2008. Body-temperature maintenance as the predominant function of the vanilloid receptor TRPV1. *Trends in Pharmacology Science* 29: 550.
5. Story, G. M., A. M. Peier, A. J. Reeve et al. 2003. ANKTM1, a TRP-like channel expressed in nociceptive neurons, is activated by cold temperatures. *Cell* 112: 819.
6. Anand, U., W. R. Otto, P. Facer et al. 2008. TRPA1 receptor localisation in the human peripheral nervous system and functional studies in cultured human and rat sensory neurons. *Neuroscience Letters* 438: 221.
7. Bautista, D. M., S. E. Jordt, T. Nikai et al. 2006. TRPA1 Mediates the Inflammatory Actions of Environmental Irritants and Proalgesic Agents. *Cell* 124: 1269.
8. Trevisani, M., J. Siemens, S. Materazzi et al. 2007. 4-Hydroxynonenal, an endogenous aldehyde, causes pain and neurogenic inflammation through activation of the irritant receptor TRPA1. *Proceedings of the National Academy of Science of the United States of America* 104: 13519.
9. McNamara, C. R., J. Mandel-Brehm, D. M. Bautista et al. 2007. TRPA1 mediates formalin-induced pain. *Proceedings of the National Academy of Science of the United States of America* 104: 13525.

10. Obata, K., H. Katsura, T. Mizushima et al. 2005. TRPA1 induced in sensory neurons contributes to cold hyperalgesia after inflammation and nerve injury. *Journal of Clinical Investigations* 115: 2393.

11. Petrus, M., A. M. Peier, M. Bandell et al. 2007. A role of TRPA1 in mechanical hyperalgesia is revealed by pharmacological inhibition. *Molecular Pain* 3: 40.

12. Chen, J., M. R. Lake, R. S. Sabet et al. 2007. Utility of large-scale transiently transfected cells for cell-based high-throughput screens to identify transient receptor potential channel A1 (TRPA1) antagonists. *Journal of Biomolecular Screening* 12: 61.

13. Finkel, A., A. Wittel, N. Yang et al. 2006. Population patch clamp improves data consistency and success rates in the measurement of ionic currents. *Journal of Biomolecular Screening* 11: 488.

14. Zheng, W., R. H. Spencer, and L. Kiss. 2004. High throughput assay technologies for ion channel drug discovery. *Assay and Drug Development Technologies* 2: 543.

15. Priest, B. T., A. M. Swensen, and O. B. McManus. 2007. Automated electrophysiology in drug discovery. *Current Pharmaceutical Design* 13: 2325.

16. Klionsky, L., R. Tamir, B. Gao et al. 2007. Species-specific pharmacology of Trichloro(sulfanyl)ethyl benzamides as transient receptor potential ankyrin 1 (TRPA1) antagonists. *Molecular Pain* 3: 39.

17. Bevan, S., S. Hothi, G. Hughes et al. 1992. Capsazepine: a competitive antagonist of the sensory neurone excitant capsaicin. *British Journal of Pharmacology* 107: 544.

18. Liu, K., M. Samuel, R. K. Harrison, and J. W. Paslay. Rb$^+$ efflux assay for assessment of non-selective cation channel activities. *Assay and Drug Development Technologies* 8: 380.

19. Bianchi, B. R., R. B. Moreland, C. R. Faltynek, and J. Chen. 2007. Application of large-scale transiently transfected cells to functional assays of ion channels: different targets and assay formats. *Assay and Drug Development Technologies* 5: 417.

20. Le Bars, D., M. Gozariu, and S. W. Cadden. 2001. Animal models of nociception. *Pharmacological Reviews* 53: 597.

21. Bridges, D., S. W. Thompson, and A. S. Rice. 2001. Mechanisms of neuropathic pain. *British Journal of Anaesthia* 87: 12.

22. Immke, D. C., and N. R. Gavva. 2006. The TRPV1 receptor and nociception. *Seminars in Cell and Developmental Biology* 17: 582.

23. Jordt, S. E., D. M. Bautista, H. H. Chuang et al. 2004. Mustard oils and cannabinoids excite sensory nerve fibres through the TRP channel ANKTM1. *Nature* 427: 260.

24. Hargreaves, K., R. Dubner, F. Brown, C. Flores, and J. Joris. 1988. A new and sensitive method for measuring thermal nociception in cutaneous hyperalgesia. *Pain* 32: 77.

25. Dieppe, P. A., and L. S. Lohmander. 2005. Pathogenesis and management of pain in osteoarthritis. *Lancet* 365: 965.

26. Fernihough, J., C. Gentry, M. Malcangio et al. 2004. Pain related behaviour in two models of osteoarthritis in the rat knee. *Pain* 112: 83.

27. Chandran, P., M. Pai, E. A. Blomme et al. 2009. Pharmacological modulation of movement-evoked pain in a rat model of osteoarthritis. *European Journal of Pharmacology* 613: 39.

28. Bennett, G. J., and Y. K. Xie. 1988. A peripheral mononeuropathy in rat that produces disorders of pain sensation like those seen in man. *Pain* 33: 87.

29. Flatters, S. J., and G. J. Bennett. 2004. Ethosuximide reverses paclitaxel- and vincristine-induced painful peripheral neuropathy. *Pain* 109: 150.

30. Chaplan, S. R., F. W. Bach, J. W. Pogrel, J. M. Chung, and T. L. Yaksh. 1994. Quantitative assessment of tactile allodynia in the rat paw. *Journal of Neuroscience Methods* 53: 55.

31. Bianchi, B. R., R. El Kouhen, J. Chen, and P. S. Puttfarcken. 2009. Binding of [(3)H] A-778317 to native transient receptor potential vanilloid-1 (TRPV1) channels in rat dorsal root ganglia and spinal cord. *European Journal of Pharmacology* 633: 15.

32. Chen, J., X. F. Zhang, M. E. Kort et al. 2008. Molecular determinants of species-specific activation or blockade of TRPA1 channels. *Journal of Neuroscience* 28: 5063.

19 Studying TRP Channels in Intracellular Membranes

Mohammad Samie and Haoxing Xu

CONTENTS

19.1 INTRODUCTION

Advances in modern cell biology and physiological techniques have dramatically improved our understanding of basic cellular functions. Together with classical genetic and biochemical approaches, these technical advances have allowed us to uncover the novel functions of a variety of proteins inside the cell. For example, recent studies have revealed intracellular functions of several TRP proteins (reviewed in Ref. 1), a family of cation non-selective ion channels that were initially thought to operate exclusively at the plasma membrane to regulate the transmembrane flux of Na^+, K^+, Ca^{2+}, and Mg^{2+}.[2,3] Initially discovered in the *Drosophila melanogaster* photoreceptor, TRPs are also found in vertebrates, widely expressed in most tissues and cell types.[2] The TRP superfamily can be divided into six subfamilies: canonical (TRPC), vanilloid (TRPV), melastatin (TRPM), polycystin (TRPP), mucolipin (TRPML), and ankyrin transmembrane proteins (TRPA). Currently, the most well-defined TRP functions are serving as cellular sensors for detecting an array of environmental stimuli including temperature, mechanical forces, and pain.[2–4] However, the list of nonsensory functions for TRPs also has expanded rapidly in recent years.[3]

Most TRPs are Ca^{2+}-permeable nonselective cation channels, generally believed to regulate intracellular Ca^{2+} ($[Ca^{2+}]_i$) levels. $[Ca^{2+}]_i$ is usually low at rest (approximately

100 nM) but may increase 10- to 100-fold to the micromolar range upon cellular stimulation.[5] Ca^{2+} is a ubiquitous second messenger and is reportedly involved in almost every single biological process.[5] Therefore, the Ca^{2+} flux pathways, i.e., channels or transporters, must be tightly regulated to ensure the functional specificity of each cellular stimulus.[5] Despite the apparent importance of TRPs, their activation mechanisms are largely unknown. Nevertheless, TRP channel dysfunction can cause human diseases, such as polycystic kidney disease and mucolipidosis type IV (MLIV), which result from mutations of human TRPP2 and TRPML1 genes, respectively.[6,7] One source for $[Ca^{2+}]_i$ increase is the Ca^{2+} in extracellular space, which is approximately 20,000 times (2 mM) more concentrated than resting $[Ca^{2+}]_i$. Hence, plasma membrane TRPs are natural candidates to mediate Ca^{2+} influx.[3] However, because most intracellular organelles also contain Ca^{2+} (at concentrations from hundreds of micromolar to millimolar),[1] activation of TRPs localized in these compartments could result in elevation of $[Ca^{2+}]_i$. The best studied example of Ca^{2+} release from intracellular membranes is the phospholipase C (PLC)-inositol-1,4,5-trisphosphate receptor (IP_3R) system.[8] Receptor-mediated activation of PLC leads to the breakdown of phosphatidylinositol 4,5-bisphosphate (PIP_2) into IP_3, which binds to the IP_3R in the membranes of endoplasmic reticulum (ER), and releases Ca^{2+} into the cytoplasm.[8] Unlike plasma membrane TRPs, the activation mechanisms of TRPs localized in intracellular organelles[1] are largely unknown. Many plasma membrane TRPs are activated or regulated by protein kinases or lipid signaling.[3] It is not clear whether intracellular TRPs are also regulated by these mechanisms and, if so, whether they exhibit electrophysiological characteristics similar to plasma membrane TRPs. Although Ca^{2+} release from intracellular compartments is important for signal transduction and membrane trafficking, our knowledge of ion channels involved in Ca^{2+} release remains very limited.[1] Because TRPs are Ca^{2+} permeable, and some localize to intracellular organelles, they are natural candidates for Ca^{2+} release from intracellular organelles. In this chapter, we will discuss techniques employed to identify and characterize TRPs localized in intracellular compartments.

19.2 STUDYING INTRACELLULAR LOCALIZATION OF TRP CHANNELS

Subcellular localization studies have revealed that many TRPs are localized on the membranes of intercellular organelles.[1] Among these, TRPV2, TRPM2, TRPML1, TRPML2, and TRPML3 are found in endosomal and/or lysosomal (endolysosomal, collectively) compartments.[1] Two strategies are commonly used to study intracellular localization of TRPs: fluorescent fusion proteins and antibody-mediated immunochemical approaches.

19.2.1 FLUORESCENT FUSION PROTEINS

Reporter genes usually encode fluorescent proteins (FP) that can emit fluorescence at specific wavelengths.[9] As such, they are often fused in-frame to the target proteins to monitor protein localization in live cells.[9] There are a number of these reporter proteins, each emitting at specific wavelengths; for example, red (RFP/

mCherry/DsRed), green (GFP), and yellow (YFP). A combination of reporter genes allows monitoring of multiple proteins simultaneously. Studies using overexpressed TRPML proteins fused with various reporter genes have revealed that TRPMLs primarily localize to a population of membrane-bound vesicles along the endosomal/lysosomal pathways in a variety of host cells.[10–14] GFP-fused TRPML (TRPML1-GFP) is mainly colocalized with Lamp-1, a late endosome and lysosome marker (Figure 19.1). Both Anti-Lamp1 and Lamp1-FP fusion proteins can be used to define late endosomes and lysosomes (LELs). The colocalization index of TRPML1 and Lamp-1 is more than 80%.[15] Thus, TRPML1 is a TRP channel specifically localized in the LEL. Using a similar approach, TRPML2, TRPML3, and TRPM2 are also

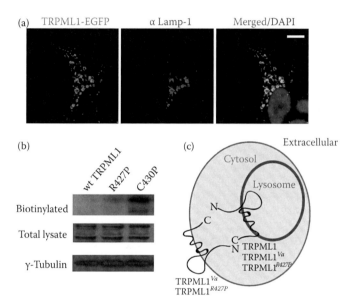

FIGURE 19.1 (**See color insert.**) While wild-type TRPML1 is localized mainly in the late endosomal and lysosomal compartments, gain of function proline-substituted TRPML1 channels are present at the plasma membrane. (a) Confocal images showing the colocalization of EGFP-TRPML1 fusion proteins and Lamp-1. EGFP reporter gene is fused to the N-terminus of TRPML1. EGFP-TRPML1-transfected HEK293T cells are fixed and immunostained with anti-Lamp-1. (b) Surface expression of TRPML1 and proline-substituted TRPML1 is demonstrated by biotinylation. Cell lysates are prepared from EGFP-TRPML1-transfected HEK293T cells. Cell-surface membrane proteins are prelabeled and pulled down with biotin. The Western blot is probed with an anti-EGFP antibody. While little or no wild-type TRPML1 is biotinylated, mutant forms of TRPML1 (TRPML1^{R427P} and TRPML1^{C430P}) exhibit significant biotinylation. The bottom panel shows the loading controls (anti-γ tubulin). (c) Illustration of the putative membrane topology and localization of TRPML1, TRPML1^{Va}, and TRPML1^{R427P} at the plasma membrane and the lysosomal membrane. Both the N- and C-termini of the channel face the cytosolic side, regardless of whether the channel is localized at the plasma membrane or on lysosomal membranes. (Modified from Dong, X.P. et al., *Journal of Biological Chemistry*, 284(46), 32040–32052, 2009; from Dong, X.P. et al., *Nature* 455(7215), 992–996, 2008. With permission.)

found in Lamp-1-labeled compartments.[11,13,14,16–18] TRPML3 and TRPV2 also are present in compartments that are positive for Rab5, an early endosome-specific small GTPase,[13,14,19] TRPML2 is also present in Arf6-positive recycling endosomes.[17] By tagging TRPMLs to different reporter genes, recent studies suggest that although TRPMLs are localized in endolysosomes, the colocalization between them is rather limited.[13,20]

Considering the heterologous nature and overexpression of the FP-fused TRPs (FP-TRPs), results obtained from fusion protein studies should be interpreted with caution and, ideally, verified by other independent approaches (see below). For example, substantial overexpression of TRPs might lead to accumulation of the proteins in intracellular compartments involved in their biosynthetic pathway, i.e., the ER and Golgi. In addition, TRPs are localized differently in different cell types. TRPV2-GFP is localized in the ER of pancreatic beta-cells,[21,22] but in J774 macrophage cells and HEK293 cells, TRPV2 is localized in early endosomes.[19,23] Thus, the intracellular localization pattern of TRPV2-GFP depends on the host cell type. Because FPs are relatively large proteins (e.g., GFP is 27 kDa), it is possible that tagging strategies could alter the localization of TRPs. For example, GFP tagged to the C-terminus of TRPML2 or TRPML3 appeared to function differently compared with the N-terminal tagged counterparts.[11,13,16] Therefore, although FPs are useful tools to investigate protein localization, conclusions drawn from these studies should be tested by other means.

19.2.2 IMMUNOSTAINING AND CELLULAR FRACTIONATION STUDIES

Because heterologously expressed FP-TRPs could potentially localize and function differently from endogenous proteins, TRP localization is analyzed using immuno-cytochemical techniques. However, there are only few effective and specific TRP antibodies available for this purpose, partially owing to the low endogenous expression levels of most TRPs.[24] In cases where the specificity of TRP antibody is not clear, immunocytochemical approaches could still yield useful information if they are used in combination with fluorescent fusion proteins. For example, if used in combination, reasonable confidence could be achieved if substantial overlapping was observed between TRP-specific immunofluorescence and FP fluorescence. By using antibodies against endogenous TRPMLs, endogenous TRPML1 was shown to be localized to the late endosomal/lysosomal vacuoles,[25] a conclusion drawn from previous studies based on FP fusion proteins.[10–14] Similarly, endogenous TRPML2 and TRPML3 exhibit a more unrestrained vesicular expression pattern,[25] consistent with observations that TRPML2-GFP and TRPML3-GFP also are found in recycling and early endosomes, respectively.[13,17,26]

Cellular fractionation has been used successfully to identify TRPML1- and TRPML3-resident vesicles with the aid of specific antibodies.[13] Cell fractionation refers to the separation of homogeneous organelles from total cell lysates by using centrifugation at different speeds. Fractionation of different cellular compartments can be achieved by filtering cell lysates on a Percoll gradient.[13] Using antibodies specific for TRPMLs and fusion tags, TRPML1 is found in the heavy fractions containing Lamp-1.[13] Unlike TRPML1, TRPML3 was found to be associated with both early endosomes and LELs.[13] These results are consistent with fluorescence studies of FP-TRPMLs.[13,14] TRPP1 (previously known as polycystin-2, PC2, or APKD2) is another intercellular TRP channel whose localization was originally studied using density gradient fractionation.[27] By using TRPP1-specific antibody on cell lysates from kidney tissues, TRPP1 was found mainly in the ER membranes, a result consistent with several other expression studies.[1,27] These studies validate the use of fluorescent fusion proteins in the heterologous systems for studying subcellular localization of TRPs, while demonstrating the need for verification by other methods.

19.3 STUDYING CHANNEL PROPERTIES OF INTRACELLULAR TRPS

19.3.1 ELECTROPHYSIOLOGICAL CHARACTERIZATION OF INTRACELLULAR TRP CHANNELS AT THE PLASMA MEMBRANE

A subset of TRPs, such as TRPML3, TRPV2, and TRPM2, have double lives; these channels are able to traffic between the plasma membrane and intracellular organelles.[1] This unique property makes it possible to characterize the channel properties of these TRPs using whole-cell recordings. Other intracellular TRPs, such as TRPML1 and TRPML2, however, are mainly localized in the intercellular organelles.[7] Fortunately, recent studies suggest that gating and/or trafficking mutations of TRPML1 and TRPML2 are present at the plasma membrane (Figure 19.1), allowing electrophysiological characterization of the pore properties of these channels.[1,7,15,20] Furthermore, TRPML2 was shown to be present at the plasma membrane of specific cell types.[28]

TRPML3 was the first TRPML family member to be characterized electrophysiologically, owing to its limited plasma membrane expression and its ability to generate whole-cell currents.[13,14,29–32] Much larger whole-cell TRPML3-mediated currents are recorded in the presence of small molecule activators.[20] A unique mutation at amino acid 419 (A419P) of TRPML3 causes pigmentation and hearing defects in Varitint-Waddler (*Va*) mice.[29,33,34] Interestingly, the *Va* mutation could dramatically increase the plasma membrane expression and whole-cell currents of TRPML3, facilitating its electrophysiological characterization.[30–32,35] Whole-cell patch clamping studies revealed that TRPML3/TRPML3[*Va*] is an inwardly rectifying non-selective cation channel.[30–32,36] Unlike TRPML3, determining the functional characteristics of plasma membrane localized TRPV2 and TRPM2 was less challenging owing to their high levels of plasma membrane expression. Whole-cell recordings revealed that TRPV2 is an outwardly rectifying nonselective cation channel activated by heat and other chemical agonists.[37] TRPM2 is a linear current channel activated by free cytosolic adenosine diphosphate ribose (ADPR), nicotinic acid adenine dinucleotide phosphate (NAADP), and Ca^{2+}.[18,38,39]

Although no significant whole-cell current can be recorded from TRPML1- and TRPML2-expressing cells, equivalent *Va* mutations of TRPML1 and TRPML2 exhibit large inwardly rectifying Ca^{2+}-permeable currents.[32,40] In addition, mutation of the dileucine lysosome-targeting motifs of TRPML1 (TRPML1-NC) results in significant plasma membrane expression of TRPML1.[11,12] A small molecule activator of TRPML3 can induce Ca^{2+} influx in cells expressing TRPML1-NC but not wild-type TRPML1.[20] These results suggest that like TRPML3, wild-type TRPML1 is likely to be a Ca^{2+}-permeable inwardly rectifying channel.

19.3.2 ENDOLYSOSOME PATCH CLAMP

Because basic properties of intracellular membranes differ significantly from the plasma membrane,[1] it is necessary to characterize the functions of intracellular TRPs in their native environment. One of the biggest hurdles to characterizing intracellular TRPs in their native membranes is the relatively small size of intracellular vesicles. For example, most endosomes and lysosomes are usually less than 0.5 µm in diameter, which is suboptimal for patch-clamping studies. However, recent studies suggest that endolysosomes can be enlarged by disrupting membrane trafficking using genetic and pharmacological approaches.[19,40]

Endosomes can be enlarged by introducing a hydrolysis-deficient mutant form of the AAA ATPase, SKD1/VPS4, into HEK293 cells.[19] Such a maneuver could lead to the formation of large endosomes (3–6 µm in diameter) in HEK293 cells. TRPV2 is localized in the early endosomes of macrophages[23] and other cell types.[19] TRPV2 could act as an endosomal Ca^{2+} channel, potentially facilitating endosomal fusion and fission.[19] The endosomal enlargement allowed Saito and colleagues to test this hypothesis by directly patch-clamping isolated endosomes.[19] Enlarged endosomes are identified by GFP-tagged endososomal markers and isolated by slicing the cell membrane using electrodes.[19] Because the endosomal current had similar pharmacological characteristics to whole-cell TRPV2 current, Saito and colleagues proposed that TRPV2 is an endosomal Ca^{2+} channel.

Dong and colleagues[40] used a chemical approach to enlarge the endosomes and lysosomes (Figure 19.2). Vacuolin-1 is a small molecule that induces the formation of enlarged endosomes and lysosomes by an unknown mechanism.[41] After exposure to 1 µM vacuolin-1 for 1 hour, large endosomes and lysosomes (up to 3–5 µm in diameter) are observed.[40] Enlarged LELs that are positive for both mCherry-Lamp1 and TRPML1-GFP are isolated by slicing the cell membrane using a patch electrode (Figure 19.2). Similar to whole-cell recordings, four distinct configurations can be used for lysosome recordings: lysosome-attached, lysosome luminal-side-out, lysosome cytoplasmic-side-out, and whole lysosome (Figure 19.2). In the whole-lysosome configurations, the extracellular solution in the patch pipette (electrode) was adjusted to pH 4.6 in order to mimic the acidic condition of the LEL.[40] On the LEL membrane, TRPML1 is positioned such that its large intraluminal loop, between trans-membrane domains one and two, faces the luminal side of the LEL, while its short N- and C-termini face the cytoplasm (Figure 19.2). Whole lysosome recordings of TRPML1-positive enlarged LELs revealed that TRPML1 and TRPML1^{Va} give rise to inwardly rectifying cationic currents similar to the whole-cell TRPML1^{Va} currents.[40] In this case, inward is defined as cations flowing out of the LEL lumen (Figure 19.2). These results demonstrate that the Va-activating mutation is a valid approach to studying TRPML1 channels at the plasma membrane. Because Dong and colleagues were also able to occasionally isolate enlarged LELs and measure TRPML1 current without any vacuolin-1 treatment,[40] they have confirmed that vacuolin-1 is unlikely to alter TRPML1 channel pore properties

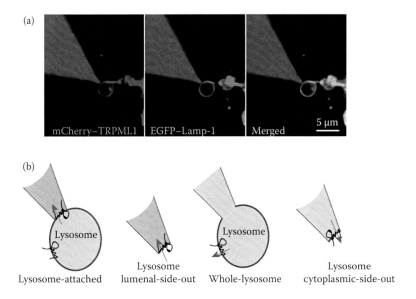

FIGURE 19.2 Lysosome patch-clamp configurations. (a) Colocalization of mCherry–TRPML1 and EGFP–Lamp-1 at the membrane of an isolated enlarged LEL. The patch pipette is filled with rhodamine B dye. (b) Diagrams of four distinct patch-clamp configurations of lysosomal recordings: lysosome-attached, lysosome luminal side-out, whole lysosome, and lysosome cytoplasmic side-out. In each configuration, the gray arrow indicates the direction of the inward current mediated by TRPML1 (at negative potentials; cations flow out of the lysosomes). (Modified from Dong, X.P. et al., *Nature* 455(7215), 992–996, 2008. With permission.)

significantly. Similar approaches can be used to characterize the channel functions of TRPML2, TRPML3, and TRPM2.

19.3.3 PLANAR LIPID BILAYER STUDIES

An alternative way to study intracellular TRPs is to reconstitute TRP proteins or TRP-resident intracellular membranes into a planar lipid bilayer. This technique is used commonly to study the electrophysiological properties of single ion channels in a defined and artificial lipid bilayer.[42] Two intracellular TRPs (TRPP1 and TRPML1) have been studied using planar lipid bilayers.[27,43] NAADP is a second messenger that induces Ca^{2+} release from acidic organelles in a variety of cell types.[44] By reconstituting lysosomal membranes into a planar lipid bilayer, NAADP activated a linear current that could be blocked by antibodies against TRPML1.[45] As this NAADP-activated current differed significantly from the aforementioned inwardly rectifying current, the significance of this study is not clear, especially because there is now compelling evidence that two-pore channels (TPCs) are likely to be the NAADP receptors.[46] In addition, ER membrane TRPP1 has been studied using planar lipid bilayers.[27,47] TRPP1 can be activated by cytoplasmic Ca^{2+} and is regulated by SNARE proteins.[27,47]

19.3.4 Monitoring Ca²⁺ Release from Intracellular Ca²⁺ Stores

Ca^{2+} imaging has been used extensively to study TRP channel function. There are two different types of Ca^{2+} indicators that are commonly used to detect intracellular Ca^{2+} levels: Ca^{2+}-sensitive fluorescent dyes or genetically encoded Ca^{2+} sensors. Fura-2 is the most commonly used fluorescent dye that can be easily loaded into most cell types.[48] Fura-2 is excited at two different wavelengths (340 and 380 nm); the ratio of light emission at these two wavelengths (F340/F380), which increases dramatically with increased Ca^{2+}, is used to estimate Ca^{2+} levels independent of dye loading.[48] Not only has Fura-2, or other Ca^{2+}-sensitive dyes such as Fluo4, been used extensively to study plasma membrane TRPs such as TRPV1,[49] but they also are frequently used to evaluate Ca^{2+} release from intracellular compartments, with modifications.[1] As many TRPs are present in both plasma membrane and intracellular membranes, to exclude the possibility of extracellular Ca^{2+} entry, experiments should be performed in a Ca^{2+}-free bath solution.[15] For instance, when TRPV1 is present in the ER or Golgi membranes;[50] in the absence of extracellular Ca^{2+}, application of the TRPV1-specific, membrane-permeable agonist capsaicin induces Ca^{2+} release from ER membranes.[50] The ER/Golgi Ca^{2+} source is confirmed using a SERCA pump inhibitor to deplete ER Ca^{2+} stores. Using a similar approach, TRPM2 was shown to mediate ADPR-induced Ca^{2+} release from LELs.[18] The LEL source of Ca^{2+} can be confirmed using the proton pump inhibitor bafilomycin A to disrupt lysosomal pH gradients and, consequently, the Ca^{2+} storage.[18] Because ADPR is a membrane-impermeable molecule, it can be delivered by a patch electrode or through photolysis of pre-injected caged-ADPR.[18] Photo un-caging has also been used to study NAADP-activated TPCs in the LEL.[46]

Genetically encoded Ca^{2+} indicators (GECIs) are a new generation of research tools for Ca^{2+} signaling that use luminescent and/or fluorescent reporter genes, whose spectral properties are modified upon binding to Ca^{2+}.[51] On the basis of their structure, there are at least three different types of GECIs. Aequorins are naturally bioluminescent reporters, whose Ca^{2+}-dependent activation requires a co-factor called coelenterazine.[52] The second class of probes, which includes Pericams and G-CaMPs, has a Ca^{2+}-binding sequence, for example, from calmodulin (CaM), inserted into a single reporter protein, so that Ca^{2+} binding will change the spectral properties of the reporter protein.[53] The third class of probes, which includes cameleon, has a Ca^{2+}-binding motif inserted between two different reporter proteins so that, upon Ca^{2+} binding, the efficiency of fluorescence resonance energy transfer (FRET) changes.[54]

One major advantage of GECIs over fluorescent dyes is that GECIs can be targeted to desired organelles by fusing the construct to organelle-specific targeting motifs.[55] Thus, GECIs can be tuned to measure Ca^{2+} levels in different intracellular compartments, such as the ER or LELs. In addition, GECIs can detect Ca^{2+} levels in microdomains in living cells. Furthermore, low-affinity (micromolar range) GECIs allow luminal Ca^{2+} measurements if they are engineered to localize on the luminal side of target proteins or TRPs. In contrast, high-affinity (several hundred nanomolar range) GECIs allow juxta-organellar Ca^{2+} measurement if they are engineered to accumulate on the cytoplasmic side of target proteins or TRPs. For example, aequorin fused to the luminal side of vesicle-associated membrane protein-2 (VAMP-2), a single transmembrane SNARE protein expressed on the surface of secretory vesicles/ granules, was used to monitor the Ca^{2+} levels of dense core vesicles in neuroendocrine cells.[56] Similar approaches could be applied to monitor Ca^{2+} levels in other organelles by generating chimeric proteins between aequorin and organelle-specific proteins, including intracellular TRPs. For example, a chimera between VAMP-7, a LEL-specific SNARE protein, and aequorin could be used to monitor the channel activities of TRPMLs.

19.4   CONCLUSION

Several intracellular TRPs have been well characterized with regard to their subcellular localization and channel properties. Further, the functions of intracellular TRPs can be revealed by correlating their subcellular localization and channel properties. Considering the Ca^{2+} permeability of most TRPs and the Ca^{2+} dependence of membrane traffic, intracellular TRPs are likely involved in regulating vesicular trafficking. Indeed, the accumulation of lipids and other macromolecules in LELs observed in MLIV patients suggests a role for TRPMLs in lipid trafficking and lysosome biogenesis.[57,58] Lysosomal biogenesis is a Ca^{2+}-dependent process involving membrane fusion/fission of vacuoles containing different materials along the endosomal–lysosomal pathway.[59] It is possible that unidentified trafficking cues activate TRPMLs to release Ca^{2+}, which then bind to Ca^{2+} sensors, for example, CaM and synaptotagmins, in the LEL and induce membrane fusion/fission events.[7] In the near future, with the TRP-resident compartments identified, more intracellular functions of TRPs will be revealed.

ACKNOWLEDGMENTS

The work in the authors' laboratory is supported by start-up funds to H. X. from the Department of MCDB and Biological Science Scholar Program, the University of Michigan, and an NIH RO1 grant (NS062792 to H. X). We appreciate the encouragement and helpful comments from other members of the Xu laboratory.

REFERENCES

1. Dong, X. P., X. Wang, and H. Xu. 2010. TRP channels of intracellular membranes. *Journal of Neurochemistry* 113(2): 313–328.
2. Montell, C. 2005. The TRP superfamily of cation channels. *Science STKE* 2005(272): re3.
3. Ramsey, I. S., M. Delling, and D. E. Clapham. 2006. An introduction to TRP channels. *Annual Review of Physiology* 68: 619–647.
4. Moran, M. M., H. Xu, and D. E. Clapham. 2004. TRP ion channels in the nervous system. *Current Opinion in Neurobiology* 14(3): 362–369.
5. Clapham, D. E. 2007. Calcium signaling. *Cell* 131(6): 1047–1058.
6. Zhou, J. 2009. Polycystins and primary cilia: primers for cell cycle progression. *Annual Review of Physiology* 71: 83–113.
7. Cheng, X., D. Shen, M. Samie, and H. Xu. 2010. Mucolipins: intracellular TRPML1-3 channels. *FEBS Letters* 17: 2013.

8. Berridge, M. J., M. D. Bootman, and H. L. Roderick. 2003. Calcium signalling: dynamics, homeostasis and remodelling. *Nature Reviews. Molecular Cell Biology* 4(7): 517–529.

9. Ibraheem, A., and R. E. Campbell. 2010. Designs and applications of fluorescent protein-based biosensors. *Current Opinion in Chemical Biology* 14(1): 30–36.

10. Pryor, P. R., F. Reimann, F. M. Gribble, and J. P. Luzio. 2006. Mucolipin-1 is a lysosomal membrane protein required for intracellular lactosylceramide traffic. *Traffic* 7(10): 1388–1398.

11. Venkatachalam, K., T. Hofmann, and C. Montell. 2006. Lysosomal localization of TRPML3 depends on TRPML2 and the mucolipidosis-associated protein TRPML1. *Journal of Biological Chemistry* 281(25): 17517–17527.

12. Vergarajauregui, S., and R. Puertollano. 2006. Two di-leucine motifs regulate trafficking of mucolipin-1 to lysosomes. *Traffic* 7(3): 337–353.

13. Kim, H. J., A. A. Soyombo, S. Tjon-Kon-Sang, I. So, and S. Muallem. 2009. The Ca²⁺ channel TRPML3 regulates membrane trafficking and autophagy. *Traffic* 10(8): 1157–1167.

14. Martina, J. A., B. Lelouvier, and R. Puertollano. 2009. The calcium channel mucolipin-3 is a novel regulator of trafficking along the endosomal pathway. *Traffic* 10(8): 1143–1156.

15. Dong, X. P., X. Wang, D. Shen et al. 2009. Activating mutations of the TRPML1 channel revealed by proline-scanning mutagenesis. *Journal of Biological Chemistry* 284(46): 32040–32052.

16. Song, Y., R. Dayalu, S. A. Matthews, and A. M. Scharenberg. 2006. TRPML cation channels regulate the specialized lysosomal compartment of vertebrate B-lymphocytes. *European Journal of Cell Biology* 85(12): 1253–1264.

17. Karacsonyi, C., A. S. Miguel, and R. Puertollano. 2007. Mucolipin-2 localizes to the Arf6-associated pathway and regulates recycling of GPI-APs. *Traffic* 8(10): 1404–1414.

18. Lange, I., S. Yamamoto, S. Partida-Sanchez, Y. Mori, A. Fleig, and R. Penner. 2009. TRPM2 functions as a lysosomal Ca²⁺-release channel in beta cells. *Science Signaling* 2(71): ra23.

19. Saito, M., P. I. Hanson, and P. Schlesinger. 2007. Luminal chloride-dependent activation of endosome calcium channels: patch clamp study of enlarged endosomes. *Journal of Biological Chemistry* 282(37): 27327–27333.

20. Grimm, C., S. Jors, S. A. Saldanha et al. 2010. Small molecule activators of TRPML3. *Chemical Biology* 17(2): 135–148.

21. Hisanaga, E., M. Nagasawa, K. Ueki, R. N. Kulkarni, M. Mori, and I. Kojima. 2009. Regulation of calcium-permeable TRPV2 channel by insulin in pancreatic beta-cells. *Diabetes* 58(1): 174–184.

22. Kanzaki, M., Y. Q. Zhang, H. Mashima, L. Li, H. Shibata, and I. Kojima. 1999. Translocation of a calcium-permeable cation channel induced by insulin-like growth factor-I. *Nature. Cell Biology* 1(3): 165–170.

23. Wainszelbaum, M. J., B. M. Proctor, S. E. Pontow, P. D. Stahl, and M. A. Barbieri. 2006. IL4/PGE2 induction of an enlarged early endosomal compartment in mouse macrophages is Rab5-dependent. *Experimental Cell Research* 312(12): 2238–2251.

24. Clapham, D. E. 2003. TRP channels as cellular sensors. *Nature* 426(6966): 517–524.

25. Zeevi, D. A., A. Frumkin, V. Offen-Glasner, A. Kogot-Levin, and G. Bach. 2009. A potentially dynamic lysosomal role for the endogenous TRPML proteins. *Journal of Pathology* 219(2): 153–162.

26. Vergarajauregui, S., J. A. Martina, and R. Puertollano. 2009. Identification of the penta-EF-hand protein ALG-2 as a Ca²⁺-dependent interactor of mucolipin-1. *Journal of Biological Chemistry* 284(52): 36357–36366.

27. Koulen, P., Y. Cai, L. Geng, Y. Maeda, S. Nishimura, R. Witzgall, B. E. Ehrlich, and S. Somlo. 2002. Polycystin-2 is an intracellular calcium release channel. *Nature. Cell Biology* 4(3): 191–197.

28. Lev, S., D. A. Zeevi, A. Frumkin, V. Offen-Glasner, G. Bach, and B. Minke. 2010. Constitutive activity of the human TRPML2 channel induces cell degeneration. *Journal of Biological Chemistry* 285(4): 2771–2782.

29. Cuajungco, M. P., and M. A. Samie. 2008. The varitint-waddler mouse phenotypes and the TRPML3 ion channel mutation: cause and consequence. *Pflügers Archiv* 457(2): 463–473.

30. Grimm, C., M. P. Cuajungco, A. F. van Aken, M. Schnee, S. Jors, C. J. Kros, A. J. Ricci, and S. Heller. 2007. A helix-breaking mutation in TRPML3 leads to constitutive activity underlying deafness in the varitint-waddler mouse. *Proceedings of the National Academy of Sciences of the United States of America* 104(49): 19583–19588.

31. Kim, H. J., Q. Li, S. Tjon-Kon-Sang, I. So, K. Kiselyov, and S. Muallem. 2007. Gain-of-function mutation in TRPML3 causes the mouse Varitint-Waddler phenotype. *Journal of Biological Chemistry* 282(50): 36138–36142.

32. Xu, H., M. Delling, L. Li, X. Dong, and D. E. Clapham. 2007. Activating mutation in a mucolipin transient receptor potential channel leads to melanocyte loss in varitint-waddler mice. *Proceedings of the National Academy of Sciences of the United States of America* 104(46): 18321–18326.

33. Di Palma, F., I. A. Belyantseva, H. J. Kim, T. F. Vogt, B. Kachar, and K. Noben-Trauth. 2002. Mutations in Mcoln3 associated with deafness and pigmentation defects in varitint-waddler (Va) mice. *Proceedings of the National Academy of Sciences of the United States of America* 99(23): 14994–14999.

34. Puertollano, R., and K. Kiselyov. 2009. TRPMLs: in sickness and in health. *American Journal of Physiology. Renal Physiology* 296(6): F1245–F1254.

35. Nagata, K., L. Zheng, T. Madathany, A. J. Castiglioni, J. R. Bartles, and J. Garcia-Anoveros. 2008. The varitint-waddler (Va) deafness mutation in TRPML3 generates constitutive, inward rectifying currents and causes cell degeneration. *Proceedings of the National Academy of Sciences of the United States of America* 105(1): 353–358.

36. Kim, H. J., T. Jackson, and K. Noben-Trauth. 2003. Genetic analyses of the mouse deafness mutations varitint-waddler (Va) and jerker (Espnje). *Journal of Association for Research in Otolaryngology* 4(1): 83–90.

37. Hu, H. Z. et al. 2004. 2-aminoethoxydiphenyl borate is a common activator of TRPV1, TRPV2, and TRPV3. *Journal of Biological Chemistry* 279(34): 35741–35748.

38. Beck, A., M. Kolisek, L. A. Bagley, A. Fleig, and R. Penner. 2006. Nicotinic acid adenine dinucleotide phosphate and cyclic ADP-ribose regulate TRPM2 channels in T lymphocytes. *Faseb Journal* 20(7): 962–964.

39. Du, J., J. Xie, and L. Yue. 2009. Intracellular calcium activates TRPM2 and its alternative spliced isoforms. *Proceedings of the National Academy of Sciences of the United States of America* 106(17): 7239–7244.

40. Dong, X. P., X. Cheng, E. Mills, M. Delling, F. Wang, T. Kurz, and H. Xu. 2008. The type IV mucolipidosis-associated protein TRPML1 is an endolysosomal iron release channel. *Nature* 455(7215): 992–996.

41. Cerny, J., Y. Feng, A. Yu et al. 2004. The small chemical vacuolin-1 inhibits Ca^{2+}-dependent lysosomal exocytosis but not cell resealing. *EMBO Report* 5(9): 883–888.

42. Mayer, M., J. K. Kriebel, M. T. Tosteson, and G. M. Whitesides. 2003. Microfabricated teflon membranes for low-noise recordings of ion channels in planar lipid bilayers. *Biophysics Journal* 85(4): 2684–2695.

43. Zhang, F., and P. L. Li. 2007. Reconstitution and characterization of a nicotinic acid adenine dinucleotide phosphate (NAADP)-sensitive Ca^{2+} release channel from liver lysosomes of rats. *Journal of Biological Chemistry* 282(35): 25259–25269.

44. Churchill, G. C., Y. Okada, J. M. Thomas, A. A. Genazzani, S. Patel, and A. Galione. 2002. NAADP mobilizes Ca^{2+} from reserve granules, lysosome-related organelles, in sea urchin eggs. *Cell* 111(5): 703–708.

45. Zhang, F., S. Jin, F. Yi, and P. L. Li. 2008. TRP-ML1 functions as a lysosomal NAADP-sensitive Ca²⁺ release channel in coronary arterial myocytes. *Journal of Cellular and Molecular Medicine* 13(9B): 3174–3185.

46. Calcraft, P. J., M. Ruas, Z. Pan et al. 2009. NAADP mobilizes calcium from acidic organelles through two-pore channels. *Nature* 459(7246): 596–600.

47. Geng, L., W. Boehmerle, Y. Maeda et al. 2008. Syntaxin 5 regulates the endoplasmic reticulum channel-release properties of polycystin-2. *Proceedings of the National Academy of Sciences of the United States of America* 105(41): 15920–15925.

48. Grynkiewicz, G., M. Poenie, and R. Y. Tsien. 1985. A new generation of Ca²⁺ indicators with greatly improved fluorescence properties. *Journal of Biological Chemistry* 260(6): 3440–3450.

49. Caterina, M. J., M. A. Schumacher, M. Tominaga, T. A. Rosen, J. D. Levine, and D. Julius. 1997. The capsaicin receptor: a heat-activated ion channel in the pain pathway. *Nature* 389(6653): 816–824.

50. Turner, H., A. Fleig, A. Stokes, J. P. Kinet, and R. Penner. 2003. Discrimination of intracellular calcium store subcompartments using TRPV1 (transient receptor potential channel, vanilloid subfamily member 1) release channel activity. *Biochemistry Journal* 371(Part 2): 341–350.

51. McCombs, J. E., and A. E. Palmer. 2008. Measuring calcium dynamics in living cells with genetically encodable calcium indicators. *Methods* 46(3): 152–159.

52. Inouye, S., M. Noguchi, Y. Sakaki, Y. Takagi, T. Miyata, S. Iwanaga, and F. I. Tsuji. 1985. Cloning and sequence analysis of cDNA for the luminescent protein aequorin. *Proceedings of the National Academy of Sciences of the United States of America* 82(10): 3154–3158.

53. Baird, G. S., D. A. Zacharias, and R. Y. Tsien. 1999. Circular permutation and receptor insertion within green fluorescent proteins. *Proceedings of the National Academy of Sciences of the United States of America* 96(20): 11241–11246.

54. Miyawaki, A., J. Llopis, R. Heim, J. M. McCaffery, J. A. Adams, M. Ikura, and R. Y. Tsien. 1997. Fluorescent indicators for Ca²⁺ based on green fluorescent proteins and calmodulin. *Nature* 388(6645): 882–887.

55. Demaurex, N. 2005. Calcium measurements in organelles with Ca²⁺-sensitive fluorescent proteins. *Cell Calcium* 38(3–4): 213–222.

56. Mitchell, K. J., P. Pinton, A. Varadi, C. Tacchetti, E. K. Ainscow, T. Pozzan, R. Rizzuto, and G. A. Rutter. 2001. Dense core secretory vesicles revealed as a dynamic Ca²⁺ store in neuroendocrine cells with a vesicle-associated membrane protein aequorin chimaera. *Journal of Cell Biology* 155(1): 41–51.

57. Piper, R. C., and J. P. Luzio. 2004. CUPpling calcium to lysosomal biogenesis. *Trends in Cell Biology* 14(9): 471–473.

58. Zeevi, D. A., A. Frumkin, and G. Bach. 2007. TRPML and lysosomal function. *Biochimica et Biophysica Acta* 1772(8): 851–858.

59. Luzio, J. P., P. R. Pryor, and N. A. Bright. 2007. Lysosomes: fusion and function. *Nature Reviews. Molecular Cell Biology* 8(8): 622–632.

20 Methods for Studying *Drosophila* TRP Channels

Elkana Kohn and Baruch Minke

CONTENTS

20.1 INTRODUCTION

Phototransduction is a process by which light is converted into electrical signals used by the central nervous system. Invertebrate phototransduction is a process mediated by the phosphoinositide signaling cascade, characterized by phospholipase C (PLC) as the effector enzyme[1] and the transient receptor potential (TRP) channels

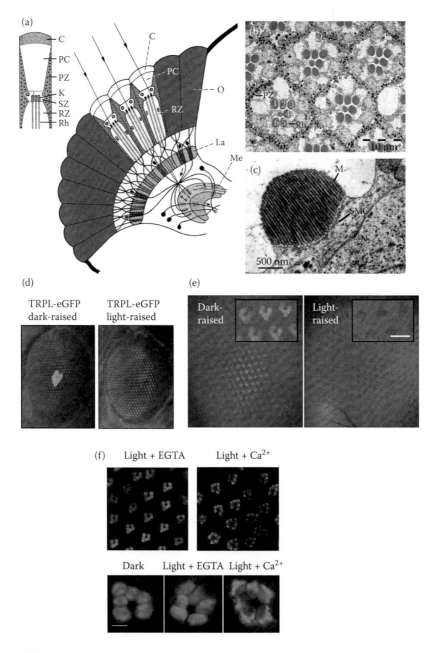

FIGURE 20.1 **(See color insert.)** The morphology of the *Drosophila* compound eye. The *Drosophila* compound eyes are made up of well-organized units termed ommatidia (a–e). (a) Morphology of the ommatidium, consisting of 20 cells. The cells in the ommatidium are the cornea (C), Semper cells (SZ), primary pigment cells (PZ), and an extracellular fluid-filled cavity called pseudocone (PC). Below are eight photoreceptor cells (RZ). (Adapted from Kirschfeld, K., *Experimental Brain Research*, 248–270, 1967. With permission.)

as its target.[2] The great advantage of using invertebrate photoreceptors is the accessibility of the preparation, the ease of light stimulation, the robust expression of key molecular components, and most importantly, the ability to apply the power of molecular genetics. This last feature is mainly attributed to *Drosophila melanogaster* as a preferred animal model. Extensive genetic studies have established the fruit fly, *Drosophila melanogaster*, as an extremely useful experimental model for genetic dissection of complex biological processes.[3] The relatively small size of the *Drosophila* genome, ease of growth, and rapid generation time make this system ideally suited for screening large numbers of mutagenized individual flies for defects in virtually any phenotypically observable or measurable trait. The creation of balancer chromosomes, containing dominant markers and multiple inversions, which prevent recombination with the native chromosomes, allows any mutation, once recognized, to be rapidly isolated and maintained. Importantly, germ line transformation, using P-element transposition, combined with the availability of tissue-specific promoters, allows for the introduction of cloned genes and subsequent expression in specific

FIGURE 20.1 (Continued) (b, c) Electron microscopic (EM) cross sections of (b) ommatidia and a (c) rhabdomere (Rh). The rhabdomere consists of a stack of ~30,000–50,000 microvilli (M). The transduction machinery is arrayed in the rhabdomere, while the nucleus (N) and cellular organelles like submicrovillar cisternae (SMC) reside at the cell body. The eight photoreceptors can be divided into two functional groups according to their position, spectral specificity, and axonal projection. The R1–6 cells (marked 1-6 in b) represent the major class of photoreceptors in the retina. They have peripherally located rhabdomeres, express the Rh1 opsin that forms a blue-absorbing rhodopsin (R, *text*), or orange-absorbing metarhodopsin (M, *text*), and project their axons to the first optic lobe, the lamina (La, green area in a). The second group includes R7 (marked 7 in b) and R8 (not shown) which have rhabdomeres located in the center of each ommatidium. Both R7 and R8 span only half the length of the retina, and they are stacked with R7 being on top of R8. Their rhabdomere expresses one of two opsins, Rh3 or Rh4 (R7) and Rh3, Rh5, or Rh6 (R8). The absorption spectrum of R is UV (for Rh3 and Rh4), blue (Rh5), or green (Rh6). The R7 and R8 cells project their axons to the second optic lobe, the medulla (Me, pink area in a). (Adapted from Minke, B., and Z. Selinger, *Current Opinion in Neurobiology*, 459–466, 1996. With permission.) (d) Green fluorescing deep pseudopupil of TRPL-eGFP expressing flies (*TRPL-eGFP*) raised in the dark (left panel). Fluorescence of the pseudopupil was not observed when the flies were raised in the light (right panel). (Adapted from Mayer, N.E. et al., *Journal of Cell Science*, 2592–2603, 2006. With permission.) (e) Optical neutralization of the cornea by water immersion displays the subcellular localization of TRPL-eGFP in (left) dark-raised and (right) light-raised (for 16 hours under orange light) intact transgenic flies (*yw; TRPL-eGFP*). The insets show the central area of the eye at higher magnification. Scale bar, 15 μm. (Adapted from Mayer, N.E. et al., *Journal of Cell Science*, 2592–2603, 2006. With permission.) (f) Removal of extracellular Ca^{2+} inhibits light-induced translocation of TRPL-eGFP. The confocal images (using a 60× water immersion objective) show optical sections of live isolated retinas of flies expressing TRPL-eGFP on a null *TRPL* background. Sliced heads were incubated in oxygenated extracellular solution supplemented with 1% FBS and 5 mM sucrose. Images were obtained from sliced heads illuminated for 4 hours, and incubation solution was supplemented with (upper left and lower middle) 1 mM EGTA or (upper and lower right) 1 mM Ca^{2+}, or (lower left) dark-raised flies. Bar, 10 and 2 μm for the upper and lower panels, respectively. (Adapted from Mayer, N.E. et al., *Journal of Cell Science*, 2592–2603, 2006. With permission.)

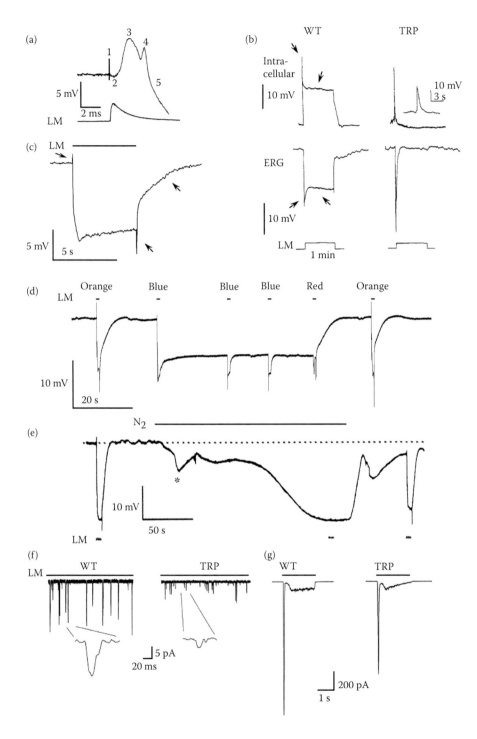

cells of the organism, thus providing a method to study in vitro modified gene products in their native cellular environment.[4] These powerful molecular genetic tools, combined with the available genome sequence, allow screening for mutants defective in critical molecules while devoid of a priori assumptions. Indeed, this methodology has produced an unequivocally large number of mutant flies with defects in novel proteins that would have been otherwise difficult to predict.[3]

Phototransduction begins by absorption of a photon by Rhodopsin (R), which belongs to the superfamily of seven transmembrane G-protein-coupled receptors. The absorbed photon isomerizes the chromophore (11-*cis* 3-hydroxy retinal[5]), resulting in a conformational change of the opsin molecule and production of the physiologically active metarhodopsin (M) state of the photopigment.[6] To ensure high sensitivity,

FIGURE 20.2 The electrophysiological responses to light of the *Drosophila eye.* (a) The metarhodopsin (M) potential. Shown is the initial part of the ERG, recorded from wild-type fly in response to intense orange flash following 5 s of intense blue light. The numbers indicate: 1, artifact at flash onset; 2, a linear response of the photoreceptor cells arising from charge displacement of the Rh1 photopigment molecules at the M state; 3, the amplified response of the M molecules by sign inverting synapse of lamina neurons; 4, the "on" transient of the lamina neurons; 5, the initial extracellular response of the photoreceptor cells due to openings of the TRP and TRPL channels. (Adapted from Minke, B., and K. Kirschfeld, *Journal of General Physiology*, 517–540, 1979. With permission.) (b) Intracellular (upper panel) and the corresponding ERG (lower panel) recordings from a (left) wild-type (WT) and a (right) *TRP* mutant fly in response to prolonged white light. (Adapted from Peretz, A. et al., *Journal of General Physiology*, 1057–1077, 1994. With permission.) (c) The typical ERG in response to medium intensity orange light. The "on" and "off" transients of the lamina neurons and the slow components of the pigment cells are indicated by arrows. (d) The induction and suppression of the PDA. An intense blue light pulse converted ~80% of the Rh1 photopigment from R to M and resulted in prolonged corneal negative response that continued in the dark, owing to maximal activation of the channels, making R1–6 cells nonresponsive. Two additional intense blue lights elicited small responses that originated from R7 and R8 cells. The following red light suppressed the PDA after light is turned off. Intense orange light pulses were applied before and after the PDA. Light monitor (LM) indicate the periods of light. (e) Anoxia induces a rapid and reversible depolarization of *Drosophila* photoreceptors in the dark and abolishes light excitation. The trace shows an extracellular voltage change recordings of intact *Drosophila* eye in response to application of anoxia (N_2, as indicated). Orange light pulses (OG590, indicated as LM below) before, during, and after demonstrate the recovery from anoxia. Note that a light pulse applied during the maximal voltage response to anoxia did not elicit any response. Asterisk indicates movement artifact, which probably resulted from movements of the fly at anoxia onset. (Adapted from Agam, K. et al., *Journal of Neuroscience*, 5748–5755, 2000. With permission.) (f) Whole-cell patch-clamp recordings of quantum bumps from photoreceptor cells of (left) wild-type (WT) and (right) *TRP^{P343}* mutant. Quantum bumps are elicited in response to very dim orange light. (Adapted from Katz, B., and B. Minke, *Frontiers in Cellular Neuroscience*, 1–18, 2009. With permission.) (g) The macroscopic light induced current in response to 100,000 folds more intense light (relative to panel f). The bumps sum up to produce a noisy LIC (note the change in scale). In WT, the LIC shows an initial fast transient that declined to a small steady-state response. In the *TRP^{P343}* mutant, the "steady-state" response declined to baseline during light. (Adapted from Katz, B., and B. Minke, *Frontiers in Cellular Neuroscience*, 1–18, 2009. With permission.)

high temporal resolution, and low dark noise of the photoresponse, the active M has to be quickly inactivated and recycled. The latter requirement is achieved by two means: the absorption of an additional photon by the dark stable M, which photo-converts M back to R, or by a multistep photochemical cycle.[6] Photoexcited R activates a heterotrimeric G protein. Combined biochemical and electrophysiological experiments revealed that the light-dependent $G_q\alpha$ activates PLC.[1] The key evidence for the actual participation of PLC in visual excitation of the fly was provided by isolation and analysis of the *Drosophila* PLC gene, designated no receptor potential A (*norpA*). The *norpA* gene encodes a β-class PLC, which is the only PLC isoform expressed in the photo-signaling compartment called rhabdomeres[7] (Figure 20.1b, c). The *norpA* mutant provides essential evidence for the critical role of inositol-lipid signaling in phototransduction, by showing that no excitation takes place in the absence of functional PLC.[7,8]

Activation of the PLCβ enzyme hydrolyzes the minor membrane phospholipid, phosphatidylinositol 4,5-bisphosphate (PIP_2) into inositol 1,4,5-trisphosphate (IP_3), and diacylglycerol (DAG).[9] Subsequently, DAG production (directly or indirectly) causes in a still unclear way, coordinated opening of TRP and TRP-like (TRPL) channels,[10–12] which results in Ca^{2+} and Na^+ influx that produce the single photon response designated quantum bump (Figure 20.2f). Summation of the quantum bumps produces the macroscopic receptor potential (Figure 20.2g).[13] To explain the brief (~25 ms) duration of an average bump, each step of the cascade requires both fast activation and fast termination. Shutoff of the activated intermediates consists of several biochemical reactions: M deactivation is achieved by binding of arrestin2 (Arr2), which prevents association of M to the G-protein.[14–16] Deactivation of PLCβ is obtained by GTP hydrolysis and subsequent breakdown of the active $G_q\alpha$-GTP-PLCβ complex. This GTPase reaction is greatly facilitated by PLC itself acting as a GTPase activating protein (GAP).[17]

The TRP channel was discovered owing to a spontaneously occurring *Drosophila* mutant, which exhibits a decline in the receptor potential to baseline during prolonged illumination (Figure 20.2a, b). This mutant was designated transient receptor potential (*TRP*) by Minke and colleagues.[18] Minke and Selinger suggested that the *TRP* gene encodes a Ca^{2+} channel/transporter mainly because application of the Ca^{2+} channel blocker La^{3+} to wild-type photoreceptors mimicked the *TRP* phenotype.[8] Subsequently, the cloning of the *TRP* locus by Montell and Rubin and by Wong and colleagues revealed a novel membrane protein.[19,20] The available sequence of the *TRP* gene led several years later to the discovery of mammalian TRPs[21,22] and the TRP superfamily. However, the significance of the *TRP* sequence, as a gene encoding a putative channel protein, was only first appreciated after a comparative patch-clamp study of the wild-type fly, and the *TRP* mutant conducted by Hardie and Minke revealed that TRP is the major route of Ca^{2+} entry into the photoreceptor cells,[2] and after a *TRP* homologue, *TRP-like* (*TRPL*), was cloned by Kelly and colleagues.[23] The use of Ca^{2+} indicator dyes and Ca^{2+}-selective microelectrodes, by Minke and colleagues, directly demonstrated that the TRP channel is a robust Ca^{2+} entry pathway into the photoreceptor cell.[24,25] The final evidence showing that TRP and TRPL are the light-activated channels came from the isolation of a null mutant of the *TRPL* gene, by Zuker and colleagues, who demonstrated that the double mutant, *TRPL;TRP*, is blind.[26]

In the following section, we describe methods exploiting the power of *Drosophila* molecular genetics, which have been used for genetic dissection of complex biological processes including activation and regulation of TRP channels.

20.2 GENETICS

A genetic screen is a procedure to identify and select mutated individuals that possess a phenotype with a specific malfunction in a specific trait. This method requires a large number of mutated individuals, usually obtained by the use of mutagens or by random DNA insertions (transposons) and a simple but powerful isolation procedure. Additionally, this method requires genetic tools to isolate and maintain the stability of the mutation through generations. These requirements make *Drosophila* an ideal organism for genetic screens. Wide genetic screens targeting the two autosomes (II and III) and the X chromosome of *Drosophila* have produced many mutants with identified phenotypes linked specifically to genes involved in the visual behavior or the photoreceptor cell function. These screens used defects in a number of visual functions such as optomotor response, phototaxis, and electrophysiological response to light. The isolation of mutants, specifically defective in the visual transduction pathway, initially made use of the electroretinogram (ERG), which measures the electrical activity of the eye at the corneal level[27] (Figure 20.2c). However, the ERG methodology failed to isolate the large number of mutants expected from a multistep and complex process such as the phototransduction cascade. The main reasons for this failure are that the phototransduction proteins are highly abundant, and the upper limit in the depolarization signal is reached even when only a small fraction of the signaling molecules are excited. Therefore, even mutations causing a significant reduction in phototransduction components could not be identified by this method. Similarly, this method could not identify mutations causing subtle malfunction in the components of the phototransduction cascade. This necessitated the employment of a more sensitive and yet simple method for isolation of mutants.

In order to find mutants that have specific malfunctions in the phototransduction cascade and to identify those with subtle phenotypes, Pak and colleagues[3] employed a method devised by Hillman, Hochstein, and Minke, designated Prolonged Depolarizing Afterpotential (PDA, Figure 20.2d).[6] This method is based on a large net photoconversion of R to its dark stable photoproduct M with a minimal conversion back to R. This method brings the capacity of phototransduction to its upper limit.[6] PDA is achieved in the fly by genetically removing the red screening pigment (Figure 20.1a, b) and by applying blue light, which is preferentially absorbed by the R state of the Rh1 photopigment. A large net photoconversion of R to M prevents phototransduction termination at the photopigment level. As a result, excitation is sustained long after the light is turned off, and the PDA producing cells are maximally activated and unable to respond further (termed inactivation, Figure 20.2d). Subsequently, application of orange light reconverts the activated M back to R, thus terminating the sustained excitation after the light is turned off.[28] Because the PDA tests the maximal capacity of the photoreceptor cell to maintain excitation for an extended period and is strictly dependent on the presence of high concentrations of R, it detects even minor defects in R biogenesis, or exhaustion of critical signaling molecules. Thus, defects in the PDA easily reveal deficiencies in the concentration of phototransduction components. Indeed, the PDA screen yielded a plethora of novel and interesting visual mutants. One group of PDA mutants exhibits a loss in several features of the PDA. They are termed *nina* mutants, which stands for neither inactivation nor afterpotential (Figure 20.2a, lower trace). Most *nina* mutants were found to be caused by reduced levels of R. The second group of PDA mutants lost the ability to produce the voltage response associated with the PDA but were still inactivated by strong blue light, and the inactivation could be relieved by orange light. These mutants are termed *ina*, which stands for inactivation but no afterpotential (no PDA voltage response).[29] The *ina* mutants were found to have normal R levels but deficiencies in proteins associated with the function of the TRP channel. The *nina* and *ina* mutants have led to the identification of most of the crucial components of *Drosophila* phototransduction, many of which are novel proteins of general importance for many cells and tissues.[3]

20.2.1 MUTAGENESIS

The ability to induce DNA lesions and the relative ease in identifying the new mutations have made *Drosophila* an important model organism. Ethyl methanesulfonate (EMS) mutagenesis is one of the most frequently used techniques for forward genetics in *Drosophila*. In a few *Drosophila* generations, one can obtain thousands of fly strains, each carrying a few point mutations in the genome.[30] For identifying a mutation in a gene of interest, in such a large stock of flies, an effective screen of the stock is needed.[31]

In EMS mutagenesis, new mutations are induced at random sites once every 150–300 kb on average (depending on EMS concentrations), i.e., about 50–100 open reading frame DNA mutations in a single chromosome, of which about 12% cause inactivation of the gene.[31] To handle such large amounts of mutations, an effective method for screening is needed. For practical considerations, only one chromosome is screened for mutations and stable or semi-stable lines are established. The screening method should be of high throughput nature, low cost, and should easily discriminate phenotypes. Such a screen can be behavioral, i.e., a stimulus that elicits typical behavior, like avoidance or attraction.[32] Another screen can be for specific phenotype, like electrical response to light, such as the ERG or PDA (see above). In the past few years, screens using expression of reporter genes, like GFP, took place to investigate protein activation and regulation. For example, a screen for mutations causing retinal degeneration was performed by expressing GFP in the rhabdomeres and flies that had lost the deep pseudopupil (DPP, see *photoreceptor imaging*) owing to retinal degeneration were analyzed for the mutation.[33] Another screen for mutations affecting TRPL translocation was performed by mutating transgenic flies expressing TRPL-GFP and screening for flies with defects in TRPL translocation.[34]

Below, we present two examples of protocols used to isolate mutations on the second and on the X chromosome. Similar schemes can be adapted to other chromosomes.

20.2.1.1 Second Chromosome Mutation Screen

The second chromosome screen exploits females that carry a dominant temperature-sensitive mutation. Starve male flies for two days and then feed them with sucrose solution containing 25 mM EMS for 30 min. Place flies in a clean vial overnight. Cross the treated males with DTS/SM5 females, which are heterozygous for the second chromosome so that one chromosome carries the dominant temperature-sensitive mutation DTS over the background of the balancer chromosome, with curly wings as a phenotypic marker. The progeny of the crossing are first generation flies (F_1) that are heterozygous for the mutated chromosome, over the background of either DTS or SM5 chromosomes. Expose the larva to heat shock (two days at 30°C), which will lead to death of all larva carrying the DTS mutation. Every surviving fly in the F_1 generation is heterozygote for the EMS mutated chromosome over the background of SM5 balancer chromosome. To increase the number of flies carrying each putative mutation, cross each F1 male with several DTS/SM5 females to generate the F_2 generation. Expose F_2 larva to heat shock. The surviving flies establish a semi-stable stock of identical mutated second chromosome.[29]

20.2.1.2 X Chromosome Mutation Screen

An X chromosome screen exploits an "attached-X" strain in which the females have two copies of X chromosome attached to each other (designated X^X) and a Y chromosome. Crossing an attached-X female with a male produces two lethal combinations: Y/Y and X^X/X and two viable combinations: X^X/Y (female) and X/Y (male). Expose males to EMS (as described above) and cross them with attached-X females. As stated above, the F_1 generation has two viable and two lethal progeny: viable *X/Y males carry the EMS-treated chromosome with putative mutations. Cross each male carrying a unique mutated X chromosome with a few attached-X females to establish a stable stock in which all the males have the same putative mutated X chromosome.[29]

20.2.2 Transposable P-Element and P-Element-Mediated Germ Line Transformation

P-element is a 2907-bp DNA sequence encoding a transposase specific for moving its own sequence within the genome. It has a pair of 31-bp inverted repeats at both ends serving as a recognition site for transposition. The principle of this method is to inject into the embryo a transposase that recognizes a P-element containing, instead of a transposase, the gene of interest to be incorporated into the genome. In a transposition event, an 8-bp duplication of the target-site DNA, adjacent to the terminal repeats, is created. Upon excision, the 8-bp duplicate is not transposed with the P-element; thus, excision of a transposable element can also affect genes.[35] The P-element can cause disruption of genes by being randomly inserted into coding regions, promoter, or regulatory sequences. The ongoing Berkeley Gene Disruption Project to disrupt and identify genes by P-element insertions has produced many deficiencies in *Drosophila* genes, as well as alleles for many genes with P-element disrupting their expression.[36,37]

P-element-mediated germ line transformation is a powerful tool to manipulate gene expression in *Drosophila*. In this technique, a DNA plasmid is injected into the posterior side of the *Drosophila* embryo. The plasmid contains: (1) the DNA sequence, which is desired to be incorporated into the genome, (2) an adequate promoter driving expression, (3) a reporter gene, and (4) two 31-bp tandem repeat P-element recognition sites on both ends. Another plasmid, encoding functional transposase, is co-injected to facilitate the incorporation of the P-element into the genome, using the P-element recognition sites.[38,39] The incorporation is random, and subsequent balancing is needed to achieve a stable line. Introduction of a specific promoter (e.g., tissue specific, heat shock, etc.) allows modification of tissue expression patterns as needed. A plasmid that generates dsRNA instead of mRNA can be used to achieve a RNAi strain, which downregulates the specific gene expression.[40]

The protocol for the P-element injection is as follows: Two days before injecting, move a few hundred flies (according to cage size) to a laying cage. The laying cage is made up of a plastic bottle in which a Petri dish with apple juice agar and some baker's yeast spread on it makes the bottom end. The apple juice attracts the females to the bottom of the cage and yeast encourages egg laying. Replace the apple juice agar dish, and add fresh yeast twice a day. On the day of injection, replace the Petri dish with a fresh dish every hour. Thirty minutes before starting the injection, replace the dish again. After 30 min with the new dish, replace it, and wash the embryos that are on the dish with distilled water. Gently brush the embryos onto a metal mesh (hole size 0.1 mm). To remove the chorion, which covers the eggs, place the mesh into 10% sodium hypochlorite (bleach) for 30–60 s, and immediately immerse the mesh in a Petri dish containing distilled water to wash off the bleach. Remove excess water by placing the mesh on a stack of tissue paper, and with a fine brush, gently transfer the embryos from the mesh to a slice of agar. Arrange the embryos in a row near the edge of the agar slice, with their anterior side (which has a small outgrowth of the micropyle) pointing toward the edge. Touch the row of embryos gently with a microscope cover-slide covered with glue, which will allow the embryos to adhere to the slide, and now the posterior side is facing the edge of the slide. In order to reduce osmotic pressure, place the slide with the embryos facing up in a small covered box containing silica gel for 30 s and then cover the embryos with mineral oil to prevent dehydration. Place the slide on a microscope stage. Using a micromanipulator, bring the tip of the injection needle just in front of the embryos. To check whether the needle tip is in place, gently move the needle to touch the posterior of the embryo. The needle should push the posterior end evenly to form a small indentation. After positioning the needle in the z axis, inject each embryo by moving the needle on the x axis (toward the embryos and away) along the row of embryos on the y axis (to place a new embryo for injection). Penetrate into the embryo with the needle; it should not penetrate more than 1/4 of the embryo's total length. Pull the needle back so it will be as close as possible to the posterior end, and inject. The method is based on injection into the syncytium stage of the embryo. Embryos that are older than embryonic stage 2 (i.e., after the syncytium stage) should be killed by penetrating with the needle all the way to the anterior end. This is to assure that only injected embryos will survive. The whole process should take no more than 30 min at 19°C. After injection, put the slide on an apple juice agar Petri dish with some dried yeast. This makes the search for new larva easier. De-chorionated embryos are very sensitive to dehydration, so put the dish in a box with water-soaked tissue paper at 18°C. From 36 to 48 hours after injection, start looking for larva under a stereo microscope. With a toothpick dipped in mineral oil, gently transfer the larva into a fresh vial with food. The P-element is incorporated into the germ cells, and its effect will be recognized in the next generation. Therefore, let the flies eclose and cross each fly with 2–3 flies of the opposite sex. The F_1 progeny of the transformant fly displays the reporter gene, usually red eyes. Eye color can vary from wild-type red eyes to faint orange. Any change in color intensity is an indication for incorporation of the P-element into the genome. Cross the progeny with balancer chromosomes to determine into which chromosome the transgene is inserted.

20.3   EXTRACELLULAR AND INTRACELLULAR VOLTAGE RECORDINGS

20.3.1   Electroretinogram (ERG)

ERG is an extracellular voltage recording from the entire eye, which reflects the total electrical activity arising from the eye in response to light stimulation. The *Drosophila* ERG light response is robust and easily obtained, thus making it a convenient method to identify defects in light response owing to mutations.

Extracellularly Recorded Electrical Activity of the Eye Resulting from Activation of the TRP and TRPL Channels: The ERG response to light arises mainly from the photoreceptor cells, pigment (glia) cells, and secondary neurons of the lamina (see Figure 20.1). The main components of the ERG light response are (1) the extracellularly recorded photoreceptor potential, (2) the "on" and "off" transients, at the beginning and the end of the light pulse, arising from the second order lamina neurons, and (3) the slow response of the pigment cells (Figure 20.2c). The photoreceptor potential is the physiological response to light arising from the light-induced openings of the TRP and TRPL channels. In response to intense light, it is composed of an initial fast transient that declines to a lower steady-state phase because of light adaptation. The depolarization of the photoreceptor cell triggers the induction of the corneal positive "on" transient, via a sign inverting synapse between the photoreceptor axon and the large monopolar neurons of the lamina (Figure 20.1a). The response of the pigment cells appears as a slow rise and slow decay of the ERG after light on and off, respectively (see Figure 20.2c).[41] These slow components arise from an increase in extracellular K^+, resulting from K^+ efflux from the photoreceptor cells via the TRP and TRPL channels, which depolarize the pigment cells surrounding the photoreceptor cells. These components largely distort the waveform of the extracellularly recorded photoreceptor potential (Figure 20.2b, c).

Extracellularly Recorded Electrical Activity of the Eye Reflecting Activation of the Photopigment: The Metarhodopsin Potential (M-Potential). Absorption changes in the photopigment molecules during illumination are accompanied by redistribution of charges in these molecules. This charge displacement induces a spread of current that can be recorded extracellularly (or intracellularly) instantaneously at light onset. Because this is a linear response proportional to the amount of activated photopigment, it is very small when recorded extracellularly, and it is masked by noise. Nevertheless, simultaneous activation of 10^7 M molecules can be detected extracellularly, especially when this electrical signal is amplified by the sign inverting synapse between the photoreceptor cells and the lamina neurons. Accordingly, this electrical signal, designated the M potential,[42] arises in the lamina neurons owing to activation of a large fraction of M molecules and appears without latency before the appearance of the "on" transient (Figure 20.2a).[43] The M potential is a useful and easy to use tool to compare the photopigment levels between wild type (WT) and mutants with reduced amount of photopigment.

The PDA: The PDA phenomenon, like the photoreceptor potential, arises from activation of the TRP and TRPL channels. However, unlike the photoreceptor potential, which quickly declines to baseline after the cessation of the light stimulus (Figure 20.2c), the PDA response, which is induced by a large photoconversion of R to M by intense blue light (see above), continues long after light offset (Figure 20.2d). With the PDA-inducing light on, all eight photoreceptor cells are activated, but after the light goes off, the peripheral photoreceptors (R1–6) remain continuously activated in the dark at their maximal capacity, reaching saturation. Additional intense blue lights do not produce any additional response in R1–6 cells for many seconds. The additional response to saturating blue light, which is superimposed on the PDA, arises from activation of the central photoreceptor cells (R7,8), in which PDA is not induced. The PDA can be quickly suppressed by photoconversion of M to R with intense orange light (Figure 20.2d).[28] Minke and Selinger explained the PDA phenomenon as follows:[44] Cellular ARR2 is present at a concentration that is insufficient to inactivate all the M generated by a large net photoconversion of R to M, leaving an excess of M persistently active in the dark. This explanation easily accounts for the elimination of the PDA response by mutations or by carotenoid deprivation, which reduces the photopigment level. It also accounts for the need to photoconvert large net amounts of R to M to elicit the PDA response.[44] Electrophysiological analysis performed by Zuker and colleagues on null ARR2 (*arr2*) and ARR1 (*arr1*) mutant alleles revealed a set of phenotypes consistent with the stoichiometric requirement of ARR binding for M inactivation in vivo.[15] In their study, they showed that a significant reduction in ARR2 levels leads to abnormally slow termination of light-activated currents and production of a PDA in mutants with low R levels.[45] The PDA has been a major tool to screen for visual mutants of *Drosophila*.[29]

20.3.2 Intracellular Voltage Recordings

The major disadvantage of the ERG is its heterogeneous cellular origin, which distorts the waveform of the typical response of the photoreceptor cells. The electrical response of the photoreceptors can be reliably recorded by intracellular recordings, which display only the light response of the photoreceptor cells (Figure 20.2b). Intracellular recordings are insensitive to responses of cells downstream of the photoreceptor cells, and thus constitute an important tool to confirm that a specific response originates in the photoreceptor cell itself.[29] A major drawback of intracellular recordings is that while ERG recordings can be maintained for hours, it is difficult to perform reliable intracellular recordings for more than 15–30 min.

20.3.3   Protocols for ERG and Intracellular Recordings

20.3.3.1   Fly Preparation

An easy to apply and a reliable protocol for voltage recordings from the intact fly is as follows: Anesthetize a fly with CO_2 and then immobilize it with low melting point wax on a fiberglass stand. Under a stereo microscope, slightly insert the glass recording pipette into the cornea and the ground pipette into the thorax or into a drop of conducting gel (such as that used for electrocardiography) placed on the thorax. The ERG setup is positioned in a Faraday cage covered with black fabric to avoid stimulation by stray light. The electrical responses are recorded using pipettes filled with normal *Drosophila* Ringer's solution (130 mM NaCl, 2 mM KCl, 5 mM $MgCl_2$, 2 mM $CaCl_2$, 10 mM Hepes, titrated to pH = 7.0). For ERG, we use the same pipettes as for whole-cell recordings (see below), but the pipette tip breaks upon contact with the cornea. The voltage signals can be recorded using an amplifier with high input resistance suitable for both extracellular and intracellular recordings (e.g., DP-311, Warner Instruments, Hamden, CT, USA) and digitized with an A/D converter and software (respectively: Digidata and pClamp, both by Molecular Devices, Sunnyvale, CA, USA). For intracellular recordings, an amplifier with a zap function is needed. The fly preparation for intracellular recordings is similar to the ERG preparation. After immobilizing the fly, a razor blade chip is used to cut a small shallow hole in the cornea (without cutting any ommatidia) within ~1/10 of the compound eye surface. Carefully remove the piece of cornea and cover the surface with petroleum jelly. Intracellular receptor potentials are measured with standard long tip sharp glass micropipettes filled with 2 M KCl (electrical resistance 80–140 MΩ), introduced through the corneal opening. Insert a ground (reference) electrode containing *Drosophila* Ringer's solution into the same cut in the cornea to avoid recording the ERG by the reference electrode. Carefully insert the recording electrode into the retina using a micromanipulator. Using the amplifier zap function, apply current through the pipette to break into the cell. Keep on lowering the electrode and pressing the zap until the pipette penetrates a cell, which will be seen as an instantaneous decrease of the electrode potential to approximately −60 mV, reflecting the cell resting potential. A test light pulse, which induces a depolarizing photoreceptor potential, confirms that the electrode has penetrated the cell.

20.3.3.2   Light Stimulation

ERG and intracellular recordings require that light stimuli be versatile in both color and intensity. We use a 100 watt Xenon high-pressure lamp. Light passes through a heat filter and a shutter and is then delivered by fiber optics to the compound eye in an electrically shielded cage (the maximal luminous intensity of orange [OG 590 Schott edge filter] is 13 mW/cm^2). The light stimulus is attenuated up to 6 orders of magnitude by neutral density filters, which are placed between the light source and the electronic shutter, and filtered by red, orange, or blue filters.

The phenotype of *Drosophila* mutants lacking the TRP channel is characterized by a decay of light response back to the dark zero baseline during prolonged (10 s) illumination of medium and above light intensities (hence the mutants were termed transient receptor potential; Figure 20.2b, g). The response to light recovers within

1 min in the dark. Accordingly, an efficient and simple screen for mutations in the *TRP* locus is obtained by an ERG recording in response to a pair of intense 5-s orange light pulses, separated by a 5 s dark interval. In the null *TRP^{p343}* mutant, the amplitude of the response to the second light is ~90% smaller than the first response. In *TRP* with reduced TRP protein levels (hypomorph alleles), the decay time during light is slowed down, and the peak amplitude of response to the second light pulse is larger.

Different mutations in the phototransduction cascade cause a decrease in sensitivity to light, which can be detected using a screen based on the intensity-response relationship. Intensity response paradigms consist of repeated light stimulations with increasing intensity, usually in a logarithmic scale. Increasing rather than decreasing intensities of orange light are required to avoid light adaptation. The orange light is required to avoid induction of a PDA, when white-eyed flies are used. Flies are dark adapted for 5–10 min and then repeated stimulations in 1 min intervals are applied with increasing light intensities. The PDA paradigm consists of illumination with an intense pulse of blue light applied to the white-eyed fly. After 10 s of darkness, apply a second pulse of intense blue light to assure full PDA. Then, apply a saturating pulse of orange light for termination of the PDA (Figure 20.2d).

20.3.3.3   Activation of the TRP and TRPL Channels by Anoxia

An efficient way to activate the TRP and TRPL channels reversibly in vivo in the dark is by inducing anoxic conditions. In a living fly, anoxia can be easily induced by blowing N_2 gas on the fly abdomen, where air inlets reside. Measurements of oxygen consumption (Q) in the retina of the blow fly *Calliphora* in the dark reveals very high Q values (Q = 90 µL of O_2 cm^{-2} min^{-1}). Accordingly, blowing N_2 over the fly is expected to dramatically reduce the partial oxygen pressure (PO_2) to anoxic level.[46,47] A two-phased corneal negative voltage change in the dark is induced by application of N_2 to intact *Drosophila* during ERG recordings (Figure 20.2e). The initial slow and small phase arises from accumulation of K^+ in the extracellular space, which depolarizes the pigment cells (as measured by K^+-sensitive microelectrodes). The subsequent faster and larger phase arises from openings of the TRP and TRPL channels. This phase is accompanied by a large Ca^{2+} influx into the photoreceptor cells (as measured by Ca^{2+}-sensitive microelectrodes) and is completely abolished in the double null mutant *TRPL;TRP*. Application of an intense light pulse during the second phase does not elicit any additional response because all the TRP and TRPL channels are already active. N_2 removal results in a fast return of the ERG response to baseline and recovery of the light response[48] (Figure 20.2e). The protocol for inducing anoxia is as follows:

After the fly is mounted and the electrodes are in place on the eye and thorax, move the end of a 10-mm-diameter metal pipe connected to a cylinder with compressed N_2 close to the abdomen and thorax of the fly. The gas flow intensity has to be carefully tuned; strong gas flow causes perturbations that vibrate the fly and causes mechanical artifacts in the recording, while poor gas flow does not induce anoxia.

20.3.3.4   Pharmacology

Application of TRP activating and inhibiting chemicals to the intact eye is a useful tool for characterizing these channels under physiological conditions. Two important antagonists of TRPs, 2-Aminoethoxydiphenyl borate[49] and carvacrol,[50] have been tested in this way. Introducing substances into the eye while recording electric responses is not an easy task and might affect the recordings. The diffusion of the substances is not uniform and the actual concentration after dilution in the native extracellular solution and hemolymph is largely unknown. In addition, injecting substances into the extracellular space of the eye has some limitations: one cannot use hydrophilic substances that are cellularly impermeable, while hydrophobic substances might not dissolve in Ringer's solution. The setup for injection consists of a PicoSpritzer (Medical System Corp., Greenvale, NY, USA) connected to an injecting pipette, in addition to the recording pipettes. As an injecting pipette, we use the same pipette type that we use for ERG recordings, but the pipette breaks on purpose when inserted into the eye. The volume of the injected drop depends on the pressure applied and the diameter of the tip after breakage, so calibration of the pressure is needed to balance between a drop that is too small to elicit an effect and a drop that is too big and will blow up the eye. The injection protocol is as follows:

To avoid blockage of the injecting pipette, first insert the pipette into the eye to create a hole in the cornea, then retract it back out. With the tip still outside the eye, inject some solution while watching the drop form to assure that the injecting pipette is not blocked. Insert the injecting pipette back into the same hole and start the injection. Control for damage should be done by injecting normal *Drosophila* Ringer's solution.

20.4 WHOLE-CELL PATCH-CLAMP CURRENT RECORDINGS FROM ISOLATED OMMATIDIA

ERG and intracellular recordings are simple to apply but cannot reveal basic parameters of the light response. Whole-cell recording is essential to complete the electrophysiological information.[51] This method uses both current and voltage clamp modes. The major advantage of whole-cell recording under the voltage clamp mode is the derivation of conductance change by measuring the current-voltage (*I-V*) relationship. This measurement reflects the opening and closing of the TRP and TRPL channels, during and following illumination (Figure 20.2g). In addition, because of the low resistance of the recording pipette (~10 MΩ) and measurement of currents (not voltage), a high signal-to-noise ratio is obtained, allowing reliable measurements of single photon responses (quantum bumps; Figure 20.2f).

20.4.1 FORMATION OF THE WHOLE-CELL CONFIGURATION

To obtain whole-cell recordings, a direct contact between the photoreceptor membrane and the recording pipette is required. This is achieved by removing the pigment (glial) cells surrounding the ommatidia (see below). However, the photoreceptor cell is unable to synthesize crucial metabolites required for ATP production and instead depends on the pigment cells for their metabolic supply.[52] Because the TRP and TRPL channels are vulnerable to metabolic stress and easily open in the dark due to metabolic stress (see above), the use of isolated ommatidia imposes great difficulty.

The following procedure is used: To achieve reliable whole-cell recordings from isolated ommatidia, always use ommatidia of newly eclosed flies. The light response induces a large consumption of ATP; therefore, the whole procedure, from the beginning of cutting the eye from the fly body, has to be done under dim red light (the fly R is not sensitive to dim 620 nm red light). The detailed protocol is as follows:

Rapidly cut the eye from the body of a newly eclosed fly in the (nominal) Ca^{2+}-free extracellular solution (see below), and separate the retina from the cornea. Move the retina into the Ca^{2+}-free extracellular solution supplemented with 10% fetal bovine serum (FBS) and 30 mM sucrose. Gently triturate the retina using a fire-polished glass pipette, reaching a diameter of 100–150 μm, in order to remove the ommatidia from the retina. During the dissociation procedure, the surrounding pigment cells disintegrate, exposing the photoreceptor membrane. The procedure also results in the ommatidia breaking off at the basement membrane; thus, the photoreceptors lack their axon terminals. Allow the isolated ommatidia to settle in a bath placed on the stage of an inverted microscope. The bottom of the bath is made up of a clean microscope cover slip. After the ommatidia settle on the cover slip, wash the bath with 7–10 bath volumes of extracellular solution. Fill a pipette for whole-cell recordings (resistance ~10 MΩ) with the intracellular solution (see below) and apply positive pressure via the electrode holder. Lower the pipette into the bath by a micromanipulator, and move it just in front of the cell membrane. Release the positive pressure and gently apply negative pressure to form a gigaseal. Use the zap function in the amplifier and, simultaneously with gentle suction, break the cell membrane and form the whole-cell configuration. Correct the capacitance transients and compensate series resistance up to 80%, and then set the membrane potential according to the experimental needs. The photoreceptor cell has a large capacitance of ~60 pF, most of which is contributed by the microvilli in the rhabdomere. In order to record the small, fast light responses to single photons, use a 5-kHz filter and a 10-kHz sampling rate.

20.4.2 SOLUTIONS

All solutions should be titrated to pH 7.15 and normal physiological osmolarity of 280 mOsmol. When the solutions are supplemented with chemicals or pharmacological agents, the final osmolarity of the solution should be maintained. Standard extracellular solution contains: 120 mM NaCl, 5 mM KCl, 10 mM N-Tris-(hydroxymethyl)-methyl-2-amino-ethanesulphonic acid, pH 7.15 (TES buffer), 4 mM $MgSO_4$, 1.5 mM $CaCl_2$, 25 mM proline, and 5 mM alanine. The intracellular solution for recording light response contains 140 mM K-gluconate, 2 mM $MgSO_4$, 10 mM TES buffer, 4 mM MgATP, 0.4 mM Na_2GTP, and 1 mM nicotinamide adenine dinucleotide (NAD). To avoid voltage-sensitive K^+ currents, during voltage steps or ramps, voltage-gated K^+ channels should be blocked. Accordingly, the intracellular solution should contain Cs^+ ions and tetraethylammonium (TEA) to block the voltage-gated K^+ channels as follows: 120 mM CsCl, 15 mM TEA-Cl, 2 mM $MgSO_4$, 10 mM TES buffer, 4 mM MgATP, 0.4 mM Na_2GTP, and 1 mM NAD.

20.4.3 LIGHT STIMULATION

A convenient way to apply light stimulation when recording from isolated ommatidia is epi-illumination via the microscope objective. A system that includes a light source, connected to the back port of the microscope via a light guide, will provide a stable, uniform light stimulation for all experiments. The effective light intensity can be calibrated by measuring the single photon responses (quantum bumps). To allow for light stimulation at different intensities, a motorized filter wheel interfaced with the light source is the most simple to use and handle. A series of ND filters, which attenuate light up to six units in logarithmic scale, provide the option to record from a single photon response of ~10 pA up to a macroscopic current of ~15 nA in response to 6 orders of magnitude of successively brighter light. Light responses in *Drosophila* have only ~30 ms delay between stimulation and response when intense light is used. Therefore, a fast shutter with opening time of less than 10 ms is required.

20.4.4 MAIN EXPERIMENTS PERFORMED BY WHOLE-CELL RECORDINGS

It is possible to monitor light responses by a continuous recording of the cell current, when the membrane potential is set to a resting potential of −70 mV. When a good seal of >1 GΩ is formed, there is only a small 10–20 pA leak current, the trace is steady and there are no sudden increases in the leak current because of changes in seal resistance. If the Faraday cage is well darkened and the cell is not exposed to any stray light, only small spontaneous bumps of ~3-pA amplitude appear at a rate of 1/s and larger bumps of <20 pA in amplitude appear at a rate of 1/min, owing to spontaneous GDP-GTP exchange[53] or thermal rhodopsin isomerization,[54] respectively.

A single light-induced bump reflects absorption of one photon by rhodopsin, leading to the synchronous opening of 10–15 channels.[54] Many parameters can be measured from a simple trace of bumps. Producing such a trace is done simply by recording responses to dim light for a period of up to 60 s. The light intensity should be calibrated so the bumps will not overlap each other, i.e., not more than three to four bumps per second. Technically, traces longer than 60 s can be recorded, but there is a slight change in bump properties after long periods of dim light stimulation, which should be considered.[54] An effective method for measuring the kinetics of the light response is by exposing the cell to a brief light pulse of 20 ms duration, with a light intensity that will produce, on average, one bump for every second light pulse (i.e., 0.5 photons per pulse). This calibration is needed to avoid overlapping of two effective photons. Response to dim light is small, of the order of 10–100 pA. Stimulation with a brighter and longer (>0.5 s) light will produce transient and steady-state phases (Figure 20.2g). Light intensities ranging over 6 orders of magnitude can be applied and the response to the brightest light reaches 10–15 nA. Beyond this amplitude of light-induced current, the light response will reach the upper limit of the amplifier. A light pulse of 1.5 s is long enough to produce a response composed of transient and steady-state phases. Longer pulses of light will produce a longer steady-state phase, but at high light intensities, the large ATP consumption usually results in irreversible openings of the channels and virtually abolishes the light response (called rundown current).[55] For measuring intensity response curves, one should increase the dark interval between light pulses to avoid light adaptation. At high light intensities, usually the full range intensity-response function cannot be measured in one cell because of exhaustion of the response to light at the higher intensities.

To determine the *I-V* relationship of the light-induced current, the voltage-gated K⁺ channels of the photoreceptor cell should be blocked. An easy way to record an *I-V* relationship is by using voltage ramps, which apply differential voltage to the cell in a short time (~1 s). From each voltage ramp sweep, the complete *I-V* relationship can be derived. Because the light response has a transient phase, the ramp should be set to start only after the transient has decayed to the steady-state level (Figure 20.2g). The membrane voltage at which the net current is zero is the reversal potential of the active channels. In a basic ramp protocol, the membrane voltage is held at zero and then the voltage is briefly dropped to its maximal negative value, generally −100 mV, gradually increased to its maximal positive value of 100 mV, and then brought back to zero again.

In *Drosophila* photoreceptors, the two light-sensitive channels, TRP and TRPL, have a reversal potential of −3 and 13 mV, respectively.[2,56] Therefore, a ramp protocol will give an intermediate reversal potential (~8 mV). To accurately measure small changes in the average reversal potential (e.g., owing to translocation of the TRPL channel upon illumination), a more accurate method is required. For this case a voltage step protocol can determine the reversal potential within 1 mV accuracy.[57] Small voltage steps around 0 mV will distinguish between the reversal potential of dark and light raised flies. The reversal potential of each channel by itself can also be derived from the *TRP* and *TRPL* mutants, where only the TRPL and TRP channels are present, respectively.

20.5 PHOTORECEPTOR IMAGING

20.5.1 Visualizing the Photoreceptor Cells in the Living Fly

In order to visualize the signaling compartment (rhabdomere) of the photoreceptor cells, there is no need to slice the fly's eye. For viewing photoreceptor cells with a fluorescent protein (i.e., GFP attached to a signaling protein) in the living fly, two methods can be used in combination with epifluorescence microscopy: (1) optical neutralization of the cornea[58] and (2) the deep pseudopupil (DPP).[58] In the first method, the cornea is optically neutralized with a drop of water (or oil) and the microscope water (or oil) immersion objective (>×40) is focused on the distal rhabdomere endings. In the second method, the DPP is measured. The DPP is a virtual image consisting of seven spots that represent the trapezoidal pattern of the rhabdomeres in each ommatidium (Figure 20.1d, e). It is generated by the superimposed images of several rhabdomeres of photoreceptor cells from different neighboring ommatidia, which have the same optical axis when focusing a low power microscope objective (×5–10) at the level of the center of eye curvature. The DPP can also be measured by reflectance or absorbance of the photopigment in the rhabdomeres.[58]

The localization of the signaling molecules in the rhabdomere is crucial for normal phototransduction. An important mechanism for long-term light adaptation in *Drosophila* arises from translocation upon illumination of the TRPL channel and $G_q\alpha$ subunit out of the rhabdomeres, while arrestin2 translocates in the opposite direction. The TRPL-eGFP chimera makes visualization of TRPL translocation in vivo possible and allows investigation of translocation under various conditions and background mutations. Expressing TRPL-eGFP using the Rh1 cell specific promoter leads to expression of TRPL-eGFP only in R1–6 cells. The central rhabdomeres of R7 and R8 photoreceptor cells do not express TRPL-eGFP and appear as a black spot, which can serve as an internal control for DPP fluorescence[59] (Figure 20.1d). The following protocol is applied for visualizing the rhabdomeres.

Anesthetize a fly with CO_2 or diethylether and mount it on an object slide using low melting point wax. Place the slide on the inverted microscope stage with the fly head toward the objective. Immerse the fly eye in a water drop placed on a water immersion objective to neutralize the optics of the cornea.

For higher magnification imaging of the rhabdomeres, isolate the retina and keep it viable in oxygenated extracellular solution containing 1% FBS and 5 mM sucrose. Move it into a standard bath (such as used for electrophysiology recordings) and use a confocal microscope to view the image of the ommatidia in optical cross sections (Figure 20.1f).

20.5.2 CALCIUM IMAGING

The *Drosophila* TRP channels are highly permeable to Ca^{2+}. Measurements of light-induced increase in cellular Ca^{2+} simultaneously with the light-induced current (LIC) is a challenging procedure. This is because light excitation of the Ca^{2+} indicator also excites the photopigment. Because the indicator fluorescence increases linearly with the intensity of the excitation light, an intense blue or UV light is required for eliciting a measurable increase in fluorescence. However, this intense light causes a massive depletion of ATP in the cell, owing to saturated excitation of the photopigment, causing metabolic stress and uncontrolled openings of the TRP and TRPL channels. Therefore, usually only one measurement can be obtained from a single cell[24] (but see below).

For single-cell Ca^{2+} imaging, two types of indicators have been used: (1) indicators for measuring relative changes in cellular Ca^{2+} such as Fluo-3, Calcium orange, or Calcium green-5N, based on single excitation-single emission spectra of the indicator[24,60] and (2) ratiometric indicators that allow measurements of absolute changes in Ca^{2+}. The indicators used for these measurements are based on measurements of single excitation-dual emission spectra, such as INDO-1 or Mag-INDO-1.[61] The indicator is membrane impermeable and therefore is applied via the whole-cell recording pipette that includes the regular intracellular solution together with the indicator. Care should be taken to ensure that the concentration of the indicator is low enough (<500 μM) to prevent reduction of intracellular Ca^{2+} concentration. Because the waveform of the LIC is strongly affected by Ca^{2+} and the indicator buffers cellular Ca^{2+}, the indicator concentration should be as low as possible in order to prevent changes in the waveform of the LIC. At least 1 min in the dark is required for the indicator to diffuse uniformly in the cell after the formation of the whole-cell configuration before the light-induced increase in fluorescence is measured. Because the intense indicator excitation light activates virtually all the TRP and TRPL channels, the increase in cellular Ca^{2+} concentration can reach more than 500 μM. This high increase in cellular Ca^{2+}, which is the major reason for the irreversibility of this method due to Ca^{2+} overload, also saturates the indicator. Therefore, in order to measure Ca^{2+} changes below saturation, one needs to use indicators with low Ca^{2+} affinity such as Mag-INDO-1 or Calcium Green-5N.

To allow repetitive measurements of light-induced increase in Ca^{2+} indicator fluorescence in a single cell, it is necessary to use an indicator that absorbs at a range of long wavelength lights, where the photopigment absorption spectrum has low absorbance. To maximize this effect, it is necessary to use photoreceptors with peak absorption spectra in the UV range (~360 nm). Because the peak absorption spectrum of the major R1–6 photoreceptors in the visible range is ~490 nm, a useful way to shift the peak absorption is to replace the normal Rh1 opsin with Rh4 opsin (which is normally expressed outside the compound eye), with peak absorption in the UV range (i.e., ~360 nm). This replacement of the photopigment can be done by P-element-mediated germ line transformation in Rh1 null mutants (Rh1+4 transgenic flies). In this system, the photopigment can be activated by UV light and the indicator (Ca^{2+} orange) by monochromatic green light.[60]

20.6   ELECTRON MICROSCOPY

The photoreceptor cell is a highly polarized cell composed of a cell body and a signaling compartment, the rhabdomere (Figure 20.1b, c). The cell body contains the nucleus and other organelles, and the rhabdomere is made of microvilli. The cell body and the rhabdomere are separated by the submicrovillar cisternae (SMC; Figure 20.1c), which is an extension of the smooth endoplasmic reticulum (ER). The microvilli are a dense membrane compartment that contains the phototransduction machinery. Visualizing the structure at the subcellular level with high resolution is possible only with electron microscopy (EM). EM can also detect degeneration of the photoreceptor and the severity of the degeneration. EM can also localize a protein with high resolution, using specific antibodies and immunogold staining. The protocol for EM of *Drosophila* eyes is described below.

Separate the head from the body and dissect it into two halves with a razor blade to have one eye in each half. Dissecting the head into two halves helps the fixative solution to penetrate into the eye and double the samples. Immerse the heads in fixative solution for two hours at room temperature (RT), at the required illumination conditions (light/dark). Continue to incubate the samples overnight (ON) at 4°C. The heads absorb the fixative solution and sink to the bottom of the tube. Heads that do not sink should be discarded. Gently take out the fixative while leaving the eyes at the bottom with some solution. Change the solution several times with a 0.1 M sodium cacodylate buffer (Electron Microscopy Sciences, Hatfield, PA, USA) by gently shaking the tube for 15 s and incubating for 10 min lying on its side at RT. This is to avoid the heads sticking to the top of the tube. Gently remove the solution as described above. Repeat 4 times.

Change the solution to 1% osmium tetroxide in a 0.1 M sodium cacodylate buffer and incubate for 4 hours at RT. Wash 4 times with 0.1 M sodium cacodylate buffer for 10 min. For dehydration, wash at RT with increasing concentrations of ethanol in distilled water: 30, 50, 70, 80, 90, and 95%, each for 10 min. Wash with 100% ethanol three times, 20 min each. Incubate the semi-heads in propylene oxide twice for 10 min each and then change to solutions with increasing concentrations (25, 50, 75, and 100%) of the epoxy resin (Agar-100, Agar Scientific, UK) diluted in propylene oxide. All incubations are overnight at 4°C in closed vials. Transfer each eye with a toothpick into a mold filled up to 2/3 with 100% resin, near the narrow end of the mold. Orient the eye so that the cornea, and therefore as many as possible rhabdomeres, will face the end of the mold. Incubate at 60°C for an hour. Fill the mold with resin to the top and incubate at 60°C for 2 hours. Adjust the orientation again and incubate at 60°C for 2–3 days.

The sample is now ready to be sectioned. Put the plastic block into the holder, and sculpture away the plastic around the eye with a blade, so it will be in the center of a trapezoid at the top of the block. Start cutting the block using an ultramicrotome with a glass knife. Remove thick sections until the plane of the rhabdomeres is reached. Using an ultramicrotome, cut 70–90 nm thin sections with a diamond knife (Ultra 2.4-mm, Diatome, Switzerland) and place them on a nickel grid. The sample is now ready for viewing in the electron microscope. High-resolution subcellular localization of components of the phototransduction system can be viewed by immunogold EM. For immunogold localization a hydrophilic acrylic resin should be used, but the protocol is similar. After placing the thin section on a nickel grid, the sample is blocked for 5 min with 5% goat serum in 10 mM Tris-HCl buffer, pH 8.2 (containing 0.9% NaCl, 0.5% bovine serum albumin, 0.1% Tween-20, and 20 mM NaN3) and then incubated for an hour with a primary antibody. After washing with the Tris-HCl buffer above, incubate for an hour with secondary antibody conjugated with 20 nm gold particles.

ACKNOWLEDGMENTS

Development of techniques described in this review was supported by grants from the National Institute of Health (RO1 EY 03529), the German Israel Foundation (GIF), The US-Israel Binational Science Foundation, and the Minerva Foundation.

REFERENCES

1. Devary, O., O. Heichal, A. Blumenfeld et al. 1987. Coupling of photoexcited rhodopsin to inositol phospholipid hydrolysis in fly photoreceptors. *Proceedings of the National Academy of Science of the United States of America* 84: 6939–6943.
2. Hardie, R. C., and B. Minke. 1992. The *TRP* gene is essential for a light-activated Ca^{2+} channel in *Drosophila* photoreceptors. *Neuron* 8: 643–651.
3. Pak, W. L. 1995. *Drosophila* in vision research. The Friedenwald Lecture. *Investigative Ophthalmology and Visual Science* 36: 2340–2357.
4. Ranganathan, R., W. A. Harris, and C. S. Zuker. 1991. The molecular genetics of invertebrate phototransduction. *Trends in Neuroscience* 14: 486–493.
5. Vogt, K., and K. Kirschfeld. 1984. Chemical identity of the chromophores of fly visual pigment. *Naturwissenschaften* 77: 211–213.
6. Hillman, P., S. Hochstein, and B. Minke. 1983. Transduction in invertebrate photoreceptors: role of pigment bistability. *Physiological Reviews* 63: 668–772.
7. Bloomquist, B. T., R. D. Shortridge, S. Schneuwly et al. 1988. Isolation of a putative phospholipase C gene of *Drosophila*, norpA, and its role in phototransduction. *Cell* 54: 723–733.
8. Minke, B., and Z. Selinger. 1991. Inositol lipid pathway in fly photoreceptors: excitation, calcium mobilization and retinal degeneration. In *Progress in Retinal Research*, eds. N. A. Osborne and G. J. Chader, 99–124. Oxford: Pergamon Press.

9. Berridge, M. J., and R. F. Irvine. 1984. Inositol trisphosphate, a novel second messenger in cellular signal transduction. *Nature* 312: 315–321.

10. Hardie, R. C. 2003. Regulation of TRP channels via lipid second messengers. *Annual Review of Physiology* 65: 735–759.

11. Minke, B., and B. Cook. 2002. TRP channel proteins and signal transduction. *Physiological Reviews* 82: 429–472.

12. Montell, C. 2001. Physiology, phylogeny, and functions of the TRP superfamily of cation channels. *Science STKE* 2001: re1.

13. Barash, S., and B. Minke. 1994. Is the receptor potential of fly photoreceptors a summation of single-photon responses? *Comments on Theoretical Biology* 3: 229–263.

14. Byk, T., M. Bar Yaacov, Y. N. Doza, B. Minke, and Z. Selinger. 1993. Regulatory arrestin cycle secures the fidelity and maintenance of the fly photoreceptor cell. *Proceedings of the National Academy of Science of the United States of America* 90: 1907–1911.

15. Dolph, P. J., R. Ranganathan, N. J. Colley, R. W. Hardy, M. Socolich, and C. S. Zuker. 1993. Arrestin function in inactivation of G protein-coupled receptor rhodopsin in vivo. *Science* 260: 1910–1916.

16. Ranganathan, R., and C. F. Stevens. 1995. Arrestin binding determines the rate of inactivation of the G protein-coupled receptor rhodopsin in vivo. *Cell* 81: 841–848.

17. Cook, B., M. BarYaacov, H. Cohen Ben-Ami et al. 2000. Phospholipase C and termination of G-protein-mediated signalling in vivo. *Nature Cell Biology* 2: 296–301.

18. Minke, B., C. Wu, and W. L. Pak. 1975. Induction of photoreceptor voltage noise in the dark in *Drosophila* mutant. *Nature* 258: 84–87.

19. Montell, C., and G. M. Rubin. 1989. Molecular characterization of the *Drosophila TRP* locus: a putative integral membrane protein required for phototransduction. *Neuron* 2: 1313–1323.

20. Wong, F., E. L. Schaefer, B. C. Roop et al. 1989. Proper function of the *Drosophila TRP* gene product during pupal development is important for normal visual transduction in the adult. *Neuron* 3: 81–94.

21. Zhu, M. X., P. B. Chu, M. Peyton, and L. Birnbaumer. 1995. Molecular cloning of a widely expressed human homologue for the *Drosophila TRP* gene. *FEBS Letters* 373: 193–198.

22. Wes, P. D., J. Chevesich, A. Jeromin, C. Rosenberg, G. Stetten, and C. Montell. 1995. TRPC1, a human homolog of a *Drosophila* store-operated channel. *Proceedings of the National Academy of Science of the United States of America* 92: 9652–9656.

23. Phillips, A. M., A. Bull, and L. E. Kelly. 1992. Identification of a *Drosophila* gene encoding a calmodulin-binding protein with homology to the *TRP* phototransduction gene. *Neuron* 8: 631–642.

24. Peretz, A., E. Suss-Toby, A. Rom-Glas, A. Arnon, R. Payne, and B. Minke. 1994. The light response of *Drosophila* photoreceptors is accompanied by an increase in cellular calcium: effects of specific mutations. *Neuron* 12: 1257–1267.

25. Peretz, A., C. Sandler, K. Kirschfeld, R. C. Hardie, and B. Minke. 1994. Genetic dissection of light-induced Ca^{2+} influx into *Drosophila* photoreceptors. *Journal of General Physiology* 104: 1057–1077.

26. Niemeyer, B. A., E. Suzuki, K. Scott, K. Jalink, and C. S. Zuker. 1996. The *Drosophila* light-activated conductance is composed of the two channels TRP and TRPL. *Cell* 85: 651–659.

27. Pak, W. L., J. Grossfield, and N. V. White. 1969. Nonphototactic mutants in a study of vision of *Drosophila*. *Nature* 222: 351–354.

28. Minke, B., C.-F. Wu, and W. L. Pak. 1975. Isolation of light-induced response of the central retinular cells from the electroretinogram of *Drosophila*. *Journal of Comparative Physiology* 98: 345–355.

29. Pak, W. L. 1979. Study of photoreceptor function using *Drosophila* mutants. In *Neurogenetics, Genetic Approaches to the Nervous System*, ed. X. Breakfield, 67–99. New York: Elsevier North-Holland.

30. Pastink, A., E. Heemskerk, M. J. Nivard, C. J. van Vliet, and E. W. Vogel. 1991. Mutational specificity of ethyl methanesulfonate in excision-repair-proficient and -deficient strains of Drosophila melanogaster. *Molecular and General Genetics* 229: 213–218.

31. Bentley, A., B. MacLennan, J. Calvo, and C. R. Dearolf. 2000. Targeted recovery of mutations in Drosophila. *Genetics* 156: 1169–1173.

32. Tracey, W. D., Jr., R. I. Wilson, G. Laurent, and S. Benzer. 2003. Painless, a *Drosophila* gene essential for nociception. *Cell* 113: 261–273.

33. Georgiev, P., I. Garcia-Murillas, D. Ulahannan, R. C. Hardie, and P. Raghu. 2005. Functional INAD complexes are required to mediate degeneration in photoreceptors of the Drosophila rdgA mutant. *Journal of Cell Science* 118: 1373–1384.

34. Meyer, N. E., C. Oberegelsbacher, T. D. Durr, A. Schafer, and A. Huber. 2008. An eGFP-based genetic screen for defects in light-triggered subcelluar translocation of the Drosophila photoreceptor channel TRPL. *Fly (Austin)* 2.

35. Castro, J. P., and C. M. Carareto. 2004. Drosophila melanogaster P transposable elements: mechanisms of transposition and regulation. *Genetica* 121: 107–118.

36. Bellen, H. J. et al. 2004. The BDGP gene disruption project: single transposon insertions associated with 40% of Drosophila genes. *Genetics* 167: 761–781.

37. Spradling, A. C., D. Stern, A. Beaton et al. 1999. The Berkeley Drosophila Genome Project gene disruption project: single P-element insertions mutating 25% of vital Drosophila genes. *Genetics* 153: 135–177.

38. Rubin, G. M., and A. C. Spradling. 1982. Genetic transformation of *Drosophila* with transposable element vectors. *Science* 218: 348–353.

39. Spradling, A. C., and G. M. Rubin. 1982. Transposition of cloned P elements into *Drosophila* germ line chromosomes. *Science* 218: 341–347.

40. Lee, Y. S., and R. W. Carthew. 2003. Making a better RNAi vector for *Drosophila*: use of intron spacers. *Methods* 30: 322–329.

41. Minke, B. 1982. Light-induced reduction in excitation efficiency in the *TRP* mutant of *Drosophila*. *Journal of General Physiology* 79: 361–385.

42. Pak, W. L., and K. J. Lidington. 1974. Fast electrical potential from a long-lived, long-wavelength photoproduct of fly visual pigment. *Journal of General Physiology* 63: 740–756.

43. Minke, B., and K. Kirschfeld. 1980. Fast electrical potentials arising from activation of metarhodopsin in the fly. *Journal of General Physiology* 75: 381–402.

44. Selinger, Z., Y. N. Doza, and B. Minke. 1993. Mechanisms and genetics of photoreceptors desensitization in *Drosophila* flies. *Biochimica et Biophysica Acta* 1179: 283–299.

45. Hochstein, S., B. Minke, and P. Hillman. 1973. Antagonistic components of the late receptor potential in the barnacle photoreceptor arising from different stages of the pigment process. *Journal of General Physiology* 62: 105–128.

46. Hamdorf, K., and A. H. Kaschef. 1964. Der Sauerstoffverbrauch des Facettenauges von *Calliphora erythrocephalla* in Abhangigkeit von der Temperatur und dem Ionenmilieu. *Zeitschrift für vergleichende Physiologie* 48: 251–265.

47. Tsacopoulos, M., S. Poitry, and A. Borsellino. 1981. Diffusion and consumption of oxygen in the superfused retina of the drone (*Apis mellifera*) in darkness. *Journal of General Physiology* 77: 601–628.

48. Agam, K., M. von Campenhausen, S. Levy et al. 2000. Metabolic stress reversibly activates the *Drosophila* light-sensitive channels TRP and TRPL in vivo. *Journal of Neuroscience* 20: 5748–5755.

49. Chorna-Ornan, I., T. Joel-Almagor, H. C. Ben-Ami et al. 2001. A common mechanism underlies vertebrate calcium signaling and *Drosophila* phototransduction. *Journal of Neuroscience* 21: 2622–2629.

50. Parnas, M., M. Peters, D. Dadon et al. 2009. Carvacrol is a novel inhibitor of *Drosophila* TRPL and mammalian TRPM7 channels. *Cell Calcium* 45: 300–309.

51. Hardie, R. C. 1991. Whole-cell recordings of the light induced current in dissociated *Drosophila* photoreceptors: evidence for feedback by calcium permeating the light-sensitive channels. *Proceedings of the Royal Society of London, Series B* 245: 203–210.

52. Dimitracos, S. A., and M. Tsacopoulos. 1985. The recovery from a transient inhibition of the oxidative metabolism of the photoreceptors of the drone (*Apis mellifera*). *Journal of Experimental Biology* 119: 165–181.

53. Elia, N., S. Frechter, Y. Gedi, B. Minke, and Z. Selinger. 2005. Excess of G_{betae} over $G_{qalphae}$ in vivo prevents dark, spontaneous activity of *Drosophila* photoreceptors. *Journal of Cell Biology* 171: 517–526.

54. Henderson, S. R., H. Reuss, and R. C. Hardie. 2000. Single photon responses in *Drosophila* photoreceptors and their regulation by Ca^{2+}. *Journal of Physiology* 524(Part 1): 179–194.

55. Hardie, R. C., and B. Minke. 1994. Spontaneous activation of light-sensitive channels in *Drosophila* photoreceptors. *Journal of General Physiology* 103: 389–407.

56. Katz, B., and B. Minke. 2009. *Drosophila* photoreceptors and signaling mechanisms. *Frontiers in Cellular Neuroscience* 3: 2.

57. Bahner, M., S. Frechter, N. Da Silva, B. Minke, R. Paulsen, and A. Huber. 2002. Light-regulated subcellular translocation of Drosophila TRPL channels induces long-term adaptation and modifies the light-induced current. *Neuron* 34: 83–93.

58. Franceschini, N., and K. Kirschfeld. 1971. Pseudopupil phenomena in the compound eye of *Drosophila*. *Kybernetik* 9: 159–182.

59. Meyer, N. E., T. Joel-Almagor, S. Frechter, B. Minke, and A. Huber. 2006. Subcellular translocation of the eGFP-tagged TRPL channel in *Drosophila* photoreceptors requires activation of the phototransduction cascade. *Journal of Cell Science* 119: 2592–2603.

60. Ranganathan, R., B. J. Bacskai, R. Y. Tsien, and C. S. Zuker. 1994. Cytosolic calcium transients: spatial localization and role in *Drosophila* photoreceptor cell function. *Neuron* 13: 837–848.

61. Hardie, R. C. 1996. INDO-1 measurements of absolute resting and light-induced Ca^{2+} concentration in *Drosophila* photoreceptors. *Journal of Neuroscience* 16: 2924–2933.

21 Studying TRP Channels in *Caenorhabditis elegans*

Rui Xiao and X. Z. Shawn Xu

CONTENTS

21.1 INTRODUCTION

21.1.1 TRP CHANNELS IN *CAENORHABDITIS ELEGANS*

Transient receptor potential (TRP) family channels are conserved from *Caenorhabditis elegans* to humans. About 28 TRP members have been identified in mammals. On the basis of their sequence homology and functional similarity, these channels are further divided into seven subfamilies. Accumulating evidence shows that mammalian TRP channels are broadly involved in regulating sensory physiology, as they are important for sensing a wide variety of physical and chemical cues from both intracellular and extracellular sides.[1]

Through genetic screening and database searching, 17 TRP channels have been identified in *C. elegans*, and they cover all of the seven TRP subfamilies.[2] The phylogenic relationship between each *C. elegans* TRP channel and some of their general functions are summarized in Figure 21.1. TRP channels in *C. elegans* may also form homo- or heterotetramers just like their mammalian homologs. Each worm TRP subunit is also believed to contain six transmembrane segments (S1–S6) with a putative pore-forming loop between S5 and S6. Interestingly, many *C. elegans* TRP channels have been shown to play important roles in sensory transduction, including chemosensation, mechanosensation, osmosensation, and proprioception,[2] suggesting that TRP channels play an evolutionarily conserved role in sensory physiology in the animal kingdom.

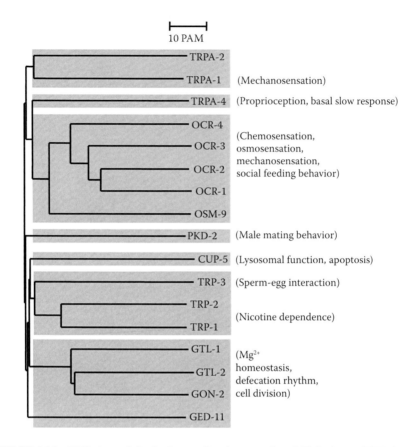

10 PAM

TRPA-2

TRPA-1 (Mechanosensation)

TRPA-4 (Proprioception, basal slow response)

OCR-4

OCR-3 (Chemosensation, osmosensation, mechanosensation, social feeding behavior)

OCR-2

OCR-1

OSM-9

PKD-2 (Male mating behavior)

CUP-5 (Lysosomal function, apoptosis)

TRP-3 (Sperm-egg interaction)

TRP-2 (Nicotine dependence)

TRP-1

GTL-1 (Mg²⁺ homeostasis, defecation rhythm, cell division)

GTL-2

GON-2

GED-11

FIGURE 21.1 TRP channels in *C. elegans*. Dendrogram plot of 17 *C. elegans* TRP channels using ClustalW algorism. The evolutionary distance between each TRP channel is indicated by the branch length in point accepted mutation (PAM) units. Some general functions of specific TRP channels are indicated.

As a powerful genetic model organism, *C. elegans* offers a number of benefits in characterizing the function and regulation of TRP channels in vivo. First, all TRP subfamilies are present in *C. elegans,* and many of them display high homology to their vertebrate counterparts. Second, multiple mutant strains are available for each *C. elegans* TRP channel. Third, the short generation time (~3 days) and availability of rich genetic tools, together with a simple and well-characterized nervous system, make *C. elegans* a valuable system to characterize gene functions in vivo, particularly those with a role in the nervous system.

21.1.2 PHYSIOLOGICAL ROLES OF *C. ELEGANS* TRP CHANNELS

Among the three *C. elegans* TRPC members, TRP-1 and TRP-2 are widely distributed in multiple types of neurons,[3,4] while TRP-3 is enriched in sperm.[5] Consistent with its expression pattern, TRP-3 is required for sperm-egg interactions, and *TRP-3* null mutant worms are infertile.[5] By contrast, *TRP-1* and *TRP-2* mutant worms are superficially wild type. However, these mutant worms exhibit impaired neuronal Ca^{2+} transients and are defective in their response to nicotine.[4] Interestingly, ectopic expression of human TRPC3 rescues the nicotine-dependent behavior in *TRP-2* mutant worms,[4] implying a conserved role of TRPC channels in nicotine-dependent behavior. As is the case with mammalian TRPC channels, the activation of *C. elegans* TRPC channels also appear to depend on $G_{q/11}$-coupled receptors and PLCβ.[4]

The TRPV channel OSM-9 was identified in a genetic screen for mutants with a defective response to osmotic shock and odorants.[3,6] Subsequent homology cloning identified four other *osm-9*/capsaicin receptor-related (*ocr*) genes, *ocr-1* to *ocr-4*.[7] *C. elegans* TRPV channels are primarily expressed in sensory neurons.[2] OSM-9 and OCR-2, the best-characterized TRPV channels in *C. elegans*, regulate chemosensation, osmosensation, and mechanosensation.[3,8,9] This polymodal feature of OSM-9 and OCR-2 is observed in their mammalian homologs, such as TRPV1. Several *C. elegans* TRPV channels have been suggested to form heterotetramers,[7,10] a feature that is also shared by many mammalian TRP channels. Furthermore, OSM-9 and OCR-2 are regulated by polyunsaturated fatty acids (PUFAs) like some *Drosophila* and mammalian TRP channels.[11–13] However, while mammalian TRPV1-4 are heat-activated channels,[14] *C. elegans* TRPV channels have not been found to display temperature sensitivity.

Four TRPM channels are present in *C. elegans*: GON-2, GTL-1, GTL-2, and CED-11. GON-2 (abnormal *gon*ad development) is mainly expressed in the gonad and intestine, where it regulates gonad development and Mg^{2+} uptake, respectively.[15,16] GTL-1 (*gon-two like*) is highly expressed in the intestine of *C. elegans*, where it plays an essential role in regulating Mg^{2+} homeostasis and defecation rhythm together with GON-2.[16]

TRP-4, the sole TRPN/NOMPC channel in *C. elegans*, is mainly expressed in a subgroup of sensory neurons and interneurons.[17,18] Similar to its *Drosophila* and zebrafish homologs, TRP-4 is involved in mechanosensation and is required for detecting mechanical attributes from the bacteria lawn on which worms navigate.[18,19] More interestingly, TRP-4 also regulates proprioception by modulating the worm's body posture through the stretch-sensitive neuron DVA.[18]

The *C. elegans* genome encodes two TRPA channels, TRPA-1 and TRPA-2. TRPA-1 displays about 40% similarity to mouse and *Drosophila* TRPA1, while TRPA-2 is more distantly related and lacks the long ankyrin-like repeats found in TRPA1.[20] TRPA-1 is widely expressed in many cell types, including neurons, muscles, and epithelium.[20] In OLQ and IL1 but not ASH neurons, TRPA-1 is involved in head withdrawal and nose touch responses.[20] Calcium imaging experiments show that in OLQ neurons TRPA-1 is involved in nose touch-evoked Ca^{2+} transients, suggesting that TRPA-1 functions in mechanosensation.[20]

PKD-2 (*polycystic kidney disease-related*), the sole TRPP channel in *C. elegans*, shares 33% sequence identity and 52% similarity with its human homolog PKD2 (also called TRPP2).[21] PKD-2 is colocalized with the TRPP1 protein LOV-1 in male-specific neurons where it regulates male mating behavior.[21] Cilia targeting of PKD-2 is required for its normal function.[21]

CUP-5 (*coelomocyte uptake-defective*) is the *C. elegans* homolog of TRPML (mucolipin) and is mainly expressed in coelomocytes and some head neurons.[22]

Similar to its mammalian homolog, CUP-5 is important for the biogenesis of late endosomes and lysosomes.[23] In addition, *cup-5* mutant worms display an abnormal accumulation of apoptotic cells,[24] a phenomenon that has also been observed in mucolipidosis type IV (MLIV) patients and a *Drosophila* model of MLIV.[25]

In summary, TRP channels in *C. elegans* are expressed in a wide range of cell types and tissues and play diverse functions, ranging from sensory perception to intracellular homeostasis. Overall, the physiological functions of many *C. elegans* TRP channels are well correlated with their vertebrate counterparts.

21.2 IDENTIFICATION OF TRP CHANNELS IN *C. ELEGANS*

A number of *C. elegans* TRP channels were cloned through classic forward genetic screens for mutants defective in behavior and development. These include OSM-9, GON-2, PKD-2, and CUP-5. For example, mutant alleles of *osm-9*, the first characterized *C. elegans* TRPV homolog, were isolated in a genetic screen for mutants that cannot avoid high osmotic stress, and the *osm-9* gene was then identified through positional cloning.[3,6] In this screen, wild-type N2 worms were mutagenized with ethyl methanesulfonate (EMS), causing random mutations across the genome at a rate of 5×10^{-4} to 5×10^{-2} per gene.[26] The progeny of thousands of F1 animals were then assayed for osmotic avoidance response. Defective F2 single worms were cloned out for further analysis. After confirmation of the phenotype, the responsible gene was mapped to a small region of a defined chromosome with phenotypic markers.[3,6] Single nucleotide polymorphism (SNP) markers are now more commonly used to map mutations. Subsequently, overlapping cosmids covering the mapped interval can be introduced individually or in combination into the mutant as a transgene through germ line transformation. Cosmids that rescue the behavioral defect must contain the gene, which can be further identified by genomic sequencing of the molecular lesions in the mutated gene followed by direct rescue of the mutant phenotype with the wild-type gene through germ line transformation.[3] Recent advances in deep-sequencing technologies have shown that mutant genes can be directly identified through whole-genome sequencing at an affordable cost for organisms like *C. elegans*, whose genome size is about one-thirtieth of that of mammals.[27] It is anticipated that this approach will be widely utilized for cloning in the future.

Subsequent database searching identified four *C. elegans* homologs of *osm-9*: *ocr-1* to *ocr-4*. The TRPM members GTL-1, GTL-2, and CED-11 were identified through the same strategy following the positional cloning of the first *C. elegans* TRPM member GON-2.[2] Members in the other three TRP subfamilies (TRPC, TRPN, and TRPA) were all identified by homology cloning.[2]

21.3 GENETIC CHARACTERIZATION OF TRP CHANNELS IN *C. ELEGANS*

Several downstream effectors of TRP channels have been identified through classic forward genetic screens. For example, an unexpected role of the TRPV channels OSM-9 and OCR-2 in serotonin synthesis has been revealed by screening EMS-mutagenized worms. These worms carry a GFP transgene driven by the promoter of *tph-1*, a gene essential for the synthesis of serotonin.[28] By screening ~6500 worms, Zhang et al. isolated several mutant strains that displayed greatly reduced GFP expression specifically in the ADF serotonin neurons. One mutant strain each was mapped to *ocr-2* and *osm-9*.[28] Thus, in ADF neurons, the TRPV channels OSM-9 and OCR-2 act upstream of serotonin synthesis. This illustrates the power of forward genetic screens in revealing new functions of TRP channels. Forward genetic screens also provide a useful tool to dissect the signaling molecules downstream and upstream of *C. elegans* TRP channels. For instance, *cup-5*-null mutant worms are lethal.[29] Schaheen et al. carried out a genetic screen for mutants that suppress the lethal phenotype of *cup-5* worms. MRP-4, an ABC transporter, has been identified as the downstream suppressor of CUP-5.[29] Using a similar approach, several suppressors of *gon-2* have been isolated. One such suppressor has been cloned and called *gem-4* (*gon-2 e*xtragenic *m*odifier), a gene that is a member of the conserved copine family of calcium-dependent phosphatidylserine-binding proteins.[30] GEM-4 is predicted to be membrane associated and may potentially inhibit GON-2 activity. Suppressor screens may identify novel genes that functionally interact with TRP channels.

Reverse genetics has also been applied to study TRP channels in *C. elegans*, including TRP-1-4, OCR-1-4, GTL-1, and TRPA-1. The first step in reverse genetics is to obtain mutant alleles of the studied gene based on its sequence information. The most commonly used method is chemical-induced gene deletion. A large pool of wild-type worms is mutagenized by EMS (~13% of EMS-induced mutations are deletions) or trimethylpsoralen (TMP)/UV (~50% TMP/UV-induced mutations are deletions).[26] The deletion events can be readily detected by PCR with primers corresponding to the genes of interest.[26] Two consortiums (http://celeganskoconsortium .omrf.org/ and http://www.shigen.nig.ac.jp/c.elegans/index.jsp) have been formed to generate deletion alleles for all worm genes, which are freely available to the research community. In addition, classic homologous recombination techniques have also been developed to knockout and knockin genes in *C. elegans*.[31,32] These approaches may become widely used in the future.

In the past decade, the development of RNA interference (RNAi) techniques has offered a powerful tool to study gene functions.[33] Sequence-defined double-stranded RNAs (dsRNA) cause potent and long-lasting degradation of endogenous mRNA of the corresponding gene.[33] Currently, dsRNA can be introduced into *C. elegans* via three routes: microinjection, soaking, and bacterial feeding.[34] The choice of method depends on the experimental requirement. In general, microinjection and soaking of dsRNA are applied to small-scale experiments and can produce a potent gene knockdown effect. Bacterial feeding is less labor intensive and thus a better fit for large-scale experiments, though its knockdown effect is sometimes less effective. In this protocol, bacteria produce dsRNA from a plasmid that carries a DNA fragment of the gene of interest under the control of an IPTG-inducible T7 promoter at both 5′ and 3′ ends, which can be directly fed to worms.[34] A brief protocol is described as follows:

- Inoculate the bacterial strain HT115 carrying an RNAi plasmid overnight.
- Seed about 300 μL liquid culture on regular NGM plates containing 25 μg/ mL carbenicillin and 1 mM IPTG.

- Let the plate sit overnight at room temperature to induce the expression of double-strand RNA.
- Transfer several (~10) young adult worms to RNAi plates to lay eggs for 5 hours.
- After eggs hatch and reach L4 stage, transfer them to new RNAi plates and let them grow for 1–2 days.

Mutant worms isolated through reverse genetic screens or worms treated with RNAi can then be subjected to phenotypic analysis using a host of behavioral and functional assays as described below.

21.4 BEHAVIORAL CHARACTERIZATION OF C. ELEGANS TRP CHANNELS

Multiple behavioral assays have been developed to test the in vivo functions of TRP channels in *C elegans*. Here we give a brief description of these assays.

Chemotaxis assay: *C. elegans* TRPV channels are mainly involved in chemosensation, osmosensation, and mechanosensation.[35] Among the best characterized TRPVs are OSM-9 and OCR-2, and several behavioral assays have been performed to study their functions. In *osm-9* or *ocr-2* mutant worms, the AWA and ASH neuron-mediated sensory functions are largely defective, including chemotaxis to odors and nose touch and high osmotic stress-induced avoidance responses.[3,7] Additionally, they are important for social feeding behavior mediated by the ASH and ADL neurons.[8] Chemotaxis assay was originally developed by Ward and later modified by Bargmann et al.[36,37] Briefly, 10 mL chemotaxis agar (1.6% agar, 5 mM K_3PO_4 [pH 6.0], 1 mM $CaCl_2$, 1 mM $MgCl_2$) is poured into a 10-cm petri dish. Test compound and control (~1–2 μL) are placed on opposing sides of the agar surface. A group of adult worms is washed three times with S Basal (0.1 M NaCl, 0.05 M K_3PO_4, 5 μg/mL cholesterol) to remove bacteria and then placed at the center of the test plate with an equal distance to the test compound and control. At different time points, the number of worms at the test compound area and control area are counted to determine the chemotaxis index (CI). In many cases, a small drop of NaN_3 (1M, 1–2 μL) is also applied to both spots to paralyze worms once they reach there, which greatly facilitates counting.

$$CI = \frac{N(\text{test}) - N(\text{control})}{N(\text{total})}$$

Nose touch response: The sensory neurons ASH, OLQ, and FLP mediate the nose touch response in *C. elegans* with ASH playing a dominant role.[38] Wild-type worms respond to gentle nose touch by initiating robust backward movement,[38] and several TRP channel mutants are defective in this response.[7,20] In this assay, an eyelash hair is placed right in front of the path of a forward-moving worm as shown in Figure 21.2b. A nose-on collision with the eyelash usually stops the worm from moving forward and also evokes backward movement. Testing plates should contain a thin

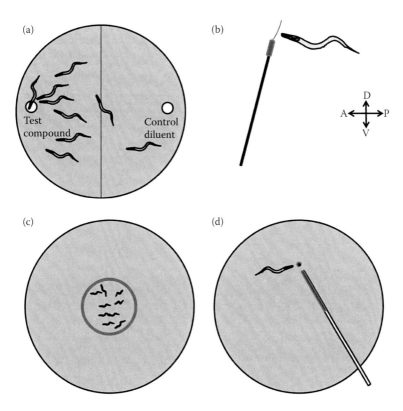

FIGURE 21.2   Behavior tests of TRP channels in *C. elegans*. (a) Chemotaxis assay in which a group of worms is placed at the middle line of the plate, while test compound and control diluent are at opposite sides. The number of worms at both sites is counted at different time points, and the chemotaxis index (CI) is calculated. (b) Nose touch assay where an eyebrow hair was put perpendicularly ahead of a forward-moving worm. (c) Osmotic ring assay where a group of worms is placed inside the ring painted with high-osmolarity liquid (1–4 M glycerol or fructose). Wild-type N2 worms should avoid the ring and consequently are retained within the ring. (d) Droplet assay for TRP channels. High-osmotic solution is sucked into a capillary pipette. A droplet of high-osmotic glycerol solution is placed in front of a forward-moving worm. Wild-type worms should initiate backward movement immediately.

layer of OP50 bacteria because the nose touch response on food is much more robust. However, too many bacteria may allow the worm to crawl through under the hair. *osm-9*, *ocr-2*, and *TRPa-1* mutant worms all show defects in this response.

   High-osmolarity avoidance assay: To test osmotic avoidance behavior, the osmotic ring assay is typically used for population analysis, while the droplet assay is usually employed for testing individual worms. When the nose of a worm reaches high-smolarity environment, it usually stops forward movement and moves backward immediately. Unseeded NGM plates are used for these assays. For the ring assay, wild-type worms are usually retained within the high-osmolarity ring for at least 10 min, while most *osm-9* and *ocr-2* mutant worms escape, as they are severely defective in this avoidance behavior. To make a high-osmolarity ring barrier, one can soak a test tube cap in 20–50 µL 1–4 M glycerol or fructose solution and then place the open end on the plate to print a ring.[39] In order to better visualize the ring, a trace amount of bromophenol blue may be included in the solution. After three washes with S basal solution, young adult worms are placed at the center of the ring as shown in Figure 21.2c. The number of worms that have crawled through or are retained within the ring is counted 10 min after the onset of the assay, and the escape rate can be calculated. In the droplet assay, a high-osmolarity solution is absorbed into a capillary pipette. A small drop of solution is then gently dipped close to the nose tip of a forward-moving worm as shown in Figure 21.2d. After contacting the droplet, wild-type worms typically pause and initiate backward movement immediately. A robust avoidance response is observed in wild-type worms but is absent in *osm-9* or *ocr-2* mutants.

Defecation cycles: In the TRPM subfamily, little is known about GTL-2 and CED-11, while GON-2 and GTL-1 are highly expressed in the intestine and required for maintaining normal defecation cycles.[40,41] Defecation occurs rhythmically every 40–45 s with little variance.[42] Each cycle consists of four well-defined stages: (1) posterior body-wall muscle contraction (pBoc), (2) relaxation, (3) anterior body-wall muscle contraction (aBoc), and (4) expulsion.[40,42] This behavior can be readily observed under a stereoscope. At the cellular level, IP$_3$ receptor (IP$_3$R)-dependent calcium oscillation in intestine cells is critical for this rhythmic contraction.[41,43] Defecation cycles are disrupted in *gon-2; gtl-1* double mutant worms. Consequently, these animals are constipated. Ca^{2+} oscillation is disrupted in the mutant.[41]

Locomotion assay: Unlike TRP-3, which is specifically expressed in sperm and required for sperm-egg fusion during fertilization, TRP-1 and TRP-2 are expressed in neurons, suggesting that they may be involved in behavioral control. Although *TRP-1* and *TRP-2* mutant worms are superficially wild type, an unexpected role of these two TRPC channels has been revealed in nicotine-dependent behavior. Similar to rodent models, worms display acute, adaptation, sensitization, and withdrawal responses to nicotine.[4] Nicotine effects can be quantified by measuring the locomotion velocity in response to nicotine treatments. This can be achieved by counting the body bending frequency through human observation or, ideally, using an automated worm tracking system. Interestingly, in *TRP-1* and *TRP-2* mutant worms, these nicotine-dependent behavioral responses are abolished.[4] Moreover, 2-APB, a nonspecific TRPC channel blocker, completely blocks nicotine-dependent behavior in *C. elegans*, supporting a critical role of TRP-1 and TRP-2 in nicotine dependence.[4] For practical assays, fresh nicotine or other types of drugs can be dissolved in warm nematode growth media (NGM) and poured into petri dishes. Alternatively, nicotine solution may be directly spread on the surface of NGM plates, which should be allowed to diffuse throughout the plate overnight before use. As nicotine is light-sensitive, assay plates should be kept in the dark and used within a week. *C. elegans* TRPN (TRP-4) channels are likely mechanosensitive. The *TRP-4* mutant worms are defective in locomotion behavior. These worms exhibit a loopy body posture during locomotion.[18] To quantify this phenotype, locomotion behavior is recorded using an automated worm tracking system, and the body posture of the worm is determined by examining the video clips.[18] To do so, the worm body is artificially divided into 12 segments, and the bending angle between each segment can be quantified accordingly. Worms reduce their locomotion velocity upon encountering a bacteria lawn, the so-called basal slowing response that is mediated by the mechanosensory dopamine neurons CEP, ADE, and PDE. These neurons sense the mechanical attributes from the bacteria lawn.[44] The *TRP-4* mutant worms are defective in basal slowing response, suggesting that TRP-4 may be the mechanosensitive channel in dopamine neurons. An automated worm tracker can also be used to quantify this behavioral response.

Foraging assay: The *TRPa-1* mutant worms display defective head foraging movement and nose touch response.[20] Foraging behavior is manifested by oscillating nose movement when worms explore their environment on food. This behavior requires the function of the OLQ and IL1 mechanosensory neurons.[45] The *TRPa-1* mutant worms display a reduced rate of foraging behavior. In response to touch delivered by a hair pick on the top of the head or anterior segment of a forward-moving worm, the worm usually initiates backward movement, and the foraging behavior is inhibited during backward movement.[20] The inhibition of foraging in wild-type worms reaches ~80%, whereas only about 30% inhibition was observed in *TRPa-1* mutants, suggesting that TRPA-1 functions in mechanosensation.[20]

Male mating behavior: PKD-2, the sole TRPP member in *C. elegans*, plays an important role in male mating behavior together with the *C. elegans* polycystin-1 homolog LOV-1. Male mating behavior consists of a series of stereotypic steps executed by male worms upon encountering hermaphrodites, including response to contact, backing, turning, location of vulva, spicule protraction, and sperm transfer.[46] This behavior can be quantified by human observation under a stereoscope. The *pkd-2* mutant males display a strong phenotype at the step of location of the vulva. PKD-2 is expressed in male-specific sensory neurons that are considered both chemosensory and mechanosensory.[21]

21.5 FUNCTIONAL CHARACTERIZATION OF TRP CHANNELS IN *C. ELEGANS*

Functional assays are needed to understand the molecular and cellular details of TRP channels that cannot be obtained through behavioral analysis. As most TRP channels are nonselective cation channels, two major assays are constantly employed to study various aspects of these channels: Ca^{2+} imaging and electrophysiological recording.

The optical transparency of the worm body makes it possible to perform calcium imaging in a noninvasive manner. To achieve noninvasive recording, genetically encoded Ca^{2+} indicators (e.g., G-CaMP and cameleon) are expressed as a transgene in defined cell types using cell-specific promoters.[47,48] Cameleon is a FRET-based sensor, and upon calcium binding, the ratio of YFP/CFP fluorescence of cameleon increases.[48] This ratiometric readout of $[Ca^{2+}]_i$ is largely insensitive to illumination intensity and uneven tissue distribution of the sensor. G-CaMP is a nonratiometric sensor that is based on a single GFP variant whose fluorescence intensity increases upon calcium binding.[47] Nonratiometric $[Ca^{2+}]_i$ measurement is sensitive to light intensity and uneven intracellular distribution of the sensor. Li et al. co-expressed a red fluorescence protein (DsRed) as an internal reference marker together with G-CaMP in the same cells, thereby enabling ratiometric imaging.

In vivo Ca^{2+} imaging experiments have been performed to record the function of many *C. elegans* TRP channels. Typically, worms are glued on a 2% agarose pad with Nexaband cyanoacrylate glue.[18] For cameleon imaging, CFP bleed-through is a concern for calculating the ratio change and should be subtracted. The ratio change is tabulated as

$$\Delta Ratio = Ratio\ (apparent) - Ratio\ (CFP)$$

where

$$Ratio\ (apparent) = \frac{I(YFP, measured) - I(YFP, background)}{I(CFP, measured) - I(CFP, background)},$$

where I indicates the fluorescence intensity and Ratio (CFP) is typically assumed to be 0.6.[19]

Several *C. elegans* TRP channels have been functionally expressed in heterologous systems, such as HEK293 cells and CHO cells. In these cases, it is plausible to study these channels with chemical Ca^{2+} indicators. For example, *C. elegans* TRPC channels have been successfully expressed in HEK293T cells. Store- and receptor-operated Ca^{2+} entry have been measured in TRPC channel-expressing HEK293T cells.[5,18]

Another important method to study ion channels is patch-clamp recording. The advantages of electrophysiological recordings over Ca^{2+} imaging include (1) outstanding temporal resolution, (2) higher sensitivity, (3) direct measurement of biophysical properties of ion channels, and (4) easy access to the intracellular content of recorded cells. However, owing to the small size of most *C. elegans* cells, especially neurons where most TRP channels are expressed, whole-cell patch-clamp recording of ion channel activity in their native environment is technically challenging. In the past decade, conventional whole-cell voltage-clamp and current-clamp techniques have been developed to record *C. elegans* neurons.[49] This approach can be applied to study TRP channels in vivo. The general setup is similar to that used for conventional brain slice recording. General information about whole-cell patch-clamp techniques has been discussed in many textbooks and reviews.[50–52] Here we just describe some special aspects pertaining to whole-cell recording of *C. elegans* cells.

Prior to recording, worms are glued with Histoacryl Blue (B. Braun, Melsungen, Germany) on a cover slip coated with Sylgard 184 (Dow Corning, Midland, MI). Glue is typically applied with a mouth pipette pulled from borosilicate glass. If the region to be recorded is on the ventral side of the worm, glue should be applied to the dorsal side, and vice versa. After the worm is glued, the cuticle is cut open with a sharp glass capillary. Intrinsic hydrostatic pressure within the worm body forces out part of the intestine, gonad, and eggs. These tissues may be removed. A small drop of collagenase (0.4 mg/mL in bath solution) is often applied to briefly digest tissues to facilitate access of recording pipettes. In most cases, cells of interest are marked by a transgene expressing fluorescent proteins. This allows for easy identification of cells for recording.

Alternatively, ion channels can be recorded from primary cultured *C. elegans* embryonic cells. Cultured embryonic cells typically maintain many in vivo features. For example, they retain the expression of some cell-specific markers,[53,54] and importantly, may undergo partial differentiation in culture.[53] A detailed protocol of culturing *C. elegans* embryonic cells has been described.[55] Briefly, a large quantity of *C. elegans* eggs can be isolated by lysing adult worms with bleach-based buffer (for 25 mL lysis buffer, mix 5 mL fresh Chlorox bleach, 1.25 mL 10 N NaOH, and 18.75 mL sterile H_2O). A seven-minute treatment is usually sufficient to break 70% of adult worms. The lysis reaction can then be terminated by adding egg buffer (118 mM NaCl, 48 mM KCl, 2 mM $CaCl_2$, 2 mM $MgCl_2$, 25 mM Hepes, pH 7.3, 340 mOsm). After several rounds of centrifugation and resuspension in egg buffer, pellets are resuspended and centrifuged in 30% sucrose to separate eggs from debris. Isolated eggs are digested for ~20 min with 4 mg/mL chitinase (Sigma) to remove eggshell. After digestion, cells are washed and dissociated using a syringe needle in L-15 culture medium (in 500 mL L-15 medium [Gibco], add 50 mL heat-inactivated Fetal Bovine Serum, 7.7 g sucrose, and 5 mL 1:100 Pen/Strep). Dissociated cells are plated on polylysine-pretreated cover slips and cultured at room temperature. Practically, transgenic worms expressing fluorescent protein markers in specific cell types are used, allowing for easy recognition of cells of interest for recording. Conventional patch-clamp techniques can be directly applied to these cultured embryonic cells. The *C. elegans* TRPM channels GON-2 and GTL-1 have been studied using this approach.[16,41] One disadvantage of this approach is that cultured neurons have lost their native synaptic connections, and some of their neuronal properties may not be preserved.

As discussed previously, several *C. elegans* TRP channels have been functionally expressed in heterogonous systems.[3,5,18,35] For these TRPs, heterologous systems offer a powerful tool to study the biophysical properties of these channels in vitro.

21.6 CONCLUSION

The power of *C. elegans* genetics provides a unique opportunity to study gene functions in vivo. All TRP channel subfamilies are present in the *C. elegans* genome, and importantly, many *C. elegans* TRP channels share similar functions and modes of regulation with their vertebrate counterparts.[2] Thus, characterization of *C. elegans* TRP channels may provide valuable insights into the function and regulation of their vertebrate homologs. Genetic screens in *C. elegans* have identified several endogenous regulators and effectors of TRP channels. More signaling molecules upstream and downstream of TRP channels are expected to be identified through genetic approaches in *C. elegans*. Owing to technical reasons, most studies of *C. elegans* TRP channels were performed using genetic and behavioral approaches. Electrophysiological recording results of TRP channels in *C. elegans* are rather limited. Thus, the biophysical properties of most *C. elegans* TRP channels remain to be determined. The recent development of in situ patch-clamp recording techniques in *C. elegans* will foster a thorough understanding of these channels.

ACKNOWLEDGMENTS

We thank the members of the Xu lab for the insightful comments. Work in the lab is supported by the NIGMS, NIDA, and Pew Scholar Award (X. Z. S. X.).

REFERENCES

1. Clapham, D. E. 2003. TRP channels as cellular sensors. *Nature* 426(6966): 517–524.
2. Xiao, R., and X. Z. Xu. 2009. Function and regulation of TRP family channels in *C. elegans*. *Pflügers Archiv* 458(5): 851–860.
3. Colbert, H. A., T. L. Smith, and C. I. Bargmann. 1997. OSM-9, a novel protein with structural similarity to channels, is required for olfaction, mechanosensation, and olfactory adaptation in *Caenorhabditis elegans*. *Journal of Neuroscience* 17(21): 259–269.
4. Feng, Z. et al. 2006. A *C. elegans* model of nicotine-dependent behavior: regulation by TRP-family channels. *Cell* 127(3): 621–633.
5. Xu, X. Z., and P. W. Sternberg. 2003. A *C. elegans* sperm TRP protein required for sperm-egg interactions during fertilization. *Cell* 114(3): 285–297.
6. Colbert, H. A., and C. I. Bargmann. 1995. Odorant-specific adaptation pathways generate olfactory plasticity in *C. elegans*. *Neuron* 14(4): 803–812.
7. Tobin, D. et al. 2002. Combinatorial expression of TRPV channel proteins defines their sensory functions and subcellular localization in *C. elegans* neurons. *Neuron* 35(2): 307–318.
8. de Bono, M. et al. 2002. Social feeding in *Caenorhabditis elegans* is induced by neurons that detect aversive stimuli. *Nature* 419(6910): 899–903.
9. Kahn-Kirby, A. H. et al. 2004. Specific polyunsaturated fatty acids drive TRPV-dependent sensory signaling in vivo. *Cell* 119(6): 889–900.
10. Jose, A. M. et al. 2007. A specific subset of transient receptor potential vanilloid-type channel subunits in *Caenorhabditis elegans* endocrine cells function as mixed heteromers to promote neurotransmitter release. *Genetics* 175(1): 93–105.
11. Chyb, S., P. Raghu, and R. C. Hardie. 1999. Polyunsaturated fatty acids activate the Drosophila light-sensitive channels TRP and TRPL. *Nature* 397(6716): 255–259.
12. Hwang, S. W. et al. 2000. Direct activation of capsaicin receptors by products of lipoxygenases: endogenous capsaicin-like substances. *Proceedings of the National Academy of Sciences of the United States of America* 97(11): 6155–6160.
13. Watanabe, H. et al. 2003. Anandamide and arachidonic acid use epoxyeicosatrienoic acids to activate TRPV4 channels. *Nature* 424(6947): 434–438.
14. Venkatachalam, K., and C. Montell. 2007. TRP channels. *Annual Review of Biochemistry* 76: 387–417.
15. Sun, A. Y., and E. J. Lambie. 1997. *gon-2*, a gene required for gonadogenesis in *Caenorhabditis elegans*. *Genetics* 147(3): 1077–1089.
16. Teramoto, T., E. J. Lambie, and K. Iwasaki. 2005. Differential regulation of TRPM channels governs electrolyte homeostasis in the *C. elegans* intestine. *Cell Metabolism* 1(5): 343–354.
17. Walker, R. G., A. T. Willingham, and C. S. Zuker. 2000. A *Drosophila* mechanosensory transduction channel. *Science* 287(5461): 2229–2234.
18. Li, W. et al. 2006. A *C. elegans* stretch receptor neuron revealed by a mechanosensitive TRP channel homologue. *Nature* 440(7084): 684–687.
19. Kindt, K. S. et al. 2007. Dopamine mediates context-dependent modulation of sensory plasticity in *C. elegans*. *Neuron* 55(4): 662–676.
20. Kindt, K. S. et al. 2007. *Caenorhabditis elegans* TRPA-1 functions in mechanosensation. *Nature Neuroscience* 10(5): 568–577.

21. Barr, M. M., and P. W. Sternberg. 1999. A polycystic kidney-disease gene homologue required for male mating behaviour in *C. elegans. Nature* 401(6751): 386–389.

22. Fares, H., and I. Greenwald. 2001. Regulation of endocytosis by CUP-5, the *Caenorhabditis elegans* mucolipin-1 homolog. *Nature Genetics* 28(1): 64–68.

23. Treusch, S. et al. 2004. *Caenorhabditis elegans* functional orthologue of human protein h-mucolipin-1 is required for lysosome biogenesis. *Proceedings of the National Academy of Sciences of the United States of America* 101(13): 4483–4488.

24. Hersh, B. M., E. Hartwieg, and H. R. Horvitz. 2002. The *Caenorhabditis elegans* mucolipin-like gene cup-5 is essential for viability and regulates lysosomes in multiple cell types. *Proceedings of the National Academy of Sciences of the United States of America* 99(7): 4355–4360.

25. Venkatachalam, K. et al. 2008. Motor deficit in a *Drosophila* model of mucolipidosis type IV due to defective clearance of apoptotic cells. *Cell* 2008. 135(5): 838–851.

26. Jansen, G. et al. 1997. Reverse genetics by chemical mutagenesis in *Caenorhabditis elegans. Nature Genetics* 17(1): 119–121.

27. Sarin, S. et al. 2008. *Caenorhabditis elegans* mutant allele identification by whole-genome sequencing. *Nature Methods* 5(10): 865–867.

28. Zhang, S. et al. 2004. *Caenorhabditis elegans* TRPV ion channel regulates 5HT biosynthesis in chemosensory neurons. *Development* 131(7): 1629–1638.

29. Schaheen, L., G. Patton, and H. Fares. 2006. Suppression of the cup-5 mucolipidosis type IV-related lysosomal dysfunction by the inactivation of an ABC transporter in *C. elegans. Development* 133(19): 3939–3948.

30. Church, D. L., and E. J. Lambie. 2003. The promotion of gonadal cell divisions by the *Caenorhabditis elegans* TRPM cation channel GON-2 is antagonized by GEM-4 copine. *Genetics* 165(2): 563–574.

31. Berezikov, E., C. I. Bargmann, and R. H. Plasterk. 2004. Homologous gene targeting in *Caenorhabditis elegans* by biolistic transformation. *Nucleic Acids Research* 32(4): e40.

32. Vazquez-Manrique, R. P. et al. 2009. Improved gene targeting in *C. elegans* using counter-selection and Flp-mediated marker excision. *Genomics* 95(1): 37–46.

33. Fire, A. et al. 1998. Potent and specific genetic interference by double-stranded RNA in *Caenorhabditis elegans. Nature* 391(6669): 806–811.

34. Ahringer, J. Reverse genetics. In *WormBook*, ed. The *C. elegans* Research Community, WormBook, doi/10.1895/wormbook.1.47.1.

35. Kahn-Kirby, A. H., and C. I. Bargmann. 2006. TRP channels in *C. elegans. Annual Review Physiology* 68: 719–736.

36. Ward, S. 1973. Chemotaxis by the nematode *Caenorhabditis elegans:* identification of attractants and analysis of the response by use of mutants. *Proceedings of the National Academy of Science of the United States of America* 70(3): 817–821.

37. Bargmann, C. I., E. Hartwieg, and H. R. Horvitz. 1993. Odorant-selective genes and neurons mediate olfaction in *C. elegans. Cell* 74(3): 515–527.

38. Kaplan, J. M., and H. R. Horvitz. 1993. A dual mechanosensory and chemosensory neuron in *Caenorhabditis elegans. Proceedings of the National Academy of Science of the United States of America* 90(6): 2227–2231.

39. Hart, A. C. 2006. Behavior. In *WormBook*, ed. The *C. elegans* Research Community, WormBook, doi/10.1895/wormbook.1.87.1.

40. Kwan, C. S. et al. 2008. TRPM channels are required for rhythmicity in the ultradian defecation rhythm *of C. elegans. BMC Physiology* 8: 11.

41. Xing, J. et al. 2008. Highly Ca2+-selective TRPM channels regulate IP3-dependent oscillatory Ca^{2+} signaling in the *C. elegans* intestine. *Journal of General Physiology* 131(3): 245–255.

42. Thomas, J. H. 1990. Genetic analysis of defecation in *Caenorhabditis elegans. Genetics* 124(4): 855–872.

43. Dal Santo, P. et al. 1999. The inositol trisphosphate receptor regulates a 50-second behavioral rhythm in C. elegans. *Cell* 98(6): 757–767.
44. Sawin, E. R., R. Ranganathan, and H. R. Horvitz. 2000. *C. elegans* locomotory rate is modulated by the environment through a dopaminergic pathway and by experience through a serotonergic pathway. *Neuron* 26(3): 619–631.
45. Hart, A. C., S. Sims, and J. M. Kaplan. 1995. Synaptic code for sensory modalities revealed by *C. elegans* GLR-1 glutamate receptor. *Nature* 378(6552): 82–85.
46. Barr, M. M., and L. R. Garcia. 2006. Male mating behavior. In *WormBook*, 1–11.
47. Nakai, J., M. Ohkura, and K. Imoto. 2001. A high signal-to-noise Ca^{2+} probe composed of a single green fluorescent protein. *Nature Biotechnology* 19(2): 137–141.
48. Miyawaki, A. et al. 1997. Fluorescent indicators for Ca^{2+} based on green fluorescent proteins and calmodulin. *Nature* 388(6645): 882–887.
49. Lockery, S. R., and M. B. Goodman. 1998. Tight-seal whole-cell patch clamping of *Caenorhabditis elegans* neurons. *Methods in Enzymology* 293: 201–217.
50. Sakmann, B., and E. Neher. 1984. Patch clamp techniques for studying ionic channels in excitable membranes. *Annual Review of Physiology* 46: 455–472.
51. Sakmann, B., and E. Neher. 1995. *Single-channel Recording*, 2nd ed. New York: Plenum Press.
52. Walz, W. 2007. Patch-clamp analysis: advanced techniques. In *Neuromethods*, 2nd ed. Totowa, NJ: Humana Press.
53. Bianchi, L., and M. Driscoll. 2006. Culture of embryonic *C. elegans* cells for electrophysiological and pharmacological analyses. In *WormBook*, 1–15.
54. Bianchi, L. et al. 2004. The neurotoxic MEC-4(d) DEG/ENaC sodium channel conducts calcium: implications for necrosis initiation. *Nature Neuroscience* 7(12): 1337–1344.
55. Christensen, M. et al. 2002. A primary culture system for functional analysis of *C. elegans* neurons and muscle cells. *Neuron* 33(4): 503–514.

Index

Page numbers followed by f and t indicate figures and tables, respectively.

Printed and bound by CPI Group (UK) Ltd, Croydon, CR0 4YY

21/10/2024

01777112-0010